Trainee Guide

Electrical

Level 3

National Center for Construction Education and Research

Wheels of Learning
Standardized Craft Training

Prentice Hall
Upper Saddle River, New Jersey   Columbus, Ohio

Prentice-Hall, Inc.
A Simon & Schuster Company
Upper Saddle River, New Jersey 07458

This information is general in nature and intended for training purposes only. Actual performance of activities described in this manual requires compliance with all applicable operations procedures under the direction of qualified personnel. References in this manual to patented or proprietary devices do not constitute a recommendation for their use.

Printed in the United States of America

10 9 8 7 6 5 4 3 2 1

ISBN: 0-13-530494-6

Prentice-Hall International (UK) Limited, *London*
Prentice-Hall of Australia Pty. Limited, *Sydney*
Prentice-Hall of Canada, Inc., *Toronto*
Prentice-Hall Hispanoamericana, S. A., *Mexico*
Prentice-Hall of India Private Limited, *New Delhi*
Prentice-Hall of Japan, Inc., *Tokyo*
Simon & Schuster Asia Pte. Ltd., *Singapore*
Editora Prentice-Hall do Brasil, Ltda., *Rio de Janeiro*

This volume is one of many in the *Wheels of Learning* craft training program. This program, covering more than 20 standardized craft areas, including all major construction skills, was developed over a period of years by industry and education specialists. Sixteen of the largest construction and maintenance firms in the U.S. committed financial and human resources to the teams that wrote the curricula and planned the national accredited training process. These materials are industry-proven and consist of competency-based textbooks and instructor guides.

Wheels of Learning was developed by the National Center for Construction Education and Research in response to the training needs of the construction and maintenance industries. The NCCER is a nonprofit educational entity supported by the following industry and craft associations:

- Associated Builders and Contractors
- American Fire Sprinkler Association
- Carolinas Associated General Contractors
- Metal Building Manufacturers Association
- National Association of Minority Contractors
- National Association of Women in Construction
- National Insulation Association
- Painting and Decorating Contractors of America

Some of the features of the *Wheels of Learning* program include:

- A proven record of success over many years of use by industry companies
- National standardization providing "portability" of learned job skills and educational credits that will be of tremendous value to trainees
- Recognition: Upon successful completion of training with an accredited sponsor, trainees receive an industry-recognized certificate and transcript from NCCER.
- Approved by the U.S. Department of Labor for use in formal apprenticeship programs
- Well illustrated, up to date, and practical. All standardized manuals are reviewed annually in a continuous improvement process.

Acknowledgments

This manual would not exist were it not for the dedication and unselfish energy of those volunteers who served on the Technical Review Committee. A sincere thanks is extended to:

Art Bishop	Jimmy Hearn	Jim Mitchem
Dennis Chelew	Allen Hill	John Motlick
Eddie Davis	Tim Langford	Bob Mueller
Al Hamilton	C. Earl Miller	Linn Newell

Contents

Load Calculations—Branch Circuits

Module 20301

Electrical Trainee Task Module 20301

LOAD CALCULATIONS – BRANCH CIRCUITS

Objectives

Upon completion of this module, the trainee will be able to:

1. Calculate loads for single-phase and three-phase branch circuits.
2. Size branch circuit overcurrent protection devices (circuit breakers and fuses) for non-continuous duty and continuous duty circuits.
3. Understand and apply derating factors to size branch circuits.
4. Calculate ampacity for single-phase and three-phase loads.
5. Use load calculations to determine branch circuit conductor sizes.
6. Use NEC Table 220-19 to calculate residential cooking equipment loads.
7. Select branch circuit conductors and overcurrent protection devices for electric heat, air conditioning equipment, motors, and welders.

Prerequisites

Successful completion of the following Task Modules is required before beginning study of this Task Module: Common Core Curricula, Electrical Level 1, and Electrical Level 2.

Required Student Material

1. Trainee Module
2. Copy of the latest edition of the National Electrical Code
3. Pencils and eraser
4. Paper and notebook
5. Hand-held calculator

COURSE MAP

This course map shows all of the *Wheels of Learning* task modules in the third level of the Electrical curricula. The suggested training order begins at the bottom and proceeds up. Skill levels increase as a trainee advances on the course map. The training order may be adjusted by the local Training Program Sponsor.

Course Map: Electrical, Level 3

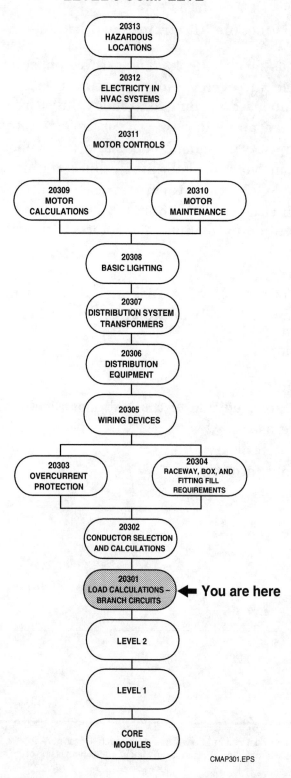

LEVEL 3 COMPLETE

20313
HAZARDOUS LOCATIONS

20312
ELECTRICITY IN HVAC SYSTEMS

20311
MOTOR CONTROLS

20309
MOTOR CALCULATIONS

20310
MOTOR MAINTENANCE

20308
BASIC LIGHTING

20307
DISTRIBUTION SYSTEM TRANSFORMERS

20306
DISTRIBUTION EQUIPMENT

20305
WIRING DEVICES

20303
OVERCURRENT PROTECTION

20304
RACEWAY, BOX, AND FITTING FILL REQUIREMENTS

20302
CONDUCTOR SELECTION AND CALCULATIONS

20301
LOAD CALCULATIONS – BRANCH CIRCUITS ← **You are here**

LEVEL 2

LEVEL 1

CORE MODULES

CMAP301.EPS

TABLE OF CONTENTS

Trade Terms Introduced In This Module

Ampacity: The current in amperes that a conductor can carry continuously under the conditions of use without exceeding its temperature rating.

Appliance: Utilization equipment, generally other than industrial, normally built-in standardized sizes or types, that is installed or connected as a unit to perform one or more functions such as clothes washing, air conditioning, food mixing, deep frying, etc.

Appliance branch circuit: A branch circuit supplying energy to one or more outlets to which appliances are to be connected. Such circuits are to have no permanently-connected lighting fixtures that are not part of an appliance.

Branch circuit: The circuit conductors between the final overcurrent device protecting the circuit and the outlet(s).

Continuous load: A load in which the maximum current is expected to continue for three hours or more.

Demand factor: The ratio of the maximum demand of a system, or part of a system, to the total connected load of a system or the part of the system under consideration.

Device: A unit of an electrical system that is intended to carry but not utilize electric energy.

General purpose branch circuit: A branch circuit that supplies a number of outlets for lighting and appliances.

Individual branch circuit: A branch circuit that supplies only one piece of utilization equipment. - Dedicated circuit

Lighting outlet: An outlet intended for the direct connection of a lampholder, a lighting fixture, or a pendant (hanging) cord terminating in a lampholder.

Multi-outlet assembly: A type of surface or flush raceway designed to hold conductors and receptacles, assembled in the field or at the factory. Ex: wiremold

Multi-wire branch circuit: A branch circuit consisting of two or more ungrounded conductors having a potential difference between them, and a grounded conductor having equal potential difference between it and each ungrounded conductor of the circuit and that is connected to the neutral or grounded conductor of the system.

Nominal voltage: A nominal value assigned to a circuit or system for the purpose of conveniently designating its voltage class (e.g., 120/240 volts, 208Y/120 volts, 480Y/277 volts, 600 volts, etc.).

Outlet: A point on the wiring system at which current is taken to supply utilization equipment.

Overcurrent: Any current in excess of the rated current of equipment or the ampacity of a conductor. It may result from overload, short circuit, or ground fault.

Receptacle: A contact device installed at an outlet for connection as a single contact device. A single receptacle is a single contact device with no other contact device on the same yoke. A multiple receptacle is a single device containing two (duplex) or more receptacles.

Receptacle outlet: An outlet where one or more receptacles are installed.

Utilization equipment: Equipment that utilizes electric energy for electronic, electromechanical, chemical, heating, lighting, or similar purposes.

Voltage to ground: For grounded circuits, the voltage between the given conductor and that point or conductor of the circuit that is grounded; for ungrounded circuits, the greatest voltage between the given conductor and any other conductor of the circuit.

1.0.0   BRANCH CIRCUITS – NEC 210 AND NEC 220

The purpose of **branch circuit** load calculations is to determine the size of branch circuit **overcurrent** protection **devices** and branch circuit conductors using National Electrical Code requirements. Once the branch circuit load is accurately calculated, branch circuit components may be sized to serve the load safely. Branch circuits supply **utilization equipment**. Utilization equipment is defined by the NEC as equipment that utilizes electric energy.

NEC Article 210 covers branch circuits (except for branch circuits that supply only motor loads). Article 210-2 provides a listing of other code articles for specific purpose branch circuits. Per NEC 210-3, branch circuits are rated by the maximum rating or setting of the overcurrent device. Except for circuits serving individual utilization equipment (dedicated circuits), branch circuits shall be rated 15, 20, 30, 40, and 50 amperes. Branch circuits designed to serve individual loads can supply any size load with no restrictions to the ampere rating of the circuit.

Per NEC 210-19, branch circuit conductors are required to be sized with an ampacity rating that is no less than the maximum load to be served. Branch circuit overcurrent protection is required to have a rating or setting not exceeding the rating specified in 240-3 for conductors, 240-2 for equipment, and 210-21 for **outlet** devices including lampholders and **receptacles**. NEC Article 430 applies to branch circuits supplying only motor loads and Article 440 applies to branch circuits supplying only air conditioning equipment, refrigerating equipment, or both.

Branch circuit conductors must have an **ampacity** rating equal to, or greater than, the non-continuous load plus 125% of the continuous load without the application of any adjustment or correction factors per NEC 210-22(c).

NEC 210-23(a) through (d) defines permissible loads for branch circuits. This is important information because it lists the types of loads which may be served according to the size of the branch circuit. Article 210-24 and Table 210-24 summarize the branch circuit requirements.

NEC Article 220 includes the requirements used to determine the number of branch circuits required and the requirements used to compute branch circuit, feeder, and service loads. NEC 220-3(a) states that the rating of a branch circuit shall not be less than the non-continuous load plus 125% of the **continuous load**. Table 220-3(b) gives general lighting loads listed by types of occupancies. These general lighting loads are expressed as a unit load per square foot in volt amperes (VA). For example, the unit lighting load for a barber shop is 3VA/sq. ft.; a store is also 3VA/sq. ft.; and a storage warehouse is ¼VA/sq. ft. Article 220-3(c) (1) through (6) lists minimum loads for outlets used in all occupancies—these outlets include general use receptacles and outlets not used for general illumination. There are five exceptions, listed in the NEC, to these minimum load requirements.

It should be noted at this point that local codes may require different values than the minimum NEC values used in this lesson. For instance, local codes may limit the number of outlets on a branch circuit to less than the calculated value or they may require dedicated circuits other than those that are listed in the NEC.

1.1.0 BRANCH CIRCUIT RATINGS

The maximum load that a branch circuit may serve is determined by multiplying the rating or setting of the overcurrent protection device (circuit breaker or fuse) by the circuit voltage. For example, the maximum load that may be supplied by a 20-amp, two-wire, 120-volt circuit is calculated by multiplying 20 amps by 120 volts.

$$20A \times 120V = 2,400VA$$

The maximum load supplied by a 20-amp, three-wire, 120/240-volt circuit is:

$$20A \times 240V = 4,800VA$$

The maximum load supplied by a 20-amp, 208-volt, three-phase circuit is:

$$20A \times 208V \times \sqrt{3} = 7205.33VA$$

Branch circuits may supply non-continuous loads, continuous loads, or a combination of non-continuous and continuous loads. Per NEC 220-3(a), the branch circuit rating shall not be less than the non-continuous load plus 125% of the continuous load. A continuous load is defined by the NEC as a load where the maximum current is expected to continue for three hours or more. Continuous loads are calculated at 125% of the maximum current rating of the load. NEC 384-16(c) states that, except where the assembly is listed for operation at 100% of its rating (exception), overcurrent devices located in panelboards shall not exceed 80% of their rating where the overcurrent device supplies a continuous duty load.

For 15-amp and 20-amp branch circuits, NEC 220-3(a) allows fastened-in-place utilization equipment to be connected in the same circuit with lighting units, cord- and plug-connected utilization equipment not fastened in place, or both. Under this condition, the fastened-in-place utilization equipment shall not exceed 50% of the branch circuit ampere rating.

Example Problems

Conditions: Store fluorescent lighting fixtures consisting of nine fluorescent ballasts rated at 1.5 amps at 120 volts. The fixtures will operate continuously during normal business hours from 9:00 A.M. until 9:00 P.M. daily.

Question: What is the minimum size circuit breaker required for a branch circuit to serve this load?

Solution: Because the lighting fixtures will stay on for more than three hours, this is a continuous load. Determine the branch circuit load:

    9 x 1.5A = 13.5A

Determine the continuous duty load:

    13.5A x 125% = 16.88 amps

Circuit breaker = 20 amps.

Question: What is the maximum continuous load that may be connected to a 30-amp, 120-volt fuse?

Solution: Per NEC 384-16(c):

    30A x 80% = 24 amps

The maximum continuous load cannot exceed 24 amps.

Question: How many **receptacle outlets** can be connected to a 20-amp, two-wire, 120-volt circuit? The receptacle outlets serve non-continuous duty loads.

Solution: Determine branch circuit capacity:

$$20A \times 120V = 2,400VA$$

Per NEC 220-3(c)(6), each outlet is assigned a load of 180VA.

$$2,400VA \div 180VA = 13.33$$

Thirteen receptacle outlets can be connected to this circuit.

Conditions: An office manager has purchased a new state-of-the-art copy machine. The nameplate rating on the copy machine is 17 amps, 120 volts.

Question: What is the minimum size branch circuit required to serve this equipment?

Solution: This is not a continuous load, however, per NEC 210-23(a) and (b), the rating of one cord-connected utilization equipment shall not exceed 80% of the branch circuit ampere rating for 15-, 20-, and 30-amp circuits. Since the load is greater than 15 amps, determine the current-carrying capacity of a 20-amp (the smallest logical size) circuit to serve this equipment:

$$20A \times 80\% = 16A$$

A 20-amp circuit is not sufficient; determine the current-carrying capacity of a 30-amp circuit:

$$30A \times 80\% = 24 \text{ amps}$$

The minimum size branch circuit required is 30 amps.

Note that the solution above assumes that only multi-receptacle 20-amp and 30-amp branch circuits exist in the office. An alternate solution would be to install a 20-amp **individual branch circuit** exclusively for the copy machine.

Question: What is the maximum lighting load that may be connected to a 20-amp branch circuit supplying a piece of fixed equipment that has a rating of 8.5 amps, 120 volts, 1 phase (Ø)?

Solution: The equipment rating is smaller than 50% of the rating of the 20-amp branch circuit. Therefore, 10 amps of lighting may be added.

$$20A \times 50\% = 10A$$

Conditions: A restaurant dishwasher has a nameplate rating of 14.7A, 208V, 3 Ø. During busy times in the restaurant, it is anticipated that the dishwasher will be turned on and operated for more than three hours at a time.

Question: What is the minimum size branch circuit required to supply this equipment?

Solution: This equipment is considered a continuous load (operated for more than three hours). Therefore, the load is to be multiplied by 125% to determine the branch circuit size.

14.7A x 125% = 18.38A

The minimum size branch circuit required is 20 amps.

1.2.0   DERATING

The current-carrying capacity (ampacity) of conductors may be derated (reduced) due to several conditions that may apply to branch circuits. Branch circuit conductors are derated when:

1. The load to be supplied is a continuous duty load per NEC 220-3(a), 210-22, and 210-23.
2. There are more than three current-carrying conductors in a raceway per NEC Note 8 of the notes to the Ampacity Tables.
3. The ambient temperature that the conductors will pass through exceeds the temperature ratings for conductors listed in NEC Table 310-16.
4. A voltage drop exists that exceeds 3% for branch circuits or 5% for the combination of the feeder and branch circuits that is caused when the distance of branch circuit conductors becomes excessive.

NEC Table 310-16 lists allowable ampacities for insulated conductors rated 0 to 2,000 volts. The ampacities listed apply when no more than three current-carrying conductors are installed in a raceway, are part of a cable assembly, or when conductors are direct buried. The ampacities are based upon an ambient temperature of 30°C (86°F). If the number of current-carrying conductors in the conduit or cable exceeds 3, the ampacity of the conductors must be derated using Note 8(a) to Table 310-16. Note 8 is titled *Adjustment Factors*. If the ambient temperature exceeds 30°C (86°F), the ampacity of the conductors must be derated using the temperature correction factors listed at the bottom of Table 310-16. Note that the multipliers listed at the bottom of Table 310-16 are called *Correction Factors*. NEC references to adjustment and correction factors mean ambient temperatures other than 30°C and more than three current-carrying conductors in a raceway or cable.

1.2.1   TEMPERATURE DERATING

NEC 110-14(c) states that the lowest temperature rating of any component in a branch circuit (circuit breaker, fuse, receptacle, conductors, etc.) must be used to determine the ampacity rating of the branch circuit conductors. Conductors that have a higher temperature rating per NEC 310-16 can be used for ampacity adjustment, correction, or both. For example, if the termination rating of a circuit breaker is 60°C, the ampacity rating of the branch circuit conductor rating cannot exceed the value given in the 60° column of Table 310-16. However, if the branch circuit has a higher temperature rating (90° THHN for example) the higher ampacity rating can be used for ampacity adjustment, correction, or both.

Example Problems

Conditions: It is determined that by combining branch circuits we can eliminate two runs of conduit. The branch circuits are to be pulled in a single conduit. All of the conductors in the conduit will be current-carrying conductors.

Question: What is the ampacity for each of eight #12 THHN copper conductors in a single conduit?

Solution: Per Table 310-16, the ampacity of #12 THHN is 30 amps. Per Note 8(a), the ampacity must be adjusted by 70% (7 to 9 current-carrying conductors).

    30A x 70% = 21A

Question: What is the maximum load that may be connected on each of six #2 current-carrying THWN copper conductors in a single conduit?

Solution: Per Table 310-16, the ampacity of #2 THWN is 115 amps. Per Note 8(a), the ampacity must be adjusted by 80% (4 to 6 current-carrying conductors).

    115A x 80% = 92A

The maximum load cannot exceed 92 amps.

Conditions: A branch circuit is required to supply a non-continuous equipment load with a nameplate rating of 55 amps. The equipment is located in a room with plastic extrusion equipment. The ambient temperature in the room is 92°F.

Question: What is the minimum size THHN copper conductors required to supply the equipment?

Solution: Per Table 310-16, #8 THHN copper conductors have an ampacity rating of 55 amps. Per correction factors to Table 310-16, the ampacity of the conductors must be multiplied by a factor of .96.

    55A x .96 = 52.8 amps

The #8 THHN copper conductors will not meet NEC requirements. A #6 THHN copper conductor has a rating of 75 amps.

75A x .96 = 72 amps

#6 THHN copper conductors will be required.

Conditions: A branch circuit is needed to supply a continuous load of 19.5 amps. The branch circuit conductors will pass through an area that has an ambient temperature of 117°F. The project specification requires XHHW copper conductors for branch circuits.

Question: What is the minimum size XHHW copper conductor required to supply this load and what is the minimum size fuse required as overcurrent protection?

Solution: Because this is a continuous load:

19.5A x 125% = 24.38 amps

The minimum size standard fuse per NEC 240-6 is 25 amps. Per Table 310-16, the ampacity of #12 XHHW copper conductors is listed as 25 amps. The temperature correction factor listed for 117°F is .82.

25A x .82 = 20.5 amps

#12 XHHW will not meet NEC requirements. #10 XHHW copper is rated for 35 amps.

35A x .82 = 28.7 amps

This equipment branch circuit will require #10 XHHW copper conductors and a 25-amp fuse.

1.2.2 VOLTAGE DROP DERATING

NEC 210-19(a) FPN No. 4 states that voltage drop for branch circuit conductors shall not exceed 3% to the farthest outlet and 5% for the combination of feeder and branch circuit distance to the farthest outlet. There are two formulas used to calculate voltage drop for single-phase circuits.

Formula 1:

$$VD = \frac{2 \times L \times R \times I}{1,000}$$

Where:

L = Load center or total length in feet
R = Conductor resistance per NEC Table 8, Chapter 9 (see *Figure 1*)
I = Current

The number 1,000 represents 1,000 feet or less of conductor.

Size AWG/ kcmil	Area Cir. Mills	Conductors				DC Resistance at 75° C (167° F)		
		Stranding		Overall		Copper		Aluminum
		Quan-tity	Diam. in.	Diam. in.	Area in.^{2}	Uncoated ohm/kFT	Coated ohm/kFT	ohm/ kFT
18	1620	1	-	0.040	0.001	7.77	8.08	12.8
18	1620	7	0.015	0.046	0.002	7.95	8.45	13.1
16	2580	1	-	0.051	0.002	4.89	5.08	8.05
16	2580	7	0.019	0.058	0.003	4.99	5.29	8.21
14	4110	1	-	0.064	0.003	3.07	3.19	5.06
14	4110	7	0.024	0.073	0.004	3.14	3.26	5.17
12	6530	1	-	0.081	0.005	1.93	2.01	3.18
12	6530	7	0.030	0.092	0.006	1.98	2.05	3.25
10	10380	1	-	0.102	0.008	1.21	1.26	2.00
10	10380	7	0.038	0.116	0.011	1.24	1.29	2.04
8	16510	1	-	0.128	0.013	0.764	0.786	1.26
8	16510	7	0.049	0.146	0.017	0.778	0.809	1.28
6	26240	7	0.061	0.184	0.027	0.491	0.510	0.808
4	41740	7	0.077	0.232	0.042	0.308	0.321	0.508
3	52620	7	0.087	0.260	0.053	0.245	0.254	0.403
2	66360	7	0.097	0.292	0.067	0.194	0.201	0.319
1	83690	19	0.066	0.332	0.087	0.154	0.160	0.253
1/0	105600	19	0.074	0.373	0.109	0.122	0.127	0.201
2/0	133100	19	0.106	0.528	0.219	0.0608	0.0626	0.100
3/0	167800	19	0.094	0.470	0.173	0.0766	0.0797	0.126
4/0	211600	19	0.106	0.528	0.219	0.0608	0.0626	0.100
250	-	37	0.082	0.575	0.260	0.0515	0.0535	0.0847
300	-	37	0.090	0.630	0.312	0.0429	0.0446	0.0707
350	-	37	0.097	0.681	0.364	0.0367	0.0382	0.0605
400	-	37	0.104	0.728	0.416	0.0321	0.0331	0.0529
500	-	37	0.116	0.813	0.519	0.0258	0.0265	0.0424
600	-	61	0.099	0.893	0.626	0.0214	0.0223	0.0353
700	-	61	0.107	0.964	0.730	0.0184	0.0189	0.0303
750	-	61	0.111	0.998	0.782	0.0171	0.0176	0.0282
800	-	61	0.114	1.03	0.834	0.0161	0.0166	0.0265
900	-	61	0.122	1.09	0.940	0.0143	0.0147	0.0235
1000	-	61	0.128	1.15	1.04	0.0129	0.0132	0.0212
1250	-	91	0.117	1.29	1.30	0.0130	0.0106	0.0169
1500	-	91	0.128	1.141	1.57	0.00858	0.00883	0.0141
1750	-	127	0.117	1.52	1.83	0.00735	0.00756	0.0121
2000	-	127	0.126	1.63	2.09	0.00643	0.00662	0.0106

E301F01.TIF

Figure 1. Conductor Properties

Formula 2:

$$VD = \frac{2 \times L \times K \times I}{CM}$$

Where:

L = Load center or total length in feet
K = Constant (12.9 for copper conductors; 21.2 for aluminum conductors)
I = Current
CM = Area of the conductor in *circular mills* (from NEC Table 8, Chapter 9)

The result of these formulas is in volts. The result is divided by circuit voltage and the second result is multiplied by 100 to determine the % voltage drop. If the branch circuit voltage drop exceeds 3%, the conductor size is increased to compensate. For a 120-volt circuit, the maximum voltage drop, in volts, is 120V x 3% = 3.6 volts. For a 240-volt 1 Ø circuit, the maximum voltage drop, in volts, is 240V x 3% = 7.2 volts.

If in commercial fixed load applications, the branch circuit load is not concentrated at the end of the branch circuit but is spread out along the circuit due to multiple outlets, the load center length of the circuit should be calculated and used in the above formulas. This is because the total current does not flow the complete length of the circuit. If the full length is used in computing the voltage drop, the drop determined would be greater than what would actually occur. The load center length of a circuit is that point in the circuit where, if the load were concentrated at that point, the voltage drop would be the same as the voltage drop to farthest load in the actual circuit. To determine the load center length of a branch circuit with multiple outlets, as shown in *Figure 2*, multiply each outlet load by its actual physical routing distance from the supply end of the circuit. Add these products for all loads fed from the circuit and divide this sum by the sum of the individual loads. The resulting distance is the load center length (L) for the total load (I) of the branch circuit.

Example Problems

Conditions: The length of a 120-volt, two-wire branch circuit is 95 feet. The non-continuous load is 14.5A.

Question: If #12 THHN solid copper conductors are used, will the voltage drop for this branch circuit exceed 3%?

Solution: Use the first voltage drop formula to determine voltage drop for this branch circuit. Look up the resistance for #12 solid copper in NEC Table 8, Chapter 9.

2 x 95 x 1.93 x 14.5 ÷ 1,000 = 5.32V

5.32V ÷ 120V = .0443 x 100 = 4.43%

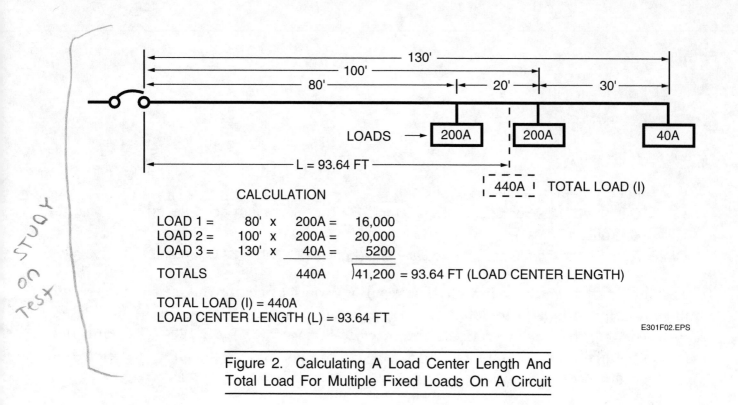

CALCULATION

LOAD 1 =   80' x   200A =   16,000
LOAD 2 =   100' x   200A =   20,000
LOAD 3 =   130' x   40A =   5200

TOTALS   440A)41,200 = 93.64 FT (LOAD CENTER LENGTH)

TOTAL LOAD (I) = 440A
LOAD CENTER LENGTH (L) = 93.64 FT

E301F02.EPS

Figure 2. Calculating A Load Center Length And Total Load For Multiple Fixed Loads On A Circuit

Since this percentage exceeds the 3% allowable, the answer to the question is yes and larger conductors would be required for this circuit. Note that this solution uses the first formula for 1 Ø voltage drop.

Question: What size THHN solid copper branch circuit conductors would be required for a non-continuous branch circuit load of 23 amps, 240 volts, 1 Ø? The length of the circuit is 130 feet.

Solution: Use the second voltage drop formula to determine the voltage drop for this branch circuit. Since #10 THHN copper would be the smallest size permitted, this size will be evaluated first. Look up the area in CM for #10 solid conductor in NEC Table 8, Chapter 9.

should be 12.9 → 7.43v

2 x 130 x 12 x 23 ÷ 10,380 = 6.91V

7.43 ÷ 240 = .0309 x 100 = 3.09%

6.91V ÷ 240V = .0288 x 100 = 2.88%

This percentage does not exceed 3%; therefore, #10 THHN copper conductors would meet NEC requirements.

Question: What size THHN copper conductors would be required for a 240V, single-phase, three-wire branch circuit with multiple outlets consisting of continuous fixed 240V loads of 30A at 60 feet, 30A at 80 feet, and at the end, a non-continuous fixed load of 20A at 100 feet?

Solution: Since there are two continuous load outlets and one non-continuous load outlet, determine the total load for these three outlets.

Outlet 1 = 30A x 125% = 37.5A

Outlet 2 = 30A x 125% = 37.5A

Outlet 3 = 20A x 100% = 20A

Determine the load center length of the circuit by multiplying the outlet loads by their distance from the circuit source and then by dividing the sum of the three products by the sum of the three loads as follows:

Outlet 1 = 60' x 37.5A = 2,250

Outlet 2 = 80' x 37.5A = 3,000

Outlet 3 = 100' x 20A = 2,000

Sum the products:

2,250 + 3,000 + 2,000 = 7,250

Divide by the sum of the loads:

37.5A + 37.5A + 20A = 95A

7,250 ÷ 95 = 76.32 feet for the load center

For a load of 95A (per NEC Table 310-16) #4 copper THHN conductors at 90°C will be selected. Use the second formula and substitute values to check voltage drop.

2 x 76.32 ft. x 12.9 x 95A ÷ 41,740 CM = 4.5V

Since the permissible voltage drop is 240V x 3%, or 7.2V, the #4 THHN conductors are satisfactory for this application.

Voltage drop for balanced three-phase circuits (with negligible reactance and a power factor of 1) can be calculated using the two formulas above by substituting √3 for the value of 2 in the formulas:

$$VD = \frac{\sqrt{3} \times L \times R \times I}{1,000} \qquad VD = \frac{\sqrt{3} \times L \times K \times I}{CM}$$

Example Problems

Question: What is the voltage drop of a 208V, 3 Ø branch circuit with a load of 32 amps, a distance from the circuit breaker to the load of 115 feet, and using #8 stranded copper conductors?

Solution: Use the voltage drop formula to determine voltage drop for this 3 Ø circuit. Look up the resistance for #8 stranded copper conductor in Table 8.

$\sqrt{3}$ x 115 x .778 x 32 ÷ 1,000 = 4.96

The maximum voltage drop permitted for a 208-volt circuit is:

208 x 3% = 6.24 volts

Since, in this problem, 4.96 volts is less than the maximum allowable voltage drop, this circuit will not have a voltage drop problem.

Question: What size THHN copper conductors would be required for a 208V, 3 Ø, four-wire branch circuit with a length of 150 feet from the source to a fixed continuous load of 100A?

Solution: Determine load by multiplying 100A x 125% to obtain 125A. Using NEC Table 310-16, determine that #2 THHN conductors at 90°C are satisfactory for a load of 125A. Use the second formula for a three-phase voltage drop and substitute values.

$\sqrt{3}$ x 150 x 12.9 x 125 ÷ 55,360 CM = 7.6V

Since the permissible voltage drop is 208 x 3%, or 6.24 volts, the next larger conductor will have to be selected. AWG #1 THHN copper conductors will be satisfactory for this application.

1.3.0   CALCULATING BRANCH CIRCUIT AMPACITY

Branch circuit ampacity for single-phase circuits is calculated by dividing VA by the circuit voltage. For example, the ampacity of a 120-volt load rated at 1,600VA is determined by dividing 1,600VA by 120V.

1,600VA ÷ 120V = 13.33 amps

The ampacity of a 3,450VA load at 277V is 12.45 amps.

3,450VA ÷ 277V = 12.45A

Branch circuit ampacity for three-phase circuits is calculated by dividing VA by the circuit voltage times $\sqrt{3}$. For example, the ampacity of a 208-volt, three-phase load rated at 5,200VA is determined by dividing 5,200VA by (208V x $\sqrt{3}$).

$$5,200VA \div (208V \times \sqrt{3}) = 14.43 \text{ amps}$$

The ampacity of a 14,000VA three-phase load at 480V is:

$$14,000VA \div (480V \times \sqrt{3}) = 16.84 \text{ amps}$$

Example Problems

Question: What is the ampacity of a single-phase load with a nameplate rating of 5.5kW, 240V?

Solution: Multiply 5.5kW by 1,000 to determine VA (watts = volt-amperes) and then divide the result by 240V.

$$5.5kW \times 1,000 = 5,500VA \div 240V = 22.92 \text{ amps}$$

The ampacity of this load is 22.92 amps.

Question: What is the ampacity of a three-phase electric water heater with a nameplate rating of 20kW, 208V?

Solution: Multiply 20kW by 1,000 and divide the result by (208V x $\sqrt{3}$).

$$20kW \times 1,000 = 20,000kW \div (208V \times \sqrt{3}) = 55.51 \text{ amps}$$

The ampacity of this load is 55.51 amps.

2.0.0 LIGHTING LOADS

Lighting load branch circuit calculations are based upon the type of lighting (incandescent or electric discharge), the branch circuit voltage, and whether or not the lighting is to be used for more than three hours without an off period (continuous duty).

Branch circuit loads for incandescent lighting are determined by adding the total incandescent load (watts = VA) and dividing the total by the circuit voltage. For example, three 500-watt quartz lamps connected on a 120-volt circuit would equal 12.5 amps.

$$3 \times 500VA = 1,500VA \div 120V = 12.5A$$

Branch circuit loads for electric discharge lighting (lighting units having ballasts, transformers, or autotransformers) are determined by multiplying the number of fixtures times the ampacity rating of the ballast. Per NEC 210-22(b), the calculated load shall be based upon the total ampere rating of the fixture (ballast or ballasts) and not the total watts of the lamps. For example, fifteen 150-watt, high-pressure sodium fixtures connected on a 277-volt circuit would equal 11.85 amps.

15 x .79A = 11.85 amps

In this example, the ampere rating of the 150-watt ballast at 277 volts is .79 amp.

Example Problems

Conditions: Incandescent lighting utilizing 150-watt medium base lamps is required for temporary lighting on a construction site. This lighting will remain on all day and all night.

Question: How many 150-watt lamps can be connected on a 20-amp, 120-volt circuit?

Solution: Since the lighting will remain on for longer than three hours, this is a continuous load. For continuous duty, multiply 150VA (watts) x 125% = 187.5VA. Determine the total capacity (in VA) for the 20-amp circuit:

20A x 120V = 2,400VA

Divide the circuit ampacity by the lamp demand load:

2,400VA ÷ 187.5VA = 12.8 lamps

Twelve 150-watt lamps may be safely connected to a 20-amp, 120-volt circuit.

Conditions: Two 400-watt metal halide high bay fixtures are required to provide additional lighting for a production machine. There is an existing, 20-amp, 277-volt lighting circuit available near the equipment and, after investigation, it is determined that the circuit has six of the same type of fixtures connected. The nameplate ampere rating for these fixtures is 2.0 amps at 277 volts. During normal operation, the fixtures are turned on for 12 hours every day.

Question: Can two 400-watt fixtures be safely added to the existing circuit?

Solution: This is a continuous lighting load. The maximum continuous load that may be connected to the 20A circuit is 16 amps.

20A x 80% = 16A

The load for eight fixtures (six existing and two new fixtures) is 16 amps.

8 fixtures x 2.0A = 16A

Per NEC requirements, two new fixtures may safely be added to this circuit.

Question: What is the difference in the load, in amps, for one 96-inch high output fluorescent strip light when the ampacity of the ballast is used compared to using the total wattage of the lamps? The ballast is rated at 1.5 amps, 277 volts. The fixture takes two lamps rated at 110 watts each.

Solution: Multiply the number of lamps by the rating of the lamps to obtain the total VA, then divide the total VA by the voltage rating to obtain amps.

$$2 \times 110 = 220VA \div 277V = .79A$$

By using the wattage of the lamps to calculate the load in the last question, we have obtained an ampacity that is approximately ½ the true ampacity of the fixture. This explains why NEC 210-22(b) requires the ampacity of the ballast to be used in the calculation. Using wattage would create a hazard by allowing the branch circuit to be overloaded.

2.1.0 RECESSED LIGHTING

The load for recessed lighting fixtures (excluding the residential general lighting load that is calculated at 3VA per square foot) is calculated per NEC 220-3(c)(3) by using the maximum VA rating of the equipment (fixture) and lamps. For example, a load of 250VA would be used to calculate the load for a recessed incandescent fixture rated at 250 watts. To calculate the load for a recessed fixture that uses compact fluorescent lamps, we would need to know the voltage and ampacity rating of the fluorescent ballast. To calculate the load for a recessed H.I.D. fixture (high pressure sodium, metal halide, etc.), we would again need to know the voltage and ampacity rating of the ballast.

Example Problems

Question: What is the load in amps for seven incandescent recessed cans that have a nameplate indicating a maximum lamp size of 150 watts and the fixtures are to be connected to a 120-volt circuit?

Solution: The number of fixtures is multiplied by the VA rating of each fixture to obtain the total VA load.

$$7 \times 150VA \text{ (watts)} = 1,050VA$$

To determine ampacity, the total VA rating is divided by the circuit voltage.

$$1,050VA \div 120V = 8.75 \text{ amps}$$

Question: What is the total load in amps for seven fluorescent recessed fixtures? Each fixture has a ballast ampacity of .20 amps at 277 volts and each fixture takes two compact fluorescent lamps rated 26 watts each.

Solution: Because these fixtures are of the electric discharge type, we need to know the ampacity of the fixture and the number of fixtures to calculate total ampacity.

$$7 \times .20A = 1.4 \text{ amps}$$

2.2.0   HEAVY DUTY LAMPHOLDER OUTLETS

Per NEC 220-3(c)(4), outlets for heavy duty lampholders are calculated at 600VA. A non-continuous load consisting of four 120-volt outlets for heavy duty lampholders could be connected to a 20-amp branch circuit

2,400VA ÷ 600VA = 4 outlets

3.0.0   RECEPTACLE LOADS

When the exact VA rating of a load that is to be cord and plug connected to a receptacle outlet is not known, a VA rating of 180VA per outlet is used per NEC 220-3(c)(6). This NEC reference does not apply to residential receptacles. Residential receptacles are included in the general illumination load **(general purpose branch circuits)** or in specific residential loads such as receptacles required for small appliance and laundry loads. For non-continuous receptacle loads, the total non-continuous load is calculated at 100%. For continuous loads, the total continuous load is calculated at 125%.

Example Problems

Conditions: A 15-amp, 120-volt circuit is to be added to supply general purpose receptacles. The receptacles are to be located above work benches that are to be positioned against a wall. The owner states that the workers using the benches will plug small tools into the outlets, but the tools will only be occasionally used for short periods of time.

Question: How many general purpose duplex receptacles can be connected to a 15-amp, 120-volt circuit when the outlets are to be rated for non-continuous duty?

Solution: Per NEC 220-3(c)(6), each receptacle is assigned a value of 180VA. The capacity of a 15-amp, 120-volt circuit is 1,800VA.

15A x 120V = 1,800VA

The total circuit capacity is then divided by the rating per receptacle to determine the total number that may be connected.

1,800VA ÷ 180VA = 10 receptacles

Question: What is the ampacity of each duplex receptacle connected in a 15-amp, 120-volt circuit with other outlets? The load for this receptacle is to be continuous.

Solution: Divide the volt-amp rating for the receptacle by the circuit voltage and take the answer (in amps) times 125% for continuous duty.

180VA ÷ 120V = 1.5A x 125% = 1.875A

4.0.0 MULTI-OUTLET ASSEMBLIES

A **multi-outlet assembly** is defined by the NEC as a type of surface or flush raceway designed to hold conductors and receptacles, assembled in the field or at the factory. Multi-outlet assemblies may consist of single outlets wired to one or more circuits and typically spaced equally apart at distances of 6", 12", 18", etc. The NEC rules for calculating loads for multi-outlet assemblies do not apply to dwelling units.

Per NEC 220-3(c) Ex. No. 1, each five feet of multi-outlet assembly is considered as one outlet of at least 180VA capacity (light duty). In locations where numerous **appliances** are likely to be used simultaneously, each one foot of multi-outlet assembly is considered as one outlet of at least 180VA capacity (heavy duty).

Example Problems

Conditions: A total of 40 feet of Plugmold is installed to serve light duty loads. The style specified for installation includes one single receptacle per foot of the assembly. The multi-outlet assembly is to be connected to a single 120-volt circuit.

Question: What is the minimum size 120-volt circuit that would safely supply this light duty multi-outlet assembly?

Solution: Since this is a light duty application, the load is to be calculated at 180VA per five feet of multi-outlet assembly. Divide 40 feet of multi-outlet assembly by 5 and multiply the result by 180VA to obtain the total VA load. Divide the total VA load by 120 volts to determine the circuit ampacity and minimum circuit size.

40 ft. ÷ 5 = 8 x 180VA =1,440VA

1,440VA ÷ 120V = 12A

The minimum 120-volt circuit size would be a 15-amp circuit.

Conditions: The conditions listed for the problem above have changed and it will now be necessary to use two-circuit Plugmold. The reason for the change is that there will now be more workers using equipment plugged into the assembly. This change means that the assembly will now be rated heavy duty.

Question: How many 20-amp, 120-volt circuits would be required to safely supply this heavy duty multi-outlet assembly?

Solution: Since this is a heavy duty application, the load is to be calculated at 180VA per foot of multi-outlet assembly. Multiply 40 feet of multi-outlet assembly by 180VA to obtain the total VA load. To determine the VA capacity for a 20-amp, 120-volt circuit:

20A x 120V = 2,400VA

Divide the total VA by 2,400VA (20A circuit capacity) to determine the number of circuits required.

40 ft. x 180VA = 7,200VA (total VA)

7,200VA ÷ 2,400VA = 3

Three 20A, 120V circuits would be required to supply this assembly.

5.0.0 SHOW WINDOW LOADS

The NEC provides two options for calculating the load for show window lighting. NEC 220-3(c) Ex. No. 3 requires a load of not less than 200VA per linear foot of show window. NEC 210-62 requires one receptacle at 180VA for each 12 linear feet of show window. Since show window lighting would likely be used for more than three hours, it is considered a continuous load when using the receptacle method. The 200VA per linear foot method is not required to be increased by 125%.

Example Problems

Question: How many receptacles would be required for a show window area that measures 67 feet in length?

Solution: Divide the length of show window by 12 to determine the number of receptacles required.

67 ÷ 12 = 5.58

Since NEC 21-62 requires one receptacle for each 12 linear feet or major fraction thereof, six receptacles would be required.

Question: What is the total load for the receptacles required in the previous problem?

Solution: Multiply the number of receptacles by 180VA per receptacle and multiply the result by 125% to determine the load.

6 x 180VA = 1,080VA

1,080 x 125% = 1,350VA

The load for these show window receptacles is 1,350VA.

Question: What is the load for 30 feet of show window area?

Solution: Multiply the total length of show window area by 200VA to determine the load.

30 ft. x 200VA = 6,000VA

6.0.0 SIGN LOADS

NEC Article 600 covers requirements for signs and outline lighting. NEC Article 600-5(b)(1) and (2) specify that the rating for sign circuits that supply lamps, ballasts, transformers, or combinations of these devices shall not exceed 20 amps. For sign circuits consisting of electric-discharge lighting transformers exclusively, the circuit size shall not exceed 30 amps. NEC 600-5(a) requires at least one 20-amp sign circuit that supplies no other load. This sign circuit must be provided for each commercial building and each commercial occupancy accessible to pedestrians. NEC 600-5(b)(3) requires the load for the sign circuit to be computed at a minimum of 1,200VA. Since signs for commercial occupancies are expected to operate for more than three hours at a time, the sign circuit is typically considered as a continuous load. Under this condition, the branch circuit calculation of 1,200VA is increased by 25%. The actual sign load cannot exceed 80% of the branch circuit rating.

Example Problem

Question: What is the maximum continuous load, in VA, which may be connected to a 20-amp, 120-volt sign circuit?

Solution: Multiply 20 amps by 120 volts to determine total VA for a 20A circuit and multiply the result by 80% to determine maximum continuous VA load.

20A x 120V = 2,400VA

2,400VA x 80% = 1,920VA

7.0.0 RESIDENTIAL BRANCH CIRCUITS

There are a number of branch circuits that are required for dwelling units. The NEC lists the requirements for these circuits including the requirements for calculating branch circuit loads for specific dwelling unit (residential) loads. These loads, which are unique to dwelling units, will be covered here.

7.1.0 SMALL APPLIANCE LOAD

NEC 210-52(b) and 220-4(b) requires at least two 20-amp small **appliance branch circuits** to be installed to supply receptacles installed in the kitchen, pantry, breakfast room, and dining room. NEC 220-16(a) requires that the small appliance branch circuits be computed at 1,500VA for each circuit.

Example Problem

Question: What is the total load for three small appliance branch circuits rated at 20 amps, 120 volts?

Solution: Multiply the number of small appliance branch circuits by 1,500VA to determine the total load.

    3 circuits x 1,500VA = 4,500VA

7.2.0 LAUNDRY CIRCUIT

NEC 210-52(f) and 220-4(c) require the installation of at least one receptacle to supply laundry equipment. NEC 220-16(b) requires the load for each laundry circuit to be computed at 1,500VA.

7.3.0 DRYERS

As specified in NEC Article 220-18, the load for household electric dryers is calculated at 5,000VA or the nameplate rating of the dryer, whichever is larger. NEC Table 220-18 lists demand factors for dryers. Note that this table applies to residential dryers used in single family dwelling units or in multi-family dwelling units. The demand factors listed do not apply to commercial dryers used in commercial facilities.

Example Problems

Question: What is the demand load for one household electric dryer with a nameplate rating of 5,500 watts?

look at. p. 74 code bk.

Solution: Per NEC Table 220-18, the demand factor for one dryer is 100%. Therefore, the load would be calculated as 5,500VA.

Conditions: A new apartment building is planned. It consists of ten dwelling units. Each unit has an electric dryer with a nameplate rating of 5,500 watts. There will be one utility transformer and one service entrance serving the building.

Question: What is the demand load, in VA, for the ten dryers listed above? *Check the demand factor in code.*

Solution: Per Table 220-18, the demand factor for ten dryers is 50%. Add the total VA for the ten dryers and multiply the result by 50%.

    10 dryers x 5,500 watts = 55,000VA

    55,000VA x 50% = 27,500VA

7.4.0 COOKING APPLIANCES

Loads for ranges, wall-mounted ovens, counter-mounted cooking units, and other household cooking appliances are calculated using the **demand factors** listed in NEC Table 220-19 and the notes to Table 220-19. The most important single piece of information required to size the circuit for residential cooking equipment is the nameplate rating of the equipment. Although Table 220-19 lists loads in kW, Article 220-19 states that kVA shall be considered equivalent to kW for loads calculated using Table 220-19.

Demand factors are calculated using Table 220-19 based upon the number of appliances, the maximum demand listed in Column A, and a demand factor percentage that is found using Columns B and C. Column A is used when the nameplate rating is over 8³/₄kW but less than 12kW. Column A is also used to calculate larger range loads per Notes 1 and 2. Column B is used when the nameplate rating is less than 3¹/₂kW. Column C is used when the nameplate rating is from 3¹/₂kW to 8³/₄kW. Note 3 provides a method to calculate the demand for multiple ranges, each of which has a nameplate rating of more than 1³/₄kW but less than 8³/₄kW. Note 4 provides a method to determine the size for a single branch circuit supplying one counter-mounted unit and no more than two wall-mounted ovens, provided the equipment is located in the same room.

Example Problems

Question: What is the demand load for one range with a nameplate rating of 11.3kW?

Solution: Since the nameplate rating is greater than 8³/₄kW and less than 12kW, Column A is used. The demand rating for one range, not over 12kW, is listed as 8kW.

Question: What is the demand load, in amps, for one range with a nameplate rating of 15.7kW, 1 Ø, 240 volts?

Solution: Since the nameplate rating exceeds 12kW, Note 1 to Table 220-19 is used. Subtract 12kW from the nameplate rating of the range to determine the number of kW exceeding 12.

> 15.7kW - 12kW = 3.7kW

Since .7 is a major fraction (≥ .5), the result is rounded off to 4kW. Multiply 5% by 4 to obtain the demand increase percentage.

> 4kW x 5% = 20%

The maximum demand listed in Column A for one range (8kW) is then multiplied by 120% to determine the demand load for this range.

> 8kW x 120% = 9.6kW

To determine the ampacity of this load, divide VA by the circuit voltage.

> 9.60 x 1,000 = 9,600VA
>
> 9.6kW = 9,600VA
>
> 9,600VA ÷ 240V = 40 amps

Conditions: One wall-mounted oven and one counter-mounted cooking unit are to be installed in the kitchen. To reduce the cost of the electrical installation, both pieces of cooking equipment are to be connected to the same 240-volt, 1 Ø circuit. The oven has a nameplate rating of 12.5kW and the counter-mounted unit has a nameplate rating of 8.0kW.

Question: What is the demand load, in amps, for a single circuit to supply this cooking equipment?

Solution: Note 4 to Table 220-19 may be used for this calculation since the conditions required by Note 4 are met. Obtain the total load by adding the nameplate ratings of the individual pieces of cooking equipment.

12.5kW + 8.0kW = 20.5kW

The equivalent of one range is 20.5kW. Using Note 1, subtract 12kW from 20.5kW to determine the kW exceeding 12.

20.5kW - 12kW = 8.5kW

Since .5 is a major fraction, the result is rounded off to 9kW. Multiply 5% by 9:

9 x 5% = 45%

The maximum demand in Column A for one range (8kW) is then multiplied by 145% to determine the demand load.

8kW x 145% = 11.6kW

Divide total VA by circuit voltage to determine branch circuit ampacity.

11,600VA ÷ 240V = 48.33 amps

Question: Using the resulting ampacity for the branch circuit in the problem above, what is the minimum size Type NM cable that may be used for this branch circuit and what size circuit breaker is required?

Solution: Per NEC 336-30(b) the rating for Type NM cable shall be 60°C. Per the 60°C column of Table 310-16, #6 copper conductor is rated at 55 amps. A 50A circuit breaker would protect this circuit; however, per NEC 240-3(b), the next higher standard size breaker may be used. A 60A circuit breaker could be used.

8.0.0 COMMERCIAL KITCHEN EQUIPMENT

Loads for commercial kitchen equipment are calculated based upon the nameplate rating of the equipment. The load is non-continuous (100%) if operated for less than three hours at a time and continuous (125%) if operated for three hours or longer at a time. NEC Table 220-20 lists demand factors for commercial kitchen equipment. NEC Article 220-20 applies to commercial kitchen equipment, including cooking equipment, dishwasher booster heaters, water heaters, and other kitchen equipment.

Example Problems

Question: What is the load, in amps, for one dishwasher booster heater? The booster heater has a nameplate rating of 15kW, 208V, 3 Ø.

Solution: Under this condition, kW = kVA. Divide total kVA by (voltage x $\sqrt{3}$).

15 kVA x 1,000 = 15,000VA

15,000VA ÷ (208V x $\sqrt{3}$) = 41.64A

Question: What is the demand load for six pieces of commercial kitchen equipment with a total VA rating of 47,000VA?

Solution: Per Table 220-20, the demand factor for six equipment units is 65%. Multiply the total VA rating by 65%.

47,000VA x 65% = 30,550VA

The demand load is 30,550VA.

9.0.0 WATER HEATERS

As specified in NEC Article 422-14(b), fixed storage water heaters having a storage capacity of 120 gallons or less shall have the branch circuit rated at no less than 125% of the nameplate rating of the water heater.

Example Problem

Question: What is the minimum size circuit breaker and branch circuit conductors using Type NM cable for a water heater with a nameplate rating of 9,000VA at 240 volts?

Solution: Divide total VA by voltage to determine ampacity; then multiply the result by 125%.

9,000VA ÷ 240V = 37.5A

37.5A x 125% = 46.88A

Per Table 310-16, 60°C column, #6 copper conductor is rated (for NM cable) at 55 amps. The circuit breaker size would be 50A.

10.0.0   ELECTRIC HEATING LOADS

NEC Article 424 covers fixed electric space heating equipment, including heating cable, unit heaters, boilers, central systems, or other approved heating equipment. This article does not apply to process heating or room air conditioning. Per 424-3, branch circuits are permitted to supply any size heating equipment. If two or more outlets for heating equipment are supplied, the branch circuit size is limited to 15, 20, or 30 amps. When space heating equipment utilizes electric resistance heating elements, protection for the resistance elements shall not exceed 60 amps. If the equipment is rated at more than 48 amps, the heating elements must be subdivided and each subdivided load must not exceed 48 amps.

For space heating equipment consisting of resistance heating elements with or without a blower motor, the ampacity of the branch circuit shall not be less than 125% of the total load. The NEC does not require an additional increase in circuit size or conductor size for continuous load.

Example Problems

Question: What is the branch circuit load, in amps, for the following baseboard electric heaters rated 240 volts, 1 Ø: one 1,500-watt unit, one 1,000-watt unit, and two 500-watt units?

Solution: Determine the total load by adding the load of each baseboard unit and then divide the total load in VA by the circuit voltage.

1,500W + 1,000W + 500W + 500W = 3,500W (VA)

3,500VA ÷ 240V = 14.58A

Question: What size NM conductors and circuit breaker would be required to supply the load for the previous question?

Solution: Multiply the load by 125% to determine circuit ampacity.

14.58A x 125% = 18.23A

Per Table 310-16, 60°C column, #12 copper conductor (for Type NM) is rated at 25 amps and may be connected to a 20A circuit breaker. The load, 18.23A, would require a 20A circuit breaker.

Question: What is the branch circuit ampacity for an electric forced air furnace with heating elements totaling 25kW and a blower motor with a nameplate full-load current of 4.9 amps? The resistance heat and the motor are rated at 240V, 1 Ø.

Solution: Divide the total of the resistance heat, in VA, by the circuit voltage to determine the ampacity for the resistance heat. Add the motor ampacity to the result and multiply the total by 125% to determine total circuit ampacity.

$$25,000VA \div 240V = 104.17A$$

$$104.17A + 4.9A = 109.07A$$

The total circuit ampacity is 109.07A.

$$109.07A \times 125\% = 136.34A$$

Note that since this equipment is rated at more than 48 amps, the heating elements must be subdivided and each subdivided load cannot exceed 48 amps.

11.0.0 AIR CONDITIONING LOADS

Branch circuit conductors are rated at least 125% of the air conditioning compressor full-load current or the branch circuit selection current, whichever is greater, per NEC 440-32. Branch circuit selection current is determined by the equipment manufacturer and this current is required to be greater than or equal to the compressor full-load current. If the branch circuit selection current is higher, its rating shall be used instead of the compressor full-load current. Per NEC 440-22(a), overcurrent protection devices shall not exceed 175% of the compressor full-load current or the branch circuit selection current. If this breaker is not sufficient to hold the starting current of the motor, the rating is permitted to be increased but cannot exceed 225% of the motor-rated load current or the branch circuit selection current, whichever is greater.

Example Problems

Question: What size THHN copper conductors are required for a branch circuit supplying an air conditioner with a nameplate rating of 45 amps, 480 volts, 3 Ø?

Solution: Multiply the nameplate ampacity by 125% and, using the result, select the proper THHN conductor size from NEC Table 310-16.

$$45A \times 125\% = 56.25A$$

Table 310-16 gives an ampacity of 75 amps for #6 THHN copper. Therefore, the conductor size would be #6 THHN copper.

Question: What size Type NM cable and circuit breaker would be required to supply an air conditioner with a nameplate rating of 16.5 amps, 240 volts, 1 Ø?

Solution: Multiply the nameplate ampacity by 125% and, using the result, select the proper NM conductor size from the 60°C column of Table 310-16.

> 16.5A x 125% = 20.63A

The 60°C column gives a rating of 25A for #12 copper conductors (for Type NM). Multiply the nameplate ampacity by 175% to determine the circuit breaker size.

> 16.5A x 175% = 28.88A

Since 440-22(a) states that the rating of the overcurrent device cannot exceed 175%, a 25-amp circuit breaker would be required.

12.0.0 MOTOR LOADS – NEC 430

Branch circuit conductors and overcurrent protection devices for motors are determined by using NEC Table 430-148 for single-phase motors and NEC Table 430-150 for three-phase motors. To use the NEC motor tables, it is necessary to know the motor horsepower, voltage, phase, and design letter. Motor information is obtained from the motor nameplate. NEC 430-6(a) requires that the values given in Tables 430-147 through 430-150 be used to determine the ampacity of branch circuit conductors and the ampere rating of switches, branch circuit short-circuit and ground fault protection, etc. The motor nameplate current data is not permitted to be used in sizing these components unless otherwise specified.

To determine the branch circuit conductor size for motor circuits, the motor full-load current taken from the appropriate table is multiplied by 125% and the resulting ampacity is used to select branch circuit conductors. To determine motor full-load current using the tables, it is necessary to know the motor size in horsepower, motor voltage, and motor phase. If the motor nameplate includes a value for amps but not horsepower, NEC 430-6(a) states that the horsepower rating shall be assumed to be the value given in the tables. To assume a horsepower value, the horsepower value that most closely corresponds to the ampere value in the table is selected, using interpolation if necessary.

Example Problems

Question: What is the motor full-load current for a 2 HP, 230-volt, 1 Ø motor?

Solution: Table 430-148 is used to determine full-load current for single-phase AC motors. The left column in the table is used to find the correct motor HP and the ampacity is then determined based upon the motor voltage. A 2 HP, 230-volt motor has a full-load current of 12A.

Question: What is the motor full-load current for a 10 HP, 460-volt, 3 Ø motor?

Solution: Table 430-150 is used to determine full-load current for three-phase AC motors. The left column is used to locate the motor HP and the ampacity is then obtained under the voltage column. A 10 HP, 460-volt, 3 Ø motor has a full-load current of 14A.

Question: What size copper THWN branch circuit conductors would be required to supply a 25 HP, 3 Ø, 460V motor?

Solution: Using Table 430-150, the full-load current for this motor is 34A. Multiply the motor full-load current by 125% to determine the minimum ampacity for branch circuit conductors.

   34A x 125% = 42.5A

Branch circuit conductor size is then determined using Table 310-16. #8 THWN copper is rated at 55 amps. The branch circuit conductors required to supply this 25 HP motor would be #8 THWN copper.

Motor short circuit and ground fault protection is sized using NEC Table 430-152. To use this table, it is helpful to know the type of motor, horsepower, phase, motor ampacity (full-load current), and the design letter, if any.

Example Problems

Question: What size dual element time delay fuse would be required for a motor with the following rating: 50 HP, 3 Ø, 460V, Design Letter D?

Solution: The motor full-load current is determined using Table 430-150. Using Table 430-152, the full-load current is multiplied by the proper percentage listed for the characteristics of the motor. From Table 430-150, the full-load current for a 50 HP, 3 Ø, 460V motor is 65 amps. From Table 430-152, the full-load current is multiplied by 175% to select a time delay fuse for a polyphase motor (3 Ø) with a design letter other than E.

   65A x 175% = 113.75A

Per NEC 430-52(c)(1), the next lower standard fuse size is 110A. A fuse with a 110A rating would be used. If the 110A fuse proves to be inadequate to carry the load, NEC 430-52(c) (1) Ex. 1 allows the next higher size fuse to be used. In this case, the next higher size would be a 125A fuse size.

Question: What size inverse time circuit breaker would be required for a motor with the following rating: 10 HP, 3 Ø, 208V, Design Letter E?

Solution: The motor full-load current is 30.8A per Table 430-150. The full-load current is multiplied by 250% per Table 430-152 to size an inverse time circuit breaker for a 3 Ø motor with Design Letter E.

30.8A x 250% = 77A

The next lower size circuit breaker is 70A and NEC 430-52(c)(1) Ex. 1 would permit an 80A circuit breaker if the 70A breaker is inadequate for the load.

13.0.0 WELDERS

Branch circuit conductors and overcurrent protection devices for welders are sized using the primary current and duty cycle for the welder. This information is obtained from the nameplate on the welder. Using the relevant NEC section, the multiplier (demand factor) is determined by using the nameplate duty cycle of the welder. The primary current is multiplied by the NEC multiplier to determine the ampacity that is then used to size branch circuit conductors and overcurrent protection. Typical welders include transformer arc welders, motor-generator welders, and resistance welders.

NEC Article 630 covers electric welders. Part B covers AC transformer and DC rectifier arc welders. Part C covers motor-generator arc welders and Part D covers resistance welders. Each NEC section lists duty cycle multipliers for individual welders based upon the type of welder. NEC 630-11(a) provides multipliers for arc welders, 630-21(a) lists multipliers for motor-generator welders, and 630-31(a) provides multipliers for resistance welders.

Per NEC 630-12(a) and 630-22(a), the overcurrent protection device for an individual welder branch circuit shall have a rating or setting not exceeding 200% of the primary current rating of the welder. These NEC sections apply to arc welders and motor-generator welders. For resistance welders, NEC 630-32(a) provides that the rating or setting for the overcurrent protection device shall not exceed 300% of the primary current rating of the welder.

Note   It is advisable to check with the local inspection authority or project specification regarding the setting or rating for welder branch circuit overcurrent protection devices. Many authorities will not permit the overcurrent device rating to exceed the rating of the branch circuit conductors.

Example Problems

Question: What size THWN copper branch circuit conductors would be required to supply an individual arc welder with a nameplate primary current of 70 amps and a duty cycle of 80%?

Solution: The nameplate primary current is multiplied by .89. This multiplier is obtained from NEC 630-11(a) for an arc welder with a duty cycle of 80%.

70A x .89 = 62.3A

The copper THWN conductor size is selected per NEC Table 310-16. #6 THWN copper conductors, with a rating of 65A, would be required.

Question: What size THWN copper branch circuit conductors and what size circuit breaker would be required to supply a resistance welder with a nameplate primary current of 125 amps and a duty cycle of 40%?

Solution: The nameplate primary current is multiplied by .63. This multiplier is obtained from NEC 630-31(a) for an arc welder with a duty cycle of 40%.

125A x .63 = 78.75A

Per Table 310-16, #4 THWN conductors (85 amps) would be used. Per NEC 630-32(a), the primary current is multiplied by 300% to size the overcurrent protection device. The nameplate rating for this resistance welder is 125A.

125A x 300% = 375A

Since this value exceeds the rating of a standard 300A overcurrent protection device, NEC 630-32 permits the next higher standard size (400A) to be used.

References

For advanced study of topics covered in this Task Module, the following works are suggested:

National Electrical Code Handbook, Latest Edition, NFPA, Quincy, MA.

SELF CHECK REVIEW / PRACTICE QUESTIONS

1. The general lighting load, expressed as a unit load per square foot, for an office building where the actual number of general purpose receptacles is *not* known is:
   a. 180.
   b. 3.
   c. 3½.
   d. 4½.

2. The general lighting load, expressed as a unit load per square foot, for a warehouse is:
   a. ½.
   b. ¼.
   c. 1.
   d. 1½.

3. An office building has 1,750 square feet of hallway and stairway space within the building. What is the total general lighting load for this space?
   a. 6,125VA
   b. 7,875VA
   c. 875VA
   d. 1,750VA

4. A continuous load is defined by the NEC as a load where the maximum current is expected to continue at least:
   a. 30 minutes.
   b. 1 hour.
   c. 8 hours.
   d. 3 hours.

5. What is the capacity, in amperes, for a 20A circuit breaker supplying a continuous load?
   a. 20
   b. 12
   c. 16
   d. 18

6. Branch circuit conductors supplying continuous duty loads are calculated at _____% of the rated load.
   a. 100
   b. 125
   c. 80
   d. 115

7. What is the ampacity of a load rated 208 volt, 3 Ø, 10kW?
   a. 48.08A
   b. 27.76A
   c. 43.48A
   d. 25.10A

8. What size copper conductors would be required for a 240 volt, 1 Ø circuit supplying a non-continuous load of 27.5 amps at a distance of 145 feet?
   a. #6
   b. #8
   c. #12
   d. #10

9. A correction factor of _____ must be used for #10 THHN copper conductors that are installed in an area with an ambient temperature of 75°F.
   a. .41
   b. 1.04
   c. 0
   d. 1.00

10. The allowable ampacity for eleven #12 THHN conductors in a cable must be adjusted by:
    a. 40%.
    b. 80%.
    c. 70%.
    d. 50%.

11. What is the current-carrying capacity for #6 THWN aluminum conductors run through an ambient temperature of 38°C?
    a. 41A
    b. 44A
    c. 45.5A
    d. 50A

12. What is the current-carrying ampacity for each of ten #10 THHN copper conductors installed in a single conduit?
    a. 20A
    b. 28A
    c. 32A
    d. 40A

13. Voltage drop shall not exceed _____% to the farthest outlet in a branch circuit.
    a. 2
    b. 3
    c. 5
    d. 10

14. What size solid copper branch circuit conductors are required to serve a 16-amp, 120-volt load? The total length of the circuit is 88 feet.
    a. #12
    b. #10
    c. #8
    d. #6

15. The voltage drop for a 208-volt, 1 Ø circuit with a load of 26.5 amps, using #8 stranded copper conductors at a circuit length of 145 feet is _____ volts.
    a. 5.87
    b. 5.98
    c. 5.08
    d. 5.18

16. The voltage drop for a 480-volt, 3 Ø circuit with a load of 26.5 amps, using #8 stranded copper conductors at a circuit length of 145 feet is _____ volts.
    a. 5.87
    b. 5.98
    c. 5.08
    d. 5.18

17. The voltage drop for an 18-amp, 208-volt, 3 Ø load with a total circuit length of 105 feet using #10 solid copper conductors is _____ volts.
    a. 3.96
    b. 4.57
    c. 4.06
    d. 4.69

18. Load calculations for circuits supplying lighting units with ballasts are based upon:
    a. the total wattage of all lamps in the fixtures.
    b. the circuit voltage times lamp wattage.
    c. the ampere ratings of the ballasts.
    d. 180VA per ballast.

19. What is the total load, in amps, for a 120-volt circuit supplying four 150-watt recessed incandescent fixtures and six recessed fluorescent fixtures? (Each fluorescent ballast is rated at .65A.)
   a. 3.9A
   b. 6.07A
   c. 12.5 A
   d. 8.9A

20. What is the total ampacity for a 120-volt branch circuit consisting of four duplex receptacles rated non-continuous duty and six duplex receptacles rated continuous duty?
   a. 17.25A
   b. 18.75A
   c. 15.00A
   d. 16.50A

21. What is the maximum number of general purpose non-continuous duty duplex receptacles that can be connected to a 20A, 120V circuit in a commercial building, and what is the NEC required load rating for each receptacle?
   a. 10 receptacles, 1.5A
   b. 10 receptacles, 1.875A
   c. 16 receptacles, 150VA
   d. 13 receptacles, 180VA

22. The load for 12 feet of multi-outlet assembly used for non-continuous duty is:
   a. 432VA.
   b. 2160VA.
   c. dependent upon the number of outlets.
   d. 180VA per outlet.

23. The load for six feet of multi-outlet assembly used at continuous duty is:
   a. 180VA per outlet.
   b. 180VA per outlet times 125%.
   c. 1,080VA.
   d. 216VA.

24. Show window lighting is calculated at _____ per linear foot.
   a. 180VA
   b. 200VA
   c. 100 watts
   d. 120 watts

25. How many receptacles are required for 90 linear feet of show window area?
   a. 4
   b. 7
   c. 9
   d. 8

26. What is the total load for 55 linear feet of show window?
   a. 9,900VA
   b. 11,000VA
   c. 10,000VA
   d. 6,600VA

27. What is the total load, in VA, for two small appliance branch circuits in a single-family dwelling?
   a. 2,400VA
   b. 1,800VA
   c. 3,000VA
   d. 1,500VA

28. What is the demand load for one household range with a nameplate rating of 11.75kW, 240V, 1 Ø?
   a. 8kW
   b. 49 amps
   c. 12kW
   d. 9.4kW

29. What is the demand load, in amps, for one household electric range with a nameplate rating of 16.75kW, 1 Ø, 240 volts?
   a. 69.79A
   b. 41.67A
   c. 50A
   d. 33.33A

30. What is the demand load, in amps, for a single circuit supplying one wall-mounted oven rated 11.75kW, 1 Ø, 240 volt, and a countertop cooking unit rated 9.6kW, 1 Ø, 240 volts?
   a. 66.67A
   b. 48.33A
   c. 50A
   d. 88.96A

31. What is the demand load, in amps, for seven pieces of commercial cooking equipment with a total nameplate rating of 44.5kVA operating at continuous duty? Each piece of equipment has a nameplate rating of 208-volt, 3 Ø.
    a. 80.29A
    b. 108.08A
    c. 100.36A
    d. 123.52A

32. What is the ampacity used to size branch circuit conductors for a 3 Ø, 208V branch circuit supplying 25kW of resistance heating elements?
    a. 120A
    b. 69A
    c. 104A
    d. 87A

33. What is the branch circuit sizing load, in amps, for four 1,000 watt, 208 volt, 1 Ø electric baseboard heaters?
    a. 24.04A
    b. 33.33A
    c. 20.83A
    d. 19.23A

34. What size THWN copper branch circuit conductors are required to supply an air conditioning unit with a nameplate rating of 33.5 amps, 208 volts, 3 Ø?
    a. 10
    b. 8
    c. 6
    d. 4

35. The dual element fuse size required for a 40 HP, 3 Ø, 460V motor with a Design Letter D is:
    a. 60A.
    b. 90A.
    c. 175A.
    d. 125A.

36. What is the motor full-load current for a motor with a nameplate rating of 50 HP, 440 volt, 3 Ø?
    a. 81.25A
    b. 130A
    c. 52A
    d. 65A

37. What size time delay fuse is required for a motor rated at 15 HP, 208-volt, 3 Ø, with a Design Letter E?
    a. 60A
    b. 80A
    c. 90A
    d. 110A

38. What size instantaneous trip breaker is required for a motor rated at 7½ HP, 460 volt, 3 Ø, with a Design Letter B?
    a. 60A
    b. 30A
    c. 90A
    d. 15A

39. What is the motor full-load current for a motor rated at 5 HP, 230 volt, 1 Ø?
    a. 28A
    b. 15.2A
    c. 16.7A
    d. 30.8A

40. An AC arc welder on an individual circuit has a nameplate primary current of 100 amps and a duty cycle of 70%. The branch circuit conductors must be a minimum size _____ THHN copper.
    a. #3
    b. #6
    c. #2
    d. #4

Conductor Selection and Calculations

Module 20302

CONDUCTOR SELECTION AND CALCULATIONS

Objectives

Upon completion of this module, the trainee will be able to:

1. Select electrical conductors for specific applications.
2. Calculate voltage drop in both single- and three-phase applications.
3. Interpret and apply NEC regulations governing conductors.
4. Understand and apply NEC parallel rules.
5. Understand and apply NEC tap rules.
6. Size conductors for the load.
7. Explain types and purposes of different conductor insulations.
8. Understand and apply NEC Tables 310-16 through 310-19.
9. Derate conductors for fill, temperature, and voltage drop.
10. Select conductors for various temperature ranges and atmospheres.

Prerequisites

Successful completion of the following Task Modules is required before beginning study of this Task Module: Core Curricula, Electrical Levels 1 and 2, Electrical Level 3, Module 20301.

Required Student Materials

1. Trainee Task Module
2. Copy of the latest edition of the National Electrical Code

COURSE MAP INFORMATION

This course map shows all of the *Wheels of Learning* Task Modules in the third level of the Electrical curricula. The suggested training order begins at the bottom and proceeds up. Skill levels increase as a trainee advances on the course map. The training order may be adjusted by the local Training Program Sponsor.

Course Map: Electrical, Level 3

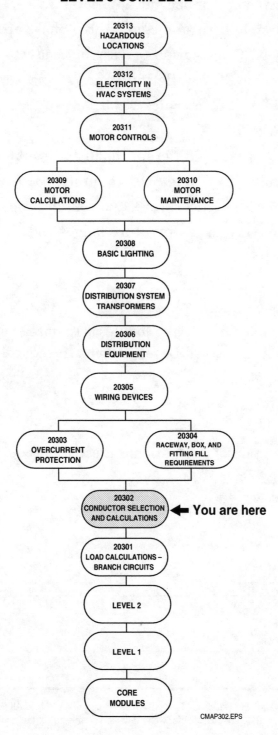

TABLE OF CONTENTS

Trade Terms Introduced in This Module

AAC: All aluminum conductor.

AASC: Aluminum alloy stranded conductors.

ACSR (aluminum, conductor, steel reinforced): A bare composite of aluminum and steel wires, usually aluminum around steel.

Al: Aluminum.

Al-Cu: An abbreviation for aluminum and copper, commonly marked on terminals, lugs, and other electrical connectors to indicate that the device is suitable for use with either aluminum or copper conductors.

ampacity: The current-carrying capacity of conductors or equipment, without exceeding its temperature rating, expressed in amperes.

armor: Mechanical protector for cables; usually a helical winding of metal tape, formed so that each convolution locks mechanically upon the previous one (interlocked armor); may be a formed metal tube or a helical wrap of wires.

AWG (American Wire Gage): The standard for measuring wires in America.

cable: An assembly of two or more wires that may be insulated or bare.

cable, aerial: An assembly of one or more conductors and a supporting messenger. Sometimes referred to as "messenger cable."

cable, armored: A cable having armor (see armor).

cable, belted: A multiconductor cable having a layer of insulation over the assembled insulated conductors.

cable, bore-hole: The term given vertical-riser cables in mines.

cable clamp: A device used to clamp around a cable to distribute mechanical strain evenly to all elements of the cable.

cable, coaxial: A cable used for high frequency, consisting of two cylindrical conductors with a common axis separated by a dielectric; normally the outer conductor is operated at ground potential for shielding.

cable, control: Used to supply voltage (usually ON or OFF) to motor controls and other controllers that operate electrical circuits or apparatus. Often used in cable trays.

cable, duplex: A twisted pair of cables.

cable, festoon: A cable draped in accordion fashion from sliding or rolling hangers, usually used to feed moving equipment such as bridge cranes.

cable, hand: A mining cable used to connect equipment to a reel truck.

cable, parkway: Designed for direct burial with heavy mechanical protection of jute, lead, and steel wires.

cable, portable: Used to transmit power to mobile equipment.

cable, power: Used to supply current (power).

cable, pressure: A cable having a pressurized fluid (gas or oil) as part of the insulation.

cable, ribbon: A flat multiconductor cable.

cable, service drop: The overhead cable from the utility line to the customer's property.

cable, signal: Used to transmit power-limiting data.

cable, spacer: An aerial distribution cable made of covered conductors held by insulated spacers; designed for wooded areas.

cable, spread room: A room adjacent to a control room to facilitate routing of cables in trays away from the control panels.

cable, tray: A multiconductor having a nonmetallic jacket, designed for use in cable trays; (not to be confused with type TC cable for which the jacket must also be flame retardant).

cable tray: A rigid structure to support cables; normally having the appearance of a ladder and accessible at the top to facilitate changes.

cable, triplexed d: Helical assembly of three insulated conductors and sometimes a bare grounding conductor.

cable, unit: A cable having pairs of cables stranded into groups (units) of a given quantity, then these groups form the core.

cable, vertical riser: Cables utilized in circuits of considerable elevation change; usually incorporate additional components for tensile strength.

A variety of materials is used to transmit electrical energy, but copper, due to its excellent cost-to-conductivity ratio, still remains the basic and most ideal conductor. Electrolytic copper, the type used in most electrical conductors, can have three general characteristics:

- Method of stranding
- Degree of hardness (temper)
- Bare, tinned, or coated

Method of Stranding: Stranding refers to the relative flexibility of the conductor and may consist of only one strand or many thousands, depending on the rigidity or flexibility required for a specific need. For example, a small-gauge wire that is to be used in a fixed installation is normally solid (one strand), whereas a wire that will be constantly flexed requires a high degree of flexibility and would contain many strands.

- Solid wire is the least flexible form of a conductor and is merely one strand of copper.
- Stranded refers to more than one strand in a given conductor and may vary from 3 to 37 depending on size. See Fig. 1.
- Flexible simply indicates that there are a greater number of strands than are found in normal stranded construction.

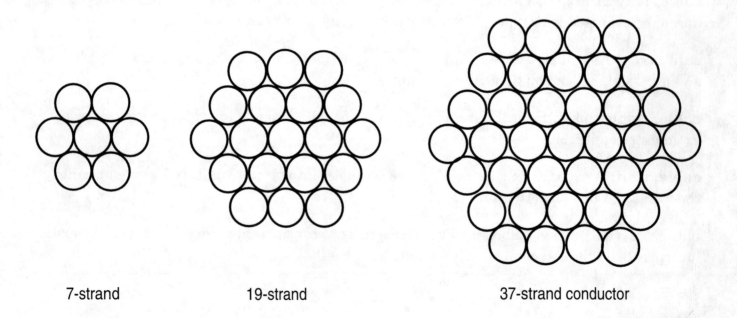

7-strand   19-strand   37-strand conductor

Figure 1. Common number of strands used in conductors.

Degree of Hardness (Temper): Temper refers to the relative hardness of the conductor and is noted as soft drawn-annealed (SD), medium hard drawn (MHD), and hard drawn (HD). Again, the specific need of an installation will determine the required temper. Where greater tensile strength is indicated, MHD would be used over SD, and so on.

Bare, Tinned, or Coated: Untinned copper is plain bare copper that is available in either solid, stranded, or flexible and in the various tempers just described. In this form it is often referred to as *red* copper.

Bare copper is also available with a coating of tin, silver, or nickel to facilitate soldering, to impede corrosion, and to prevent adhesion of the copper conductor to rubber or other types of conductor insulation. The various coatings will also affect the electrical characteristics of copper.

Conductor Size: The American Wire Gage (AWG) is used in the United States to identify the sizes of wire and cable up to and including No. 4/0 (0000), which is commonly pronounced in the electrical trade as "four-aught" or "four-naught." These numbers run in reverse order as to size; that is, No. 14 AWG is smaller than No. 12 AWG and so on up to size No. 1 AWG. To this size (No. 1 AWG), the larger the gauge number, the smaller the size of the conductor. However, the next larger size after No. 1 AWG is No. 1/0 AWG, then 2/0 AWG, 3/0 AWG, and 4/0 AWG. At this point, the AWG designations end and the larger sizes of conductors are identified by circular mils (CM or cmil). From this point, the larger the size of wire, the larger the number of circular mils. For example, 300,000 cmil is larger than 250,000 cmil. In writing these sizes in circular mils, the "thousand" decimal is replaced by the letter k, and instead of writing, say, 500,000 cmil, it is usually written 500 kcmil — pronounced *five-hundred kay-cee-mil*. See Fig. 2 for a comparison of the different wire sizes.

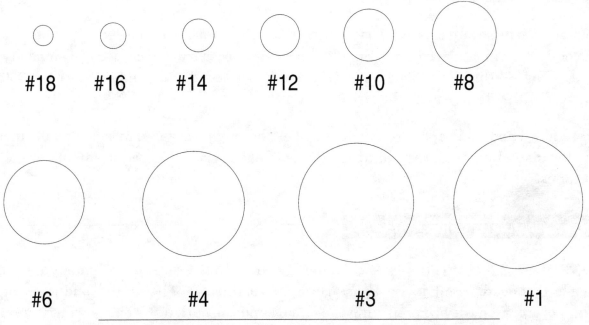

Figure 2. Comparison of some different wire sizes (not to scale).

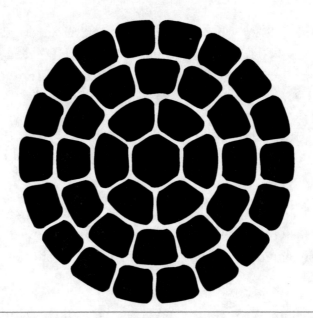

Figure 3. Cross-section of a 37-strand compressed conductor.

1.1.0 COMPRESSED CONDUCTORS

Compressed aluminum conductors are those which have been compressed so as to reduce the air space between the strands. Figure 3 shows a cross-section of a 37-strand compressed conductor. Notice that the once round strands are compressed into a sort of rectangular shape that reduces the air space between strands.

The purpose of compressed conductors is to reduce the overall diameter of the cable so that it may be installed in a conduit that is smaller than that required for standard conductors of the same wire size. Compressed conductors are especially useful when increasing the ampacity of an existing service or feeder circuits.

For example, let's say an existing service is rated at 250 amperes and is fed with four 350 kcmil THW conductors in 3-inch conduit. Should it become necessary to increase the ampacity of the service to, say, 300 amperes, 500 kcmil THW compressed conductors may replace the 350 kcmil conductors without increasing the size of conduit.

Both standard and compressed conductors will be thoroughly covered in this module, including conductor insulation and practical applications of each type. NEC requirements are followed at all times.

2.0.0 CONDUCTOR APPLICATIONS

The NEC defines *feeder* as the circuit conductors between the service equipment and the final branch-circuit overcurrent device. *Branch circuit* is defined as "The circuit conductors between the final overcurrent device protecting the circuit and the outlet(s)." The power-riser diagram in Fig. 4 shows examples of both feeders and branch circuits.

ELECTRICAL TRAINEE TASK MODULE 20302

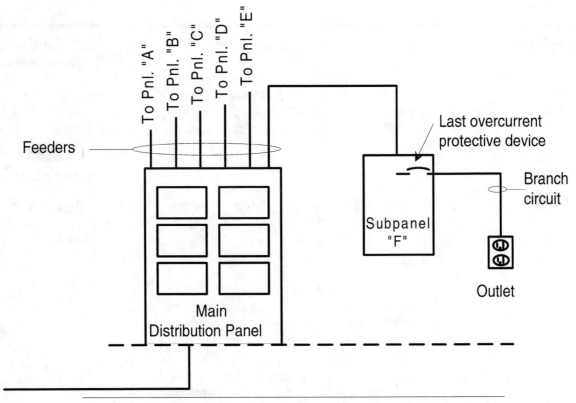

Figure 4. Power-riser diagram showing feeders and branch circuits.

When current-carrying conductors are used in an electrical system, the NEC requires that each ungrounded conductor be protected from damage by an overcurrent protective device such as a fuse or circuit breaker. The conductors must also be identified so that the ungrounded conductors may be distinguished from the grounded and grounding conductors. Minimum sizes or required ampacity for any of these conductors used in a circuit are selected based on NEC rules and tables. The NEC also provides tables that list the physical and electrical properties of conductors to allow the selection of the proper conductor for any application.

Article 240

The table in Fig. 5 summarizes the NEC requirements for protecting conductors from excess current caused by overloads, short circuits, or ground faults. The setting or sizes of the protective device are based on the ampacity of the conductors as listed in appropriate NEC tables. Under certain conditions, the overcurrent device setting may be larger than the ampacity rating of the conductors as listed in the exceptions to the basic rules. For convenience, a standard rating of a fuse or circuit breaker may be used even if this rating exceeds the ampacity of the conductor if the rating does not exceed 800 amperes. For example, a branch-circuit with a load of 56 amperes may be protected with a 60-amp fuse, because 56 amperes is not a standard fuse size.

Section 240-3

Application	NEC Regulation	NEC Section or Comments
Protection of conductors	Conductors, other than flexible cords and fixture wires, must be protected against overcurrent in accordance with their ampacities as specified in Section 310-15.	240-3
	Exceptions	240-3 (a) through (m)
	a) Conductor overload protection is not required where the interruption of the circuit would create a hazard (material handling magnet circuits, fire-pump circuits, etc.)	Note: short-circuit protection must be provided
	b) Conductors that are not part of a multioutlet branch circuit supplying receptacles for cord- and plug-connected portable loads and where the ampacity of the conductors does not correspond with the standard ampere rating of a fuse or circuit breaker without overload trip adjustments above its rating, the next higher rating is permitted provided this rating does not exceed 800 amperes.	
	c) Where the overcurrent device is rated over 800 amperes, the ampacity of the conductors it protects must be equal to or greater than the rating of the overcurrent device as defined in NEC Section 240-6.	
	d) Tap conductors may be protected according to Sections 210-19(c), 240-21, 364-11, 364-12, and 430-53(d).	
	e) Motor-operated appliance circuit conductors are permitted to be protected against overcurrent as specified in Parts B and D of NEC Article 422.	
	f) Motor and motor-control circuit conductors are permitted to be protected against overcurrent as permitted in Parts C, D, E, and F of NEC Article 430.	
	g) Supply conductors for phase converters for both motor and nonmotor loads are permitted to be protected against overcurrent as specified in NEC Section 455-7.	

Figure 5. Overcurrent protection of conductors.

Application	NEC Regulation	NEC Section or Comment
	h) air conditioning and refrigeration equipment circuit conductors are permitted to be protected against overcurrent as specified in Parts C and F of NEC Article 440.	
	i) Single-phase, (other than two-wire) and multiphase (other than delta-delta, three-wire) transformer secondary conductors are not considered to be protected by the primary overcurrent protection. However, conductors supplied by the secondary side of a single-phase transformer having a two-wire (single-voltage) secondary or a three phase delta-delta connected transformer having a three-wire (single-voltage) secondary are permitted to be protected by overcurrent protection provided on the primary side of the transformer, provided the protection complies with NEC Section 450-3 and does not exceed the value determined by multiplying the secondary conductor ampacity by the secondary to primary transformer voltage ratio.	
	j) Capacitor circuit conductors are permitted to be protected against overcurrent according to NEC Sections 460-8(b) and 460-25(a) through (d).	
	k) Welder circuit conductors are permitted to be protected against overcurrent according to NEC Sections 630-12, 630-22, and 630-32.	
	l) Remote-control circuits, transformer secondary conductors, capacitor circuits, circuits for welders, and circuits that cannot allow a power loss have special NEC rules that apply.	
	m) Fire protective signaling system circuit conductors shall be protected against overcurrent in accordance with NEC Sections 760-23, 760-24, 760-41, and Chapter 9, Tables 12(a) and 12(b).	
Overcurrent device required	A fuse or overcurrent trip unit of a circuit breaker is required to be connected in series with each ungrounded conductor.	240-20(a)

Figure 5. Overcurrent protection of conductors. *(Cont.)*

Application	NEC Regulation	NEC Section or Comment
Location in circuit	An overcurrent device must be connected at the point where the conductor receives its supply. Exceptions: a) Feeder and branch-circuit conductors must be protected by overcurrent-protective devices connected at the point the conductors receive their supply, unless otherwise permitted in (b) through (m) below. b) Conductors are permitted to be tapped—without overcurrent protection at the tap—to a feeder or transformer secondary where all the following conditions are met: (1) The length of the tap conductors does not exceed 10 feet; (2) The ampacity of the tap conductor is not less than the combined computed load on the circuit supplied by the tap conductor, and not less than the rating of the overcurrent-protective device at the termination of the tap conductors. Furthermore, (3) the tap conductors must not extend beyond the switchboard, panelboard, disconnecting means or control devices they supply. c) Conductors may be tapped, without overcurrent protection at the tap, to a feeder where the length of the tap conductors does not exceed 25 feet; where the ampacity of the tap conductors is not less than $\frac{1}{3}$ the rating of the overcurrent device protecting the feeder conductors; where the tap conductors terminate in a single circuit breaker or a single set of fuses that will limit the load to the ampacity of the tap conductors. This device may supply any number of additional overcurrent devices on its load side; and, the tap conductors are suitably protected from physical damage or are enclosed in a raceway.	240-21

Figure 5. Overcurrent protection of conductors. *(Cont.)*

Application	NEC Regulation	NEC Section or Comment
	d) Conductors supplying a transformer are permitted to be tapped, without overcurrent protection at the tap, from a feeder where all of the following conditions are met: (1) The conductors supplying the primary of a transformer must have an ampacity at least $\frac{1}{3}$ the rating of the overcurrent device protecting the feeder conductors; (2) The conductors supplied by the secondary of the transformer must have an ampacity that, when multiplied by the ratio of the secondary-to-primary voltage, is at least $\frac{1}{3}$ the rating of the overcurrent device protecting the feeder conductors; (3) The total length of one primary plus one secondary conductor, excluding any portion of the primary conductor that is protected at its ampacity, is not over 25 feet;(4) The primary and secondary conductors are suitably protected from physical damage; (5) The secondary conductors terminate in a single circuit breaker or set of fuses that will limit the load current to no more than the conductor ampacity allowed in NEC Section 310-15. e) Conductors over 25 feet long are permitted to be tapped from feeders in high bay manufacturing buildings which has walls over 35 feet high. Conductors tapped, without overcurrent protection at the tap, to a feeder must not be over 25 feet long horizontally and not over 100 feet total length where all of the conditions in NEC Section 240-21(e)(1) through (7) are met. f) Taps to individual outlets and circuit conductors supplying a single electric household range are permitted to be protected by the branch-circuit overcurrent devices where installed according to NEC Sections 210-19, 210-20, and 210-24. g) Busways and busway taps are permitted to be protected against overcurrent as specified in NEC Sections 364-10 through 364-13.	

Figure 5. Overcurrent protection of conductors. *(Cont.)*

Application	NEC Regulation	NEC Section or Comment
	h) Motor feeder and branch-circuit conductors are permitted to be protected against overcurrent according to Sections 430-28 and 430-53 respectively. i) Conductors from generator terminals are permitted to be protected against overcurrent as specified in NEC Section 445-5. j) Conductors connected to a transformer secondary of a separately derived system for industrial installations may be considered to be protected against overcurrent where the conditions in NEC Section 240-21(j)(1) through (4) are met. m) Outside conductors tapped to a feeder or connected to a transformer secondary are permitted to be protected by complying with all of the conditions in NEC Section 240-21(m)(1) through (5). n) Service-entrance conductors are permitted to be protected by overcurrent devices as specified in NEC Section 230-91.	

Figure 5. Overcurrent protection of conductors. *(Cont.)*

Section 240-21

Section 200-2

In most cases, an overcurrent device must be connected at the point where the conductor to be protected receives its supply. The most common situations are shown in Fig. 6, which illustrate the basic rule and several exceptions, including the 10-foot tap rule. Figure 7 illustrates the 25-foot tap rule.

Wiring systems require a grounded conductor in most installations. A grounded conductor, such as a neutral, or a grounding conductor must be identified either by the color of its insulation, by markings at the terminals, or by other suitable means. In general, a grounded conductor must have a white or natural gray finish. When this is not practical for conductors larger than No. 6 AWG, marking the terminations with white color is an acceptable method of identifying the conductor. Tagging is also an acceptable method.

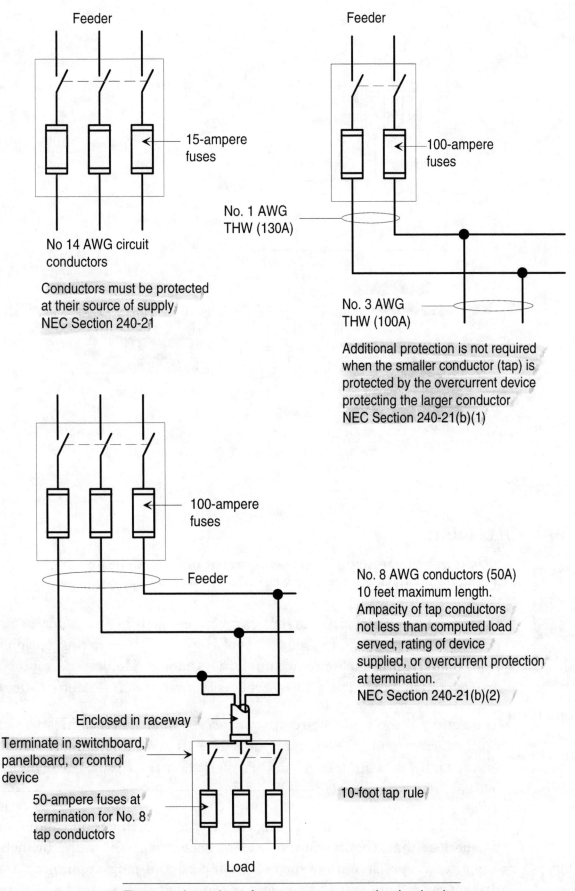

Feeder

15-ampere fuses

No 14 AWG circuit conductors

Conductors must be protected at their source of supply. NEC Section 240-21

Feeder

100-ampere fuses

No. 1 AWG THW (130A)

No. 3 AWG THW (100A)

Additional protection is not required when the smaller conductor (tap) is protected by the overcurrent device protecting the larger conductor NEC Section 240-21(b)(1)

100-ampere fuses

Feeder

No. 8 AWG conductors (50A) 10 feet maximum length. Ampacity of tap conductors not less than computed load served, rating of device supplied, or overcurrent protection at termination. NEC Section 240-21(b)(2)

Enclosed in raceway

Terminate in switchboard, panelboard, or control device

50-ampere fuses at termination for No. 8 tap conductors

10-foot tap rule

Load

Figure 6. Location of overcurrent protection in circuits.

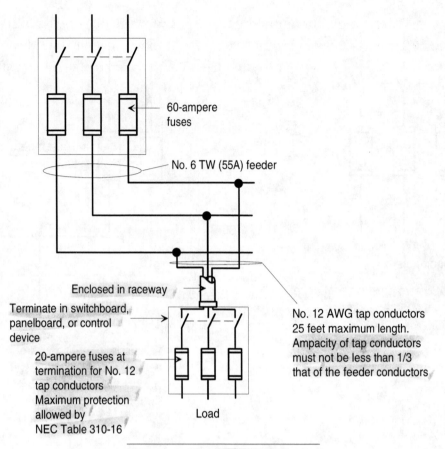

60-ampere fuses

No. 6 TW (55A) feeder

Enclosed in raceway

Terminate in switchboard, panelboard, or control device

No. 12 AWG tap conductors 25 feet maximum length. Ampacity of tap conductors must not be less than 1/3 that of the feeder conductors

20-ampere fuses at termination for No. 12 tap conductors Maximum protection allowed by NEC Table 310-16

Load

Figure 7. NEC 25-foot tap rule.

2.1.0 BRANCH CIRCUITS

National Electrical Code 1996

Now that an overview of NEC overcurrent protection has been presented, let's dig deeper into the subject of NEC installation requirements for conductors.

In general, the ampacity (current-carrying capacity) of a conductor must not be less than the maximum load served. However, there are exceptions to this rule; namely, a branch circuit supplying a motor. Motors and motor circuits are covered in NEC Article 430 as well as Module 20309 — Motor Calculations.

Section 210-19

Sections 210-3 and 210-19

The rating of the branch-circuit overcurrent device determines the rating of the branch circuit. For example, if a No. 10 AWG, 30-ampere conductor is protected by a 20-ampere circuit breaker, then the circuit is considered to be a 20-ampere branch circuit.

Sections 210-3 and 210-19

Furthermore, the current-carrying capacity of branch-circuit conductors must not be less than the maximum load to be served. Where the branch circuit supplies receptacle outlets for use with cord-and-plug appliances and other utilization equipment, the conductor's ampacity must not be less than the rating of the branch-circuit overcurrent device.

As mentioned previously, when the ampacity of the conductor does not match up with a standard rating of fuses or circuit breakers, the next higher standard size overcurrent device may be used, provided the overcurrent device does not exceed 800 amperes. This exception is not permitted, however, when the branch circuit supplies receptacles where cord-and-plug connected appliances, and similar electrical equipment could be used, because too many loads plugged into the circuit could result in an overload condition. The next standard size fuse or circuit breaker may be used only when the circuit supplies a fixed load.

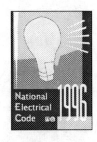

Section 240-3

Section 210-19(a)

The allowable ampacity of conductors used on most electrical systems is found in NEC Tables 310-16 through 310-19. However, the ampacities in these tables are subject to correction factors that must be applied where high ambient temperatures are encountered; that is, when the ambient temperature for the conductor location exceeds 30° C (86° F). This reduction is required even if the reduction for more than three conductors in a raceway is also applied. For example, if six No. 10 AWG, TW current-carrying conductors are installed in a single raceway, and where the ambient temperature is 40°C, the ampacity of 30 amperes must be derated or reduced to 80% because of conduit fill and then reduced again by a correction factor of .82 because of the ambient temperature. Therefore, when more than three current-carrying conductors are installed in a single raceway or cable, the allowable ampacity for this condition is calculated as follows:

See footnotes to NEC Table 310-16

$$30 A \times .8 \times .82 = 19.68 \text{ amperes}$$

In this situation, the listed ampacities must be reduced because of the heating effect of many current-carrying conductors in proximity. Grounding conductors are not counted as current-carrying conductors.

The rating of the branch-circuit overcurrent device serving continuous loads must be not less than the noncontinuous load plus 125% of the continuous load.

See footnotes to NEC Table 310-16

NOTE The NEC defines continuous load as a load where the maximum current is expected for three hours or more.

210-22(c)

2.2.0 CONDUCTOR PROTECTION

Article 100

Conductors must be installed and protected from damage—both physically and electrically—according to several NEC sections. Additional requirements specify the use of boxes or fittings for certain connections, specify how connections are made to terminals, and restrict the use of parallel conductors. When conductors are installed in enclosures or raceways, additional rules apply. Finally, if conductors are installed underground, the burial depth and other installation requirements are specified by the NEC.

Task Module 20303 — Overcurrent Protection — deals with overcurrent protection for conductors to some extent. However, a brief review is warranted here.

All conductors must be protected against overcurrents in accordance with their ampacities as set forth in the NEC Section. They must also be protected against short-circuit current damage.

Section 240-3

Sections 240-1 & 110-10

Sections 364-10 & 365-5

Ampere ratings of overcurrent-protective devices must not be greater than the ampacity of the conductor. There is, however, an exception: NEC Section 240-3 states that if such conductor rating does not correspond to a standard size overcurrent-protective device, the next larger size overcurrent-protective device may be used provided its rating does not exceed 800 amperes and when the conductor is not part of a multi-outlet branch circuit supplying receptacles for cord-and-plug connected portable loads. When the ampacity of busway or cablebus does not correspond to a standard overcurrent-protective device, the next larger standard rating may be used only if the rating does not exceed 800 amperes.

Section 240-6

Sections 430-40 & 430-52

Standard overcurrent-device sizes are: 1, 3, 6, 10, 15, 20, 25, 30, 35, 40, 45, 50, 60, 70, 80, 90, 100, 110, 125, 150, 175, 200, 225, 250, 300, 350, 400, 450, 500, 600, 700, 800, 1000, 1200, 1600, 2000, 2500, 3000, 4000, 5000, and 6000 amperes. An additional standard rating for fuses is 601 amperes.

NOTE   The small fuse ampere ratings of 1, 3, 6, and 10 have recently been added to the NEC to provide more effective short-circuit and ground-fault protection for small loads.

Protection of conductors under short-circuit conditions is accomplished by obtaining the maximum short-circuit current available at the supply end of the conductor, the short-circuit withstand rating of the conductor, and the short-circuit let-through characteristics of the overcurrent device.

When a non-current-limiting device is used for short-circuit protection, the conductor's short-circuit withstand rating must be properly selected based on the overcurrent protective device's ability to protect. See Fig. 8.

It is necessary to check the energy let-through of the overcurrent device under short-circuit conditions. Select a wire size of sufficient short-circuit withstand ability.

In contrast, the use of a current-limiting fuse permits a fuse to be selected which limits short-circuit current to a level less than that of the conductor's short-circuit withstand rating — doing away with the need of oversized ampacity conductors. See Fig. 9.

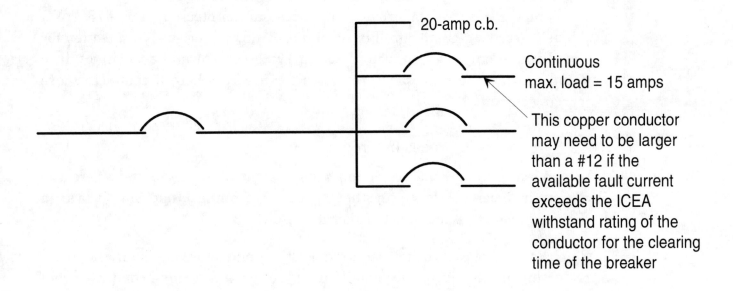

20-amp c.b.

Continuous
max. load = 15 amps

This copper conductor
may need to be larger
than a #12 if the
available fault current
exceeds the ICEA
withstand rating of the
conductor for the clearing
time of the breaker

Figure 8. Non-current-limiting device.

In many applications, it is desirable to use the convenience of a circuit breaker for a disconnecting means and general overcurrent protection, supplemented by current-limiting fuses at strategic points in the circuits.

Flexible cords, including tinsel cords and extension cords, must be protected against overcurrent in accordance with their ampacities. Supplementary fuse protection is an acceptable method of protection. For #18 AWG fixture wire of 50 feet length or more, a 6-ampere fuse will

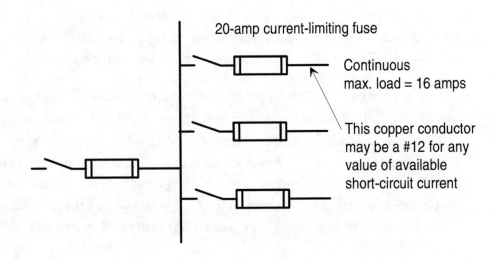

20-amp current-limiting fuse

Continuous
max. load = 16 amps

This copper conductor
may be a #12 for any
value of available
short-circuit current

Figure 9. Current-limiting device.

provide the necessary protection. For #16 AWG fixture wire of 100 feet or more, an 8-ampere fuse would provide the necessary protection. For #18 AWG extension cords, a 10-ampere fuse would provide the necessary protection for a cord where only two conductors are carrying current and a 7-ampere fuse would provide the necessary protection for a cord where 3 conductors are carrying current.

Section 240-4

2.2.1 Location Of Fuses In Circuits

Section 240-21 In general, overcurrent protection must be installed at points where the conductors receive their supply; that is, at the beginning or lineside of a branch circuit or feeder. Exceptions to this rule follow:

Rule No. 1: Fuses are not required at the conductor supply if the fuses protecting one conductor are small enough to protect a small conductor connected thereto.

Rule No. 2: Fuses are not required at the conductor supply if a feed tap conductor is not over ten feet long; is enclosed in raceway, does not extend beyond the switchboard, panelboard or control device which it supplies, and has an ampacity not less than the combined computed loads supplied and not less than the rating of the device supplied unless the tap conductors are terminated in a fuse not exceeding the tap conductors' ampacities. For field installed taps, the ampacity of the tap conductor must be at least 10% of the overcurrent device rating.

Rule No. 3: Fuses are not required at the conductor supply if a feeder tap conductor is not over 25 feet long; is suitably protected from physical damage; has an ampacity not less than 1/3 that of the feeder conductors or fuses from which the tap conductors receive their supply; and terminates in a single set of fuses sized not more than the tap conductor ampacity.

Rule No. 8: Fuses are not required at the conductor supply if a transformer feeder tap has primary conductors at least 1/3 ampacity and/or secondary conductors at least 1/3 ampacity when multiplied by the approximate transformer turns ratio of the fuse or conductors from which they are tapped; the total length of one primary plus one secondary conductor (excluding any portion of the primary conductor that is protected at its ampacity) is not over 25 feet in length; the secondary conductors terminate in a set of fuses rated at the ampacity of the tap conductors; and if the primary and secondary conductors are suitably protected from physical damage.

Rule No. 10: Fuses are not required at the conductor supply if a feeder tap is not over 25 feet long horizontally and not over 100 feet long total length in high bay manufacturing buildings when only qualified persons will service such a system, and the ampacity of the tap conductors is not less than ⅓ of the fuse rating from which they are supplied, that will limit the load to the ampacity of the tap are at least No. 6 AWG copper or No. 4 AWG aluminum, do not penetrate walls, floors, or ceilings, and are made no less than 30 feet from the floor.

WARNING!   Smaller conductors tapped to larger conductors can be a serious hazard. If not protected against short-circuit conditions, these unprotected conductors can vaporize or incur severe insulation damage.

Rule No. 11: Transformer secondary conductors of separately derived systems do not require fuses at the transformer terminals when all of the following conditions are met:

- Must be an industrial location
- Secondary conductors must be less than 25 feet long
- Secondary conductor ampacity must be at least equal to secondary full-load current of transformer and sum of terminating, grouped, overcurrent devices
- Secondary conductors must be protected from physical damage

NOTE   Switchboard and panelboard protection, along transformer protection must still be observed.   **Sections 384-16 & 450-3**

3.0.0   PROPERTIES OF CONDUCTORS

Various NEC tables define the physical and electrical properties of conductors. Electricians use these tables to select the type of conductor and the size of conduit, or other raceway, to enclose the conductors in specific applications.

NEC tables tabulate properties of conductors as follows:

- Name
- Operating temperature
- Application
- Insulation
- Physical properties
- Electrical resistance
- AC resistance and reactance

NEC Table 310-13 gives the name, operating temperature, application, and insulation of various types of conductors, while tables in NEC Chapter 9 (5, 6, 7, 8, and 9) give the physical properties and electrical resistance.

To gain an understanding of these tables and how they are used in practical applications, let's take a 4/0 THHN copper conductor and see what properties may be determined from NEC tables.

Step 1.   Turn to NEC Table 310-13 and scan down the second column from the left (Type Letter) until "THHN" is found. Scan to the left in this row to see that the trade name of this conductor is Heat-Resistant Thermoplastic.

Step 2.   Scanning to the right in this row, note that the maximum operating temperature for this wire type is 90°C (194°F). Continuing to the right in this row, under the column headed "Application Provisions," we find that this wire type is suitable for use in dry and damp locations. The next column reveals that the insulation is flame-retardant, heat-resistant thermoplastic.

Step 3.   Continuing to the right in this row, the next column lists insulation thickness for various AWG or kcmil wire sizes. The insulation thicknesses for Type THHN wire, from No. 14 AWG to 1000 kcmil, are as follows:

> AWG or
> kcmil size
> 14 - 12 .... 15 mils
> 10 .... 20 mils
> 8 - 6 .... 30 mils
> 4 - 2 .... 40 mils
> 1 - 4/0 .... 50 mils
> 250 - 500 .... 60 mils
> 501 - 1000 .... 70 mils

Consequently, the insulation thickness for 4/0 THHN is 50 mils.

Step 4.   Looking in the very right-hand column ("Outer Covering"), this wire type has a nylon jacket or equivalent.

If it is desired to find the maximum current-carrying capacity of this size and type conductor, when used in a raceway, turn to NEC Table 310-16 and proceed as following:

Step 1.   Scan down the left-hand column until the wire size is found.

Step 2.   Scan to the right in this row until the 90°C column is found. This column covers the insulation types that are rated for 90°C maximum operating temperature, and include type THHN conductor insulation.

ELECTRICAL TRAINEE TASK MODULE 20302

Step 3. Note that the current-carrying rating for this size and type of conductor is 260 amperes. This rating, however, is for not more than three conductors in a raceway. If more than three conductors, such as 4 conductors in a three-phase, 4-wire feeder to a subpanel, this figure (260 amperes) must be derated as follows:

Step 4. Turn to "Notes to Ampacity Tables of 0 to 2000 volts" which immediately follows the NEC Ampacity Tables 310-16 through 310-19. Look under Note 8(a), "Adjustment Factors" which states:

Where the number of current-carrying conductors in a raceway or cable exceed three, the allowable ampacites shall be reduced as shown in the following table:

Number of Current-Carrying Conductors	Percent of Values in Tables as Adjusted for Ambient Temperature if Necessary
4 through 6	80
7 through 9	70
10 through 20	50
21 through 30	45
31 through 40	40
41 and above	35

Step 5. Since we want to know the allowable maximum current-carrying capacity of four 4/0 THHN conductors in one raceway, it is necessary to multiply the previous amperage (260 amperes) by 80% or 0.80.

260 x .80 = 208 amperes

These amperage tables are also based on the conductors being installed in areas where the ambient air temperature is 30°C (86°F). If the conductors are installed in areas with different ambient temperatures, a further deduction is required. For example, if this same set of four 4/0 THHN conductors were installed in an industrial area where the ambient temperature averaged, say, 35°C, look in the correction-factor tables at the bottom of NEC Table 310-16 through 310-19. In doing so, we find that a correction or derating factor for our situation is .96. Consequently, our present current-carrying capacity of 208 amperes must be multiplied by 0.96 to obtain the actual current-carrying capacity of the four conductors.

208 x .96 = 199.68 amperes

Other sizes and types of conductors are handled in a similar manner; that is, find the appropriate table, determine the listed ampacity and then multiply this ampacity by the appropriate factors in the correction-factor tables.

It sometimes becomes necessary to know additional properties of conductors for some conductor calculations, especially for voltage-drop calculations which will appear in a later section in this module. There are many useful tables in NEC Chapter 9. Examples of their practical use will be presented under Section 4.0.0 Voltage Drop.

3.1.0 IDENTIFYING CONDUCTORS

The NEC specifies certain methods of identifying conductors used in wiring systems of all types. For example, the high leg of a 120/240-volt grounded three-phase delta system must be marked with an orange color for identification; a grounded conductor must be identified either by the color of its insulation, by markings at the terminals, or by other suitable means. Unless allowed by NEC exceptions, a grounded conductor must have a white or natural gray finish. When this is not practical for conductors larger than No. 6 AWG, marking the terminals with white color is an acceptable method of identifying the conductors.

3.1.1 Color Coding

Conductors contained in cables are color-coded so that identification may be easily made at each access point. The following table lists the color-coding for cables up through four-wire cable. Although some control-wiring and communication cables contain 60, 80, or more pairs of conductors — using a combination of colors — the ones listed are the most common and will be encountered the most on electrical installations.

Number of Conductors in Cable	Color of Conductors
Two-wire cable	One black (ungrounded phase conductor) One white (grounded conductor)
Two-wire cable with ground	One black (ungrounded phase conductor) One white (grounded conductor) One bare (equipment grounding conductor)
Three-wire cable	One black (ungrounded phase conductor) One white (grounded conductor) One red (ungrounded phase conductor)
Three-wire cable with ground	One black (ungrounded phase conductor) One white (grounded conductor) One red (ungrounded phase conductor) One bare (equipment grounding conductor)

Number of Conductors in Cable	Color of Conductors
Four-wire cable	One black (ungrounded phase conductor) One white (grounded conductor) One red (ungrounded phase conductor) One blue (ungrounded phase conductor)
Four-wire cable with ground	One black (ungrounded phase conductor) One white (grounded conductor) One red (ungrounded phase conductor) One blue (ungrounded phase conductor) One bare (equipment grounding conductor)

When conductors are installed in raceway systems, any color insulation is permitted for the ungrounded phase conductors except the following:

White or gray .. reserved for use as the grounded circuit conductor
Green............... reserved for use as a grounding conductor only

3.1.2 Changing Colors

Should it become necessary to change the actual color of a conductor to meet NEC requirements or to facilitate maintenance on circuits and equipment, the conductors may be reidentified with colored tape or paint.

For example, assume that a two-wire cable containing a black and white conductor is used to feed a 240-volt, two-wire single-phase motor. Since the white colored conductor is supposed to be reserved for the grounded conductor, and none is required in this circuit, the white conductor may be marked with a piece of black tape at each end of the circuit so that everyone will know that this wire is not a grounded conductor.

4.0.0 VOLTAGE DROP

Sections 210-19 & 215-2 FPN

In all electrical systems, the conductors should be sized so that the voltage drop never exceeds 3 percent for power, heating, and lighting loads or combinations of these. Furthermore, the maximum total voltage drop for conductors, feeders and branch circuits combined should never exceed 5 percent. These percentages are recommended by the NEC, but are not requirements. However, it is considered to be good practice to incorporate these percentages into every electrical installation.

In some applications, such as for circuits feeding hospital x-ray equipment, the voltage drop is even more critical — requiring a minimum of 2% voltage drop throughout. With higher rating on new insulations, it is extremely important to keep volt loss in mind, otherwise some very unsatisfactory problems are likely to be encountered.

For example, the resistance and voltage drop on long conductor runs may be great enough to seriously interfere with the efficient operation of the connected equipment. Resistance elements, such as those used in incandescent lamps and electric heating units are particularly critical in this respect; a drop of just a few volts greatly reduces their efficiency.

Electric motors are not affected by small voltage variations quite as much as pure resistance loads, but motors will not give their rated horsepower if the voltage is below that at which they are rated. When loaded motors are operated at reduced voltage, the current flow actually increases, as it requires more amperes to produce a given wattage and horsepower at low voltage than at normal voltage. This current increase is also caused by the fact that the opposition of the motor windings to current flow reduces as their speed reduces.

From the foregoing, we can see that it is very important to have all conductors of the proper size, to avoid excessive heating and voltage drop; and that, in the case of long runs, it is necessary to determine the wire size by consideration of resistance and voltage drop, rather than by the heating effect or NEC Tables 310-16 — 310-19 alone.

To solve the ordinary problems of voltage drop requires only a knowledge of a few simple facts about the areas and resistance of conductors and the application of a few mathematical equations.

The actual conductor used must also meet the other sizing requirements such as full-load current, ambient temperature, number in a raceway, etc.

4.1.0   WIRE SIZES BASED ON RESISTANCE

Earlier modules covered wire sizes and how conductors are normally specified in kcmil or AWG sizes. This numbering system was originated by the Brown & Sharpe Company and was originally called *B & S gauge*. However, the B & S gauge quickly evolved into *American Wire Gauge* (AWG) and is now standard in the United States for indicating sizes of round wires and conductors.

These gauge numbers are arranged according to the resistance of the wires —the larger numbers being for the wires of greatest resistance and smallest area. A very handy rule to remember is that decreasing the gauge by three numbers gives a wire of approximately twice the area and half the resistance.

NOTE   Decreasing the AWG wire gauge by three numbers gives a wire of approximately twice the area and half the resistance.

For example, if we increase the wire gauge from No. 3 AWG, which has resistance of 0.1931 ohms per 1000 feet, to a No. 6 AWG, we find it has .3872 ohms per 1000 feet which is almost double.

In using this rule-of-thumb, remember that it is three numbers difference; that is, 3 + 3 = 6. An AWG size 5 wire is not listed in most tables, but it must be counted as if it were there.

The American Wire Gauge numbers range from 0000 (4/0), down in size to number 60. The 4/0 conductor is nearly ½″ in diameter and the No. 60 is as fine as a small hair.

The most common sizes used for light and power installations range from 4/0 to No. 14 AWG. Lighting-fixture wires are frequently size 16 or 18 AWG, and low-voltage control wiring sometimes drop down to size No. 22 AWG.

4.1.1 Circular Mil — Unit of Conductor Area

In addition to AWG gauge numbers, a unit is used called the *mil*, for measuring the diameter and area of conductors. The mil is equal to $\frac{1}{1000}$ of an inch, so it is small enough to measure and express these sizes very accurately. For example, instead of saying a wire has a diameter of .055″, or fifty-five thousandths of an inch, we can simply call it "55 mils." So a wire of 250 mils diameter is also .250″, or ¼ inch, in diameter.

Since the resistance and current-carrying capacity of conductors both depend on their cross-sectional area, a unit is necessary to express this area. For square conductors, such as busbars, the square mil is used, which is a square $\frac{1}{1000}$ of an inch on each side. For round conductors, the circular mil unit is used, which is the area of a circle with a diameter of $\frac{1}{1000}$ of an inch.

These units greatly simplify conductor calculations. For example, to determine the area of a square conductor (see "B" in Fig. 10), multiply one side by the other, measuring the sides in either mils or thousandths of an inch.

To obtain the area of a round conductor in circular mils, square the diameter in mils or thousandths of an inch.

NOTE   To square a number, multiply the number by itself.

4.1.2 Conversion of Square Mils to Circular Mils

In comparing round and square conductors, remember that the square mil and the circular mil are not quite the same size units of area. For a comparison, see Fig. 10. At "B" a circle with a square is shown. While the circle has the same diameter as the square, the corners of the square make it the larger in area. From this, we can say that the area of one circular mil is less than that of one square mil. The actual ratio between the two is .7854, or the circle has only .7854 percent of the area of a square of the same diameter. Consequently, if it is desired to find the

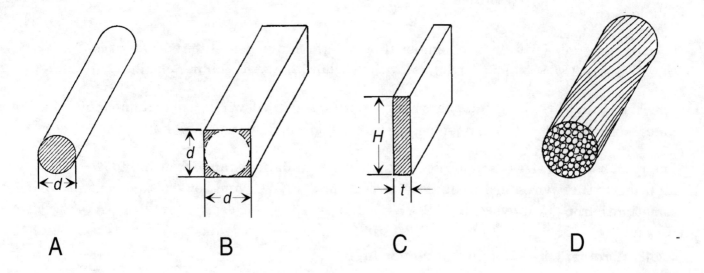

Figure 10. Electrical conductors are commonly made in several different shapes.

circular mil area from the number of square mils, divide the square mils by .7854. If the reverse were true, that is, finding the square mil area from circular mils, multiply the circular mils by .7854.

For example, if the conductor "A" in Fig. 10 is a No. 4/0 conductor with a diameter of 460 mils, what is its area both in circular mils and in square mils?

Circular mil area = 460 × 460 = 211,600 circular mils

Square mil area = 211,600 × .7854 = 166,190.64 square mils

If the busbar at "C" in Fig. 10 is 1½ inches high and ¼ inch thick, what is its area in square mils, and what size of round conductor would be necessary to carry the same current as this busbar?

First, the dimensions of a ¼″ × 1½″ busbar, stated in mils, are 250 mils × 1500 mils. Therefore, the area in square mils may be determined by the following equation:

250 × 1500 = 375,000 square mils

To find what this area would be in circular mils, divide the square mil area (375,000) by .7854; the results are as follows:

$$\frac{375,000}{.7854} = 477,463.7 \text{ circular mils}$$

The nearest standard size to this is a round conductor 500,000 circular mils in size. This is called five-hundred thousand circular mils or written *500 kcmil*.

Busbars of the shape shown at "C" in Fig. 10 are commonly used in panelboards and switchgear. These bars commonly range in thickness from .250″ to .375″ or more; and in heights from 1″ to 12″.

Stranded conductors ("D" in Fig. 10) are used on all conductor sizes No. 6 AWG and larger. Since these conductors are not solid throughout, their area cannot be determined accurately by squaring their diameter. This diameter also varies somewhat with the twist or "lay" of the strands.

To determine the cross-sectional area of such conductors, first determine the area of each strand, either from a wire table or by calculation from its diameter, and then multiply this by the number of strands, to get the total area of the cable in circular mils. However, Table 8 in Chapter 9 of the NEC (Conductor Properties) gives all the information necessary for most wiring calculations concerning voltage drop. Consequently, this table should be used to obtain the properties of conductors for use in conductor calculations.

4.2.0 RESISTANCE OF CONDUCTORS

It is often necessary to determine the exact resistance of a conductor of a certain length, in order to calculate the voltage drop under a certain current load.

The resistance per 1000 feet of various conductors can be obtained from Table 8 in NEC Chapter 9 which is reproduced in Fig. 11. These conductor specifications are necessary to accurately determine the voltage drop for various sizes of conductors. For example, assume that it is necessary to find the total resistance of a 120-volt, two-wire circuit consisting of two No. 10 AWG solid copper conductors, each 150 feet long.

Step 1. The length of one conductor (150 feet) must first be multiplied by 2 to obtain the entire length of both conductors:

$$2 \times 150 = 300 \ feet$$

Step 2. Refer to Table 8 of NEC Chapter 9 and find the resistance of No. 10 solid copper wire. The table gives a resistance of 1.21 ohms per 1000 feet for uncoated copper wire at 75°C temperature.

Step 3. Since our circuit is less than 1000 feet, we must determine the resistance of 300 feet. This is accomplished by dividing the actual footage (300 feet in this case) by 1000.

$$\frac{300}{1000} = .30 \times 1.21 = 0.363 \ ohms$$

| Size AWG/ kcmil | Area Cir. Mils | Conductors | | | | DC Resistance at 75°C (167°F) | | |
| | | Stranding | | Overall | | Copper | | Aluminum |
		Quantity	Diam. In.	Diam. In.	Area In. 2	Uncoated ohm/kFT	Coated ohm/kFT	ohm/kFT
18	1620	1	—	0.040	0.001	7.77	8.08	12.8
18	1620	7	0.015	0.046	0.002	7.95	8.45	13.1
16	2580	1	—	0.051	0.002	4.89	5.08	8.05
16	2580	7	0.019	0.058	0.003	4.99	5.29	8.21
14	4110	1	—	0.064	0.003	3.07	3.19	5.06
14	4110	7	0.024	0.073	0.004	3.14	3.26	5.17
12	6530	I	—	1	0.081	0.005	1.93	2.01
12	6530	7	0.030	0.092	0.006	1.98	2.05	3.25
10	10380	I	—	1	0.102	0.008	1.21	1.26
10	10380	7	0.038	0.116	0.011	1.24	1.29	2.04
8	16510	1	—	0.128	0.013	0.764	0.786	1.26
8	16510	7	0.049	0.146	0.017	0.778	0.809	1.28
6	26240	7	0.061	0.184	0.027	0.491	0.510	0.808
4	41740	7	0.077	0.232	0.042	0.308	0.321	0.508
3	52620	7	0.087	0.260	0.053	0.245	0.254	0.403
2	66360	7	0.097	0.292	0.067	0.194	0.201	0.319
1	83690	19	0.066	0.332	0.087	0.154	0.160	0.253
1/0	105600	19	0.074	0.373	0.109	0.122	0.127	0.201
2/0	133100	19	0.084	0.419	0.138	0.0967	0.101	0.159
3/0	167800	19	0.094	0.470	0.173	0.0766	0.0797	0.126
4/0	211600	19	0.106	0.528	0.219	0.0608	0.0626	0.100
250	—	37	0.082	0.575	0.260	0.0515	0.0535	0.0847
300	—	37	0.090	0.630	0.312	0.0429	0.0446	0.0707

Figure 11. Properties of conductors.

| Size AWG/ kcmil | Area Cir. Mils | Conductors | | | | DC Resistance at 75°C (167°F) | | |
| | | Stranding | | Overall | | Copper | | Aluminum |
		Quantity	Diam. In.	Diam. In.	Area In. 2	Uncoated ohm/kFT	Coated ohm/kFT	ohm/kFT
350	—	37	0.097	0.681	0.364	0.0367	0.0382	0.0605
400	—	37	0.104	0.728	0.416	0.0321	0.0331	0.0529
500	—	37	0.116	0.813	0.519	0.0258	0.0265	0.0424
600	—	61	0.099	0.893	0.626	0.0214	0.0223	0.0353
700	—	61	0.107	0.964	0.730	0.0184	0.0189	0.0303
750	—	61	0.111	0.998	0.782	0.0171	0.0176	0.0282
800	—	61	0.114	1.03	0.834	0.0161	0.0166	0.0265
900	—	61	0.122	1.09	0.940	0.0143	0.0147	0.0235
1000	—	61	0.128	1.15	1.04	0.0129	0.0132	0.0212
1250	—	91	0.117	1.29	1.30	0.0103	0.0106	0.0169
1500	—	91	0.128	1.41	1.57	0.00858	0.00883	0.0141
1750	—	127	0.117	1.52	1.83	0.00735	0.00756	0.0121
2000	—	127	0.126	1.63	2.09	0.00643	0.00662	0.0106

Figure 11. Properties of conductors. (*Cont.*)

In another situation, it is desired to install an outside a 120-volt, two-wire line between two buildings, a distance of 1650 feet, and using No. 1 AWG copper wire. What would be the total resistance of this circuit?

Step 1. Determine the total length of both conductors by multiplying the length (one way) by 2.

$$1650 \times 2 = 3300 \text{ feet}$$

Step 2. Referring again to NEC Table 8 in Chapter 9, No. 1 AWG uncoated copper wire has a resistance of 0.154 ohms per 1000 feet.

Step 3.   Since 3300 feet is not exactly 1000 feet, the total length must be divided by 1000 to determine the total ohms in the circuit.

$$\frac{3300}{1000} = 3.3$$

Step 4.   Multiply this result (3.3) times the resistance found in Table 8 (0.154).

$$3.3 \times 0.154 = .5082 \text{ ohms}$$

Now let's see what happens if we apply the full current to these conductors allowed by NEC Table 310-16. Assuming that Type THHN conductors are used, NEC Table 310-16 allows a maximum load on these conductors of 150 amperes. If this much current flowed through this circuit for a distance of 3300 feet (both ways), the voltage drop in the circuit would be:

$$I \times R = \text{voltage drop}$$

$$150 \times .5082 = 76.2 \text{ volts}$$

If the initial voltage is only 120 volts, this means that the voltage at the far end of the circuit will be only (120 - 76.2 =) 43.8 volts and few, if any, 120-volt loads will operate at this low voltage. Consequently, the load will have to be reduced to make this circuit of any use.

Our goal is to keep the voltage drop within 3% of the original voltage. Therefore, since the original voltage is 120 volts, the allowable voltage drop may be found by using the following equation:

$$120 \times .03 = 3.6 \text{ volts}$$

Now, what size load may be applied to this circuit to stay within this 3% range? The equation for finding the maximum current on this circuit to keep the voltage drop with 3% or 3.6 volts is as follows:

Maximum current $\times$ *total resistance in ohms = original voltage x allowable voltage drop%*

Substituting our known values in this equation, we have:

$$x \times .5082 = 3.6 \text{ volts}$$

To solve for x, we divide both sides of the equation by .5082, which results in the following:

$$x = \frac{3.6}{.5082} = 7.083 \text{ amperes}$$

Therefore, to keep the voltage drop within 3% for this length of circuit with only 120 volts applied, the amperage must be held to 7.083 amperes or below.

If a larger load is connected to this circuit, and it is desired to hold the voltage drop to within 3%, then either a larger size of conductor will have to be used, or else the voltage will have to be increased. For example, assume that this circuits feeds a 120/240-volt dual-voltage pump motor. If the motor connections were rewired to accept a 240-volt branch circuit, then the allowable voltage-drop (at 3%) would be:

$$.03 \times 240 = 7.2 \: volts$$

Continuing as before,

$$x \times \frac{7.2}{.5082} = 14.17 \: amperes$$

If 480 volts were applied to the circuit, the amperage would double again. These samples should show why that long electric transmission lines utilize extremely high voltages — up to 250,000 volts or more —to keep the current, resulting voltage drop, and conductor size to the bare minimum.

4.3.0   RESISTANCE OF COPPER PER MIL FOOT

In many cases, it may be necessary to calculate the resistance of a certain length of wire or busbar of a given size.

This can be done very easily if the unit resistance of copper is known. In doing so, a unit called the *mil foot* may be used. A mil foot represents a piece of round wire 1 mil in diameter and 1 foot in length, and is a small enough unit to be very accurate for all practical calculations. A round wire of 1 mil diameter has an area of 1 circular mil, as the diameter multiplied by itself or "squared", is $1 \times 1 = 1$ circular mil area.

The resistance of ordinary copper is 12.9 ohms per mil foot. This figure or "constant" is important and should be remembered.

Suppose it is desired to determine the resistance of a piece of No. 12 copper wire, 50 feet long. We know that the resistance of any conductor increases as its length increases, and decreases as its area increases. So, for a wire 50 feet long, we first multiply, and get 50 x 12.9 = 645, which would be the resistance of a wire 1 circular mil. in area and 50 feet long. Then we find in NEC Table 8, Chapter 9 that the area of a No. 12 wire is 6530 circular mils, which will reduce the resistance in proportion. So we now divide:

$$\frac{645}{6530} = .098775 \: ohms$$

In another case we wish to find the resistance of a coil containing 3000 ft. of No. 18 stranded wire. Then, $3000 \times 12.9 = 38,700$; and, as the area of No. 18 wire is 1620 circular mils., the total resistance may be found by the following equation:

$$\frac{38,700}{1620} = 23.88888 \ ohms$$

Checking this with NEC Table 8, Chapter 9, we find that the table gives a resistance of 7.95 ohms per 1000 feet for No. 18 stranded AWG wire. Then for 3000 feet, we use the following equation:

$$3 \times 7.95 = 23.85 \ ohms$$

The small difference in this figure and the one obtained by the first calculation, is caused by using approximate figures instead of lengthy decimal places.

The mil foot unit and its resistance of 12.9 for copper may also be used to calculate the resistance of square busbars, by simply using the figure .7854 to change from square mils to circular mils.

Suppose we wish to find the resistance of a square bus bar $\frac{1}{4}''$ x $2''$, and 100 feet long. The dimensions in mils will be 250×2000, or 500,000 square mils area. Then, to find the circular mil area, we divide 500,000 by .7854 and get 636,618.3 circular mil area. Then, 100 feet $\times$ 12.9 = 1290 ohms, or the resistance of 100 feet of copper, 1 mil in area. As the area of this bar is 636,618 C.M., we divide:

$$\frac{1290}{636,618} = .002026 \ ohms \ total \ resistance$$

NEC Tables 310-16 through 310-19 give the allowable current-carrying capacities of conductors with various types of insulation. These tables, however, do not take into consideration the length of the conductors or voltage drop. Consequently, it is often necessary to use a larger size conductor than the tables allow.

4.4.0 EQUATIONS FOR VOLTAGE DROP USING CONDUCTOR AREA OR CONDUCTOR RESISTANCE

The size of conductor required to connect an electrical load to the source of supply is determined by several factors which include the following:

- The load current in amperes
- The permissible voltage drop between source and load
- The total length of the conductor
- The type of wire; that is, copper, aluminum, copper-clad aluminum, etc. and its permissible load-carrying capability based on NEC Tables 310-16 though 310-19.

To perform voltage-drop calculations, it is also necessary to recall the following:

- The resistance of a wire varies directly with its length or:

$$R = resistance \ per \ foot \times length \ in \ feet$$

- The resistance varies inversely with its cross-sectional area or:

$$R = \frac{1}{A}$$

Combining both statements, we obtain the following equations:

$$R = \frac{L \times K}{A}$$

Where   A = Wire size (First from NEC Tables 310-16 through 310-19 for a given load, and then, if necessary, from NEC Table 8, Chapter 9 for circular mils)

L = Length of wire in feet

R = Total resistance of wire

K = Resistance per mil foot

K is a constant whose value depends upon units chosen and the type of wire (12.9 for copper, 21.2 for aluminum). Using the foot as the unit of length and the circular mil as the unit of area, the values for K represent the resistance in ohms per mil foot.

The length (L) in the above equations is the total length of only a single current-carrying conductor. For a two conductor, 120V circuit or a 240V balanced three-wire circuit (neutral current equals zero), the length in the equations must be multiplied by 2. The equation now becomes:

$$R = \frac{2 \times L \times K}{A} \text{ for single-phase circuits}$$

For a three-phase, four-wire balanced circuit with a power factor near the value of 1, the equation must be multiplied by √3 ÷ 2. The equation now becomes:

$$R = (\sqrt{3} \div 2) \times (2 \times L \times K \div A)$$

The terms of 2 cancel each other and the equation reduces to:

$$R = \frac{\sqrt{3} \times L \times K}{A} \text{ for three-phase balanced circuits}$$

Since R = E ÷ I, where E = voltage drop (VD), substitute VD ÷ I for R in the above equations:

$$\frac{VD}{I} = \frac{2 \times L \times K}{A}$$

$$\frac{VD}{I} = \frac{\sqrt{3} \times L \times K}{A}$$

Multiply each side of both equations by I:

$$VD = \frac{2 \times L \times K \times I}{A}$$

$$VD = \frac{\sqrt{3} \times L \times K \times I}{A}$$

Substitute CM for A where CM is initially determined from the wire size ampacities listed in NEC Tables 310-16 through 310-19 for the desired load and where NEC Table 8, Chapter 9 is used to convert the wire size so determined to circular mils, if required. The equations become:

$$VD = \frac{2 \times L \times K \times I}{CM} \quad \text{for single-phase circuits}$$

and

$$VD = \frac{\sqrt{3} \times L \times K \times I}{CM} \quad \text{for three-phase balanced circuits}$$

You should recognize the above equations as the same ones used to calculate voltage drop for branch circuits in the previous module. Using an exercise similar to the one above, the following equations also used in the previous module can be derived for a given wire size resistance and load:

$$VD = \frac{2 \times L \times R \times I}{1000} \quad \text{for single-phase circuits}$$

and

$$VD = \frac{\sqrt{3} \times L \times R \times I}{1000} \quad \text{for three-phase balanced circuits}$$

4.5.0   USE OF VOLTAGE DROP EQUATIONS

If it is desired to determine the voltage drop on either an existing 120V, two-wire installation or a proposed project, the voltage drop of the circuit may be found by using one of the above single-phase formulas:

$$VD = \frac{2 \times L \times K \times I}{CM}$$

For example, assume a 120-volt, two-wire circuit feeding a total load of 30 amperes and the branch-circuit length is 120 feet.

NEC Table 310-16 allows a No. 10 AWG conductor at 60°C to carry this load. Referring to NEC Table 8, Chapter 9, we find that the area (A) of a No. 10 AWG conductor is 10,380 circular mils. Then, substituting these values and a value of 12.9 (copper) for K in the equation, we have:

$$VD = \frac{2 \times 120 \; ft. \times 12.9 \times 30A}{10{,}380}$$

$$VD = 8.95 \; volts$$

The voltage at the load = 120 - 8.95 volts = 111.05 volts which is usually not acceptable on most installations since it exceeds the 3% recommended maximum voltage drop; that is, 3% of 120 = 3.6 volts. In fact, it is more than double. There are three ways to correct this situation:

- Reduce the load
- Increase the conductor size
- Increase the voltage

4.5.1 Miscellaneous Voltage Drop Equations

The final voltage drop (VD) equations given in paragraph 4.4.0 can be solved for length (L) yielding the following equations:

$$L = \frac{VD \times CM}{2 \times K \times I} \quad or \quad L = \frac{1000 \times VD}{2 \times R \times I} \; for \; single\text{-}phase$$

$$L = \frac{VD \times CM}{\sqrt{3} \times K \times I} \quad or \quad L = \frac{1000 \times VD}{\sqrt{3} \times R \times I} \; for \; three\text{-}phase$$

These equations can be used to determine the maximum circuit length for a specified voltage drop at a given load using a wire size selected from NEC Tables 310-16 through 310-19 for the ampacity of the load. NEC Table 8, Chapter 9 may be used to convert the selected wire size to circular mils, if required.

For example, determine the maximum circuit length for a feeder serving a 208V balanced three-phase load of 400A at a voltage drop not to exceed 3%. From NEC Table 310-16, determine that a 750 kcmil conductor at 60°C is rated to carry 400A. Using the three-phase length equation above containing the CM term, substitute values and calculate length:

$$L = \frac{208V \times 3\% \times 750{,}000 \; CM}{\sqrt{3} \times 12.9 \times 400A}$$

$$L = 523.6 \; feet$$

Therefore, in this example the 400A load may be positioned up to 523.6 feet from the source and the voltage drop will be 3% or less using 750 kcmil conductors at 60°C.

The two final voltage drop (VD) equations given in paragraph 4.4.0 may also be solved for wire size (CM) to directly determine, *under certain conditions*, the wire size for a given voltage drop, wire length, and load. Solving for CM yields:

$$CM = \frac{2 \times L \times K \times I}{VD} \text{ for single-phase circuits}$$

and

$$CM = \frac{\sqrt{3} \times L \times K \times I}{VD} \text{ for three-phase circuits}$$

The results from these two CM equations must be used in accordance with the following two rules:

Rule 1.   If the wire size calculated using the CM equation results in a wire size with less ampacity than the given load as determined from NEC Tables 310-16 through 310-19, discard the calculated result and use the wire size with the rated ampacity for the given load.

Rule 2.   If the wire size calculated using the CM equation results in a wire size equal to or greater than the wire size determined from NEC Tables 310-16 through 310-19 for the given load, use the calculated wire size or the next larger standard wire size.

The reason for these rules is that the equation results are for non-derated bare wire. The wire sizes determined from the NEC tables are derated for number of wires, temperatures, and other factors. Calculated sizes that are smaller than those sizes rated at the desired load ampacity cannot be used. However, calculated wire sizes that are larger may be used when excessive voltage drop occurs that is a function of only the size of wire (that has already been derated) and that is caused by additional length beyond the maximum length for a specified voltage drop and load. [See (L) equations above.]

To illustrate, let's solve for CM using the terms of the example given above for determining length (L); i.e., a 400A, three-phase load at 523.6 feet:

$$CM = \frac{\sqrt{3} \times L \times K \times I}{VD}$$

Substituting values we have:

$$CM = \frac{\sqrt{3} \times 523.6 \text{ ft.} \times 12.9 \times 400A}{208V \times 3\%}$$

$$CM = 749,938 \text{ or } 750 \text{ kcmil}$$

In this instance, of course, the length and load correlated exactly with the 3% voltage drop and a 750 kcmil conductor was obtained from the calculation. In accordance with Rule 2, the size obtained equaled the size of a 60°C conductor with a rated ampacity of 400A. Note that longer lengths for this same load will result in larger conductor sizes, each of which will be proven correct if cross-checked using the VD equations.

Now take the same problem, but change the 523.6 feet to 100 feet and solve for CM.

$$CM = \frac{\sqrt{3} \times 100 \; ft. \times 12.9 \times 400A}{208\,V \times 3\%}$$

$$CM = 143{,}227$$

From NEC Table 8, Chapter 9, the next largest standard conductor is 167,800 CM or a No. 3/0 AWG conductor. However, when NEC Table 310-16 is checked, the maximum permissible ampacity of a No. 3/0 AWG conductor is found to be only 165A — far below the load of 400A.

Therefore, in accordance with Rule 1, the 3/0 AWG conductor solution would be discarded and 750 kcmil 60°C conductors, determined from NEC Table 310-16 for the 400A load, would be used for the 100 foot run.

Even when sizing wire for low-voltage (24-V) control circuits, the voltage drop should be limited to 3% because excessive voltage drop causes:

- Failure of control coil to activate
- Control contact chatter
- Erratic operation of controls
- Control coil burnout
- Contact burnout

The voltage drop calculations described previously may also be used for low-voltage wiring, but tables are quite common and can save much calculation time.

To use the Table in Fig. 13, for example, assume a load of 35 VA with a 50-ft. run for a 24-V control circuit. Referring to the table, scan the 50-ft. column. Note that No. 18 AWG wire will carry 29 VA and No. 16 wire will carry 43 VA while still maintaining a maximum of 3% voltage drop. In this case, No. 16 wire would be the size to use.

AWG Wire Size	Length of circuit, one way in feet											
	25	50	75	100	125	150	175	200	225	250	275	300
20	29	14	10	7.2	5.8	4.8	4.1	3.8	3.2	2.9	2.8	2.4
18	58	29	19	14	11	9,6	8.2	7.2	6.4	5.8	5.2	4.8
16	86	43	29	22	17	14	12	11	9.6	8.7	7.8	7.2
14	133	67	44	33	27	22	19	17	15	13	12	11

Figure 13. Table for calculating voltage-drop in low-voltage wiring.

When the length of wire is other than listed in the table in Fig. 13, the capacity may be determined by the following equation:

$$VA \ capacity = \frac{length \ of \ circuit \ (from \ table) \times VA \ (from \ table)}{length \ of \ circuit \ (actual)}$$

The 3% voltage drop limitation is imposed to assure proper operation when the power supply is below the rated voltage. For example, if the rated 240-V supply is 10% low (216V), the transformer does not produce 24 V but rather 21.6 V. When normal voltage drop is taken from this 21.6 V, it approaches the lower operating limit of most controls. If it is assured that the primary voltage to the transformer will always be at rated value or above, the control circuit will operate satisfactorily with more than 3% voltage drop.

In most installations, several lines connect the transformer to the control circuit. One line usually carries the full load of the control circuit from the secondary side of the transformer to one control, with the return perhaps through several lines of the various other controls. Therefore, the line from the secondary side of the transformer is the most critical regarding voltage drop and VA capacity and must be properly sized.

When low-voltage lines are installed, it is suggested that one extra line be run for emergency purposes. This can be substituted for any one of the existing lines that may be defective. Also, it is possible to parallel this extra line with the existing line carrying the full load of the control circuit if the length of run affects control operation because of a voltage drop. In many cases this will reduce the voltage drop and permit satisfactory operation.

Summary

A great deal of valuable information about conductors — dimensions, resistance, and current-carrying capacity—can be obtained from convenient NEC tables; and these tables should be used whenever possible as they are great time-saving devices.

There are times, however, when appropriate tables are not available, or do not give the exact information needed for a certain project or situation. This is when a knowledge of simple conductor calculations pays off.

For example, NEC Tables 310-16 through 310-19 give the allowable current-carrying capacities of conductors with various types of insulation based on the heating of the conductors, but these tables do not take in account for voltage drop due to resistance of long circuit runs. Both of these considerations are very important and should always be kept in mind when planning or installing any electrical installation.

The information contained in this module will give the trainee a good basic knowledge of conductor calculations and NEC installation requirements. It is also a good "refresher reference" for those who are already active in the trade.

References

For a more advanced study of topics covered in this Task Module, the following works are suggested:

American Electricians Handbook, Latest Edition, Croft, McGraw-Hill, New York, NY

National Electrical Code Handbook, Latest Edition, NFPA, Quincy, MA

SELF-CHECK REVIEW/PRACTICE QUESTIONS

1. All things considered, from what material is the basic and most ideal conductor made?

    a. Bronze
    b. Silver
    c. Aluminum
    d. Copper

2. Which of the following is not a standard configuration for stranded wire?

    a. 5-strand
    b. 7-strand
    c. 19-strand
    d. 37-strand

3. What is the largest standard AWG wire size?

    a. No. 60 AWG
    b. No. 1 AWG
    c. No. 1/0 AWG
    d. No. 4/0 AWG

4. Which of the following best describes a branch circuit?

    a. The circuit conductors between the service equipment and the final branch-circuit overcurrent device
    b. The circuit conductors between the final overcurrent device protecting the circuit and the outlet(s)
    c. A conductor tapped onto another conductor
    d. Two conductors run in parallel

5. Which of the following is a situation where the NEC does not require overcurrent protection?

    a. Where the conductor run is of sufficient length to be too far away from the overcurrent device to do any good
    b. Where the interruption of the circuit would create a hazard
    c. On motor and motor-control circuit conductors
    d. On motor-operated appliance circuit conductors

6. How must overcurrent protective devices be connected to each ungrounded conductor in a circuit?

    a. In series
    b. In parallel
    c. In a series-parallel configuration (compound circuit)
    d. One conductor in series; all others in parallel

7. What determines the rating of a branch circuit?

    a. The type and size of the branch-circuit conductors
    b. The rating of the branch-circuit overcurrent protective device
    c. The total connected load
    d. The characteristics of the supply circuit

8. Which of the following best describes a *continuous load*?

    a. Where the maximum current is expected for 1 hour or more
    b. Where the maximum current is expected for 2 hours or more
    c. Where the maximum current is expected for 3 hours or more
    d. Where the maximum current is expected for 4 hours or more

9. Which of the following is not a standard rating for fuses?

    a. 600 amperes
    b. 601 amperes
    c. 650 amperes
    d. 700 amperes

10. Where must overcurrent protection be provided in a branch circuit or feeder?

    a. At the outlet
    b. At points where the conductors receive their supply
    c. At the equipment grounding terminal
    d. At the point where the conductors terminate

PERFORMANCE/LABORATORY EXERCISE

1. Use a wire gauge (supplied by instructor) and identify the size of various conductors by AWG number.

2. Calculate the voltage drop in Transparency PLE-1 which will be shown by your instructor.

3. Calculate the required conductor size in circular mils for the circuit shown in Transparency PLE-2 (furnished by instructor) to keep the voltage drop in the circuit to 3%.

Answers to Self-Check Questions

1. d

2. a

3. d

4. b

5. b

6. a

7. b

8. c

9. c

10. b

Overcurrent Protection

Module 20303

Electrical Trainee Task Module 20303

OVERCURRENT PROTECTION

Objectives

Upon completion of this module, the trainee will be able to:

1. Understand that above all, *life safety* is the most important issue when designing or installing electrical systems.

2. Understand the importance of overcurrent protection.

3. Understand the meaning and importance of commonly used electrical terms relating to overcurrent protection.

4. Understand and discuss the key NEC requirements regarding overcurrent protection.

5. Check electrical drawings to conformance to NEC sections that cover short-circuit current, fault currents, interrupting ratings, and other sections relating to overcurrent protection.

6. Determine let-through current values (peak and RMS) when current-limiting overcurrent devices are used.

7. Understand that the major sources of short-circuit currents are motors and generators.

8. Understand that transformers are not a source of short-circuit current.

9. Select and size overcurrent protection for specific applications.

10. Know how to ask the right questions concerning overcurrent protection.

Prerequisites

Successful completion of the following Task Modules is required before beginning study of this Task Module: Core Curricula, Electrical Levels 1 and 2, Electrical Level 3, Modules 20301 and 20302.

Required Student Materials

1. Trainee Task Module
2. Copy of the latest edition of the National Electrical Code

COURSE MAP INFORMATION

This course map shows all of the *Wheels of Learning* Task Modules in the third level of the Electrical curricula. The suggested training order begins at the bottom and proceeds up. Skill levels increase as a trainee advances on the course map. The training order may be adjusted by the local Training Program Sponsor.

Course Map: Electrical, Level 3

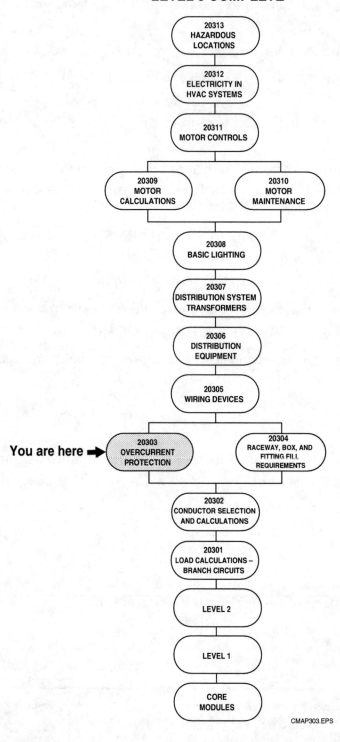

LEVEL 3 COMPLETE

20313
HAZARDOUS LOCATIONS

20312
ELECTRICITY IN HVAC SYSTEMS

20311
MOTOR CONTROLS

20309
MOTOR CALCULATIONS

20310
MOTOR MAINTENANCE

20308
BASIC LIGHTING

20307
DISTRIBUTION SYSTEM TRANSFORMERS

20306
DISTRIBUTION EQUIPMENT

20305
WIRING DEVICES

You are here ➤ 20303
OVERCURRENT PROTECTION

20304
RACEWAY, BOX, AND FITTING FILL REQUIREMENTS

20302
CONDUCTOR SELECTION AND CALCULATIONS

20301
LOAD CALCULATIONS – BRANCH CIRCUITS

LEVEL 2

LEVEL 1

CORE MODULES

CMAP303.EPS

ELECTRICAL TRAINEE TASK MODULE 20303

TABLE OF CONTENTS

Trade Terms Introduced in This Module

ampere rating: The current-carrying capacity of an overcurrent protective device. The fuse or circuit breaker is subjected to a current above its ampere rating, it will open the circuit after a predetermined period of time.

ampere squared seconds, I^2t: The measure of heat energy developed within a circuit during the fuse's clearing. It can be expressed as "melting I^2t," "arcing I^2t," or the sum of them as "Clearing I^2t." "I" stands for effective let-through current (RMS), which is squared, and "t" stands for time of opening, in seconds.

arcing time: The amount of time from the instant the fuse link has melted until the overcurrent is interrupted, or cleared.

clearing time: The total time between the beginning of the overcurrent and the final opening of the circuit at rated voltage by an overcurrent protective device. Clearing time is the total of the melting time and the arcing time.

current limitation: A fuse operation relating to short-circuits only. When a fuse operates in its current-limiting range, it will clear a short circuit in less than ½ cycle. Also, it will limit the instantaneous peak let-through current to a value substantially less than that obtainable in the same circuit if that fuse were replaced with a solid conductor of equal impedance.

electrical load: That part of the electrical system that actually uses the energy or does the work required.

fast-acting fuse: A fuse that opens on overloads and short circuits very quickly. This type of fuse is not designed to withstand temporary overload currents associated with some electrical loads (inductive loads).

high-speed fuses: Fuses with no intentional time-delay in the overload range and designed to open as quickly as possible in the short-circuit range. These fuses are often used to protect solid-state devices.

inductive load: An electrical load that pulls a large amount of current—an inrush current—when first energized. After a few cycles or seconds, the current "settles down" to the full-load running current.

interrupting capacity: The maximum short-circuit current that a circuit breaker can safely interrupt.

melting time: The amount of time required to melt the fuse link during a specified overcurrent.

"NEC" Dimensions: These are dimensions once referenced in the National Electrical Code®. They are common to Class H and K fuses and provide interchangeability between manufacturers for fuses and fusible equipment of given ampere and voltage ratings.

overload: Can be classified as an overcurrent that exceeds the normal full-load current of a circuit.

peak let-through current, Ip: The instantaneous value of peak current let-through by a current limiting fuse, when it operates in its current-limiting range.

resistive load: An electrical load which is characteristic of not having any significant inrush current. When a resistive load is energized, the current rises instantly to its steady state value, without first rising to a higher value.

RMS Current: The effective value of an ac sine wave which is calculated as the square root of the average of the squares of all the instantaneous values of the current throughout one cycle. RMS alternating current is that value of an alternating current that produces the same heating effect as a given dc value.

semiconductor fuses: Fuses used to protect solid-state devices.

short circuit: Can be classified as an overcurrent which exceeds the normal full-load current of a circuit by a factor many times greater than normal. Also characteristic of this type of overcurrent is that it leaves the normal current-carrying path of the circuit—it takes a "short cut" around the load and back to the source.

single phasing: That condition which occurs when one phase of a three-phase system opens, either in a low voltage or high voltage distribution system. Primary or secondary single phasing can be caused by any number of events. This condition results in unbalanced loads in polyphase motors and unless protective measures are taken, causes overheating and failure.

threshold current: The symmetrical rms available current at the threshold of the current-limiting range, where the fuse becomes current limiting when tested to the U.L. Standard. This value can be read off of a peak let-through chart where the fuse curve intersects the A-B line. A threshold ratio is the relationship of the threshold current to the fuse's continuous current rating.

U.L. Classes: Underwriters' Laboratories has developed basic physical specifications and electrical performance requirements for fuses with voltage ratings of 600 volts or less. These are known as U.L. Standards. If a type of fuse meets with the requirements of a standard, it can fall into that U.L. Class. Typical U.L. Classes are K, RK1, RK5, G, L, H, T, CC, J.

voltage rating: The maximum value of system voltage in which a fuse can be used, yet safely interrupt an overcurrent. Exceeding the voltage rating of a fuse impairs its ability to clear an overload or short circuit safely.

withstand rating of components: The maximum short-circuit current that a component or device is capable of carrying without requiring replacement because of extensive damage.

Electrical distribution systems are often quite complicated. They cannot be absolutely fail-safe. Circuits are subject to destructive overcurrents. Harsh environments, general deterioration, accidental damage or damage from natural causes, excessive expansion or overloading of the electrical distribution system are factors which contribute to the occurrence of such overcurrents. Reliable protective devices prevent or minimize costly damage to transformers, conductors, motors, and the other many components and loads that make up the complete distribution system. Reliable circuit protection is essential to avoid the severe monetary losses which can result from power blackouts and prolonged downtime of facilities. It is the need for reliable protection, safety, and freedom from fire hazards that has made overcurrent protective devices absolutely necessary in all electrical systems—both large and small.

Overcurrent protection of electrical circuits is so important that the National Electrical Code® devotes an entire Article to this subject. NEC Article 240—Overcurrent Protection—provides the general requirements for overcurrent protection and overcurrent protective devices; that is, NEC Article 240, Parts A through G covers systems 600 volts, nominal and under, while Part H covers overcurrent protection over 600 volts, nominal. This entire NEC Article will be thoroughly covered in this module with practical examples.

All conductors must be protected against overcurrents in accordance with their ampacities as set forth in NEC Section 240-3. They must also be protected against short-circuit current damage as required by NEC Sections 110-10 and 240-1. Two basic types of overcurrent protective devices that are in common use include:

- Fuses
- Circuit breakers

1.1.0   OVERCURRENTS

An overcurrent is either an overload current or a short-circuit current. The overload current is an excessive current relative to normal operating current but one which is confined to the normal conductive paths provided by the conductor and other components and loads of the distribution system. As the name implies, a short-circuit current is one which flows outside the normal conducting paths.

1.2.0   OVERLOADS

Overloads are most often between one and six times the normal current level. Usually, they are caused by harmless temporary surge currents that occur when motors are started-up or transformers are energized. Such overload currents or transients are normal occurrences. Since they are of brief duration, any temperature rise is trivial and has no harmful effect on the circuit components.

Continuous overloads can result from defective motors (such as worn motor bearings), overloaded equipment, or too many loads on one circuit. Such sustained overloads are destructive and must be cut-off by protective devices before they damage the distribution system or system loads. However, since they are of relatively low magnitude compared to short-circuit currents, removal of the overload current within a few seconds will generally prevent equipment damage. A sustained overload current results in overheating of conductors and other components and will cause deterioration of insulation which may eventually result in severe damage and short-circuits if not interrupted.

1.3.0 SHORT-CIRCUITS

The amperes interrupting capacity (AIC) rating of a circuit breaker or fuse is the maximum short circuit current which the breaker will interrupt safely. This AIC rating is at rated voltage and frequency.

Whereas overload currents occur at rather modest levels, the short-circuit or fault current can be many hundreds of times larger than the normal operating current. A high level fault may be 50,000 amperes (or larger). If not cut off within a matter of a few thousands of a second, damage and destruction can become rampant—there can be severe insulation damage, melting of conductors, vaporization of metal, ionization of gases, arcing, and fires. Simultaneously, high-level short-circuit currents can develop huge magnetic-field stresses. The magnetic forces between bus bars and other conductors can be many hundreds of pounds per lineal foot; even heavy bracing may not be adequate to keep them from being warped or distorted beyond repair.

NEC Section 110-9 clearly states that "...equipment intended to break current at fault levels (fuses and circuit breakers) must have an interrupting rating sufficient for the nominal circuit voltage and the current that is available at the line terminals of the equipment."

Equipment intended to break current at other than fault levels must have an interrupting rating at nominal circuit voltage sufficient for the current that must be interrupted.

These NEC statements mean that fuses and circuit breakers (and their related components) designed to break fault or operating currents (open the circuit) must have a rating sufficient to withstand such currents. This NEC section emphasizes the difference between clearing fault level currents and clearing operating currents. Protective devices such as fuses and circuit breakers are designed to clear fault currents and therefore must have short-circuit interrupting ratings sufficient for fault levels. Equipment such as contactors and safety switches have interrupting ratings for currents at other than fault levels. Thus, the interrupting rating of electrical equipment is now divided into two parts:

- Current at fault (short-circuit) levels.
- Current at operating levels.

Most people are familiar with the normal current-carrying ampere rating of fuses and circuit breakers. For example, if an overcurrent protective device is designed to open a circuit when the circuit load exceeds 20 amperes for a given time period, as the current approaches 20 amperes, the overcurrent protective device begins to overheat. If the current barely exceeds 20 amperes, the circuit breaker will open normally or a fuse link will melt after a given period of time with little, if any, arcing. If, say, 40 amperes of current were instantaneously applied to the circuit, the overcurrent protective device will open instantly, but again with very little arcing. However, if a ground fault occurs on the circuit that ran the amperage up to, say, 5000 amperes, an explosion effect would occur within the protective device. One simple indication of this "explosion effect" is blackened windows of plug fuses.

If this fault current exceeds the interrupting rating of a fuse or circuit breaker, the protective device can be damaged or destroyed; such current can also cause severe damage to equipment and injure personnel. Therefore, selecting overcurrent protective devices with the proper interrupting capacity is extremely important in all electrical systems.

For better understanding of interrupting rating, consider the following analogy. Let's use a dammed stream (Fig. 1) as an example. Consider the reservoir capacity to be the available fault current in an electrical circuit; the flood gates (located downstream from the dam) to be the overcurrent protective device in the circuit rated at 10,000 gallons per minute (10,000 AIC) and the stream of water coming through the discharge pipes in the dam to be the normal load current. Our drawing shows a normal flow of 100 gallons per minute. Also note the bridge crossing the stream, downstream from the flood gates. This bridge will represent downstream circuit components or equipment connected to the circuit.

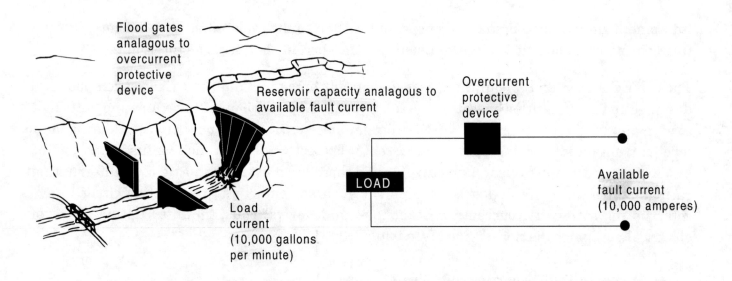

Figure 1. Normal operating current.

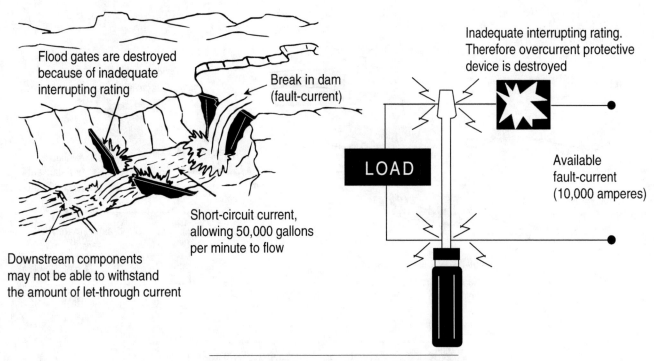

Flood gates are destroyed because of inadequate interrupting rating

Break in dam (fault-current)

Short-circuit current, allowing 50,000 gallons per minute to flow

Downstream components may not be able to withstand the amount of let-through current

Inadequate interrupting rating. Therefore overcurrent protective device is destroyed

LOAD

Available fault-current (10,000 amperes)

Figure 2. Inadequate interrupting rating.

Figure 2 shows this same diagram with a fault in the dam—creating a water "short-circuit" that allows 50,000 gallons per minute to flow (50,000 fault-circuit amperes). Such a situation destroys the flood gates because of inadequate interrupting rating. The overcurrent protective device in the circuit will also be destroyed. With the flood gates damaged, this surge of water continues downstream, wrecking the bridge. The downstream components may not be able to withstand the let-through current in an electrical circuit either.

Figure 3 shows the same situation, but with adequate interrupting capacity. Note that the flood gates have adequately contained the surge of water and restricted the let-through current to an amount that can be withstood by the bridge, or the components downstream.

There are several factors that must be considered when calculating the required interrupting capacity of an overcurrent protective device. NEC Section 110-10 states the following:

"...The overcurrent protective devices, the total impedance, the component short-circuit withstand ratings, and other characteristics of the circuit to be protected shall be selected and coordinated to permit the circuit protective devices used to clear a fault to do so without extensive damage to the electrical components of the circuit. This fault shall be assumed to be either between two or more of the circuit conductors, or between any circuit conductor or the grounding conductor or enclosing metal raceway."

Component short-circuit rating is a current rating given to conductors, switches, circuit breakers and other electrical components, which, if exceeded by fault currents, will result in "extensive" damage to the component. The rating is expressed in terms of time intervals and/or

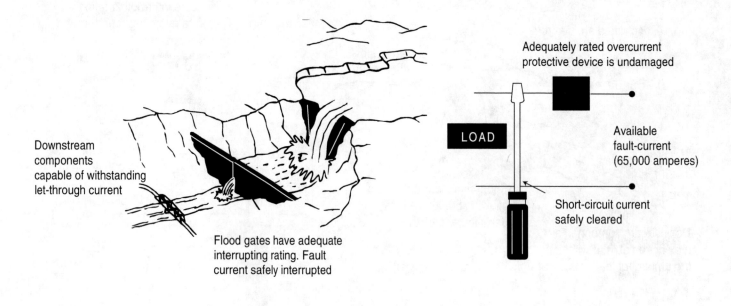

Figure 3. Adequate interrupting rating.

current values. Short-circuit damage can be the result of heat generated or the electro-mechanical force of high-intensity, magnetic field.

The NEC's intent is that the design of a system must be such that short-circuit currents cannot exceed the withstand ratings of the components selected as part of the system. Given specific system components, and level of "available" short-circuit currents that could occur, overcurrent protective devices (mainly fuses and/or circuit breakers) must be used which will limit the energy let-through of fault currents to levels within the withstand ratings of the system components.

2.0.0 FUSEOLOGY

The fuse is a reliable overcurrent protective device. A "fusible" link or links encapsulated in a tube and connected to contact terminals comprise the fundamental elements of the basic fuse. Electrical resistance of the link is so low that it simply acts as a conductor. However, when destructive currents occur, the link very quickly melts and opens the circuit to protect conductors and other circuit components and loads. Fuse characteristics are stable. Fuses do not require periodic maintenance or testing. Fuses have three unique performance characteristics.

2.1.0 VOLTAGE RATING

Most low-voltage power distribution fuses have 250-volt or 600-volt ratings. The voltage rating of a fuse must be at least equal to or greater than the circuit voltage. It can be higher, but never lower. For example, a 600-volt fuse can be used in a 240-volt circuit.

The voltage rating of a fuse is a function of or depends upon its capability to open a circuit under an overcurrent condition. Specifically, the voltage rating determines the ability of the fuse to suppress the internal arcing that occurs after a fuse link melts and an arc is produced. If a fuse is used with a voltage rating lower than the circuit voltage, arc suppression will be impaired and, under some fault-current conditions, the fuse may not safely clear the overcurrent. Special consideration is necessary for semiconductor fuse application, where a fuse of a certain voltage rating is used on a lower-voltage circuit.

2.2.0   AMPERE RATING

Every fuse has a specific ampere rating. In selecting the ampere rating of a fuse, consideration must be given to the type of load and NEC requirements. The ampere rating of a fuse should normally not exceed current-carrying capacity of the circuit. For instance, if a conductor is rated to carry 20 amperes, a 20-ampere fuse is the largest that should be used. However, there are specific circumstances where the ampere rating is permitted to be greater than the current-carrying capacity of the circuit. A typical example is the motor circuit; dual-element fuses generally are permitted to be sized up to 175% and non-time delay fuses up to 300% of the motor full-load amperes. Generally, the ampere rating of a fuse and switch combination should be selected at 125% of the continuous load current (this usually corresponds to the circuit capacity which is also selected at 125% of the load current). There are exceptions such as when the fuse-switch combination is approved for continuous operation at 100% of its rating.

2.3.0   INTERRUPTING RATING-SAFE OPERATION

A protective device must be able to withstand the destructive energy of short-circuit currents. If a fault current exceeds a level beyond the capability of the protective device, the device may actually rupture, causing additional damage. Therefore, it is important in applying a fuse or circuit breaker to use one which can sustain the largest potential short-circuit currents. The rating which defines the capacity of a protective device to maintain its integrity when reacting to fault currents is termed its "interrupting rating."

NEC Section 110-9 requires equipment intended to break current at fault levels to have an interrupting rating sufficient for the current that must be interrupted. The subjects of interrupting rating and interrupting capacity have been thoroughly treated in your Level 2 training. More advanced material is presented in this module.

2.4.0   SELECTIVE COORDINATION

The coordination of protective devices prevents system power outages or blackouts caused by overcurrent conditions. When only the protective device nearest a faulted circuit opens and larger upstream fuses remain closed, the protective devices are "selectively" coordinated (they discriminate). The word "selective" is used to denote total coordination...isolation of a faulted circuit by the opening of only the localized protective device.

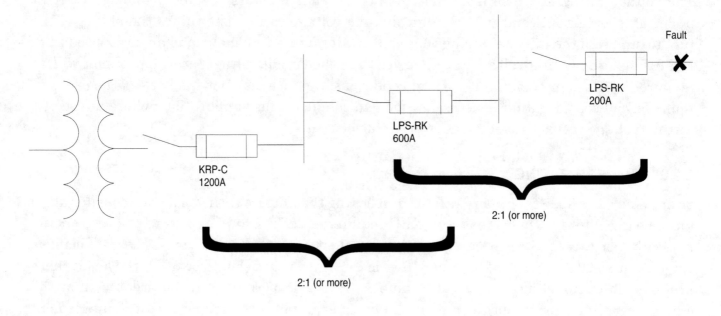

Figure 4. Minimum ratios of ampere ratings of fuses that will provide selective coordination.

The diagram in Fig. 4 shows the minimum ratios of ampere rating of low-peak fuses that are required to provide "selective coordination" of upstream and downstream fuses.

2.5.0 CURRENT LIMITATION

If a protective device cuts off a short-circuit current in less than one-half cycle, before it reaches its total available (and highly destructive) value, the device is a "current limiting" device. Most modern fuses are current limiting. They restrict fault currents to such low values that a high degree of protection is given to circuit components against even very high short-circuit currents. They permit breakers with lower interrupting ratings to be used. They can reduce bracing of bus structures. They minimize the need of other components to have high short-circuit current "withstand" ratings. If not limited, short-circuit currents can reach levels of 30,000 or 40,000 amperes or higher in the first half cycle (.008 seconds at 60 hz) after the start of a short-circuit. The heat that can be produced in circuit components by the immense energy of short-circuit currents can cause severe insulation damage or even explosion. At the same time, huge magnetic forces developed between conductors can crack insulators and distort and destroy bracing structures. Thus, it is important that a protective device limit fault currents before they can reach their full potential level.

A non-current limiting protective device, by permitting a short-circuit current to build up to its full value, can let an immense amount of destructive short-circuit heat energy through before opening the circuit as shown in Fig. 5. On the other hand, a current-limiting fuse has such a high speed of response that it cuts off a short-circuit long before it can build up to its full peak value as shown in Fig. 6.

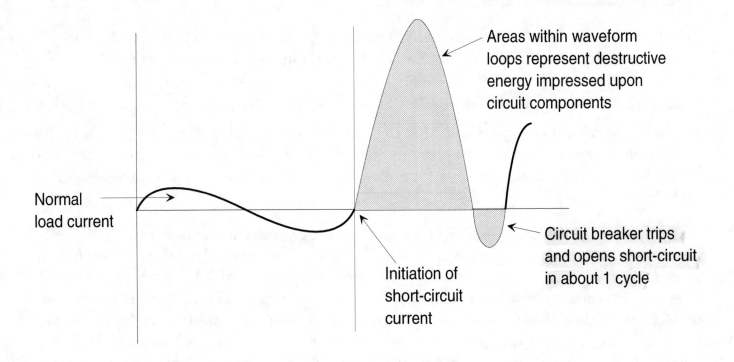

Figure 5. Characteristics of non-current limiting protective device.

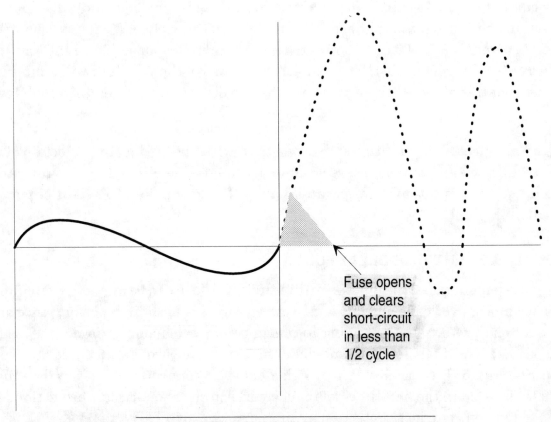

Figure 6. Characteristics of current-limiting fuse.

3.0.0 OPERATING PRINCIPLES OF FUSES

There are several different types of fuses, and although all operate in a similar fashion, all have slightly different characteristics. Each is described in the paragraphs to follow.

3.1.0 NON-TIME DELAY FUSES

The basic component of a fuse is the link. Depending upon the ampere rating of the fuse, the single-element fuse may have one or more links. They are electrically connected to the end blades (or ferrules) and enclosed in a tube or cartridge surrounded by an arc quenching filler material.

Under normal operation, when the fuse is operating at or near its ampere rating, it simply functions as a conductor. However, as illustrated in Fig. 7, if an overload current occurs and persists for more than a short interval of time, the temperature of the link eventually reaches a level which causes a restricted segment of the link to melt; as a result, a gap is formed and an electric arc established. However, as the arc causes the link metal to burn back, the gap becomes progressively larger. Electrical resistance of the arc eventually reaches such a high level that the arc cannot be sustained and is extinguished; the fuse will have then completely cut off all current flow in the circuit. Suppression or quenching of the arc is accelerated by the filler material.

Overload current normally falls within the region of between one and six times normal current—resulting in currents that are quite high. Consequently, a fuse may be subjected to short-circuit currents of 30,000 or 40,000 amperes or higher. Response of current-limiting fuses to such currents is extremely fast. The restricted sections of the fuse link will simultaneously melt within a matter of two or three-thousandths of a second in the event of a high level fault current.

The high resistance of the multiple arcs, together with the quenching effects of the filler particles, results in rapid arc suppression and clearing of the circuit. Again, refer to Fig. 7. Short-circuit current is cut-off in less than a half-cycle—long before the short-circuit current can reach its full value.

3.2.0 DUAL-ELEMENT TIME-DELAY FUSES

Unlike single-element fuses, the dual-element time-delay fuse can be applied in circuits subject to temporary motor overloads and surge currents to provide both high performance short-circuit and overload protection. Oversizing to prevent nuisance openings is not necessary with this type of fuse. The dual-element time-delay fuse contains two distinctly separate types of elements. See Fig. 5. Electrically, the two elements are connected in series. The fuse links similar to those used in the non-time-delay fuse perform the short-circuit protection function; the overload element provides protection against low-level overcurrents or overloads and will

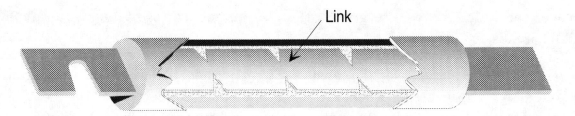

Cut-away view of single-element fuse.

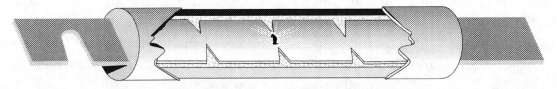

Under sustained overload a section of the link melts and an arc is established.

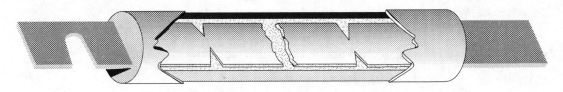

The "open" single-element fuse after opening a circuit overload.

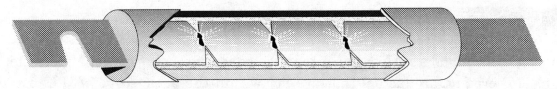

When subjected to a short-circuit, several sections of the fuse link melt almost instantly.

The appearance of an "open" single-element fuse after opening a short-circuit.

Figure 7. Characteristics of a single-element fuse.

hold an overload which is five times greater than the ampere rating of the fuse for a minimum time of 10 seconds.

As shown in Fig. 8, the overload section consists of a copper heat absorber and a spring-operated trigger assembly. The heat absorber bar is permanently connected to the heat absorber extension and to the short-circuit link on the opposite end of the fuse by the S-shaped connector of the trigger assembly. The connector electrically joins the short-circuit link to the heat absorber in the overload section of the fuse. These elements are joined by a "calibrated" fusing alloy. An overload current causes heating of the short-circuit link connected to the trigger assembly. Transfer of heat from the short-circuit link to the heat absorbing bar in the mid-section of the fuse begins to raise the temperature of the heat absorber. If the overload is sustained, the temperature of the heat absorber eventually reaches a level which permits the trigger spring to "fracture" the calibrated fusing alloy and pull the connector free of the short-circuit link and the heat absorber. As a result, the short-circuit link is electrically disconnected from the heat absorber, the conducting path through the fuse is opened, and overload current is interrupted. A critical aspect of the fusing alloy is that it retains its original characteristic after repeated temporary overloads with degradation. The main purpose of dual-element fuses are as follows:

- Provide motor overload, ground-fault and short-circuit protection.
- Permit the use of smaller and less costly switches.
- Give a higher degree of short-circuit protection (greater current limitation) in circuits in which surge currents or temporary overloads occur.
- Simplify and improve blackout prevention (selective coordination).

4.0.0   U.L. FUSE CLASSES

Safety is the U.L. mandate. However, proper selection, overall functional performance and reliability of a product are factors that are not within the basic scope of U.L. activities. To develop its safety test procedures, U.L. does develop basic performance and physical specifications of standards of a product. In the case of fuses, these standards have culminated in the establishment of distinct classes of low-voltage (600 volts or less) fuses, Classes FK 1, RK 5, G, L, T, J. H, and CC being the more important.

Class R fuses: U.L. Class R (rejection) fuses are high performance $\frac{1}{10}$ to 600 ampere units, 250 volt and 600 volt, having a high degree of current limitation and a short-circuit interrupting rating of up to 200,000 amperes (rms symmetrical). This type of fuse is designed to be mounted in rejection type fuseclips to prevent older type Class H fuses from being installed. Since Class H fuses are not current limiting and are recognized by U.L. as having only a 10,000 ampere interrupting rating, serious damage could result if a Class H fuse were inserted in a system designed for Class R fuses. Consequently, NEC Section 240-60(b) requires fuseholders for current-limiting fuses to reject non-current limiting type fuses.

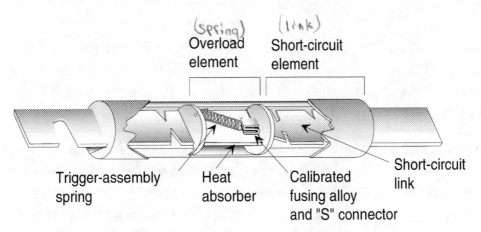

(spring)
Overload
element

(link)
Short-circuit
element

Trigger-assembly
spring

Heat
absorber

Calibrated
fusing alloy
and "S" connector

Short-circuit
link

The true dual-element fuse has distinct and separate overload and short-circuit elements.

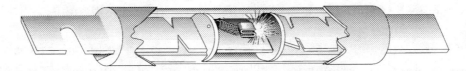

Under sustained overload conditions, the trigger spring fractures the calibrated fusing alloy and releases the "connector."

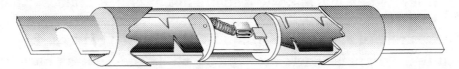

The "open" dual-element fuse after opening under an overload

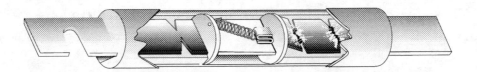

Like the single-element fuse, a short-circuit current causes the restricted portions of the short-circuit elements to melt and arcing to burn back the resulting gaps until the arcs are suppressed by the arc-quenching material and increased are resistance.

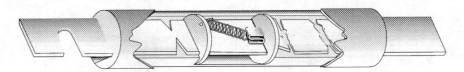

The "open" dual-element fuse after opening under a short-circuit condition.

Figure 8. Characteristics of dual-element time-delay fuses.

Figure 9 shows a standard Class H fuse (left) and a Class R fuse (right). A grooved ring in one ferrule of the Class R fuse provides the rejection feature of the Class R fuse in contrast to the lower interrupting capacity, non-rejection type. Figure 10 shows Class R type fuse rejection clips that accept only the Class R rejection type fuses.

Class CC fuses: 600-volt, 200,000-ampere interrupting rating, branch circuit fuses with overall dimensions of $^{15}/_{32}'' \times 1^1/_2''$. Their design incorporates rejection feature that allow them to be inserted into rejection fuse holders and fuse blocks that reject all lower voltage, lower interrupting rating $^{15}/_{32}'' \times 1^1/_2''$ fuses. They are available from $^1/_{10}$ ampere through 30 amperes.

Class G fuses: 300-volt, 100,000 ampere interrupting rating branch circuit fuses that are size rejecting to eliminate overfusing. The fuse diameter is $^{13}/_{32}''$ while the length varies from $1^5/_{16}''$ to $2^1/_4''$. These are available in ratings from 1 amp through 60 amps.

Class H fuses: 250-volt and 600-volt, 10,000 ampere interrupting rating branch circuit fuses that may be renewable or non-renewable. These are available in ampere ratings of 1 amp through 600 amps.

Class J fuses: These fuses are rated to interrupt 200,000 amperes ac. They are U.L. labeled as "current limiting", are rated for 600 volts ac, and are not interchangeable with other classes.

Class K fuses: These are fuses listed by U.L. as K-1, K-5, or K-9 fuses. Each subclass has designated I^2t and Ip maximums. These are dimensionally the same as Class H fuses, (NEC dimensions) and they can have interrupting ratings of 50,000, 100,000, or 200,000 amps. These fuses are current limiting, however, they are not marked "current limiting" on their label since they do not have a rejection feature.

Class L fuses: These fuses are rated for 601 through 6000 amperes, and are rated to interrupt 200,000 amperes ac. They are labeled "current limiting" and are rated for 600 volts ac. They

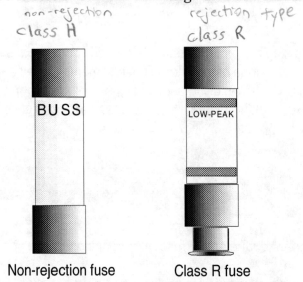

Figure 9. Comparison of Class H and Class R fuses.

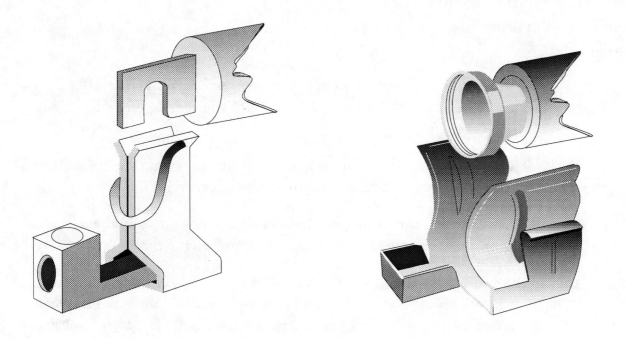

Figure 10. Class R fuse rejection clips that accept only Class R fuses.

are intended to be bolted into their mountings and are not normally used in clips. Some Class L fuses have designed in time-delay features for all purpose use.

Class T fuses: A U.L. classification of fuses in 300 volt and 600 volt ratings from 1 amp through 1200 amps. They are physically very small and can be applied where space is at a premium. They are fast acting fuses, with an interrupting rating of 200,000 amps RMS.

4.1.0   BRANCH-CIRCUIT LISTED FUSES

Branch circuit listed fuses are designed to prevent the installation of fuses that cannot provide a comparable level of protection to equipment. The characteristics of branch-circuit fuses are as follows:

- They just have a minimum interrupting rating of 10,000 amperes.
- They must have a minimum voltage rating of 125 volts.
- They must be size rejecting such that a fuse of a lower voltage rating cannot be installed in the circuit.
- They must be size rejecting such that a fuse with a current rating higher than the fuseholder rating cannot be installed.

4.2.0   MEDIUM-VOLTAGE FUSES

Fuses above 600 volts are classified under one of three classifications as defined in ANSI/IEEE 40-1981:

- General purpose current-limiting fuse
- Back-up current-limiting fuse
- Expulsion fuse

General purpose current-limiting fuse: A fuse capable of interrupting all currents from the rated interrupting current down to the current that causes melting of the fusible element in one hour.

Back-up current-limiting fuse: A fuse capable of interrupting all currents from the maximum rated interrupting current down to the rated minimum interrupting current.

Expulsion fuse: A vented fuse in which the expulsion effect of gasses produced by the arc and lining of the fuseholder, either alone or aided by a spring, extinguishes the arc.

One should note that in the definitions just given, the fuses are defined as either expulsion or current limiting. A current limiting fuse is a sealed, non-venting fuse that, when melted by a current within its interrupting rating, produces arc voltages exceeding the system voltage which in turn forces the current to zero. The arc voltages are produced by introducing a series of high resistance arcs within the fuse. The result is a fuse that typically interrupts high fault currents with the first ½ cycle of the fault.

In contrast an expulsion fuse depends on one arc to initiate the interruption process. The arc acts as a catalyst causing the generation of de-ionizing gas from its housing. The arc is then elongated either by the force of the gasses created or a spring. At some point the arc elongates far enough to prevent a restrike after passing through a current zero. Therefore, it is not atypical for an expulsion fuse to take many cycles to clear.

4.2.1 Application of Medium-Voltage Fuses

Many of the rules for applying expulsion fuses and current-limiting fuses are the same, but because the current-limiting fuse operates much faster on high-fault, some additional rules must be applied.

Three basic factors must be considered when applying any fuse:

- Voltage
- Continuous current-carrying capacity
- Interrupting rating

Voltage: The fuse must have a voltage rating equal to or greater than the normal frequency recovery voltage which will be seen across the fuse under all conditions. On three-phase systems it is a good rule-of-thumb that the voltage rating of the fuse be greater than or equal to the line-to-line voltage of the system.

Continuous current-carrying capacity: Continuous current values that are shown on the fuse represent the level of current the fuse can carry continuously without exceeding the temperature rises as specified in ANSI C37.46. An application that exposes the fuse to a current slightly above its continuous rating but below its minimum interrupting rating may damage the fuse due to excessive heat. This is the main reason overload relays are used in series with back-up current-limiting fuses for motor protection.

Interrupting rating: All fuses are given a maximum interrupting rating. This rating is the maximum level of fault current that the fuse can safely interrupt. Back-up current-limiting fuses are also given a minimum interrupting rating. When using back-up current-limiting fuses, it is important that other protective devices are used to interrupt currents below this level.

When choosing a fuse, it is important that the fuse be properly coordinated with other protective devices located upstream and downstream. To accomplish this, one must consider the melting and clearing characteristics of the devices. Two curves, the minimum melting curve and the total clearing curve, provide this information. To insure proper coordination, the following rules should be used:

- The total clearing curve of any downstream protective device must be below a curve representing 75% of the minimum melting curve of the fuse being applied.
- The total clearing curve, of the fuse being applied, must lie below a curve representing 75% of the minimum melting curve for any upstream protective device.

4.3.0 CURRENT-LIMITING FUSES

To insure proper application of a current-limiting fuse, it is important that the following additional rules be applied.

1. Current-limiting fuses produce arc voltages that exceed the system voltage. Care must be taken to make sure that the peak voltages do not exceed the insulation level of the system. If the fuse voltage rating is not permitted to exceed 140% of the system voltage, there should not be a problem. This does not mean that a higher rated fuse cannot be used, but points out that one must be assured that the system insulation level (BIL) will handle the peak arc voltage produced. BIL stands for basic impulse level which is the reference impulse insulation strength of an electrical system.

2. As with the expulsion fuse, current-limiting fuses must be properly coordinated with other protective devices on the system. For this to happen, the rules for applying an expulsion fuse must be used at all currents that cause the fuse to interrupt in 0.01 seconds or greater.

When other current-limiting protective devices are on the system, it becomes necessary to use I^2t values for coordination at currents causing the fuse to interrupt in less than 0.01 seconds.

These values may be supplied as minimum and maximum values or minimum melting and total clearing I^2t curves. In either case, the following rules should be followed.

1. The minimum melting I^2t of the fuse should be greater than the total clearing I^2t of the downstream current-limiting device.

2. The total clearing I^2t of the fuse should be less than the minimum melting I^2t of the upstream current-limiting device.

The fuse-selection chart in Fig. 11 should serve as a guide for selecting fuses on circuits of 600 volts or less. The dimensions of various Buss fuses are shown in Fig. 12. Both of these charts will prove invaluable on all types of projects—from residential to heavy industrial applications. Other valuable information may be found in catalogs furnished by manufacturers of overcurrent protective devices. These are usually obtainable from electrical supply houses or from manufacturers' reps. You may also write the various manufacturers for a complete list (and price, if any) for all reference materials offered by them.

4.4.0 FUSES FOR SELECTIVE COORDINATION

The larger the upstream fuse is relative to a downstream fuse (feeder to branch, etc.), the less possibility there is of an overcurrent in the downstream circuit causing both fuses to open. Fast action, non-time-delay fuses require at least a 3:1 ratio between the ampere rating of a large upstream, line-side time-delay fuse to that of the downstream, load-side fuse in order to be selectively coordinated. In contrast, the minimum selective coordination ratio necessary for dual-element fuses is only 2:1 when used with Low Peak loadside fuses. See Fig. 13.

The use of time-delay, dual-element fuses affords easy selective coordination—coordination hardly requires anything more than a routine check of a tabulation of required selectively ratios. As shown in Fig. 14, close sizing of dual-element fuses in the branch circuit for motor overload protection provides a large difference (ratio) in the ampere ratings between the feeder fuse and the branch fuse compared to the single-element, non-time-delay fuse.

4.5.0 FUSE TIME-CURRENT CURVES

When a low-level overcurrent occurs, a long interval of time will be required for a fuse to open (melt) and clear the fault. On the other hand, if the overcurrent is large, the fuse will open very quickly. The opening time is a function of the magnitude of the level of overcurrent. Overcurrent levels and the corresponding intervals of opening times are logarithmically plotted in graph form as shown in Fig. 15. Levels of overcurrent are scaled on the horizontal axis, time intervals on the vertical axis. The curve is therefore called a "time-current" curve.

The plot in Fig. 15 reflects the characteristics of a 200 ampere, 600 volt, dual-element fuse. Note that at the 1,000 ampere overload level, the time interval which is required for the fuse to open is 10 seconds. Yet, at approximately the 2200 ampere overcurrent level, the opening

Circuit	Load	Ampere Rating	Fuse Type	Symbol	Voltage Rating (ac)	UL Class	Interrupting Rating (K)	Remarks
Main, Feeder and Branch (Conventional dimensions)	All type load (optimum overcurrent protection)	0-600A	LOW-PEAK® (dual-element, time-delay)	LPN-RK	250V	RK1	200	All-purpose fuses. Unequaled for combined short-circuit and overload protection.
				LPS-RK	600V			
		601 to 6000A	LOW-PEAK® time delay	KRP-C	600V	L	200	
	Motors, welders, transformers, capacitor banks (circuits with heavy inrush currents)	0 to 600A	FUSETRON® (dual-element, time-delay)	FRN-R	250V	RK5	200	Moderate degree of current limitation. Time-delay passes surge currents.
				FRS-R	600V			
		601 to 4000A	LIMITRON® (time delay)	KLU	600V	L	200	All-purpose fuse. Time-delay passes surge-currents.
	Non-motor loads (circuits with no heavy inrush currents)	0 to 600A	LIMITRON® (fast-acting)	KTN-R	250V	RK1	200	Same short-circuit protection as LOW-PEAK fuses but must be sized larger for circuits with surge-currents; i.e., up to 300%.
				KTS-R	600V			
	LIMITRON fuses particular suited for circuit breaker protection	601 to 6000A	LIMITRON® (fast-acting)	KTU	600V	L	200	A fast-acting, high performance fuse.
	All type loads (optimum overcurrent protection)	0 to 600A	LOW-PEAK® (dual-element time-delay)	LPJ	600V	J	200	All-purpose fuses. Unequaled for combined short-circuit and overload protection.
	Non-motor loads (circuits with no heavy inrush currents)	0 to 600A	LIMITRON® (quick-acting)	JKS	600V	J	200	Very similar to KTS-R LIMITRON, but smaller.
		0 to 1200A	T-TRON™	JJN	300V	T	200	The space saver (⅓ the size of KTN-R/KTS-R).
				JJS	600V			

Note: All shaded areas represent fuses with reduced dimensions for installation in restricted space

Figure 11. Fuse selection chart (600 volts of less).

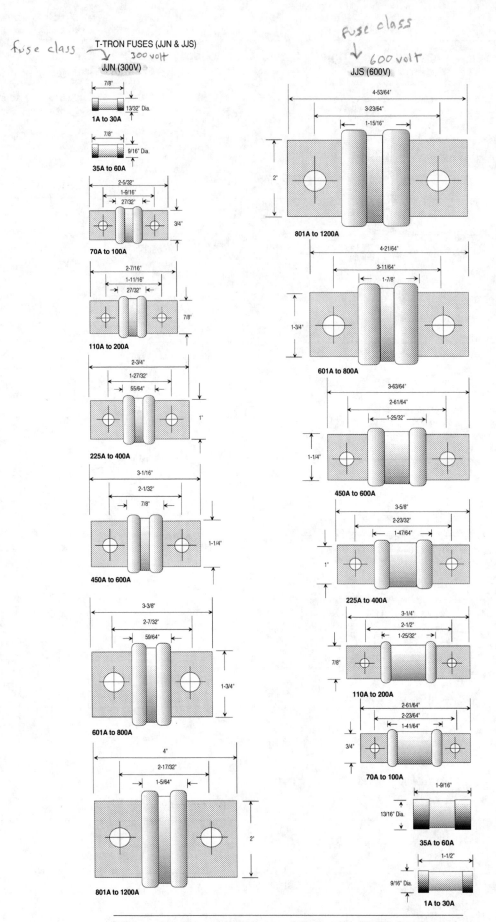

Figure 12. Buss fuse dimensional data (T-TRON fuses).

ELECTRICAL TRAINEE TASK MODULE 20303

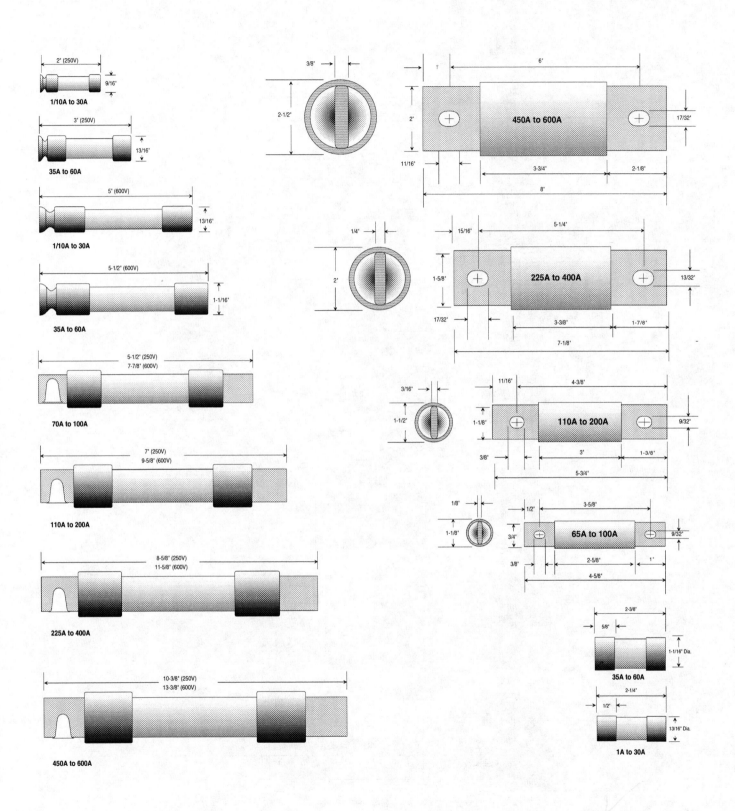

Figure 12. *(Cont.)* Buss fuse dimensional data (FUSETRON, LOW-PEAK, and LIMITRON fuses).

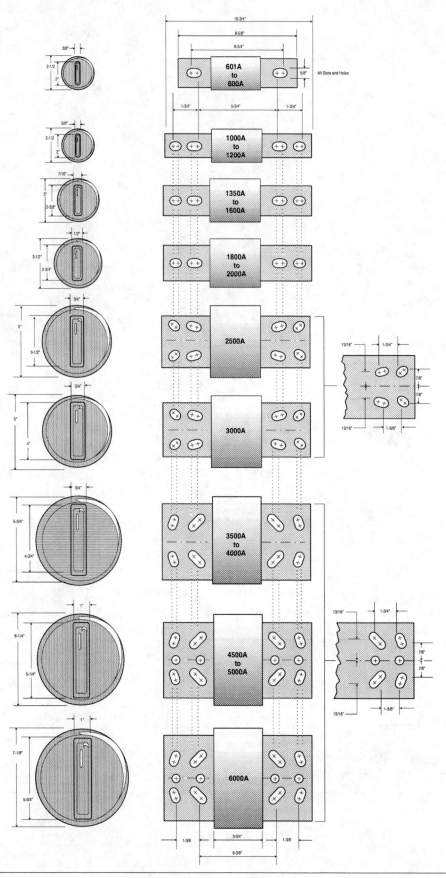

Figure 12 *(Cont.)*. Buss fuse dimensional data (LOW PEAK & LIMITRON KRP-C, KTU, & KLU)

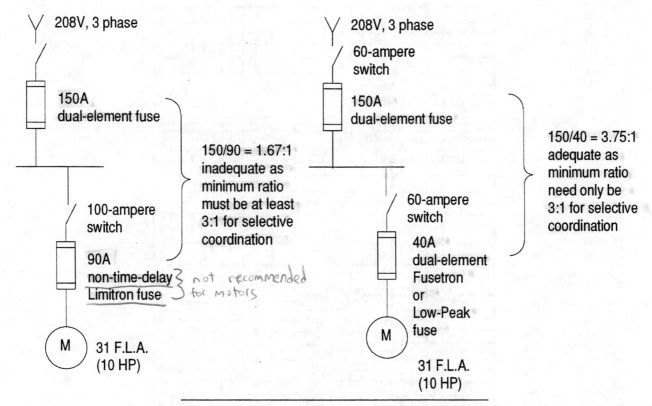

Figure 13. Fuses used for selective coordination.

(melt) time of a fuse is only 0.01 seconds. It is apparent that the time intervals become shorter and shorter as the overcurrent levels become larger. This relationship is termed an inverse time-to-current characteristic. Time-current curves are published or are available on most commonly used fuses showing "minimum melt," "average melt" and/or "total clear" characteristics. Although upstream and downstream fuses are easily coordinated by adhering to

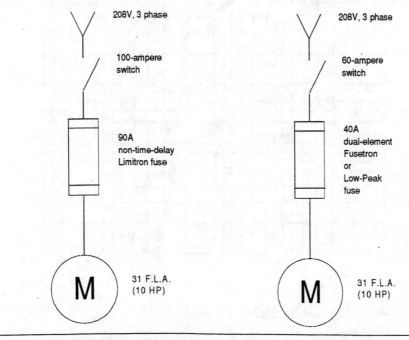

Figure 14. Dual-element fuses permit the use of smaller and less costly switches.

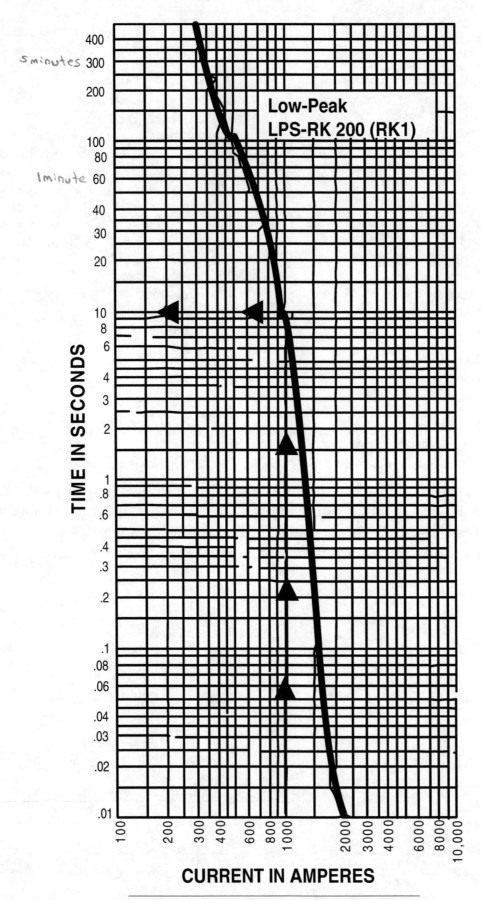

Figure 15. Typical time-current curve of a fuse.

simple ampere ratios, these time-current curves permit close or critical analysis of coordination.

4.5.1 Peak Let-Through Charts

Peak let-through charts enable you to determine both the peak let-through current and the apparent prospective rms symmetrical let-through current. Such charts are commonly referred to as *current limitation curves*. Figure 16 shows a simplified chart with explanations of the various functions. Figures 17 through 22 show current limitation curves for popular fuses in current use.

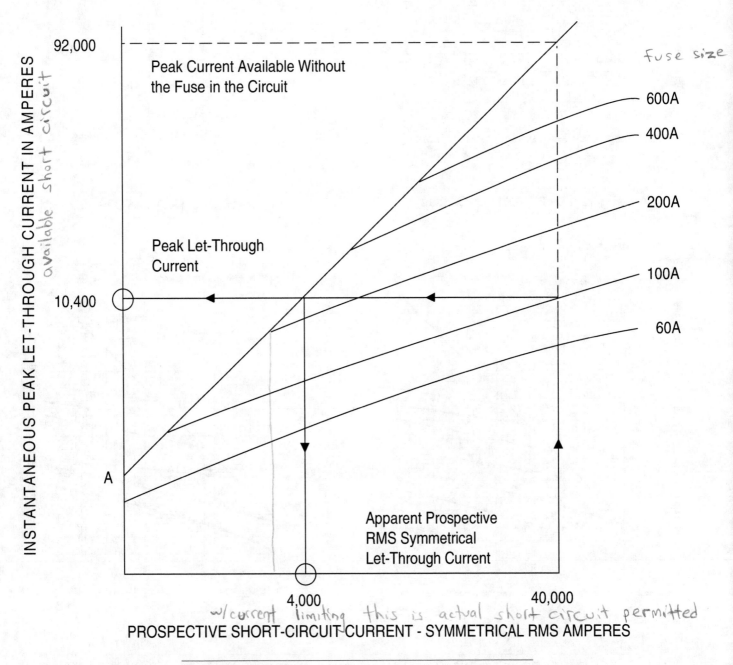

Figure 16. Principles of forming current limitation curves.

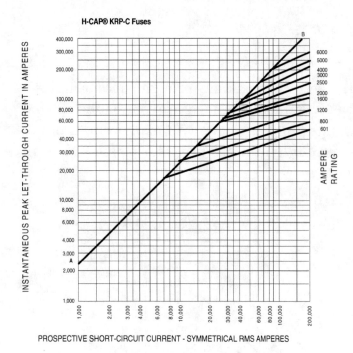

Figure 17. HI-CAP KRP-C fuses.

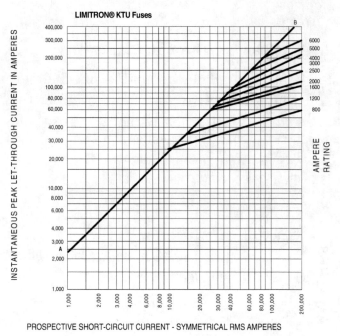

Figure 18. LIMITRON KTU fuses.

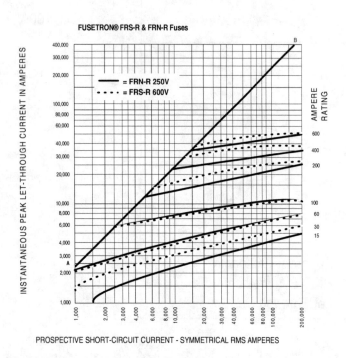

Figure 19. FUSETRON FRS-R & FRN-R fuses.

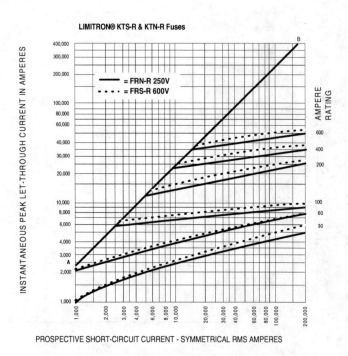

Figure 20. LIMITRON KTS-R & KTN-R fuses.

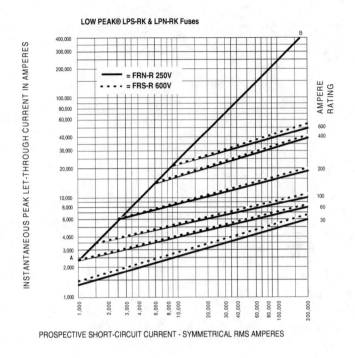

Figure 21. LOW PEAK LPS-RK & PLN-RK fuses.

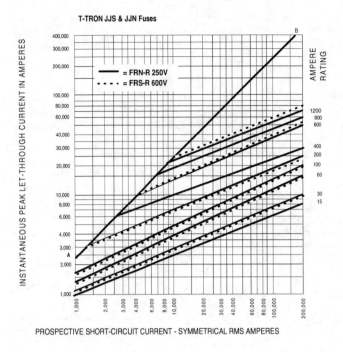

Figure 22. T-TRON JJS & JJN fuses.

5.0.0 MOTOR OVERLOAD AND SHORT-CIRCUIT PROTECTION

When used in circuits with surge currents such as those caused by motors, transformers, and other inductive components, dual-element, time-delay fuses can be sized close to full-load amperes to give maximum overcurrent protection. For example, let's assume that a 10 HP, 208-volt motor has a full-load current of 31 amperes. NEC Sections 430-32 and 430-52 require the following fuse types and sizes, as well as the switch size:

Fuse and Switch Sizing for 10 HP Motor (208V, 3Ø, 31 FLA)

Fuse Type	Maximum Fuse Size (Amperes)	Required Switch Size (Amperes)
Dual-element, time-delay	40A	60A
Single-element, non-time delay	90A	100A

The preceding table shows that a 40-ampere, dual-element fuse will protect the 31 ampere motor compared to the much larger, necessary 90-ampere, single-element fuse. It is apparent that if a sustained, harmful overload of 300% occurred in the motor circuit, the 90-ampere, single-element fuse would never open and the motor could be damaged. The non-time-delay

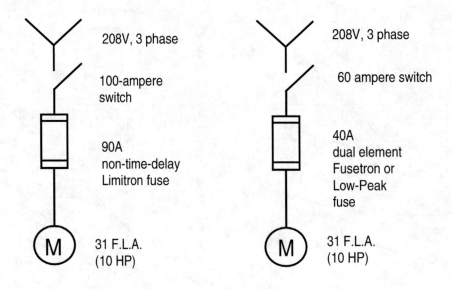

Figure 23. Dual-element fuses permit the use of smaller and less costly switches.

fuse provides only ground fault and short-circuit protection—requiring separate overload protection as per the NEC.

In contrast, the 40-amperes dual-element fuse provides ground-fault, short-circuit protection plus a high degree of back-up protection against motor burnout from overload or single-phasing, should other overload protective devices fail. If thermal overloads, relays, or contacts should fail to operate, the dual-element fuses will act independently to protect the motor.

Aside from providing only short-circuit protection, the single-element fuse also makes it necessary to use larger size switches since a switch rating must be equal to or larger than the ampere rating of the fuse; as a result, the larger switch may cost two or three times more than would be necessary were a dual-element fuse used. See Fig. 23.

When secondary single-phasing occurs, the current in the remaining phases increases to a value of 170% to 200% of rated full-load current. When primary single-phasing occurs, unbalanced voltages that occur in the motor circuit cause excessive current. Dual-element fuses sized for motor overload protection can protect motors against the overload damage caused by single-phasing.

The non-time-delay, fast-acting fuse must be oversized in circuits in which surge or temporary overload currents occur. Response of the oversized fuse to short-circuit currents is slower. Current builds up to a high level before the fuse opens—causing the current-limiting action of the oversized fuse to be less than a fuse whose ampere rating is closer to the normal full-load current of the circuit. Consequently, oversizing sacrifices some component protection and although it is permitted by the NEC, the practice is not recommended. In actual practice, dual-element fuses used to protect motors keep short-circuit currents to approximately half the value of the non-time-delay fuses, since the non-time-delay fuses must be oversized to carry the temporary starting current of motors.

6.0.0   CIRCUIT BREAKERS

Circuit breakers were thoroughly covered in your Level 2 training. However, some of the more important points are worth repeating here. Then, practical applications of both fuses and circuit breakers will be covered.

Basically, a circuit breaker is a device for closing and interrupting a circuit between separable contacts under both normal and abnormal conditions. This is done manually (normal condition) by use of its "handle" by switching to the ON or OFF positions. However, the circuit breaker also is designed to open a circuit automatically on a predetermined overload or ground-fault current without damage to itself or its associated equipment. As long as a circuit breaker is applied within its rating, it will automatically interrupt any "fault" and therefore must be classified as an inherently safe, overcurrent, protective device.

The internal arrangement of a circuit breaker is shown in Fig. 24 while its external operating characteristics are shown in Fig. 25. Note that the handle on a circuit breaker resembles an ordinary toggle switch. On an overload, the circuit breaker opens itself or *trips*. In a tripped position, the handle jumps to the middle position (Fig. 25). To reset, turn the handle to the OFF position and then turn it as far as it will go beyond this position; finally, turn it to the ON position.

A standard molded case circuit breaker usually contains:

- A set of contacts
- A magnetic trip element
- A thermal trip element
- Line and load terminals
- Bussing used to connect these individual parts
- An enclosing housing of insulating material

The circuit breaker handle manually opens and closes the contacts and resets the automatic trip units after an interruption. Some circuit breakers also contain a manually operated "push-to-trip" testing mechanism.

Circuit breakers are grouped for identification according to given current ranges. Each group is classified by the largest ampere rating of its range. These groups are:

- 15-100 amperes
- 125-225 amperes

- 250-400 amperes
- 500-1,000 amperes
- 1,200-2,000 amperes

Therefore, they are classified as 100, 225, 400, 1,000 and 2,000-ampere frames. These numbers are commonly referred to as "frame classification" or "frame sizes" and are terms applied to groups of molded case circuit breakers which are physically interchangeable with each other.

6.1.0 INTERRUPTING CAPACITY RATING

In most large commercial and industrial installations, it is necessary to calculate available short-circuit currents at various points in a system to determine if the equipment meets the requirements of NEC Sections 110-9 and 110-10. There are a number of methods used to determine the short-circuit requirements in an electrical system. Some give approximate values; some require extensive computations and are quite exacting.

The breaker interrupting capacity is based on tests to which the breaker is subjected. There are two such tests; one set up by U.L. and the other by NEMA. The NEMA tests are self-certification while U.L. tests are certified by unbiased witnesses. U.L. tests have been limited to a maximum of 10,000 amperes in the past, so the emphasis was placed on NEMA tests with higher ratings. U.L. tests now include the NEMA tests plus other ratings. Consequently, the emphasis is now being placed on U.L. tests.

The interrupting capacity of a circuit breaker is based on its rated voltage. Where the circuit breaker can be used on more than one voltage, the interrupting capacity will be shown for each voltage level. For example, the LA type circuit breaker has 42,000 amperes, symmetrical interrupting capacity at 240 volts, 30,000 amperes symmetrical at 480 volts, and 22,000 amperes symmetrical at 600 volts.

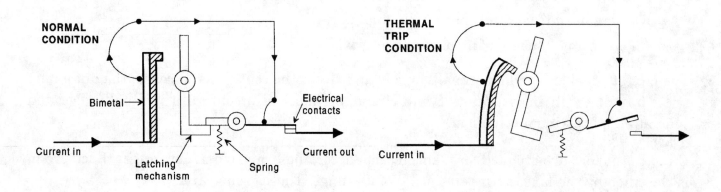

Figure 24. Characteristics of a thermal-trip circuit breaker.

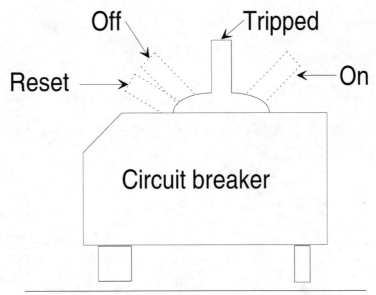

Figure 25. External characteristics of a circuit breaker.

7.0.0 CONDUCTOR PROTECTION

All conductors are to be protected against overcurrents in accordance with their ampacities as set forth in NEC Section 240-3. They must also be protected against short-circuit current damage as required by NEC Sections 240-1 and 110-10.

Ampere ratings of overcurrent-protective devices must not be greater than the ampacity of the conductor. There is, however, an exception. NEC Section 240-3 states that if such conductor rating does not correspond to a standard size overcurrent-protective device, the next larger size overcurrent-protective device may be used provided its rating does not exceed 800 amperes and when the conductor is not part of a multi-outlet branch circuit supplying receptacles for cord-and-plug connected portable loads. When the ampacity of busway or cablebus does not correspond to a standard overcurrent-protective device, the next larger stand rating may be used only if the rating does not exceed 800 amperes (NEC Sections 364-10 and 365-5).

Standard fuse sizes stipulated in NEC Section 240-6 are: 1, 3, 6, 10, 15, 20, 25, 30, 35, 40, 45, 50, 60, 70, 80, 90, 100, 110, 125, 150, 175, 200, 225, 250, 300, 350, 400, 450, 500, 600, 700, 800, 1000, 1200, 1600, 2000, 2500, 3000, 4000, 5000, and 6000 amperes.

Note   The small fuse ampere ratings of 1, 3, 6, and 10 have recently been added to the NEC to provide more effective short-circuit and ground-fault protection for motor circuits in accordance with Sections 430-40 and 430-52 and U.L. requirements for protecting the overload relays in controllers for very small motors.

Protection of conductors under short-circuit conditions is accomplished by obtaining the maximum short-circuit current available at the supply end of the conductor, the short-circuit

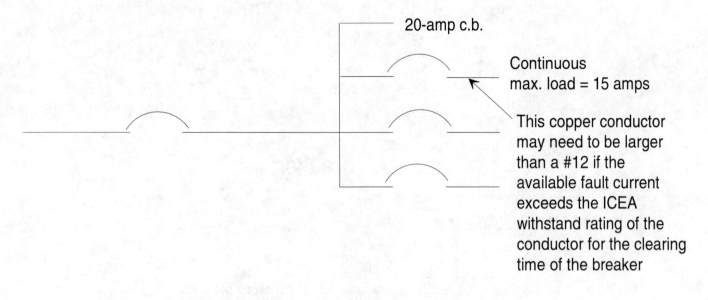

Figure 26. Conductor protection with non-current-limiting device.

withstand rating of the conductor, and the short-circuit let-through characteristics of the overcurrent device.

When a non-current-limiting device is used for short-circuit protection, the conductor's short-circuit withstand rating must be properly selected based on the overcurrent protective device's ability to protect. See Fig. 26.

It is necessary to check the energy let-through of the overcurrent device under short-circuit conditions and select a wire size of sufficient short-circuit withstand ability.

In contrast, the use of a current-limiting fuse permits a fuse to be selected which limits short-circuit current to a level less than that of the conductors short-circuit withstand rating — doing away with the need of oversized ampacity conductors. See Fig. 27.

In many applications, it is desirable to use the convenience of a circuit breaker for a disconnecting means and general overcurrent protection, supplemented by current-limiting fuses at strategic points in the circuits.

Flexible cord, including tinsel cord and extension cords, must be protected against overcurrent in accordance with their ampacities. See Task Module 20302—Conductor Selections and Calculations.

7.1.0   LOCATION OF FUSES IN CIRCUITS

In general, fuses must be installed at points where the conductors receive their supply; that is, at the beginning or lineside of a branch circuit or feeder. Exceptions to this rule are given in NEC Section 240-21.

ELECTRICAL TRAINEE TASK MODULE 20303

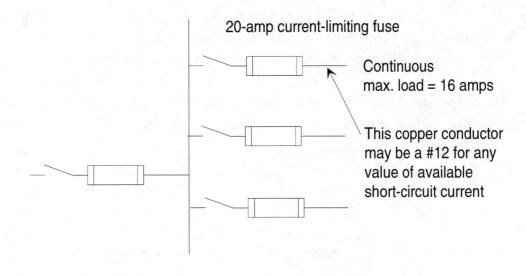

20-amp current-limiting fuse

Continuous
max. load = 16 amps

This copper conductor
may be a #12 for any
value of available
short-circuit current

Figure 27. Circuits protected by current-limiting devices.

Exception No. 1: Fuses are not required at the conductor supply if the fuses protecting one conductor are small enough to protect a small conductor connected thereto.

Exception No. 2: Fuses are not required at the conductor supply if a feed tap conductor is not over ten feet long; is enclosed in raceway, does not extend beyond the switchboard, panelboard disconnecting means or control device which it supplies, and has an ampacity not less than the combined computed loads supplied and not less than the rating of the device supplied unless the tap conductors are terminated in a fuse not exceeding the tap conductors' ampacities. For field installed taps, the ampacity of the tap conductor must be at least 10% of the overcurrent device rating.

Exception No. 3: Fuses are not required at the conductor supply if a feeder tap conductor is not over 25 feet long; is suitably protected from physical damage; has an ampacity not less than ⅓ that of the feeder conductors or fuses from which the tap conductors receive their supply; and terminate in a single set of fuses sized not more than the tap conductor ampacity.

Exception No. 8: Fuses are not required at the conductor supply if a transformer feeder tap has primary conductors at least ⅓ ampacity of the fuses protecting the feeder; the secondary conductors have at least ⅓ ampacity when multiplied by the approximate transformer turns ratio of the fuses protecting the feeder conductors; the total length of one primary plus one secondary conductor (excluding any portion of the primary conductor that is protected at its ampacity) is not over 25 feet in length; the secondary conductors terminate in a set of fuses that will limit the load current to no more than the conductor ampacity that is permitted by NEC Section 310-15; and if the primary and secondary conductors are suitably protected from physical damage.

Exception No. 10: Fuses are not required at the conductor supply if a feeder tap is not over 25 feet long horizontally and not over 100 feet long total length in high bay manufacturing buildings when only qualified persons will service such a system, and the ampacity of the tap conductors is not less than ⅓ of the fuse rating from which they are supplied, that will limit the load to the ampacity of the tap are at least No. 6 AWG copper or No. 4 AWG aluminum, do not penetrate walls, floors, or ceilings, and are made no less than 30 feet from the floor.

| *WARNING!* | Smaller conductors tapped to larger conductors can be a serious hazard. If not protected against short-circuit conditions, these unprotected conductors can vaporize or incur severe insulation damage. |

Exception No. 11: Transformer secondary conductors of separately derived systems do not require fuses at the transformer terminals when all of the following conditions are met:

- Must be an industrial location
- Secondary conductors must be less than 25 feet long
- Secondary conductor ampacity must be at least equal to secondary full-load current of transformer and sum of terminating, grouped, overcurrent devices
- Secondary conductors must be protected from physical damage

Note: Switchboard and panelboard protection (NEC Section 384-16) and transformer protection (NEC Section 450-3) must still be observed.

7.1.1   Lighting/Appliance Loads

The branch-circuit rating must be classified in accordance with the rating of the overcurrent protective device. Classifications for those branch circuits other than individual loads must be: 15, 20, 30, 40, and 50 amperes as specified in NEC Section 210-3.

Branch-circuit conductors must have an ampacity of the rating of the branch circuit and not less than the load to be served (NEC Section 210-19). The minimum size branch-circuit conductor that can be used is No. 14 (NEC Section 210-19). However, there are some exceptions as specified in NEC Section 210-19.

Branch-circuit conductors and equipment must be protected by a fuse whose ampere rating conforms to NEC Section 210-20. Basically, the branch circuit conductor and fuse must be sized for the actual non-continuous load and 125% for all continuous loads. The fuse size must not be greater than the conductor ampacity. Branch circuits rated 15 through 50 amperes with two or more outlets (other than receptacle circuits) must be fused at their rating and the branch-circuit conductor sized according to NEC Table 210-24.

7.1.2 Feeder Circuits With No Motor Load

The feeder fuse ampere rating and feeder conductor ampacity must be at least 100% of the non continuous load plus, 125% of the continuous load as calculated per NEC Article 220. The feeder conductor must be protected by a fuse not greater than the conductor ampacity. Motor loads shall be computed in accordance with Article 430.

7.1.3 Service Equipment

Each ungrounded service entrance conductor must have a fuse in series with a rating not higher than the ampacity of the conductor. There service fuses shall be part of the service disconnecting means or be located immediately adjacent thereto. (NEC Section 230-91).

Service disconnecting means can consist of one to six switches or circuit breakers for each service or for each set of service-entrance conductors permitted in NEC Section 230-2. When more than one switch is used, the switches must be grouped together (NEC Section 230-71).

7.1.4 Transformer Secondary Conductors

Field installations indicate nearly 50% of transformers installed do not have secondary protection. The NEC recommends that secondary conductors be protected from damage by the proper overcurrent protective device. For example, the primary overcurrent device protecting a 3-wire transformer cannot offer protection of the secondary conductors. Also see NEC exception in Section 240-3 for 2-wire primary and secondary circuits.

7.1.5 Motor Circuit Protection

Motors and motor circuits have unique operating characteristics and circuit components. Therefore, these circuits must be dealt with differently from other types of loads. Generally, two levels of overcurrent protection are required for motor branch circuits:

- Overload protection—Motor running overload protection is intended to protect the system components and motor from damaging overload currents.
- Short-circuit protection (includes ground-fault protection)—Short-circuit protection is intended to protect the motor circuit components such as the conductors, switches, controllers, overload relays, motor, etc. against short-circuit currents or grounds. This level of protection is commonly referred to as motor branch-circuit protection applications. Dual-element fuses are designed to provide this protection provided they are sized correctly.

There are a variety of ways to protect a motor circuit—depending upon the user's objective. The ampere rating of a fuse selected for motor protection depends on whether the fuse is of the dual-element time-delay type or the non-time-delay type.

In general, non-time-delay fuses can be sized at 300% of the motor full-load current for ordinary motors so that the normal motor starting current does not affect the fuse. Dual-element, time-delay fuses are able to withstand normal motor-starting current and can be sized closer to the actual motor rating than can non-time-delay fuses.

A summary of NEC regulations governing overcurrent protection is covered in the table in Fig. 28, while the table in Fig. 29 gives generalized fuse application guidelines for motor branch circuits. Figure 30 may be used to select fuses for motor protection.

Application	Rule	NEC Section
Scope (FPN)	Overcurrent protection for conductors and equipment is provided to open the circuit if the current reaches a value that will cause an excessive or dangerous temperature in conductors or conductor insulation. See also Sections 110-9 and 110-10 for requirements for interrupting capacity and protection against fault currents.	Section 240-1
Protection Required	Each ungrounded service-entrance conductor must have overcurrent protection. Device must be in series with each ungrounded conductor.	Section 230-90(a)
Number of Devices	Up to six circuit breakers or sets of fuses may be considered as the overcurrent device.	Section 230-90(a)
Location in Building	The overcurrent device must be part of the service disconnecting means or be located immediately adjacent to it.	Section 230-91
Accessibility	In a property comprising more than one building under single management, the ungrounded conductors supplying each building served shall be protected by overcurrent devices, which may be located in the building served or in another building on the same property, provided they are accessible to the occupants of the building served. In a multiple-occupancy building each occupant shall have access to the overcurrent protective devices.	Section 230-91
Location in Circuit	The overcurrent device must protect all circuits and devices, except equipment which may be connected on the supply side including: 1) Service switch, 2) Special equipment, such as surge arresters, 3) Circuits for emergency supply and load management (where separately protected), 4) Circuits for fire alarms or fire-pump equipment (where separately protected), 5) Meters, with all metal housing grounded, (600 volts or less), 6) Control circuits for automatic service equipment if suitable overcurrent protection and disconnecting means are provided.	Section 230-94

Figure 28. NEC regulations concerning overcurrent protection.

Application	Rule	NEC Section
Installation and Use	Listed or labeled equipment shall be used or installed in accordance with any instructions included in the listing or labeling.	Section 110-3(b)
Interrupting Rating	Equipment intended to break current at fault levels shall have an interrupting rating sufficient for the system voltage and the current which is available at the line terminals of the equipment.	Section 110-9
Circuit Impedance and Other Characteristics	The overcurrent protective devices, the total impedance, the component short-circuit withstand ratings, and other characteristics of the circuit to be protected shall be so selected and coordinated as to permit the circuit protective devices used to clear a fault without the occurrence of extensive damage to the electrical components of the circuit.	Section 110-10
Available Short-Circuit Current	Service equipment shall be suitable for the short-circuit current available at its supply terminals.	Section 230-65
Effective Grounding Path	The path to ground from circuits, equipment, and conductor enclosures shall: 2) have capacity to conduct safely any fault current likely to be imposed on it.	Section 250-51
General	Bonding shall be provided where necessary to assure electrical continuity and the capacity to conduct safely any fault current likely to be imposed.	Section 250-70
Bonding Other Enclosures	Metal raceways, cable trays, cable armor, cable sheath, enclosures, frames, fittings, and other metal noncurrent-carrying parts that are to serve as grounding conductors with or without the use of supplementary equipment grounding conductors shall be effectively bonded where necessary to assure electrical continuity and the capacity to conduct safely any fault current likely to be imposed on them. Any non conductive paint, enamel, or similar coating shall be removed at threads, contact points, and contact surfaces or be connected by means of fittings so designed as to make such removal unnecessary.	Section 250-75

Figure 28. NEC regulations concerning overcurrent protection. (*Cont.*)

Type of Motor	Dual-Element, Time-Delay Fuses			Non-Time-Delay Fuses
	Desired Level of Protection			
	Motor Overload and Short-Circuit	Backup Overload and Short-Circuit	Short-Circuit Only (Based on NEC Tables 430-147 through 150 current ratings)	Short-Circuit Only (Based on NEC Tables 430-147 through 150 current ratings)
Service Factor 1.15 or Greater or 40°C Temp. Rise or Less	125% or less of motor nameplate current	125% or next standard size not to exceed 140%	150% to 175%	150% to 300%
Service Factor Less Than 1.15 or Greater Than 40°C Temp. Rise	115% or less of motor nameplate current	115% or next standard size not to exceed 130%	150% to 175%	150% to 300%

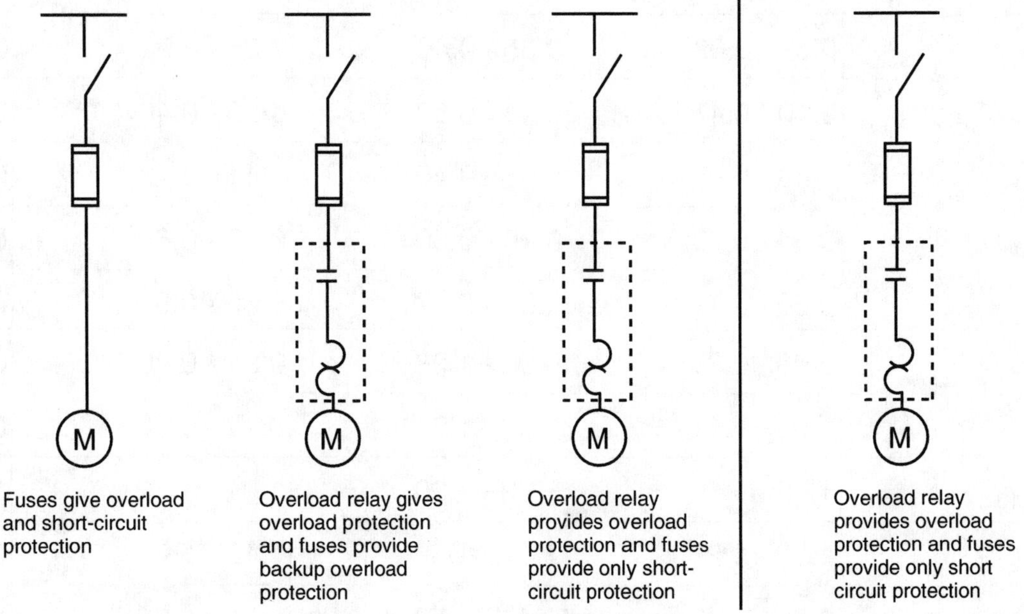

Fuses give overload and short-circuit protection

Overload relay gives overload protection and fuses provide backup overload protection

Overload relay provides overload protection and fuses provide only short-circuit protection

Overload relay provides overload protection and fuses provide only short circuit protection

Figure 29. Sizing of fuses as a percentage of motor full-load current.

Dual-Element Fuse Size	Motor Protection (Used without properly sized overload relays). Motor Full-Load Amps		Back-up Motor Protection (Used with properly sized overload relays). Motor Full-load Amps	
	Motor Service Factor of 1.15 or Greater or With Temp. Rise Not Over 40° C.	Motor Service Factor Less Than 1.15 or With Temp. Rise Not Over 40° C.	Motor Service Factor of 1.15 or Greater or With Temp. Rise Not Over 40° C.	Motor Service Factor of Less Than 1.15 or With Temp. Rise Not Over 40° C.
$\frac{1}{10}$	0.08 - 0.09	0.09 - 0.10	0 - 0.08	0 - 0.09
$\frac{1}{8}$	0.10 - 0.11	0.11 - 0.125	0.09 - 0.10	0.10 - 0.11
$\frac{5}{100}$	0.12 - 0.15	0.14 - 0.15	0.1 l - 0.12	0.12 - 0.13
$\frac{2}{10}$	0.16 - 0.19	0.18 - 0.20	0.13 - 0.16	0.14 - 0.17
$\frac{1}{4}$	0.20 - 0.23	0.22 - 0.25	0.17 - 0.20	0.18 - 0.22
$\frac{3}{10}$	0.24 - 0.30	0.27 - 0.30	0.21 - 0.24	0.23 - 0.26
$\frac{4}{10}$	0.32 - 0.39	0.35 - 0.40	0.25 - 0.32	0.27 - 0.35
$\frac{1}{2}$	0.40 - 0.47	0.44 - 0.50	0.33 - 0.40	0.36 - 0.43
$\frac{6}{10}$	0.48 - 0.60	0.53 - 0.60	0.41 - 0.48	0.44 - 0.52
$\frac{8}{10}$	0.64 - 0.79	0.70 - 0.80	0.49 - 0.64	0.53 - 0.70
1	0.80 - 0.89	0.87 - 0.97	0.65 - 0.80	0.71 - 0.87
$1\frac{1}{8}$	0.90 - 0.99	0.98 - 1.08	0.81 - 0.90	0.88 - 0.98
$1\frac{1}{4}$	1.00 - 1.11	1.09 - 1.21	0.91 - 1.00	0.99 - 1.09
$1\frac{4}{10}$	1.12 - 1.19	1.22 - 1.30	1.01 - 1.12	1.10 - 1.22
$1\frac{1}{2}$	1.20 - 1.27	1.31 - 1.39	1.13 - 1.20	1.23 - 1.30
$1\frac{6}{10}$	1.28 - 1.43	1.40 - 1.56	1.21 - 1.28	1.31 - 1.39
$1\frac{8}{10}$	1.44 - 1.59	1.57 - 1.73	1.29 - 1.44	1.40 - 1.57
2	1.60 - 1.79	1.74 - 1.95	1.45 - 1.60	1.58 - 1.74
$2\frac{1}{4}$	1.80 - 1.99	1.96 - 2.17	1.61 - 1.80	1.75 - 1.96
$2\frac{1}{2}$	2.00 - 2.23	2.18 - 2.43	1.81 - 2.00	1.97 - 2.17

Figure 30. Selection of fuses for motor protection.

Dual-Element Fuse Size	Motor Protection (Used without properly sized overload relays). Motor Full-Load Amps		Back-up Motor Protection (Used with properly sized overload relays). Motor Full-load Amps	
	Motor Service Factor of 1.15 or Greater or With Temp. Rise Not Over 40° C.	Motor Service Factor Less Than 1.15 or With Temp. Rise Not Over 40° C.	Motor Service Factor of 1.15 or Greater or With Temp. Rise Not Over 40° C.	Motor Service Factor of Less Than 1.15 or With Temp. Rise Not Over 40° C.
$2^6/_{10}$	2.24 - 2.39	2.44 - 2.60	2.01 - 2.24	2.18 - 2.43
3	2.40 - 2.55	2.61 - 2.78	2.25 - 2.40	2.44 - 2.60
$3^2/_{10}$	2.56 - 2.79	2.79 - 3.04	2.41 - 2.56	2.61 - 2.78
$3^1/_2$	2.80 - 3.19	3.05 - 3.47	2.57 - 2.80	2.79 - 3.04
4	3.20 - 3.59	3.48—3.91	2.81 - 3.20	3.05 - 3.48
$4^1/_2$	3.60 - 3.99	3.92 - 4.34	3.21 - 3.60	3.49 - 3.91
5	4.00 - 4.47	4.35 - 4.86	3.61 - 4.00	3.92 - 4.35
$5^6/_{10}$	4.48 - 4.79	4.87 - 5.21	4.01 - 4.48	4.36 - 4.87
6	4.80 - 4.99	5.22 - 5.43	4.49 - 4.80	4.88 - 5.22
$6^1/_4$	5.00 - 5.59	5.44 - 6.08	4.81 - 5.00	5.23 - 5.43
7	5.60 - 5.99	6.09 - 6.52	5.01 - 5.60	5.44 - 6.09
$7^1/_2$	6.00 - 6.39	6.53 - 6.95	5.61 - 6.00	6.10 - 6.52
8	6.40 - 7.19	6.96 - 7.82	6.01 - 6.40	6.53 - 6.96
9	7.20 - 7.99	7.83 - 8.69	6.41 - 7.20	6.97 - 7.83
10	8.00 - 9.59	8.70 - 10.00	7.21 - 8.00	7.84 - 8 70
12	9.60 - 11.99	10.44 - 12.00	8.01 - 9.60	8.71 - 10.43
15	12.00 - 13.99	13.05 - 15.00	9.61 - 12.00	10.44 - 13.04
$17^1/_2$	14.00 - 15.99	15.22 - 17.39	12.01 - 14.00	13.05 - 15.21
20	16.00 - 19.99	17.40 - 20.00	14.01 - 16.00	15.22 - 17.39
25	20.00 - 23.99	21.74 - 25.00	16.01 - 20.00	17.40 - 21.74
30	24.00 - 27.99	26.09 - 30.00	20.01 - 24.00	21.75 - 26.09
35	28.00 - 31.99	30.44 - 34.78	24.01 - 28.00	26.10 - 30.43

Figure 30. Selection of fuses for motor protection.

Dual-Element Fuse Size	Motor Protection (Used without properly sized overload relays). Motor Full-Load Amps		Back-up Motor Protection (Used with properly sized overload relays). Motor Full-load Amps	
	Motor Service Factor of 1.15 or Greater or With Temp. Rise Not Over 40° C.	Motor Service Factor Less Than 1.15 or With Temp. Rise Not Over 40° C.	Motor Service Factor of 1.15 or Greater or With Temp. Rise Not Over 40° C.	Motor Service Factor of Less Than 1.15 or With Temp. Rise Not Over 40° C.
40	32.00 - 35.99	34.79 - 39.12	28.01 - 32.00	30.44 - 37.78
45	36.00 - 39.99	39.13 - 43.47	32.01 - 36.00	37.79 - 39.13
50	40.00 - 47.99	43.48 - 50.00	36.01 - 40.00	39.14 - 43.48
60	48.00 - 55.99	52.17 - 60.00	40.01 - 48.00	43.49 - 52.17
70	56.00 - 59.99	60.87 - 65.21	48.01 - 56.00	52.18 - 60.87
75	60.00 - 63.99	65.22 - 69.56	56.01 - 60.00	60.88 - 65.22
80	64.00 - 71 .99	69.57 - 78.25	60.01 - 64.00	65.23 - 69.57
90	72.00 - 79.99	78.26 - 86.95	64.01 - 72.00	69.58 - 78.26
100	80.00 - 87.99	86.96 - 95.64	72.01 - 80.00	78.27 - 86.96
110	88.00 - 99.99	95.65 - 108.69	80.01 - 88.00	86.97 - 95.65
125	100.00 - 119.99	108.70 - 125.00	88.01 - 100.00	95.66 - 108.70
150	120.00 - 139.99	131.30 - 150.00	100.01 - 1 20.00	108.71 - 30.43
175	140.00 - 159.99	152.17 - 173.90	120.01 - 140.00	130.44 - 152.17
200	160.00 - 179.99	173.91 - 195.64	140.01 - 160.00	152.18 - 173.91
225	180.00 - 199.99	195.65 - 217.38	160.01 - 180.00	173.92 - 195.62
250	200.00 - 239.99	217.39 - 250.00	180.01 - 200.00	195.63 - 217.39
300	240.00 - 279.99	260.87 - 300.00	200.01 - 240.00	217.40 - 260.87
350	280.00 - 319.99	304.35 - 347.82	240.01 - 280.00	260.88 - 304.35
400	320.00 - 359.99	347.83 - 391.29	280.01 - 320.00	304.36 - 347.83
450	360.00 - 399.99	391.30 - 434.77	320.01 - 360.00	347.84 - 391.30
500	400.00 - 479.99	434.78 - 500.00	360.01 - 400.00	391.31 - 434.78
600	480.00 - 600.00	521.74 - 600.00	400.01 - 480.00	434.79 - 521.74

Figure 30. Selection of fuses for motor protection.

Summary

Reliable overcurrent-protective devices prevent or minimize costly damage to transformers, conductors, motors, and the other many components and electrical loads that make up the complete electrical distribution system. Consequently, reliable circuit protection is essential to avoid the severe monetary losses which can result from power blackouts and prolonged downtime of various types of facilities. Knowing these facts, the NFPA—via the NEC—has set forth various minimum requirements dealing with overcurrent devices, and how they should be installed in various types of electrical circuits.

Knowing how to select and size the type of overcurrent devices for specific applications is one of the basic requirements of every electrician. The sooner this knowledge is learned and put to practical use, the closer the Trainee becomes to being a first-class electrician.

Figure 31 gives a summary of fuse applications in all types of electrical systems. You will want to keep this chart handy for reference in your daily work routine.

References

For a more advanced study of topics covered in this Task Module, the following works are suggested:

American Electricians Handbook, Latest Edition, Croft, McGraw-Hill, New York, NY

National Electrical Code Handbook, Latest Edition, NFPA, Quincy, MA

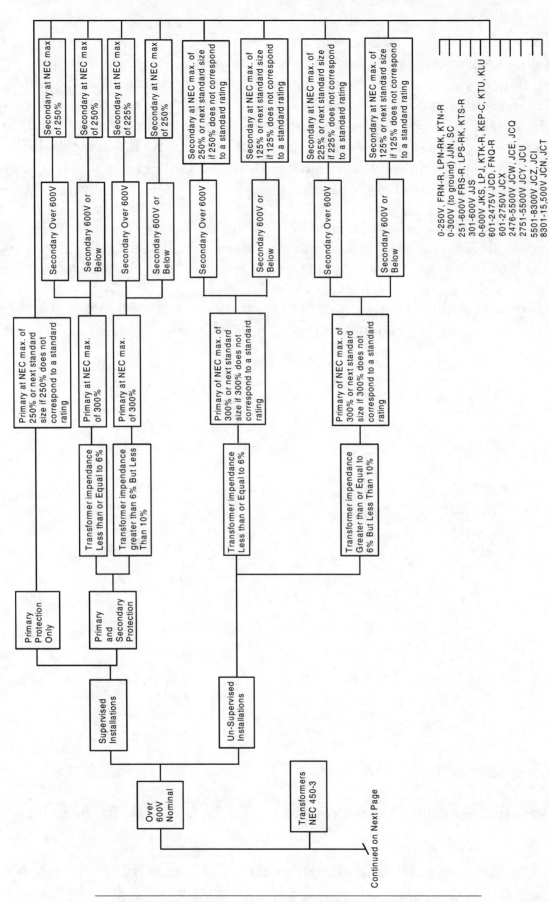

Figure 31. Summary of fuse applications in electrical systems.

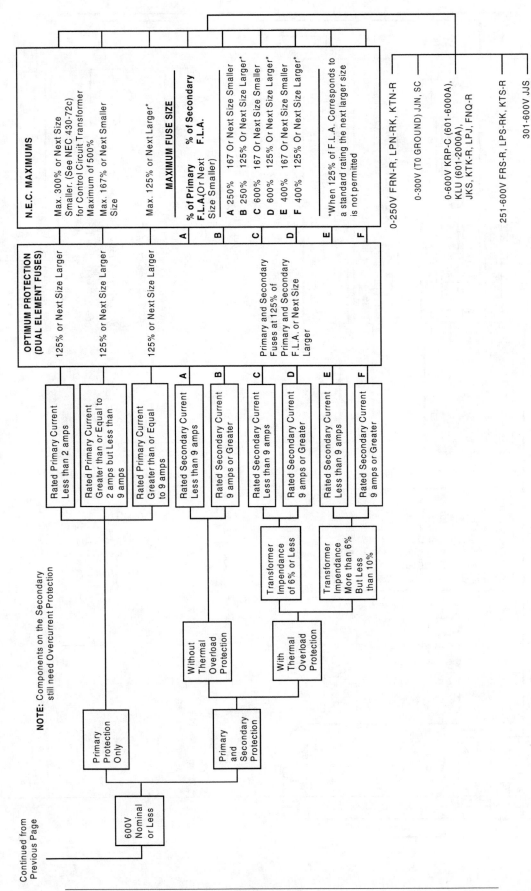

Figure 31. Summary of fuse applications in electrical systems. *(Cont.)*

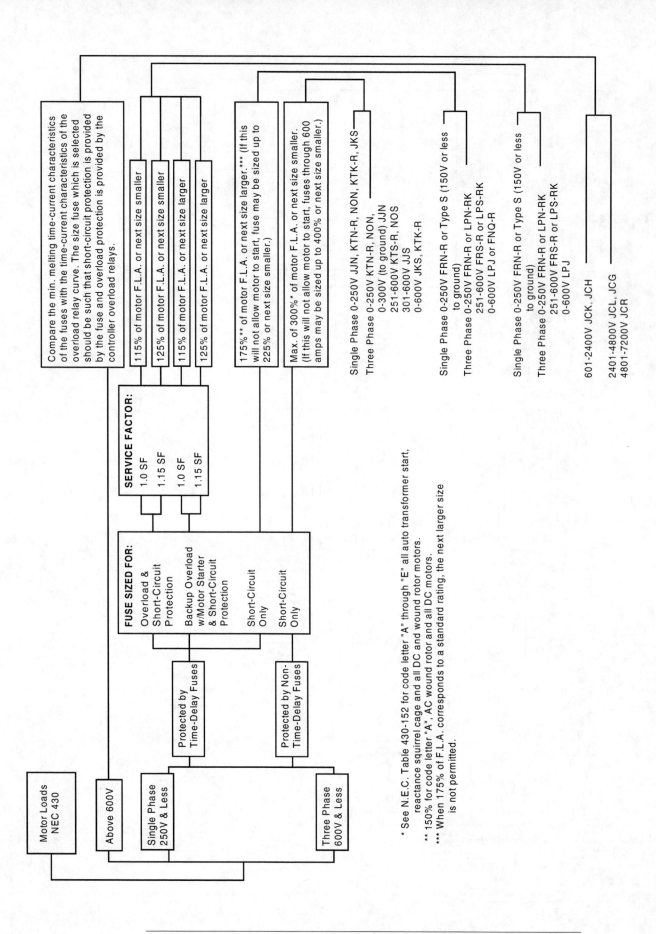

Figure 31. Summary of fuse applications in electrical systems. *(Cont.)*

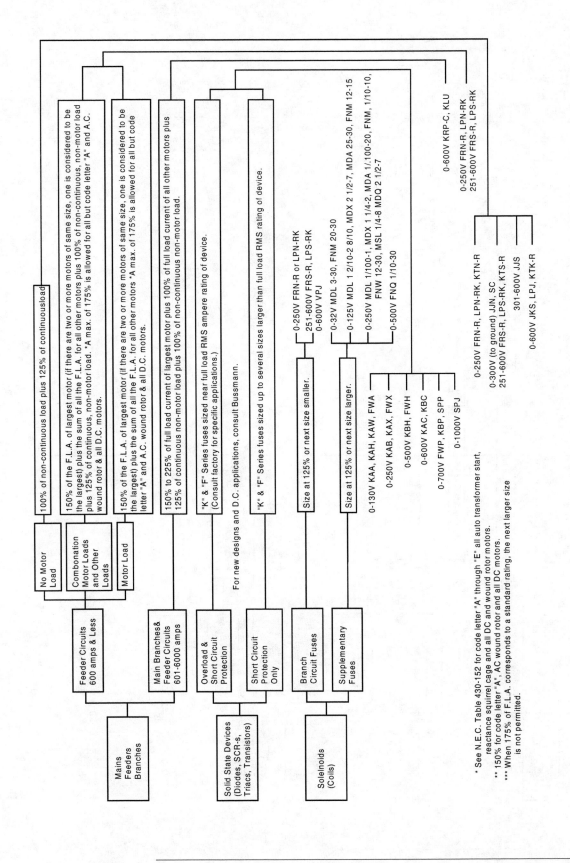

Figure 31. Summary of fuse applications in electrical systems. *(Cont.)*

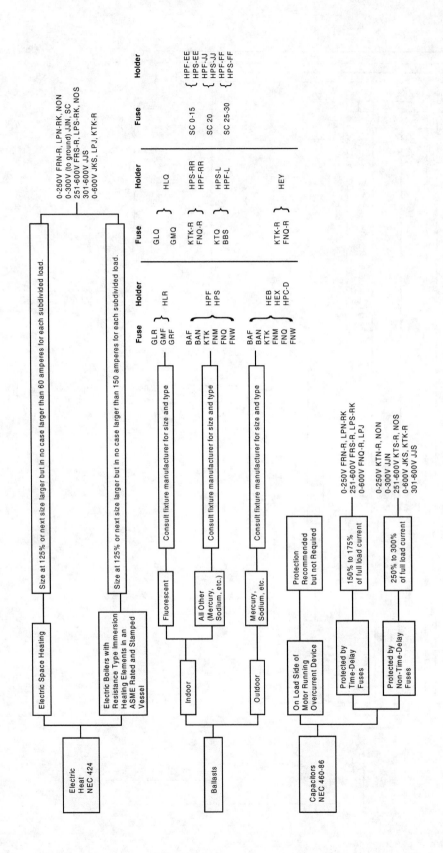

Figure 31. Summary of fuse applications in electrical systems. *(Cont.)*

SELF-CHECK REVIEW/PRACTICE QUESTIONS

1. What is the range of current-interrupting capacity of Class K fuses?

   a. 50,000 to 200,00 amperes
   b. 10,000 to 15,000 amperes
   c. 15,000 to 20,000 amperes
   d. 25,000 to 50,000 amperes

2. What is the minimum interrupting rating of branch-circuit listed fuses?

   a. 5,000 amperes
   b. 10,000 amperes   class H
   c. 15,000 amperes
   d. 20,000 amperes

3. What is the minimum voltage rating of branch-circuit fuses?

   a. 24 volts
   b. 120 volts
   c. 125 volts
   d. 240 volts

4. Which of the following is not a classification of medium-voltage fuses?

   a. General-purpose current-limiting fuses
   b. Back-up current-limiting fuses
   c. Expulsion fuses
   d. Proportion fuses

5. Which of the following best describes an expulsion fuse?

   a. A fuse capable of interrupting all currents from the rated interrupting current down to the current that causes melting of the fusible element in an hour
   b. A fuse capable of interrupting all currents from the maximum rated interrupting current down to the rated minimum interrupting current
   c. Strap-mounted devices
   d. A vented fuse in which the expulsion effect of gasses produced by the arc and lining of the fuseholder, either alone or aided by spring, extinguishes the arc

6. Which of the following is **not** a basic factor to consider when applying any fuse?

    a. Voltage
    b. Continuous current-carrying capacity
    c. Interrupting rating
    d. Manufacturer's brand name

7. The total clearing time of any downstream protective device must be below a curve representing what percentage of the minimum melting curve of the fuse being applied?

    a. 10%
    b. 20%
    c. 50%
    d. 75%

8. When a common circuit breaker trips, in what position is the handle?

    a. OFF position
    b. ON position
    c. Middle position
    d. No change in handle position

9. Circuit breakers are grouped for identification according to given current ranges. How is each group classified?

    a. The largest ampere rating of its range
    b. The smallest ampere rating of its range
    c. The absolute lowest ampere rating of its range
    d. The overall average ampere rating of its range

10. When circuit breakers are classified as 100 through 2,000-ampere frames, what are these numbers normally referred to?

    a. Frame size
    b. Overload protective current
    c. Maximum voltage allowed on the circuit
    d. Physical size of circuit breaker

PERFORMANCE/LABORATORY EXERCISE

1.   Using several types of fuses, and several different types of fuse holders, insert the proper fuse in the correct fuse holder.

2.   Disassemble a renewable fuse and install a new fuse link.

3.   In the spaces provided in the drawing below, write the type and size of fuse you would use for the various applications.

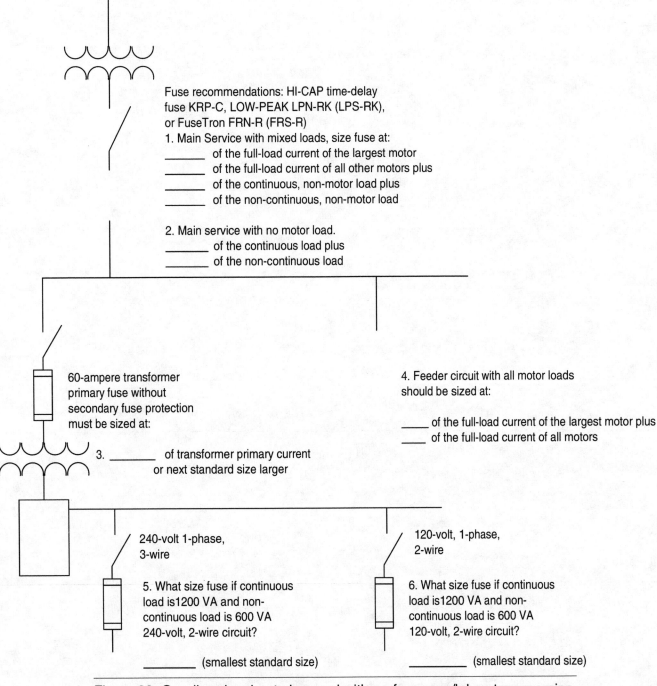

Fuse recommendations: HI-CAP time-delay fuse KRP-C, LOW-PEAK LPN-RK (LPS-RK), or FuseTron FRN-R (FRS-R)

1. Main Service with mixed loads, size fuse at:
_____ of the full-load current of the largest motor
_____ of the full-load current of all other motors plus
_____ of the continuous, non-motor load plus
_____ of the non-continuous, non-motor load

2. Main service with no motor load.
_____ of the continuous load plus
_____ of the non-continuous load

60-ampere transformer primary fuse without secondary fuse protection must be sized at:

3. _____ of transformer primary current or next standard size larger

4. Feeder circuit with all motor loads should be sized at:

____ of the full-load current of the largest motor plus
____ of the full-load current of all motors

240-volt 1-phase, 3-wire

5. What size fuse if continuous load is 1200 VA and non-continuous load is 600 VA 240-volt, 2-wire circuit?

_____ (smallest standard size)

120-volt, 1-phase, 2-wire

6. What size fuse if continuous load is 1200 VA and non-continuous load is 600 VA 120-volt, 2-wire circuit?

_____ (smallest standard size)

Figure 32. One-line drawing to be used with performance/laboratory exercise.

Answers to Self-Check Questions

1. a

2. b

3. c

4. d

5. d

6. d

7. d

8. c

9. a

10. a

Raceways/Boxes & Fill Requirements

Module 20304

Electrical Trainee Task Module 20304

RACEWAYS/BOXES: FILL REQUIREMENTS

Objectives

Upon completion of this module, the trainee will be able to:

1. Size raceways according to conductor fill and NEC installation requirements.
2. Size and install outlet boxes according to NEC installation requirements.
3. Size, select, and install pull and junction boxes according to NEC regulations.
4. Size conduit and conduit bodies using tables in NEC Chapter 9.
5. Calculate conduit fill using a percentage of the trade size conduit inside diameter (ID.)
6. Calculate required bending radius in boxes and cabinets.
7. Know when cable must be racked in pull or junction boxes.

Prerequisites

Successful completion of the following Task Modules is required before beginning study of this Task Module: Core Curricula, Electrical Levels 1 and 2, Electrical Level 3, Modules 20301 and 20302.

Required Student Materials

1. Trainee Task Module
2. Copy of the latest edition of the National Electrical Code

COURSE MAP INFORMATION

This course map shows all of the *Wheels of Learning* Task Modules in the third level of the Electrical curricula. The suggested training order begins at the bottom and proceeds up. Skill levels increase as a trainee advances on the course map. The training order may be adjusted by the local Training Program Sponsor.

Course Map: Electrical, Level 3

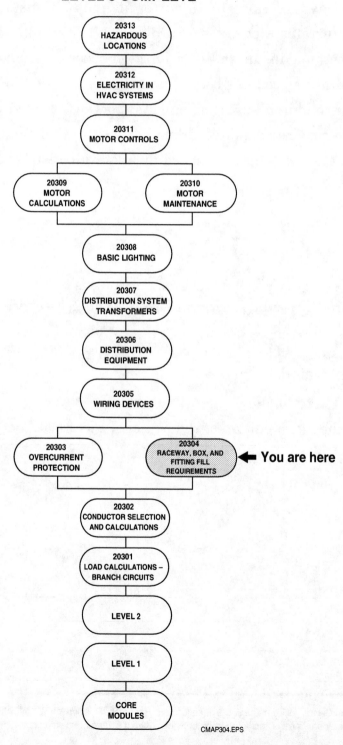

LEVEL 3 COMPLETE

20313
HAZARDOUS LOCATIONS

20312
ELECTRICITY IN HVAC SYSTEMS

20311
MOTOR CONTROLS

20309
MOTOR CALCULATIONS

20310
MOTOR MAINTENANCE

20308
BASIC LIGHTING

20307
DISTRIBUTION SYSTEM TRANSFORMERS

20306
DISTRIBUTION EQUIPMENT

20305
WIRING DEVICES

20303
OVERCURRENT PROTECTION

20304
RACEWAY, BOX, AND FITTING FILL REQUIREMENTS ◄ **You are here**

20302
CONDUCTOR SELECTION AND CALCULATIONS

20301
LOAD CALCULATIONS – BRANCH CIRCUITS

LEVEL 2

LEVEL 1

CORE MODULES

CMAP304.EPS

TABLE OF CONTENTS

Trade Terms Introduced in This Module

accessible: Capable of being removed or exposed without damaging the building structure or finish, or not permanently closed in by the structure or finish of the building.

American National Standards Institute (ANSI): An organization that publishes nationally recognized standards.

American Wire Gage (AWG): The standard for measuring wires in America.

ampacity: The current-carrying capacity of conductors or equipment, expressed in amperes.

ANSI: See American National Standards Institute.

AWG: See American Wire Gage.

cable: An assembly of two or more conductors which may be insulated or bare.

clearance: The vertical space between a cable and its conduit.

concealed: Rendered inaccessible by the structure or finish of the building. Wires in concealed raceways are considered concealed, even though they may become accessible by being withdrawn.

concentricity: The measurement of the center of the conductor with respect to the center of the insulation.

conduit fill: Amount of cross-sectional area used in a raceway.

configuration, triangular: The geometric pattern which cables will take in a conduit when the cables are triplexed or are pulled in parallel with the ratio of the conduit inside diameter (ID) to the 1/C cable outside diameter (OD) less than 2.5.

jacket: A nonmetallic polymeric close-fitting protective covering over conductor insulation; the cable may have one or more conductors.

junction box: Group of electrical terminals housed in a protective box or container.

knockout: A portion of an enclosure designed to be readily removed for the installation of a raceway.

load losses: Those losses incidental to distributing power.

Neoprene: An oil-resistant synthetic rubber used for jackets; originally a DuPont trade name, now a generic term for polychloroprene.

phase conductor: Any conductor other than the neutral or grounded conductors.

1.0.0 INTRODUCTION

A raceway is any channel used for holding wires, cables, or busbars, which is designed and used solely for this purpose. Types of raceways include rigid metal conduit, intermediate metal conduit (IMC), rigid nonmetallic conduit, flexible metal conduit, liquid-tight flexible metal conduit, electrical metallic tubing (EMT), underfloor raceways, cellular metal floor raceways, cellular concrete floor raceways, surface metal raceways, wireways, and auxiliary gutters. Raceways are constructed of either metal or insulating material.

Raceways provide mechanical protection for the conductors that run in them and also prevent accidental damage to insulation and the conducting metal. They also protect conductors from the harmful chemical attack of corrosive atmospheres and prevent fire hazards to life and property by confining arcs and flame due to faults in the wiring system.

One of the most important functions of metal raceways is to provide a path for the flow of fault current to ground, thereby preventing voltage build-up on conductor and equipment enclosures. This feature, of course, helps to minimize shock hazards to personnel and damage to electrical equipment. To maintain this feature, it is extremely important that all metal raceway systems be securely bonded together into a continuous conductive path and properly connected to a grounding electrode such as a water pipe or a ground rod.

A box or fitting must be installed at:

- Each conductor splice point
- Each outlet, switch point, or junction point
- Each pull point for the connection of conduit and other raceways

Furthermore, boxes or other fittings are required when a change is made from conduit to open wiring. Electrical workers also install pull boxes in raceway systems to facilitate the pulling of conductors.

In each case — raceways, outlet boxes, pull and junction boxes — the NEC specifies specific maximum fill requirements; that is, the area of conductors in relation to the box, fitting, or raceway system. This module is designed to cover these NEC requirements and apply these rules to practical applications.

2.0.0 CONDUIT FILL REQUIREMENTS

The NEC provides rules on the maximum number of conductors permitted in raceways. In conduits, for either new work or rewiring of existing raceways, the maximum fill must not exceed 40 percent of the conduit cross-sectional area. In all such cases, fill is based on using the actual cross-sectional areas of the particular types of conductors used. Other derating rules specified by the NE Code may be found in Article 310. For example, if more than three

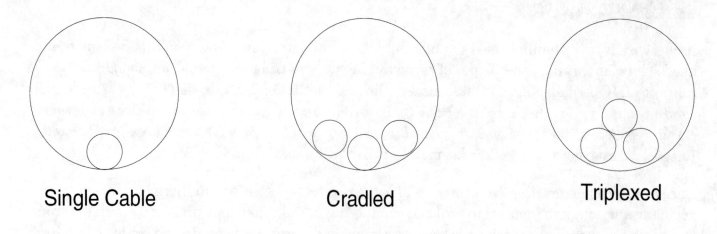

Single Cable   Cradled   Triplexed

Figure 1. Configurations of cable in raceway.

conductors are used in a single conduit, a reduction in current-carrying capacity is required. Ambient temperature is another consideration that may call for derating of wires below the values given in NEC tables.

2.1.0   FACTORS INFLUENCING CABLE FIT

Besides the NEC requirements, five factors influence a cable's mechanical fit in a raceway:

- Its configuration
- Weight
- Clearance
- Jam ratio
- Coefficient of friction

2.1.1   Configuration

The configuration of the cable in the raceway is measured by the ratio of the inner diameter of the conduit to the overall diameter of one of the cables within the conduit, expressed:

$$\frac{D}{d}$$

A cradled configuration (Fig. 1) occurs when cables with a ratio of 2.5 or greater are pulled in parallel from individual reels. A triangular configuration (Fig. 1) occurs when cables with a ratio of less than 2.5 are pulled in parallel from individual reels or when cables are assembled.

Configuration directly affects drag which is computed as Weight Correction Factor (w) in the following section.

2.1.2   Weight

When making installation calculations, use the total weight per unit length of the cables being pulled. Cabled assemblies will weigh more than paralleled cables unless the cables were specially ordered to have several paralleled cables wound on a reel.

Weight Correction Factor (w): Due to its geometric configuration, a cable is subjected to uneven force when it is pulled into a conduit. This imbalance results in additional frictional drag which must be calculated and allowed for if the installation is to be a success. See Fig. 2. The graph in Fig. 3 illustrates the Weight Correction Factor (w).

When computing Weight Correction Factor, ensure all cable diameters are equal. When in doubt, use the cradled configuration equation.

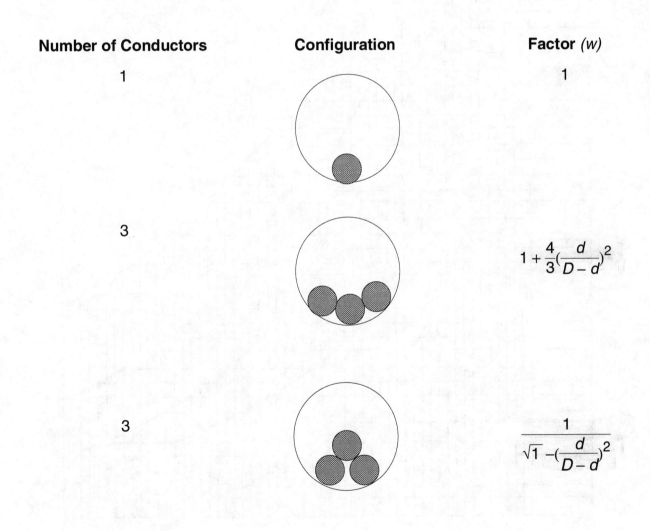

Number of Conductors	Configuration	Factor *(w)*
1		1
3		$1 + \dfrac{4}{3}\left(\dfrac{d}{D-d}\right)^2$
3		$\dfrac{1}{\sqrt{1} - \left(\dfrac{d}{D-d}\right)^2}$

Figure 2. Weight correction factors *(w)* for various cable configurations.

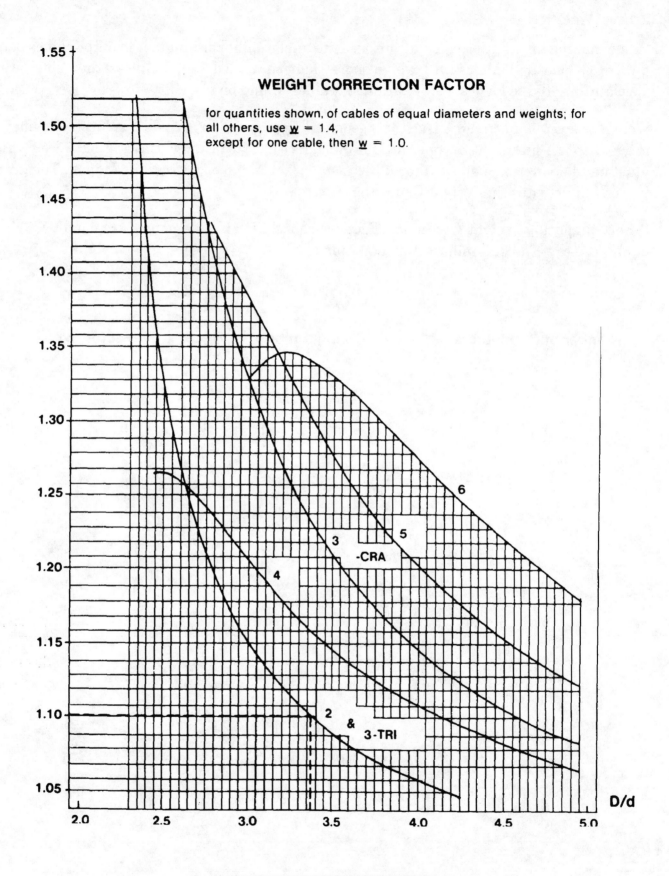

Figure 3. Weight correction factors.

The cables may be in either a single or multiple conductor construction. But, when only one cable (whether single conductor or multiple conductors under common jacket) is being pulled, no weight correction factor is needed.

2.1.3 Clearance

Clearance refers to the distance between the uppermost cable in the conduit and the inner top of the conduit. Clearance should be ¼ inch at minimum and up to one inch for large cable installations or insulations involving numerous bends. It is calculated as follows:

When calculating clearance, ensure all cable diameters are equal. Use the triplexed configuration equation (Fig. 4) if you are in doubt. Again, the cables may be of single or multiple conductor construction.

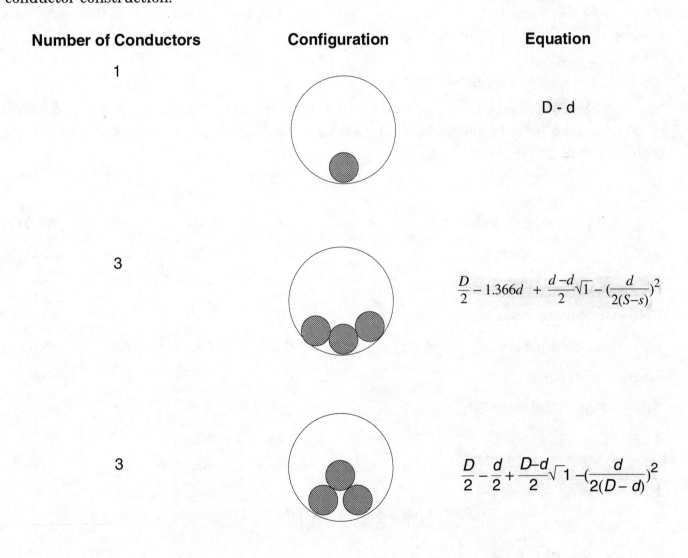

Number of Conductors	Configuration	Equation
1		$D - d$
3		$\dfrac{D}{2} - 1.366d + \dfrac{d-d}{2}\sqrt{1 - \left(\dfrac{d}{2(S-s)}\right)^2}$
3		$\dfrac{D}{2} - \dfrac{d}{2} + \dfrac{D-d}{2}\sqrt{1 - \left(\dfrac{d}{2(D-d)}\right)^2}$

Figure 4. Principles of calculating cable clearances.

2.1.4 Jam Ratio

Jamming is the wedging of three cables lying side-by-side in a conduit. This usually occurs when cables are being pulled around bends or when cables twist.

Jam Ratio is calculated by slightly modifying the ratio used to measure configuration (D/d). A value of 1.05D is used for the inner diameter of the conduit because bending a cylinder creates an oval cross-section in the bend (1.05D/d).

- If 1.05D/d is larger than 3.0, jamming is impossible.
- If 1.05D/d is between 2.8 and 3.0, serious jamming is probable.
- If 1.05D/d is less than 2.5, jamming is impossible but clearance should be checked.

Since there are manufacturing tolerances on cable, the actual overall diameter should be measured prior to computing jam ratio.

2.1.5 Coefficient of Dynamic Friction

The coefficient of dynamic friction is a measure of the friction between the cable and the conduit or roller, and can vary from 0.03 to 0.8, even with lubrication. Typical values are shown in the table in Fig. 5.

Cable Exterior	Type of Conduit			
	M	PVC	FIB	ASB
PVC - Polyvinyl Chloride	0.4	0.35	0.5	0.5
PE - Low Density HMW Polyethylene	0.35	0.35	0.5	0.5
CPE - Chlorinated Polyethylene	0.35	0.35	0.5	0.5
Hypalon - CSPE (Chlorosulfonated PE)	0.5	0.5	0.7	0.6
Neoprene - N (Chloroprene)	0.5	0.5	0.7	0.6
FREP - Flame Retardant EP	0.4	0.4	0.5	0.5
XLPE - Cross Linked PE	0.35	0.35	0.5	0.5
Lead	0.5	0.5	0.5	0.5

Figure 5. Typical coefficients of dynamic of friction (f).

- M = Metallic, Steel or Aluminum
- PVC = Polyvinyl Chloride, Thinwall or Heavy Schedule 40
- FIB = Fiber Conduit - Orangeburg or Nocrete
- ASB = Asbestos Cement - Transite or Korduct

The coefficient of friction of a duct or conduit varies with the type of cable covering, condition of duct or conduit internal surface, type and amount of pulling lubricant used and ambient installation temperature.

High ambient temperatures (80 deg. F and over) can increase the coefficient of dynamic friction for cable having a nonmetallic jacket.

Pulling lubricants must be compatible with the cable's components and be applied while the cable is being pulled.

2.2.0 CONDUIT FILL

The allowable number of conductors in a raceway system is calculated as percentage of fill as specified in Table 1 of NEC Chapter 9 (Fig. 6). When using this table, remember that equipment grounding or bonding conductors, where installed, must be included when calculating conduit or tubing fill. The actual dimensions of the equipment grounding or bonding conductor (insulated or bare) must be used in the calculation.

Number of Conductors	1	2	Over 2
All conductor types	53%	31%	40%

Figure 6. Percent of cross-section of conduit for conductors.

Conduit fill is calculated as follows:

If all of the conductors are of the same size, you can refer to Tables C1 through C12 of Appendix C, being careful to identify the type of raceway being used and the conductor type being used.

Note: If you are using compact aluminum conductors, refer to Tables C1A through C12A.

Example: How many #6 THW wires can you install in a 2" EMT conduit?

Step 1   Refer to Appendix C, locating Table C1 (Maximum Number of Conductors and Fixture Wires in Electrical Metallic Tubing).

Step 2   Locate the section of the table listing THW wire.

Step 3   Move right across the table until you reach the 2" Trade Size column.

Step 4   Move down this column until you are lined up with conductor size #6.

Step 5   The total quantity of #6 THW wire in a 2" EMT is determined to be 18.

You can also use these tables if you know the size and type of wire and need to know the size of a raceway required.

Example: What size of rigid metallic conduit is required for four 500 kcmil THHN conductors?

Step 1   Refer to Appendix C, locating Table C8 (Maximum Number of Conductors and Fixture Wires in Rigid Metallic Conduit).

Step 2   Locate the section of Table C8 listing THHN insulation.

Step 3   Move down the table until you reach 500 kcmil.

Step 4   Move right across the table from 500 kcmil until you reach the column with the number 4, indicating 4 conductors. Look up to the top of this column to find that you would need a trade diameter of 3". (If the quantity of conductors is not listed in the row, you must use the next higher listed quantity.)

If you have more than one size of conductor installed in the same raceway (or if you prefer) use Chapter 9, Tables 4 and 5 to determine the proper conduit size and fill requirements. Follow the procedure shown below.

Step 1   Using Chapter 9, Table 5, select the type and size of conductors being used.

Step 2   Move right to the column marked "Approx. Area Sq. In." to find the cross-section square inches required for each conductor size.

Step 3   Multiply the number of each size of conductors by the quantity to determine the total square inches required for each conductor size. Add the total square inches required for each conductor size together to determine the total cross-section square inches required.

Step 4   Determine the type and size of raceway being used (i.e., emt, imc, rigid, etc.).

Step 5   Refer to Chapter 9, Table 4 to locate the Type of Raceway subtable.

Step 6   Once the appropriate subtable of Table 4 is identified, move right across the row to the column that corresponds to the number of conductors in the raceway.

Step 7   Go down this column until you reach a value that is equal to or greater than the amount calculated in step 3.

Step 8   Move left across the table to find the trade size of conduit required.

Step 9   Move right to the column marked "Approx. Area Sq. In." to find the cross-section square inches required for each conductor.

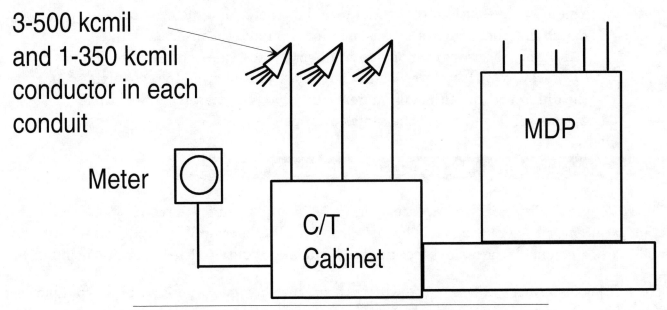

3-500 kcmil and 1-350 kcmil conductor in each conduit

Meter

C/T Cabinet

MDP

Figure 7. Power-riser diagram of a 1200-ampere service-entrance.

Example: Let's take a situation where a 1200 ampere service entrance is to be installed utilizing three parallel conduits, each containing three 500 kcmil THHN ungrounded conductors and one 350 kcmil THHN grounded conductor (neutral) as shown in Fig. 7. What size of rigid conduit is required?

Step 1   Refer to Chapter 9, Table 5 to determine the area (in square inches) of 500 kcmil THHN conductor. The area is found to be .7073 sq. in.

Step 2   Repeat step 1 to find the area for 350 kcmil THHN. The area is found to be .5242 sq. in.

Step 3   Multiply the number of conductors of each size by their corresponding areas.

.7073 x 3 = 2.1219 (Total area required for 500 kcmil conductors)

.5242 x 1 = .5242 (Total area required for 350 kcmil conductors)

Step 4   Add the totals together to obtain the total area required for all conductors.

2.1219 + .5242 + 2.6461 sq. in.

Step 5   Refer to Chapter 9, Table 4 and locate the subtable for Rigid Metal Conduit.

Step 6   Looking in the column marked "Over 2 Wires", scan down until you find a number equal to or greater than 2.6461. The value found is 3.000. Now scan back to the left to determine that the conduit size required is 3".

Therefore, each of the three conduits containing three 500 kcmil and one 350 kcmil THHN conductors must be at least 3 inches trade size to comply.

Note: You may have noticed that the internal diameter of a raceway does not match the trade size in all cases, nor does it match the internal diameter of different raceways in the same trade size. The trade size is a size given to a raceway with an internal diameter close to that size. In recognizing the difference in internal diameters, the 1996 NEC has expanded Table 4, Chapter 9, and added the tables in Appendix C.

3.0.0 CONDUIT BODIES, PULL BOXES, AND JUNCTION BOXES

The NEC specifically states that at each splice point, or pull point for the connection of conduit or other raceways, a box or fitting must be installed. The NEC specifically considers conduit bodies, pull boxes, and junction boxes and specifies the installation rules as listed in the table in Fig. 8.

Conduit bodies provide access to the wiring through removable covers. Typical examples are Types T, C, X, L, and LB. Conduit bodies enclosing No. 6 or smaller conductors must have an area twice that of the largest conduit to which they are attached, but the number of conductors within the body must not exceed that allowed in the conduit. If a conduit body has entry for three or more conduits such as Type T or X, splices may be made within the conduit body. Splices may not be made in conduit bodies having one or two entries unless the volume is sufficient to qualify the conduit body as a junction box or device box.

When conduit bodies or boxes are used as junction boxes or as pull boxes, a minimum size box is required to allow conductors to be installed without undue bending. The calculated dimensions of the box depend on the type of conduit arrangement and on the size of the conduits involved.

Application	Installation Requirements	NEC Reference or Comment
Conduit bodies	Conduit bodies enclosing No. 6 conductors or smaller must have a cross-sectional area not less than twice the cross-sectional area of the largest conduit or tubing to which it is attached. The maximum number of conductors allowed in the conduit body must not exceed the allowable fill for the attached conduit. Conduit bodies must not contain splices, taps, or devices unless they are durably and legibly marked by the manufacturer with their cubic inch capacity. Conduit bodies must be supported in a rigid and secure manner.	370-16(c)

Figure 8. Installation requirements for conduit bodies, pull and junction boxes.

Application	Installation Requirements	NEC Reference or Comment
Minimum sizes	Boxes and conduit bodies used as pull or junction boxes must comply with (a) through (d) below: (a) For raceways containing conductors of No. 4 or larger, the minimum dimensions of pull or junction boxes installed in a raceway or cable run must comply with the following: In straight pulls, the length of the box must not be less than eight times the trade diameter of the largest conduit. Where angle or U pulls are made, the distance between each raceway entry inside the box and the opposite wall of the box must not be less than six times the trade diameter of the largest raceway in a row. This distance must be increased for additional entries by the amount of the sum of the diameters of all other raceway entries in the same row on the same wall of the box. Each row must be calculated individually and the single row that provides the maximum distance must be used. The distance between raceway entries enclosing the same conductor must not be less than six times the trade diameter of the largest raceway. (b) In pull boxes or junction boxes having any dimension over 6 feet, all conductors must be cabled or racked up in an approved manner. (c) All boxes and fittings must be provided with covers compatible with the box, and if metal, must comply with NEC Section 250-42. (d) Where permanent barriers are installed in a box, each section must be considered as a separate box.	Section 370-28(a)
Accessibility	Junction, pull, and outlet boxes must be accessible without removing any part of a building or digging.	370-29
Over 600 V	Special requirements apply to boxes used on systems of over 600 volts.	Article 370, Part D

Figure 8. Installation requirements for conduit bodies, pull and junction boxes. *(Cont.)*

3.1.0 SIZING PULL AND JUNCTION BOXES

Figure 9 shows a junction box with several conduits entering it. Since 4-inch conduit is the largest size in the group, the minimum length required for the box can be determined by the following calculation:

Trade size of conduit x 8 (as per NEC) = minimum length of box

4″ x 8 = 32″

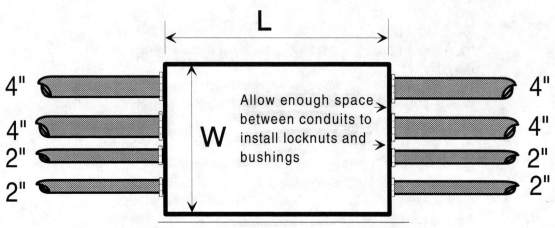

Figure 9. Pull box used on straight pulls.

Therefore, this particular pull box must be at least 32″ in length. The width of the box, however, need be only of sufficient size to enable locknuts and bushings to be installed on all the conduits or connectors entering the enclosure.

Junction or pull boxes in which the conductors are pulled at an angle as shown in Fig. 10 must have a distance of not less than six times the trade diameter of the largest conduit. The distance must be increased for additional conduit entries by the amount of the sum of the diameter of all other conduits entering the box on the same side, that is, the wall of the box. The distance between raceway entries enclosing the same conductors must not be less than six times the trade diameter of the largest conduit.

Since the 4-inch conduit is the largest of the lot in this case,

$$L_1 = 6 \times 4 + (3 + 2) = 29″$$

Since the same number and sizes of conduit are located on the adjacent wall of the box, L_2 is calculated in the same way; therefore, $L_2 = 29″$.

The distance (D) = 6 x 4 or 24" and this is the minimum distance permitted between conduit entries enclosing the same conductor.

The depth of the box need only be of sufficient size to permit locknuts and bushings to be properly installed. In this case, a 6-inch deep box would suffice.

ELECTRICAL TRAINEE TASK MODULE 20304

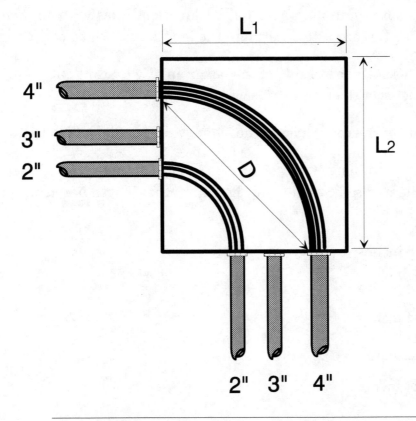

Figure 10. Junction box with conduit runs entering at right angles.

If the conductors are smaller than No. 4, the length restriction does not apply.

Figure 11 shows another straight-pull box. What is the minimum length if the box has one 3-inch conduit and two 2-inch conduits entering and leaving the box? Again, refer to NEC Section 370-28(a)(1) and find that the minimum length is 8 times the largest conduit size which in this case is:

8 x 3 inches = 24 inches

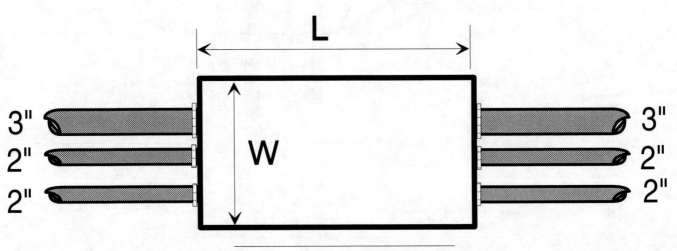

Figure 11. Typical straight-pull box.

Let's review the installation requirements for pull or junction boxes with angular or U-pulls. Two conditions must be met in order to determine the length and width of the required box.

The minimum distance to the opposite side of the box from any conduit entry must be at least six times the trade diameter of the largest raceway.

The sum of the diameters of the raceways on the same wall must be added to this figure.

Figure 12 shows the minimum length of a box with two 3-inch conduits, two 2-inch conduits, and two 1½-inch conduits in a right-angle pull. The minimum length based on this configuration is:

6 x 3 inches =	18 inches
1 x 3 inches =	3 inches
2 x 2 inches =	4 inches
2 x 1½ inches =	<u>3 inches</u>
	28 inches

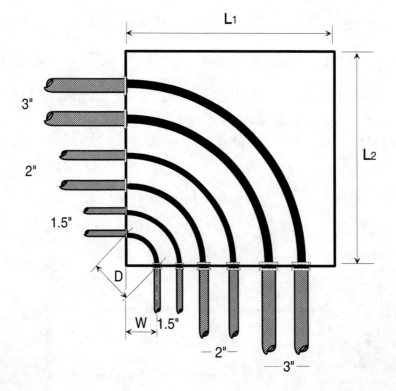

Figure 12. Minimum size pull box for angle conduit entries.

Since the number and size of conduits on the two sides of the box are equal, the box is square and has a minimum size dimension of 28 inches. However, the distance between conduit entries must now be checked to ensure that *all* NEC requirements are met; that is, the spacing (D) between conduits enclosing the same conductor must not be less than six times the conduit diameter. Again refer to Fig. 12 and note that the 1½″ conduits are the closest to the left-hand corner of the box. Therefore, the distance (D) between conduit entries must be:

$$6 \times 1\tfrac{1}{2}'' = 9''$$

The next group is the two 2″ conduits which is calculated in a similar fashion; that is:

$$6 \times 2 = 12''$$

The remaining raceways in this example are the two 3″ conduits and the minimum distance between the 3″ conduit entries must be:

$$6 \times 3 = 18''$$

A summary of the conduit-entry distances is presented in Fig. 13. However, some additional math is required to obtain the spacing (w) between the conduit entries. For example, the distance from the corner of the pull box to the center of the conduits (w) may be found by the following equation:

$$Spacing = \frac{Diagonal\ distance\ (D)}{\sqrt{2}}$$

Consequently, the spacing (w) for the 1½″ conduit may be determined using the following equation:

$$\frac{9}{\sqrt{2}} = \frac{9}{1.414} = 6.4''$$

Therefore, the spacing (w) is 6.4 inches. This distance is measured from the left lower corner of the box in each direction — both vertically and horizontally — to obtain the center of the first set of 1½″ conduits. This distance must be added to the spacing of the other conduits including locknuts or bushings.

Note: A rule-of-thumb is to allow ½″ clearance between locknuts.

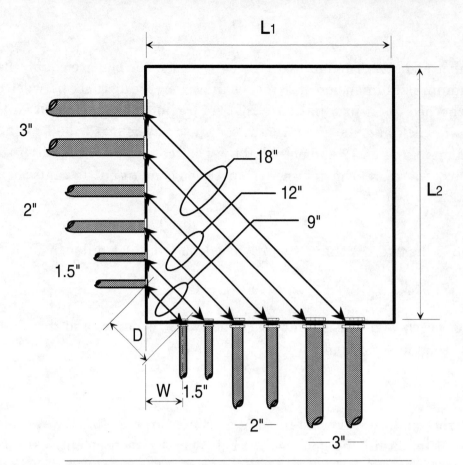

Figure 13. Required distances between conduit entries.

Using all information calculated thus far, and using Fig. 13 as reference, the required measurements of the pull box may be further calculated as follows:

Step 1.   Calculate space (w):

$$D = 6 \times 1\frac{1}{2}'' = 9''$$

Step 2.   Divide this number (9″) by the square root of 2 (1.414) and make the following calculation:

$$w = \frac{9''}{1.414} = 6.4''$$

Step 3.   Measure from the left, lower corner of the pull box over 6.4″ to obtain the center of the knockout for the first 1½″ conduit. Measure up (from the lower, left corner) to obtain the center of the knockout for this same cable run on the left side of the pull box.

Step 4.   Since there are two 3-inch (inside diameter) conduits, each with an outside diameter of approximately 4.25 inches, the space for these two conduits can be found by the following equation:

$$2 \times 4.25 = 8.5 \text{ inches}$$

ELECTRICAL TRAINEE TASK MODULE 20304

Step 5. The space required for the two 2-inch (inside diameter) conduits, each with an outside diameter of approximately 3.12 inches, may be determined in a similar manner; that is:

$$2 \times 3.12 = 6.24 \text{ inches}$$

Step 6. The space required for the two 1.5-inch (inside diameter) conduits, each with an outside diameter of approximately 2.62 inches, may be determined using the same equation:

$$2 \times 2.62 = 5.24 \text{ inches}$$

Step 7. To find the required space for locknuts and bushings, multiply 0.5″ by the total number of conduit entries on one side of the box. Since there are a total of 6 conduit entries, use the following equation:

$$6 \times .5 = 3.0 \text{ inches}$$

Step 8. Add all figures obtained in Steps 2 through 7 together to obtain the total required length of the pull box.

Clear space (w)	=	6.4 inches
1.5-inch conduits	=	5.24 inches
2-inch conduits	=	6.24 inches
3-inch conduits	=	8.5 inches
Space between locknuts	=	3.0 inches
Total length of box	=	29.2 inches

Since the same number and size of conduits enter on the bottom side of the pull box and leave, at a right angle, on the left side of the pull box, the box will be square. Furthermore, although a box exactly 29.2 inches will suffice for this application, the next larger standard size is 30 inches; this should be the size pull box selected. Even if a "custom" pull box is made in a sheet-metal shop, the workers will still probably make it an even 30 inches unless specifically ordered otherwise.

3.2.0 CABINETS AND CUTOUT BOXES

NEC Article 373 deals with the installation requirements for cabinets, cutout boxes, and meter sockets. In general, where cables are used, each cable must be secured to the cabinet or cutout box by an approved method. Furthermore, the cabinets or cutout boxes must have sufficient space to accommodate all conductors installed in them without crowding.

NEC Table 373-6(a), which is reproduced in Fig. 14, gives the minimum wire-bending space at terminals along with the width of sizing gutter in inches. Figure 15 gives a summary of NEC requirements for the installation of cabinets and cutout boxes.

AWG or Circular-Mil Size of Wire	Wires per Terminal				
	1	2	3	4	5
14-10	Not Specified	—	—	—	—
8-6	1½	—	—	—	—
4-3	2	—	—	—	—
2	2½	—	—	—	—
1	3	—	—	—	—
1/0-2/0	3½	5	7	—	—
3/0-4/0	4	6	8	—	—
250 kcmil	4½	6	8	10	—
300-350 kcmil	5	8	10	12	—
400-500 kcmil	6	8	10	12	14
600-700 kcmil	8	10	12	14	16
750-900 kcmil	8	12	14	16	18
1000-1250 kcmil	10	—	—	—	—
1500-2000 kcmil	12	—	—	—	—

Figure 14. Minimum wire-bending space in inches.

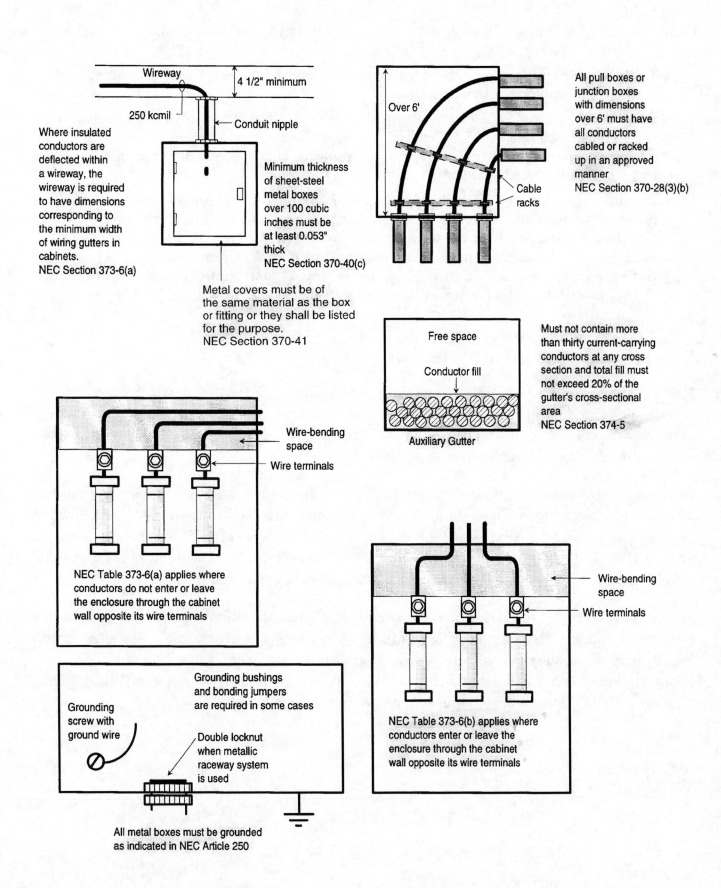

Figure 15. NEC installation requirements for cabinets, cutout boxes, and auxiliary gutters.

Other basic NEC requirements for cabinets and cutout boxes are as follows:

- Table 373-6(a) must apply where the conductor does not enter or leave the enclosure through the wall opposite its terminal.

- Exception No. 1 states: A conductor must be permitted to enter or leave an enclosure through the wall opposite its terminal provided the conductor enters or leaves the enclosure where the gutter joins an adjacent gutter that has a width that conforms to Table 373-6(b) for that conductor.

- Exception No. 2 states: A conductor not larger than 350 kcmil must be permitted to enter or leave an enclosure containing only a meter socket(s) through the wall opposite its terminal, provided the terminal is a lay-in type where either: (a) The terminal is directly facing the enclosure wall and offset is not greater than 50 percent of the bending space specified in Table 373-6(a), or (b) The terminal is directed toward the opening in the enclosure and is within a 45-degree angle of directly facing the enclosure wall.

- Table 373-6(b) must apply where the conductor enters or leaves the enclosure through the wall opposite its terminal.

NEC Article 374 covers the installation requirements for auxiliary gutters, which are permitted to supplement wiring spaces at meter centers, distribution centers, and similar points of wiring systems and may enclose conductors or busbars but must not be used to enclose switches, overcurrent devices, appliances, or other similar equipment.

In general, auxiliary gutters must not contain more than thirty (30) current-carrying conductors at any cross section. The sum of the cross-sectional areas of all contained conductors at any cross section of an auxiliary gutter must not exceed 20 percent of the interior cross-sectional area of the auxiliary gutter. We discussed earlier that conductors installed in conduits and tubing must not exceed 40 percent fill. Auxiliary gutters are limited to only 20 percent.

When dealing with auxiliary gutters, always remember the number "thirty." This is the maximum number of conductors allowed in any auxiliary gutter regardless of the cross-sectional area. This question will be found on almost every electrician's examination in the country. Consequently, this number should always be remembered. Refer to Fig. 15 for a summary of NEC installation requirements for auxiliary gutters.

On every job, a great number of boxes is required for outlets, switches, pull and junction boxes. All of these must be sized, installed and supported to meet current NEC requirements. Since the NEC limits the number of conductors allowed in each outlet or switch box — according to its size — electricians must install boxes large enough to accommodate the number of conductors that must be spliced in the box or fed through. Therefore, a knowledge of the various types of boxes and the volume of each is essential.

4.1.0 SIZING OUTLET BOXES

In general, the maximum number of conductors permitted in standard outlet boxes is listed in Table 370-16(a) of the NEC. These figures apply where no fittings or devices such as fixture studs, cable clamps, switches, or receptacles are contained in the box and where no grounding conductors are part of the wiring within the box. Obviously, in all modern residential wiring systems there will be one or more of these items contained in the outlet box. Therefore, where one or more of the above mentioned items are present, the number of conductors shall be one less than shown in the tables. For example, a deduction of two conductors must be made for each strap containing a device such as a switch or duplex receptacle; a further deduction of one conductor shall be made for one or more grounded conductors entering the box. A 3-inch x 2-inch x 2¾-inch box for example, is listed in the table as containing a maximum number of six No. 12 wires. If the box contains cable clamps and a duplex receptacle, two wires will have to be deducted from the total of six — providing for only four No. 12 wires. If the box contains cable clamps and a duplex receptacle, two wires will have to be deducted from the total of six providing for only four No. 12 wires. If a ground wire is used, only three no. 12 wires may be used, which might be the case when a three-wire cable with ground is used to feed a 3-way wall switch.

Figure 16 illustrates one possible wiring configuration for outlet boxes and the maximum number of conductors permitted in them as governed by Section 370-16 of the NEC. This example shows two single-gang switch boxes joined or "ganged" together to hold a single-pole toggle switch and a duplex receptacle. This type of arrangement is likely to be found above kitchen countertops whereas the duplex receptacle is provided for small appliances and the single-pole switch could be used to control a garbage disposal. This arrangement is also useful above a workbench — the receptacle for small power tools and the switch to control lighting over the bench.

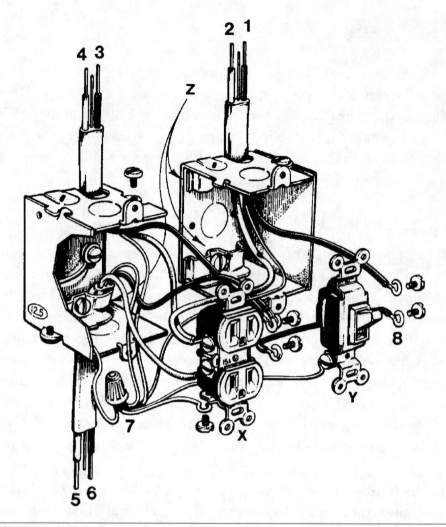

Figure 16. One possible box-sizing configuration that may appear on an electrician's examination.

Since Table 370-16(a) gives the capacity of one 3 x 2 x 2¼-inch device box as 12.5 cubic inches, the total capacity of both boxes in Fig. 16 is 25 cubic inches. These two boxes have a capacity to allow 10 No. 12 AWG conductors, or 12 No. 14 AWG conductors, less the deductions as listed below.

- Per NEC Section 370-16(b)(4), two conductors must be deducted for each strap-mounted device based on the largest size conductor connected to the device. Since there is one duplex receptacle (X) and one single-pole toggle switch (Y), four conductors must be deducted from the total number stated in the above paragraph.

- Per NEC Section 370-16(b)(2), the combined boxes contain one or more cable clamps (Z), another conductor must be deducted based on the largest size conductor in the box. Note that only one deduction is made for similar clamps, regardless of the number. However, any unused clamps may be removed to facilitate the electrical worker's job; that is, allowing for more work space.

- Per NEC Section 370-16(b)(5), the equipment grounding conductors, regardless of the number, count as one conductor only. Therefore, deduct one conductor based on the largest size grounding conductor in the box.

Therefore, to comply with the NEC, and considering the combined deduction of six conductors, only four No. 12 AWG conductors (six No. 14 AWG conductors) may be installed in the outlet-box configuration in Fig. 16.

Figure 16 shows three types of nonmetallic-sheathed (NM) cables, designated 12/2 with ground, entering the ganged outlet boxes. This is a total of six current-carrying conductors and three ground wires, for a total of nine. Is this arrangement in violation of the NEC? Yes, because the total number of conductors exceed the NEC limits. However, if No. 14 AWG conductors were installed rather than No. 12, the configuration will comply with the 1993 NEC. Another alternative is to go to 3 x 2 x 3½" device boxes which would then have a total of 36 cubic inches for the two boxes.

Also note the jumper wire in Fig. 16; this is numbered "8" in the drawing. Conductors that both originate and end in the same outlet box are exempt from being counted against the allowable capacity of an outlet box [NEC Section 370-16(b)(1)]. This jumper wire (8) taps off one terminal of the duplex receptacle to furnish a "hot wire" to the single-pole toggle switch. Therefore, this wire originates and terminates in the same set of ganged boxes and is not counted against the total number of conductors. By the same token, the three grounding conductors extending from the wire nut to the individual grounding screws on the devices originate and terminate in the same set of boxes. These conductors are also exempt from being counted with the total. Incidentally, the wire nut has a crimp connector beneath; wire nuts alone are not allowed to connect equipment grounding conductors.

A pictorial definition of stipulated conditions as they apply to Section 370-16 of the NEC is shown in the following illustrations. Figure 17 illustrates an assortment of raised covers and outlet box extensions. These components, when combined with the appropriate outlet boxes, serve to increase the usable work space. Each type is marked with their cubic-inch capacity which may be added to the figures in Table 370-16(a) of the NEC to calculate the increased number of conductors allowed.

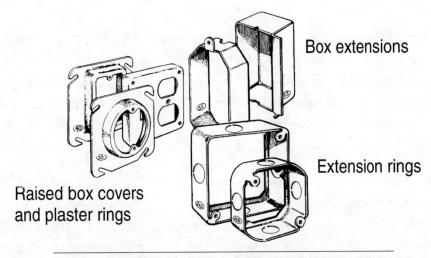

Box extensions

Extension rings

Raised box covers
and plaster rings

Figure 17. Raised box covers add to the box capacity.

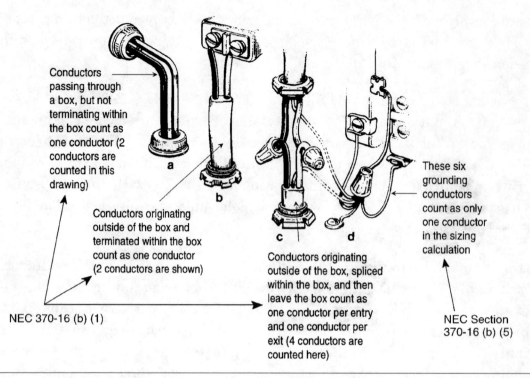

Conductors passing through a box, but not terminating within the box count as one conductor (2 conductors are counted in this drawing)

Conductors originating outside of the box and terminated within the box count as one conductor (2 conductors are shown)

NEC 370-16 (b) (1)

Conductors originating outside of the box, spliced within the box, and then leave the box count as one conductor per entry and one conductor per exit (4 conductors are counted here)

These six grounding conductors count as only one conductor in the sizing calculation

NEC Section 370-16 (b) (5)

Figure 18. Wiring configurations that must be counted as conductors when calculating box capacity.

Figure 18 shows typical wiring configurations which must be counted as conductors when calculating the total capacity of outlet boxes. A wire passing through the box without a splice or tap is counted as one conductor. Therefore, a cable containing two wires that passes in and out of an outlet box with a splice or tap is counted as two conductors. However, wires which enter a box and are either spliced or connected to a terminal, and then exit again, are counted as two conductors. In the case of two 2-wire cables, the total conductors charged will be four. Wires that enter and terminate in the same box are charged as individual conductors and in this case, the total charge would be two conductors. Remember, when one or more grounding wires enter the box and are joined, a deduction of only one conductor is required, regardless of their number.

Further components that require deduction adjustments from those specified in Table 370-16(a) include fixture studs, hickeys, and fixture stud extensions [NEC Section 370-16(b)(3)]. One conductor must be deducted from the total for each type of fitting used. Two conductors must be deducted for each strap-mounted device, like duplex receptacles and wall switches; a deduction of one conductor is made when one or more internally mounted cable clamps are used.

Figure 19 shows components which may be used in outlet boxes without affecting the total number of conductors. Such items include grounding clips and screws, wire nuts and cable connectors when the latter is inserted through knock-out holes in the outlet box and secured with lockouts. Pre-wired fixture wires are not counted against the total number of allowable conductors in an outlet box; neither are conductors originating and ending in the box.

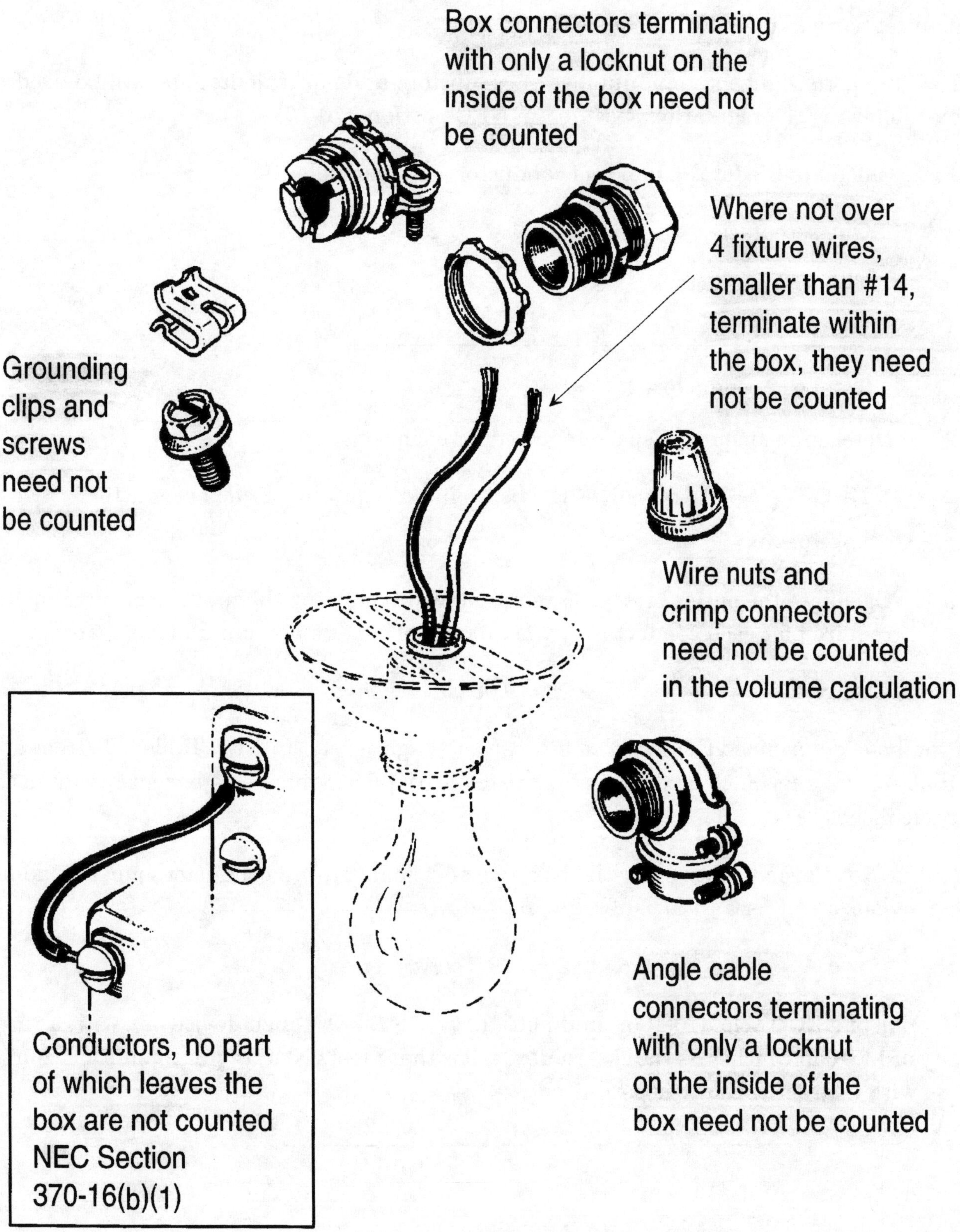

Box connectors terminating
with only a locknut on the
inside of the box need not
be counted

Where not over
4 fixture wires,
smaller than #14,
terminate within
the box, they need
not be counted

Grounding
clips and
screws
need not
be counted

Wire nuts and
crimp connectors
need not be counted
in the volume calculation

Conductors, no part
of which leaves the
box are not counted
NEC Section
370-16(b)(1)

Angle cable
connectors terminating
with only a locknut
on the inside of the
box need not be counted

Figure 19. Components that do not affect the capacity of an outlet box.

To better understand how outlet boxes are sized, let's take two No. 12 AWG conductors installed in ½″ EMT and terminating into a metallic outlet box containing one duplex receptacle. What size of outlet box will meet NEC requirements?

The first step is to count the total number of conductors and equivalents that will be used in the box—following the regulation specified in NEC Section 370-16.

Step 1.   Calculate the total number of conductors and equivalents.

One receptacle	=	2
Two #12 conductors	=	2
Total #12 conductors	=	4

Step 2.   Determine amount of space required for each conductor.

NEC Table 370-16(b) gives the box volume required for each conductor:

No. 12 AWG   =   2.25 cubic inches

Step 3.   Calculate the outlet box space required by multiplying the number of cubic inches required for each conductor by the number of conductors found in No. 1 above.

$$4 \times 2.25 = 9.00 \text{ cubic inches}$$

Once you have determined the required box capacity, again refer to NEC Table 370-16(a) and note that a $3 \times 2 \times 2\frac{1}{4}$-inch box comes closest to our requirements. This box size is rated for 10.5 cubic inches.

Where four No. 12 conductors enter the box, two additional No. 12 conductors must be added to our previous count for a total of (4 + 2 =) 6 conductors.

$$6 \times 2.25 = 13.5 \text{ cubic inches}$$

Again, refer to NEC Table 370-16(a) and note that a $3 \times 2 \times 2\frac{3}{4}$-inch device box with a rated capacity of 14.0 cubic inches, is the closest device box that meets NEC requirements. Of course, any box with a larger capacity is permitted.

Summary

The NEC specifies certain fill requirements for raceways, outlet boxes, pull and junction boxes, cabinets, cutout boxes, auxiliary gutters and similar conductor-containing housings; that is, the area of conductors in relation to the box, fitting, or raceway system. In some cases, NEC tables may be used to determine the proper size of housing; in other cases, calculations are required in conjunction with tables and manufacturers' specifications.

This module should provide a firm foundation for determining the proper size of raceway, box, or fitting for any given application. However, it is recommended that the trainee carefully review the NEC requirements pertaining to all such fill requirement. Then, frequently refer to the NEC when a certain application is encountered.

References

For a more advanced study of topics covered in this Task Module, the following works are suggested:

American Electricians Handbook, Latest Edition, Croft, McGraw-Hill, New York, NY

National Electrical Code Handbook, Latest Edition, NFPA, Quincy, MA

SELF-CHECK REVIEW/PRACTICE QUESTIONS

1. Any channel used for holding wires, cables, or busbars is called which of the following?

    a. A knockout
    b. A raceway
    c. A junction box
    d. A conduit fill

2. Which of the following is not a factor influencing a cable's mechanical fit in a raceway?

    a. Clearance
    b. Jam ratio
    c. Age
    d. Weight

3. To find the allowable number of conductors in a raceway system one can look in the NE Code book. Where?

    a. Appendix A
    b. Section 380-6(d)
    c. Article 310
    d. Chapter 9

4. What is the maximum number of conductors allowed in any auxiliary gutter?

    a. Thirty
    b. Twenty
    c. Forty
    d. Twenty five

5. What is the standard amount of clearance one should allow between locknuts when calculating pull-box size?

    a. One inch
    b. $\frac{1}{2}$ inch
    c. $\frac{1}{4}$ inch
    d. $\frac{1}{8}$ inch

6. If a straight-pull box contains two 4″ conduits entering one end and leaving the other end, what is the minimum length that the box must be?

    a. 12 inches
    b. 24 inches
    c. 32 inches
    d. 36 inches

7. If a straight-pull box contains one 4″ conduit and one 3″ conduit (both sizes entering and leaving from opposite ends), what is the minimum length that the box must be?

    a. 24 inches
    b. 32 inches
    c. 36 inches
    d. 38 inches

8. What special procedures must be taken with cables installed in a pull box that is 6 feet high or over?

    a. The cables must be racked in the box
    b. The cables must terminate within the box
    c. The cables must not terminate within the box
    d. Cable clamps of any kind are not allowed

9. What is the maximum percentage of fill allowed in an auxiliary gutter?

    a. 10%
    b. 20%
    c. 30%
    d. 40%

10. What is the minimum wire-bending space that must be provided in a cabinet or cutout box in which three #1/0 THWN conductors terminate?

    a. 7 inches
    b. 12 inches
    c. 14 inches
    d. 18 inches

PERFORMANCE/LABORATORY EXERCISE

1.   Install cable racks in a junction box provided by your instructor and secure four 350 kcmil conductors onto the rack.

2.   Calculate the minimum straight pull-box size for the condition shown in Fig. 20.

3.   Calculate the minimum pull-box size for the condition in Fig. 21.

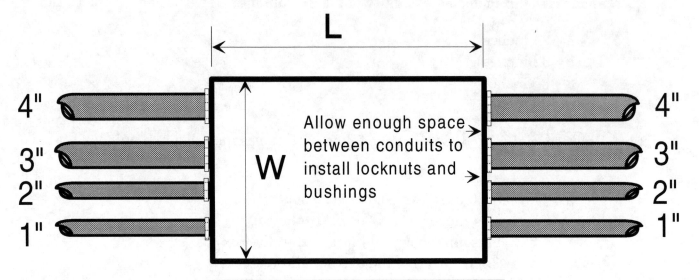

Figure 20. Pull-box with straight conduit entries.

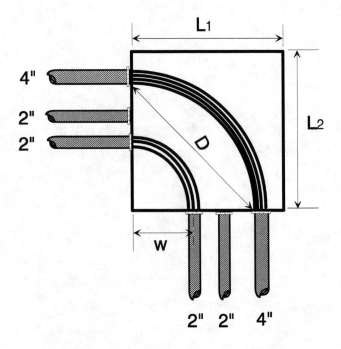

Figure 21. Pull box with angular conduit entries.

Answers to Self-Check Questions

1. b

2. c

3. d

4. a

5. b

6. c

7. b

8. a

9. b

10. a

ELECTRICAL TRAINEE TASK MODULE 20304

Wiring Devices

Module 20305

Electrical Trainee Task Module 20305

WIRING DEVICES

Objectives

Upon completion of this module, the trainee will be able to:

1. Select wiring devices according to NEMA classifications.
2. Size wiring devices according to NEC and NEMA requirements.
3. Select and install the proper box or enclosure for wiring devices.
4. Connect wiring devices.
5. Follow NEC regulations governing the installation of wiring devices.
6. Explain types and purposes of grounding wiring devices.
8. Connect a 480-volt receptacle.
9. Size the maximum load allowed on wiring devices.

Prerequisites

Successful completion of the following Task Modules is required before beginning study of this Task Module: Core Curricula, Electrical Levels 1 and 2, Electrical Level 3, Modules 20301 through 20304.

Required Student Materials

1. Trainee Task Module
2. Copy of the latest edition of the National Electrical Code

COURSE MAP INFORMATION

This course map shows all of the *Wheels of Learning* Task Modules in the third level of the Electrical curricula. The suggested training order begins at the bottom and proceeds up. Skill levels increase as a trainee advances on the course map. The training order may be adjusted by the local Training Program Sponsor.

Course Map: Electrical, Level 3

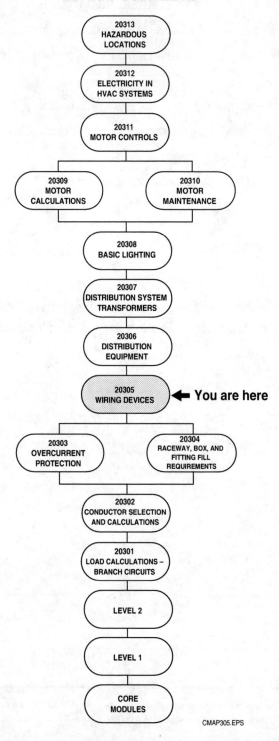

LEVEL 3 COMPLETE

20313 HAZARDOUS LOCATIONS

20312 ELECTRICITY IN HVAC SYSTEMS

20311 MOTOR CONTROLS

20309 MOTOR CALCULATIONS

20310 MOTOR MAINTENANCE

20308 BASIC LIGHTING

20307 DISTRIBUTION SYSTEM TRANSFORMERS

20306 DISTRIBUTION EQUIPMENT

20305 WIRING DEVICES ← You are here

20303 OVERCURRENT PROTECTION

20304 RACEWAY, BOX, AND FITTING FILL REQUIREMENTS

20302 CONDUCTOR SELECTION AND CALCULATIONS

20301 LOAD CALCULATIONS – BRANCH CIRCUITS

LEVEL 2

LEVEL 1

CORE MODULES

CMAP305.EPS

TABLE OF CONTENTS

Trade Terms Introduced in This Module

attachment plug or cap: The male connector for electrical cords.

convenience outlet: A point on the wiring system at which current is taken to supply portable 120-volt appliances such as TV sets, toasters, etc.

cord: A small flexible conductor assembly, usually jacketed.

cord set: A cord having a wiring connector on one or more ends.

device: An item intended to carry or help carry, but not utilize, electrical energy.

four-way switch: A device which, when used in conjunction with two three-way switches, offers control of an electrical outlet (usually lighting) at three or more locations.

gang switch: A unit of two or more switches to give control of two or more circuits from one point. The entire mechanism is mounted in one box under one cover.

jack: A plug-in type terminal.

leg: A portion of a circuit, such as a switch leg or switch loop.

receptacle: A contact device installed at an outlet for the connection of an attachment plug and flexible cord to supply portable equipment.

SP: Single pole.

SPDT: Single-pole, double-throw.

split-wire: A way of wiring a duplex receptacle outlet with a 3-wire, 120/240 volt single-phase circuit so that one hot leg and the grounded conductor (neutral) feed one of the receptacle outlets, and the other hot leg and grounded conductor (neutral) feed the other receptacle outlet. This gives the capacity of two separate circuits to one duplex receptacle.

SPST: Single-pole, single-throw.

switch: A device for opening and closing or for changing the connection of a circuit.

switch-leg: That part of a circuit run from a lighting outlet box where a luminaire or lampholder is installed down to an outlet box that contains the wall switch that turns the light or other load on or off; it is a control leg of the branch circuit.

three-way switch: A switch that is used to control a light, or set of lights, from two different points.

1.0.0 INTRODUCTION TO WIRING DEVICES

A *device*—by NEC definition—is a unit of an electrical system that is intended to carry, but not utilize, electric energy. This covers a wide assortment of system components that include, but are not limited to, the following:

- Switches
- Relays
- Contactors
- Receptacles
- Conductors

However, for our purpose, we will deal with switching devices and also those items used to connect utilization equipment to electrical circuits; namely, *receptacles*. Both are commonly known as *wiring devices*.

Other devices, such as relays, contactors, and the like, are covered elsewhere in the NCCER electrical curricula. Therefore, items covered elsewhere will not be repeated in this module.

2.0.0 RECEPTACLES

A receptacle is a contact device installed at the outlet for the connection of a single attachment plug. Several types and configurations are available for use with many different attachment plug caps—each designed for a specific application. For example, receptacles are available for two-wire 120-volt 15- and 20-ampere circuits; others are designed for use on two- and three-wire, 240-volt, 20-, 30-, 40-, and 50-ampere circuits. There are also many other types, most of which are discussed in this module.

Receptacles are rated according to their voltage and amperage capacity. This rating, in turn, determines the number and configuration of the contacts — both on the receptacle and the receptacle's mating plug. Figure 1 shows the most common configurations, along with their applications. This chart was developed by the Wiring Device Section of NEMA and illustrates 75 various configurations, which cover 38 voltage and current ratings. The configurations represent existing devices as well as suggested standards (shown with an asterisk in the chart) for future design. Note that all configurations in Fig. 1 are for general-purpose nonlocking devices. Locking-type receptacles and caps are covered later in this module.

As indicated in the chart, unsafe interchangeability has been eliminated by assigning a unique configuration to each voltage and current rating. All dual ratings have been eliminated, and interchangeability exists only where it does not present an unsafe condition.

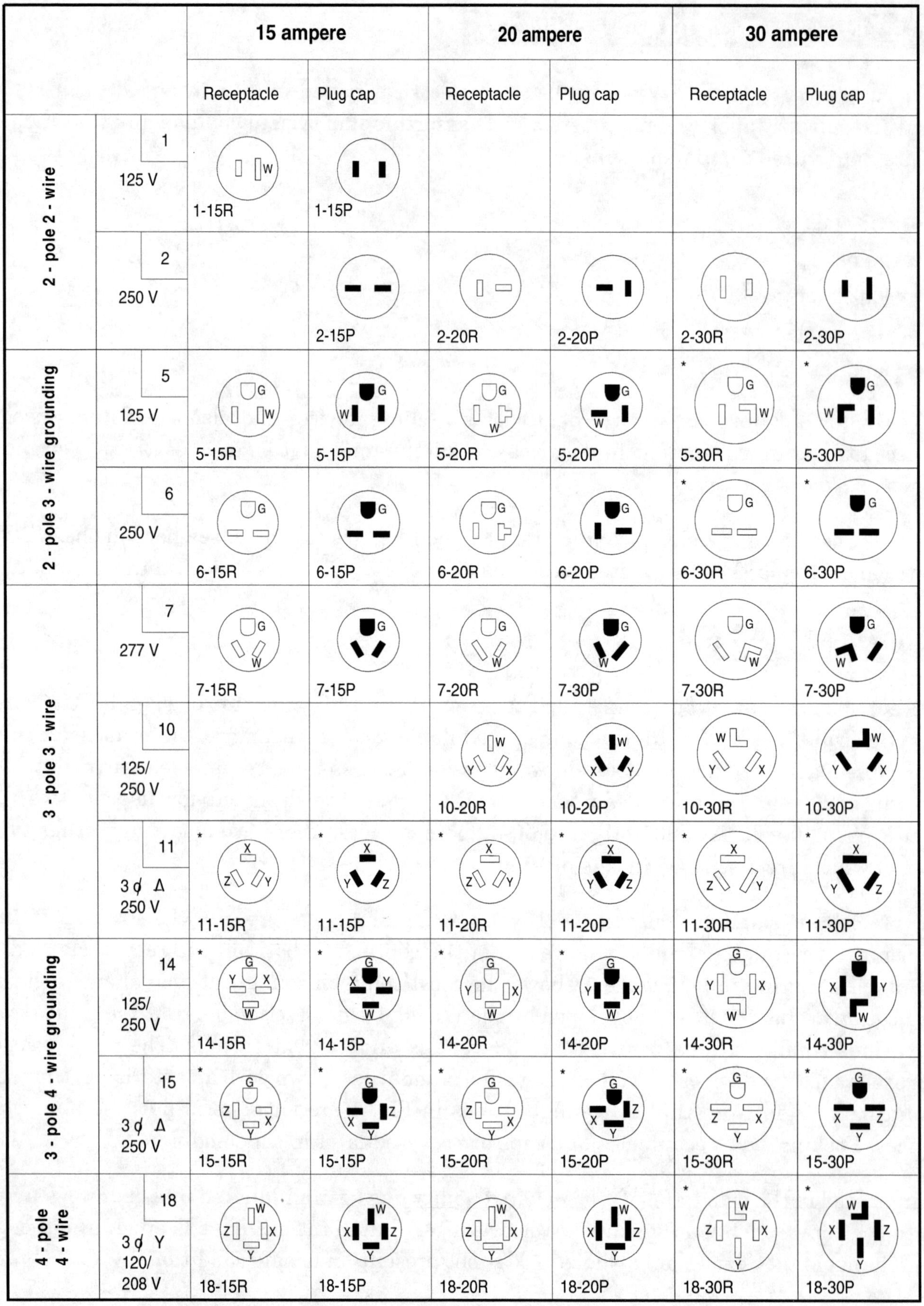

		15 ampere		20 ampere		30 ampere	
		Receptacle	Plug cap	Receptacle	Plug cap	Receptacle	Plug cap
2 - pole 2 - wire	**1** 125 V	1-15R	1-15P				
	2 250 V		2-15P	2-20R	2-20P	2-30R	2-30P
2 - pole 3 - wire grounding	**5** 125 V	5-15R	5-15P	5-20R	5-20P	5-30R	5-30P
	6 250 V	6-15R	6-15P	6-20R	6-20P	6-30R	6-30P
3 - pole 3 - wire	**7** 277 V	7-15R	7-15P	7-20R	7-30P	7-30R	7-30P
	10 125/ 250 V			10-20R	10-20P	10-30R	10-30P
	11 3φ Δ 250 V	11-15R	11-15P	11-20R	11-20P	11-30R	11-30P
3 - pole 4 - wire grounding	**14** 125/ 250 V	14-15R	14-15P	14-20R	14-20P	14-30R	14-30P
	15 3φ Δ 250 V	15-15R	15-15P	15-20R	15-20P	15-30R	15-30P
4 - pole 4 - wire	**18** 3φ Y 120/ 208 V	18-15R	18-15P	18-20R	18-20P	18-30R	18-30P

Figure 1. NEMA configurations for general-purpose nonlocking receptacles and plug caps.

50 ampere		60 ampere	
Receptacle	Plug cap	Receptacle	Plug cap
5-50R	5-50P		
6-50R	6-50P		
7-50R	7-50P		
10-50R	10-50R		
11-50R	11-50P		
14-50R	14-50P	14-60R	14-60P
15-50R	15-50P	15-60R	15-60P
18-50R	18-50P	18-60R	18-60P

Figure 1. *Continued*

Each configuration is designated by a number composed of the chart line number, the amperage, and either "R" for receptacle or "P" for plug cap. For example, a 5-15R is found in line 5 and represents a 15-ampere receptacle.

A clear distinction is made in the configurations between "system grounds" and "equipment grounds." System grounds, referred to as grounded conductors, normally carry current at ground potential, and terminals for such conductors are marked "W" for "White" in the chart. Equipment grounds, referred to as grounding conductors, carry current only during ground-fault conditions, and terminals for such conductors are marked "G" for "grounding" in the chart.

2.1.0 RECEPTACLE CHARACTERISTICS

Receptacles have various symbols and information inscribed on them that help to determine their proper use and ratings. For example, Fig. 2 shows a standard duplex receptacle and contains the following printed inscriptions:

- The testing laboratory label
- The CSA (Canadian Standards Association) label
- Type of conductor for which the terminals are designed
- Current and voltage ratings, listed by maximum amperage, maximum voltage, and current restrictions

The testing laboratory label is an indication that the device has undergone extensive testing by a nationally recognized testing lab and has met with the minimum safety requirements. The label does not indicate any type of quality rating.

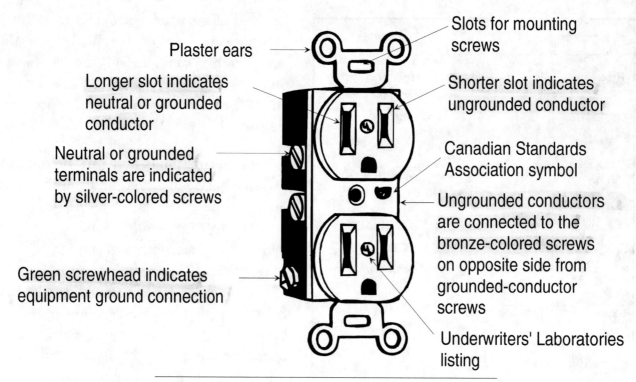

Figure 2. Characteristics of typical duplex receptacles.

The receptacle in Fig. 2 is marked with the "UL" label which indicates that the device type was tested by Underwriters' Laboratories, Inc. of Northbrook, IL. ETL Testing Laboratories, Inc. of Cortland, NY is another nationally recognized testing laboratory. They provide a labeling, listing and follow-up service for the safety testing of electrical products to nationally recognized safety standards or specifically designated requirements of jurisdictional authorities.

The CSA (Canadian Standards Association) label is an indication that the material or device has undergone a similar testing procedure by the Canadian Standards Association and is acceptable for use in Canada.

Current and voltage ratings are listed by maximum amperage, maximum voltage and current restriction. On the device shown in Fig. 2, the maximum current rating is 15 amperes at 125 volts — the latter of which is the maximum voltage allowed on a device so marked.

Conductor markings are also usually found on duplex receptacles. Receptacles with quick-connect wire clips will be marked "Use #12 or #14 solid wire only." If the inscription "CO/ALR" is marked on the receptacle, either copper, aluminum, or copper-clad aluminum wire may be used. The letters "ALR" stand for "aluminum revised." Receptacles marked with the inscription "CU/AL" should be used for copper only, although they were originally intended for use with aluminum also. However, such devices frequently failed when connected to 15- or 20-ampere circuits. Consequently, devices marked with "CU/AL" are no longer acceptable for use with aluminum conductors.

The remaining markings on duplex receptacles may include the manufacturer's name or logo, "Wire Release" inscribed under the wire-release slots, and the letters "GR." beneath or beside of the green grounding screw.

The screw terminals on receptacles are color-coded. For example, the terminal with the green screw head is the equipment ground connection and is connected to the U-shaped slots on the receptacle. The silver-colored terminal screws are for connecting the grounded or neutral conductors and are associated with the longer of the two vertical slots on the receptacle. The brass-colored terminal screws are for connecting the ungrounded or "hot" conductors and are associated with the shorter vertical slots on the receptacle.

Note: The long vertical slot accepts the grounded or neutral conductor while the shorter vertical slot accepts the ungrounded or hot conductor.

2.2.0   MOUNTING RECEPTACLES

Although no actual NEC requirements exist on mounting heights and positioning receptacles, there are certain installation methods that have become "standard" in the electrical industry. Figure 3 shows mounting heights of duplex receptacles used on conventional residential and small commercial installations. However, do not take these dimensions as gospel; they are frequently varied to suit the building structure. For example, ceramic tile might be placed above a kitchen or bathroom countertop. If the dimensions in Fig. 3 puts the receptacle part of the way out of the tile, say, half in and half out, the mounting height should be adjusted to either place the receptacle completely in the tile or completely out of the tile as shown in Fig. 4.

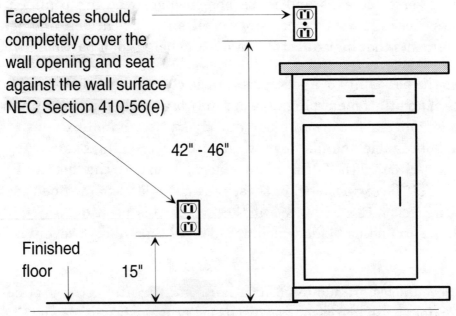

Figure 3. Recommended mounting heights of duplex receptacles.

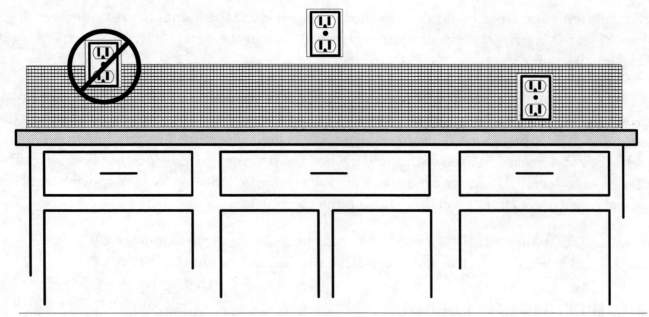

Figure 4. Adjust mounting heights so receptacles will either be completely in or completely out of tile.

Refer again to Fig. 3 and note that the mounting heights are given to the bottom of the outlet box. Many dimensions on electrical drawings are given to the center of the outlet box or receptacle. However, during the actual installation, workers installing the outlet boxes can mount them more accurately (and in less time) by lining up the bottom of the box with a chalk mark rather than trying to "eyeball" this mark to the center of the box.

A decade or so ago, most electricians mounted receptacle outlets 12 inches from the finished floor to the center of the outlet box. However, a recent survey taken of over 500 homeowners shows that they prefer a mounting height of 15 inches from the finished floor to the bottom of the outlet box. It is easier to plug and unplug the cord-and-plug assemblies at this height — especially among senior citizens and those homeowners who are confined to wheelchairs. However, always check the working drawings, written specifications, and details of construction for measurements that may affect the mounting height of a particular receptacle outlet.

There is always the possibility of a metal receptacle cover coming loose and falling downward onto the blades of an attachment plug cap that may be loosely plugged into the receptacle. By the same token, a hairpin, fingernail file, metal fly-swatter handle, or any other metal object may be knocked off a table and fall downward onto the the plug blades. Any of these objects could cause a short-circuit if the falling metal object fell on both the "hot" and grounded neutral blades of the plug at the same time. For these reasons, it is recommended that the equipment grounding slot in receptacles be placed at the top. In this position, any falling metal object would fall onto the grounding blade which would more than likely prevent a short-circuit. See Fig. 5.

When duplex receptacles are mounted in a horizontal position, the grounded neutral slots should be on top for the same reasons as discussed previously. Again, see Fig. 5.

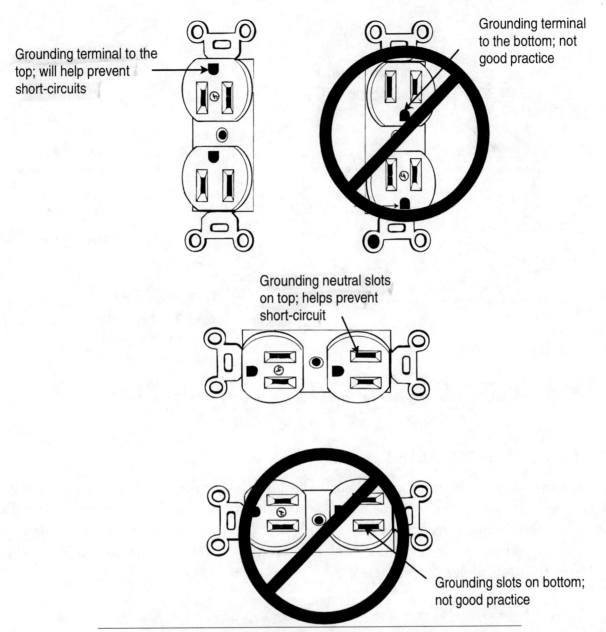

Figure 5. Recommended mounting positions for duplex receptacles.

NEC Section 370-20 requires all outlet boxes installed in walls or ceilings of concrete, tile, or other noncombustible material such as plaster or drywall to be installed in such a matter that the front edge of the box or fitting will not set back of the finished surface more than $\frac{1}{4}$". Where walls and ceilings are constructed of wood or other combustible material, outlet boxes and fittings must be flush with the finished surface of the wall. See Fig. 6.

Wall surfaces such as drywall, plaster, etc. that contain wide gaps or are broken, jagged, or otherwise damaged must be repaired so there will be no gaps or open spaces greater than $\frac{1}{8}$" between the outlet box and wall material. These repairs should be made prior to installing the faceplate. Such repairs are best made with a noncombustible caulking or spackling compound. See Fig. 7.

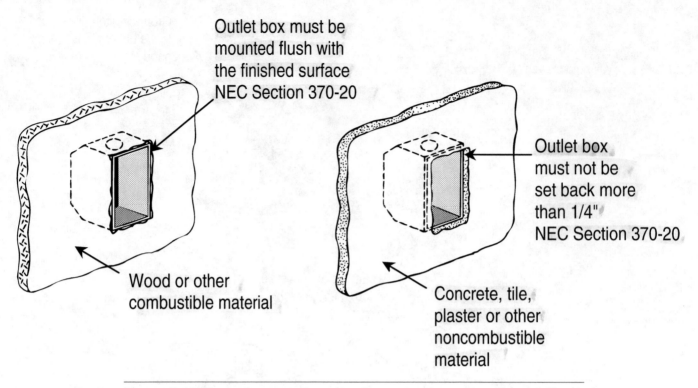

Outlet box must be mounted flush with the finished surface NEC Section 370-20

Wood or other combustible material

Outlet box must not be set back more than 1/4" NEC Section 370-20

Concrete, tile, plaster or other noncombustible material

Figure 6. NEC requirements for mounting outlet boxes in walls or ceilings.

2.3.0   TYPES OF RECEPTACLES

There are many types of receptacles. For example, the duplex receptacles that have been discussed are the straight-blade type which accepts a straight blade connector or plug. This is the most common type of receptacle and such receptacles are found on virtually all electrical projects from residential to large industrial installations. Refer again to Fig. 1 for types of receptacles that fall under this category.

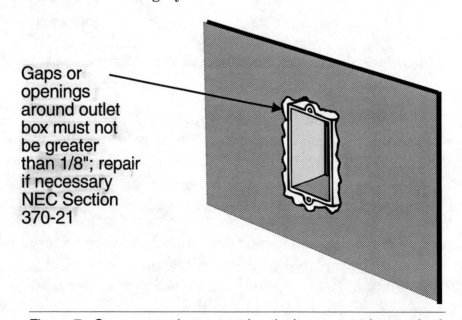

Gaps or openings around outlet box must not be greater than 1/8"; repair if necessary NEC Section 370-21

Figure 7. Gaps or openings around outlet boxes must be repaired.

Twist lock receptacles: Twist lock receptacles are designed to accept a somewhat "curved blade" connector or plug. The plug/connector and the receptacle will lock together with a slight twist. The locking prevents accidentally unplugging the equipment.

Pin-and-sleeve receptacles: Pin-and-sleeve devices have a unique locking feature. These receptacles are made with an extremely heavy-duty plastic housing that makes them highly indestructible. They are manufactured with long brass pins for long life and are color-coded according to voltage for easy identification.

Low-voltage receptacles: These receptacles are designed for both AC and DC systems where the maximum potential is 50 volts. Receptacles used for low-voltage systems must have a minimum current-carrying rating of 15 amperes.

440-volt receptacles: Portable electrical equipment operating at 440 to 460 volts is common on many industrial installations. Such equipment includes welders, battery chargers, and other types of portable equipment. Special "440-volt" plugs and receptacles are used to connect and disconnect such equipment from a power source. 440-volt receptacles are available in 2-wire, single-phase; 3-wire, three-phase, and 4-wire, three-phase. Equipment grounding is required in all cases, and provisions are provided in each receptacle for such grounding.

WARNING!   Make certain that the plug-and-cord assembly is compatible with both the equipment and receptacle before connecting to a 440-volt receptacle. Polarity and equipment-grounding checks on the plug-and-cord assembly should be made on a monthly basis and sooner if subject to hard usage.

3.0.0   LOCATING RECEPTACLES

Several NEC Sections specify specific requirements for locating receptacles in all types of installations. A summary of these requirements is presented herein.

3.1.0   RESIDENTIAL OCCUPANCIES

NEC Section 210-52 should be referred to when laying out outlets for residential and some commercial installations. This section details the general provisions along with small-appliance circuit requirements, laundry requirements, unfinished basements, attached garages, and other areas of the home.

In general, every dwelling — regardless of its size — must have receptacles located in each habitable area so that no point along the floor line in any wall space (2 feet wide or wider) is more than 6 feet in that space. The purpose of this requirement is to prevent the need for extension cords longer than 6 feet and to minimize the use of cords across doorways, fireplaces, and similar openings. See Fig. 8.

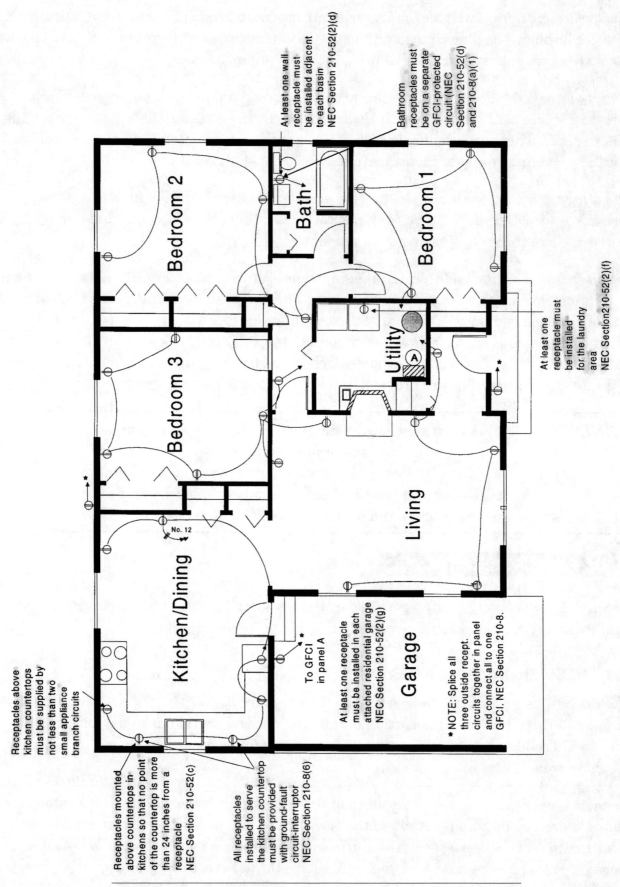

At least one wall receptacle must be installed adjacent to each basin NEC Section 210-52(2)(d)

Bathroom receptacles must be on a separate GFCI-protected circuit (NEC Section 210-52(d) and 210-8(a)(1)

At least one receptacle must be installed for the laundry area NEC Section210-52(2)(f)

Bedroom 2

Bath

Bedroom 1

Bedroom 3

Utility

A

Living

No. 12

Kitchen/Dining

To GFCI in panel A

At least one receptacle must be installed in each attached residential garage NEC Section 210-52(2)(g)

Garage

★ NOTE: Splice all three outside recept. circuits together in panel and connect all to one GFCI. NEC Section 210-8.

Receptacles above kitchen countertops must be supplied by not less than two small appliance branch circuits

Receptacles mounted above countertops in kitchens so that no point of the countertop is more than 24 inches from a receptacle NEC Section 210-52(c)

All receptacles installed to serve the kitchen countertop must be provided with ground-fault circuit-interruptor NEC Section 210-8(6)

Figure 8. Summary of NEC requirements for locating receptacles.

In addition, a minimum of two 20-ampere small appliance branch circuits are required to serve all receptacle outlets, including refrigeration equipment, in the kitchen, pantry, breakfast room, dining room, or similar area of the dwelling unit. Such circuits, whether two or more are used, must have no other outlets connected to them.

At least one receptacle is required in each laundry area, on the outside of the building at the front and back, in each basement, in each attached and detached garage, in each hallway 10 feet or more in length, and at an accessible location for servicing any HVAC equipment. Figures 9 and 10 summarize these and other NEC requirements regarding the installation of receptacles in dwelling units.

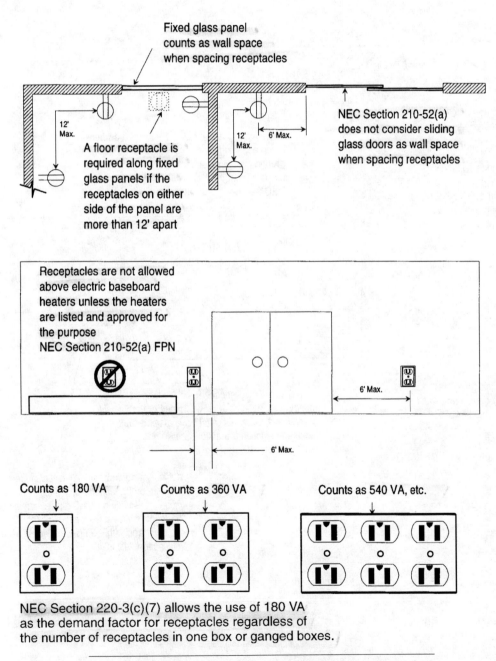

Figure 9. Additional NEC requirements for receptacles.

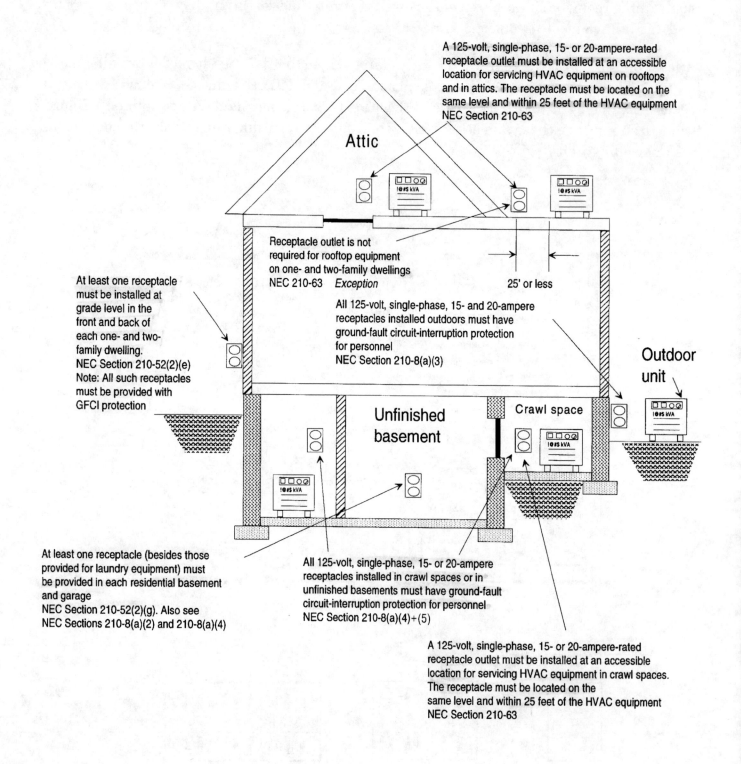

A 125-volt, single-phase, 15- or 20-ampere-rated receptacle outlet must be installed at an accessible location for servicing HVAC equipment on rooftops and in attics. The receptacle must be located on the same level and within 25 feet of the HVAC equipment NEC Section 210-63

Attic

Receptacle outlet is not required for rooftop equipment on one- and two-family dwellings NEC 210-63 *Exception*

25' or less

At least one receptacle must be installed at grade level in the front and back of each one- and two-family dwelling. NEC Section 210-52(2)(e) Note: All such receptacles must be provided with GFCI protection

All 125-volt, single-phase, 15- and 20-ampere receptacles installed outdoors must have ground-fault circuit-interruption protection for personnel NEC Section 210-8(a)(3)

Outdoor unit

Unfinished basement

Crawl space

At least one receptacle (besides those provided for laundry equipment) must be provided in each residential basement and garage NEC Section 210-52(2)(g). Also see NEC Sections 210-8(a)(2) and 210-8(a)(4)

All 125-volt, single-phase, 15- or 20-ampere receptacles installed in crawl spaces or in unfinished basements must have ground-fault circuit-interruption protection for personnel NEC Section 210-8(a)(4)+(5)

A 125-volt, single-phase, 15- or 20-ampere-rated receptacle outlet must be installed at an accessible location for servicing HVAC equipment in crawl spaces. The receptacle must be located on the same level and within 25 feet of the HVAC equipment NEC Section 210-63

Figure 10. More NEC requirements concerning placement of receptacles.

ELECTRICAL TRAINEE TASK MODULE 20305

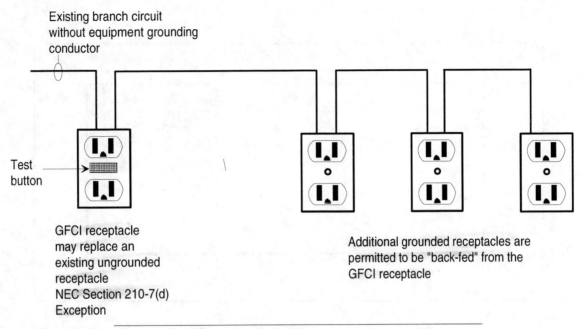

Figure 11. A GFCI may replace an ungrounded receptacle.

When upgrading existing electrical systems, the NEC permits the use of a GFCI receptacle in place of a grounded receptacle. With such an arrangement, additional grounded receptacles may be connected on the downstream side of the GFCI as shown in Fig. 11.

Other receptacles and related circuits are provided as needed according to the load to be served. For example, receptacles are normally provided in residential occupancies for electric ranges, clothes dryers, and similar appliances. Most operate on 120/240-volt branch circuits using 30- to 60-ampere receptacles. See Fig. 1.

3.2.0 COMMERCIAL APPLICATIONS

Receptacle requirements for commercial installations follow the same general requirements set forth for residential occupancies, with some exceptions. For example, guest rooms in hotels, motels, and similar occupancies must have receptacle outlets installed in accordance with Section 210-52. However, some leeway is given commercial installations. NEC Section 210-60 permits receptacle outlets to be located conveniently for permanent furniture layout.

The only other "must" requirement for commercial installations deals with the placement of receptacle outlets in show windows. NEC Section 210-62 requires at least one receptacle for each 12 linear feet of show window area measured horizontally at its maximum width. See Fig. 12. To calculate the number of receptacles required at the top of any show window, measure the total linear feet, and then divide this figure by 12 and any remainder or "fraction thereof" requires an additional receptacle. For example, the show window in Fig. 12 is 18 feet in length. Consequently, the number of receptacles required may be calculated as follows:

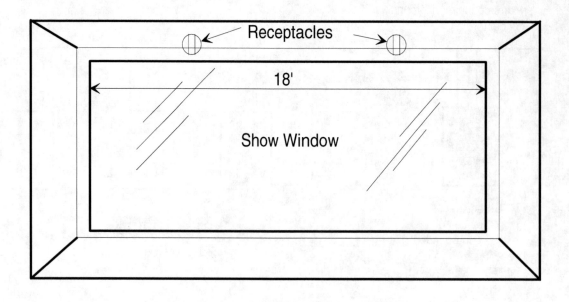

Figure 12. NEC Section 210-62 deals with commercial show windows.

$$\frac{18 \ feet}{12 \ feet} = 1.5 \ receptacles$$

Since "1.5" is more than one, to comply with NEC Section 210-62, "or major fraction thereof," two receptacles are required in this area. Had the calculation resulted in a figure of, say, 1.01, the local inspection authorities would probably require only one receptacle in the show widow.

Of course, GFCIs are required on all 15- and 20-ampere receptacles installed in commercial bathrooms or toilets, in commercial garages, receptacles installed outdoors where there is direct grade-level access (below 6 feet 6 inches), crawl spaces, boathouses, and all receptacles installed on roofs.

Other receptacles and related circuits are provided as needed according to the load to be served.

4.0.0   SWITCHES

The purpose of a switch is to make and break an electrical circuit, safely and conveniently. In doing so, a switch may be used to manually control lighting, motors, fans, and other various items connected to an electrical circuit. Switches may also be activated by light, heat, chemicals, motion, and electrical energy for automatic operation. NEC Article 380 covers the installation and use of switches.

Although there is some disagreement concerning the actual definitions of the various switches that might fall under the category of *wiring devices*, the most generally accepted ones are as follows:

ELECTRICAL TRAINEE TASK MODULE 20305

Bypass isolation switch: This is a manually operated device used in conjunction with a transfer switch to provide a means of directly connecting load conductors to a power source, and of disconnecting the transfer switch. See Fig. 13.

General-use switch: A switch intended for use in general distribution and branch circuits. It is rated in amperes, and it is capable of interrupting its rated current at its rated voltage.

General-use snap switch: A form of general-use switch so constructed that it can be installed in flush device boxes or on outlet box covers, or otherwise used in conjunction with wiring systems recognized by the NEC.

Isolating switch: A switch intended for isolating an electric circuit from the source of power. It has no interrupting rating, and it is intended to be operated only after the circuit has been opened by some other means.

Motor-circuit switch: A switch, rated in horsepower, capable of interrupting the maximum operating overload current of a motor of the same horsepower rating as the switch at its rated voltage.

Transfer switch: A transfer switch is a device for transferring one or more load conductor connections from one power source to another. This type of switch may be either automatic or nonautomatic.

Each of these switches will be thoroughly covered in this module.

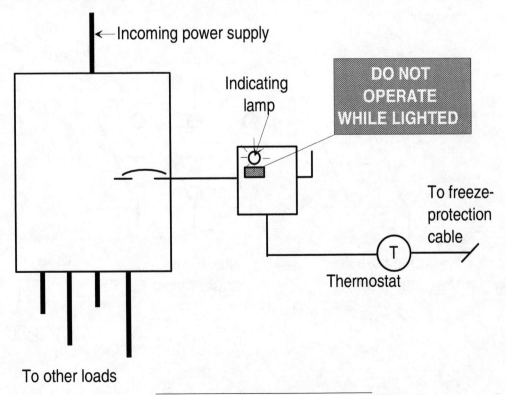

Figure 13. Bypass isolation switch.

4.1.0 COMMON TERMS

Although basic switch terms are covered to some extent in earlier modules, a brief review of these terms is warranted here. In general, the major terms used to identify the characteristics of switches are:

- Pole or poles
- Throw

The term *pole* refers to the number of conductors that the switch will control in the circuit. For example, a single-pole switch breaks the connection on only one conductor in the circuit. A double-pole switch breaks the connection to two conductors, and so forth.

The term *throw* refers to the number of internal operations that a switch can perform. For example, a single-pole, single-throw switch will "make" one conductor when thrown in one direction — the "ON" direction — and "break" the circuit when thrown in the opposite direction; that is, the "OFF" position. The commonly used ON/OFF toggle switch is an SPST switch (single-pole, single-throw). A two-pole, single-throw switch opens or closes two conductors at the same time. Both conductors are either open or closed; that is, in the ON or OFF position. A two-pole, double-throw switch is used to direct a two-wire circuit through one of two different paths. One application of a two-pole, double-throw switch is in an electrical transfer switch where certain circuits may be energized from either the main electric service, or from an emergency standby generator. The double-throw switch "makes" the circuit from one or the other and prevents the circuits from being energized from both sources at once. Figure 14 shows common switch configurations.

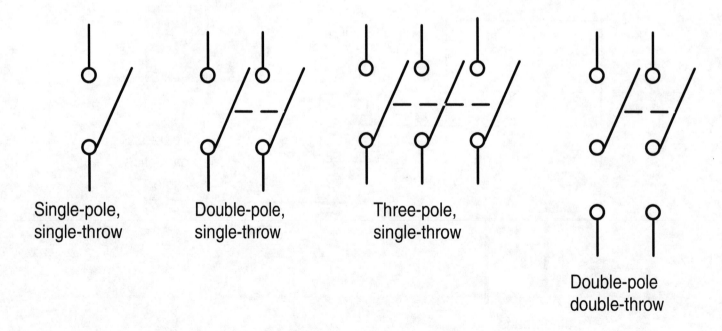

Single-pole,
single-throw

Double-pole,
single-throw

Three-pole,
single-throw

Double-pole
double-throw

Figure 14. Common switch configurations.

ELECTRICAL TRAINEE TASK MODULE 20305

4.2.0   SWITCH IDENTIFICATION

Switches vary in grade, capacity, and purpose. It is very important that proper types of switches are selected for the given application. For example, most single-pole toggle switches used for the control of lighting are restricted to AC use only. This same switch is not suitable for use on, say, a 32-volt DC emergency lighting circuit. A switch rated for AC only will not extinguish a DC arc quickly enough. Not only is this a dangerous practice (causing arcing and heating of the device), the switch contacts would probably burn up after only a few operations of the handle, if not the first time.

Figure 15 shows a typical single-pole toggle switch — the type most often used to control AC lighting in all installations. Note the identifying marks. They are similar to those on the duplex receptacle discussed previously. The main difference is the "T" rating which means that the switch is rated for switching lamps with tungsten filaments (incandescent lamps).

Screw terminals are also color-coded on conventional toggle switches. Switches are typically constructed with a ground screw attached to the metallic strap of the switch. The ground screw is usually a green-colored hex-head screw. This screw is for connecting the equipment-grounding conductor to the switch. On three-way switches, the common or pivot terminal usually has a black or bronze screw head.

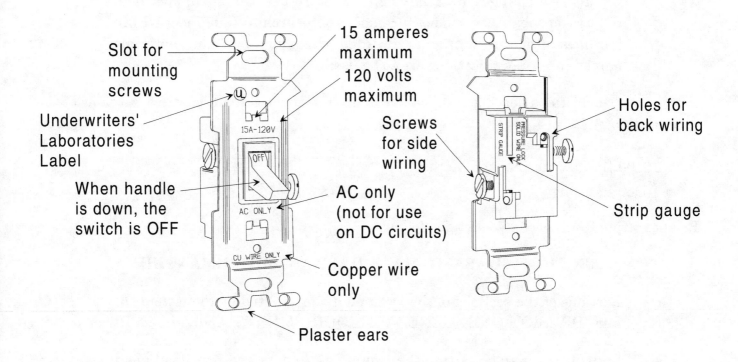

Figure 15. Typical identifying marks on a single-pole switch.

The switch shown is the type normally used for residential construction. Heavier-duty switches are usually the type used on commercial wiring — some of which are rated for use on 277-volt circuits with current-carrying ratings up to 30 amperes. Therefore, it is important to check the rating of each switch before it is installed.

The exact type and grade of switch to be used on a specific installation is often dictated by the project drawings or written specifications. Sometimes wall switches are specified by manufacturer and catalog number; other times they are specified by type, grade, voltage, current rating, and the like, leaving the contractor or electrician to select the manufacturer. The naming of a certain brand of switch for a particular project, does not necessarily mean that this brand must be used. A typical paragraph from an electrical specification (concerning the substitution of materials) may read as follows:

> The naming of a certain brand or make or manufacturer in the specifications is to establish a quality standard for the article desired. The contractor is not restricted to the use of the specific brand of the manufacturer named unless so indicated in the specifications. However, where a substitution is requested, a substitution will be permitted only with the written approval of the engineer. No substitute material or equipment shall be ordered, fabricated, shipped, or processed in any manner prior to the approval of the architect-engineer. The contractor shall assume all responsibility for additional expenses as required in any way to meet changes from the original material or equipment specified. If notice of substitution is not furnished to the architect-engineer within ten days after the contract is awarded, the equipment and materials named in the specifications are to be used.

Electrical specifications dealing with wall switches are covered in at least two sections of the specifications:

- 16100 Basic Materials and Methods
- 16500 Lighting

Brief excerpts from these two sections follow:

SECTION 16B - BASIC MATERIALS AND WORKMANSHIP

1. Portions of the sections of the Documents designated by the letters "A", "B" & "C" and "DIVISION ONE - GENERAL REQUIREMENTS" apply to this Division.

2. Consult Index to be certain that set of Documents and Specifications is complete. Report omissions or discrepancies to the architect.

c. SWITCH OUTLET BOXES: Wall switches shall be mounted approximately 54 inches above the finished floor (AFF) unless otherwise noted. When the switch is mounted in a masonry wall the bottom of the outlet box shall be in line with the bottom of a masonry unit. Where more than two switches are located, the switches shall be mounted in a gang outlet box with gang cover. Dimmer switches shall be individually mounted unless otherwise noted. Switches with pilot lights, switches with overload motor protection and other special switches that will not conveniently fit under gang wall plates may be individually mounted.

13. EQUIPMENT AND INSTALLATION WORKMANSHIP:

a. All equipment and material shall be new and shall bear the manufacturer's name and trade name. The equipment and material shall be essentially the standard product of a manufacturer regularly engaged in the production of the required type of equipment and shall be the manufacturer's latest approved design.

b. The Electrical Contractor shall receive and properly store the equipment and material pertaining to the electrical work. The equipment shall be tightly covered and protected against dirt, water, chemical or mechanical injury and theft. The manufacturer's directions shall be followed completely in the delivery, storage, protection and installation of all equipment and materials.

c. The Electrical Contractor shall provide and install all items necessary for the complete installation of the equipment as recommended or as required by the manufacturer of the equipment or required by code without additional cost to the Owner, regardless whether the items are shown on the plans or covered in the Specifications.

d. It shall be the responsibility of the Electrical Contractor to clean the electrical equipment, make necessary adjustments and place the equipment into operation before turning equipment over to the Owner. Any paint that was scratched during construction shall be "touched-up" with factory color paint to the satisfaction of the Architect. Any items that were damaged during construction shall be replaced.

6. **WIRING DEVICES:**

   a. **GENERAL:** The wiring devices specified below with ARROW HART numbers may also be the equivalent wiring device as manufactured by BRYANT ELECTRIC, HARVEY HUBBELL or PASS & SEYMOUR. All other items shall be as specified.

   b. **WALL SWITCHES:** Where more than one flush wall switch is indicated in the same location, the switches shall be mounted in gangs under a common plate.

      (1) Single Pole   AH#1991
      (2) Three-Way   AH#1993
      (3) Four-Way   AH#1994
      (4) Switch with pilot light   AH#2999-R
      (5) Motor Switch - Surface   AH#6808
      (6) Motor Switch - Flush   AH#6808-F

   c. **WALL PLATE:** Stainless steel wall plates with satin finish minimum .030 inches shall be provided for all outlets and switches.

In general, the preceding electrical specifications give the grade of materials to be used on the project and the manner in which the electrical system must be installed. Most specification writers use an abbreviated language; although it is relatively difficult for beginners to understand, experience makes possible a proper interpretation with little difficulty. However, electricians involved with any project should make certain that everything is clear. If it is not, contact the architectural or engineering firm and clarify the problem prior to installing the work, not after a system has been completed.

5.0.0   NEC REQUIREMENTS FOR SWITCHES

There are many NEC requirements for installing light switches and they are scattered in various locations in the NEC book. For example, wall-switch controlled lighting outlets are required in each habitable room of all residential occupancies. Wall-switch controlled lighting is also required in each bathroom, hallways, stairways, attached garages, and at outdoor entrances. A wall-switch controlled receptacle may be used in place of the lighting outlet in habitable rooms other than the kitchen and bathrooms. Providing a wall switch for room lighting is intended to prevent an occupant's groping in the dark for table lamps or pull chains. In stairways with six or more steps, the stairway lighting must be controlled at two locations — at the top and also the bottom of the stairway. This is accomplished by using two 3-way switches and connected as discussed in previous modules. See Fig. 16.

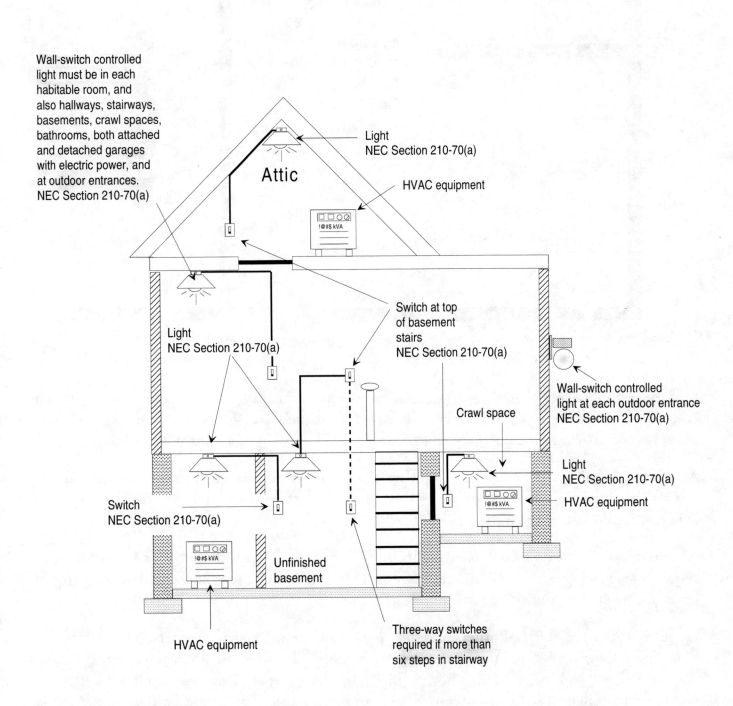

Wall-switch controlled light must be in each habitable room, and also hallways, stairways, basements, crawl spaces, bathrooms, both attached and detached garages with electric power, and at outdoor entrances. NEC Section 210-70(a)

Attic

Light
NEC Section 210-70(a)

HVAC equipment

Switch at top of basement stairs
NEC Section 210-70(a)

Light
NEC Section 210-70(a)

Wall-switch controlled light at each outdoor entrance
NEC Section 210-70(a)

Crawl space

Light
NEC Section 210-70(a)

HVAC equipment

Switch
NEC Section 210-70(a)

Unfinished basement

HVAC equipment

Three-way switches required if more than six steps in stairway

Figure 16. NEC requirements for residential wall switches.

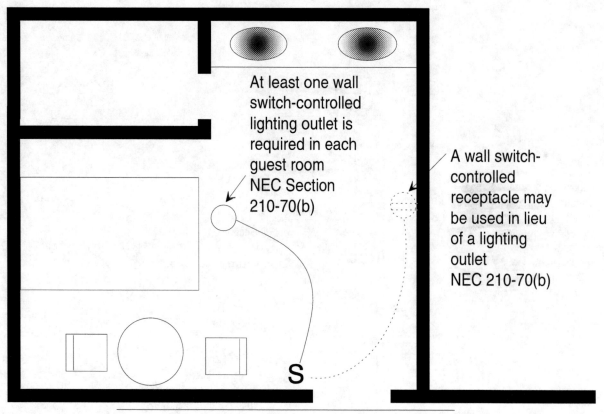

Figure 17. NEC switch requirements for guest rooms.

Lighting outlets are also required in attics, crawl spaces, utility rooms, and basements when these spaces are used for storage or contain equipment such as HVAC equipment. Again, if the basement or attic stairs have more than 6 steps, a 3-way switch is required at each landing.

At least one wall switch-controlled lighting outlet is required in each guest room in hotels, motels, or similar locations as shown in Fig. 17. Note that a wall switch-controlled receptacle is permitted in lieu of the lighting outlet.

At least one wall switch-controlled lighting outlet must be installed at or near equipment requiring servicing such as HVAC equipment. The wall switch must be located at the point of entry to the attic or underfloor space.

In many commercial installations, circuit breakers in panelboards are permitted to control main-area lighting where the areas are constantly illuminated during operating hours. Consequently, wall switches are not required in these areas. However, wall switches are normally installed at outdoor entrances, entrances to store rooms, small offices, toilets, and similar locations.

Wiring diagrams of switch circuits — single-pole, three-way, and four-way switches — were thoroughly discussed in previous modules and these diagrams will not be repeated here. However, it is recommended that the trainee review these diagrams at this time if deemed necessary.

Enclosed single-throw safety switches are manufactured to meet industrial, commercial, and residential requirements. See Fig. 18. The two basic types of safety switches are:

- General duty
- Heavy duty

Double-throw switches are also manufactured with enclosures and features similar to the general and heavy-duty single-throw designs.

The majority of safety switches have visible blades and safety handles. The switch blades are in full view when the enclosure door is open and there is visually no doubt when the switch is OFF. The only exception is Type 7 and 9 enclosures; these do not have visible blades. Switch handles, on all types of enclosures are an integral part of the box, not the cover, so that the handle is in control of the switch blades under normal conditions.

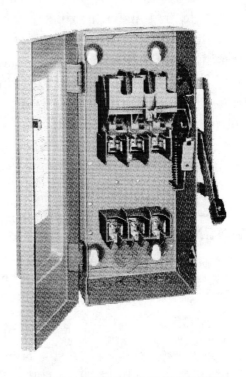

Figure 18. Typical safety switch.

6.1.0   HEAVY DUTY SWITCHES

Heavy duty switches are intended for applications where ease of maintenance, rugged construction, and continued performance are primary concerns. They can be used in atmospheres where general duty switches would be unsuitable, and are therefore widely used in industrial applications. Heavy duty switches are rated 30 through 1200 amperes and 240 to 600 volts AC or DC. Switches with horsepower ratings are capable of opening a circuit up to six times the rated current of the switch. When equipped with Class J or Class R fuses for 30 through 600 ampere switches, or Class L fuses in 800 and 1200 ampere switches, many heavy duty safety switches are UL listed for use on systems with up to 200,000 RMS symmetrical amperes available fault current. This, however, is about the highest short-circuit rating available for any heavy-duty safety switch. Applications include use where the required enclosure is NEMA TYPE 1, 3R, 4, 4X, 5, 7, 9, 12 or 12K.

6.1.1   Switch Blade and Jaws

Two types of switch contacts are used by the industry in today's safety switches. One is the "butt" contact; the other is a knife-blade and jaw type. On switches with knife-blade construction, the jaws distribute a uniform clamping pressure on both sides of the blade contact surface.

In the event of a high-current fault, the electromagnetic forces which develop tend to squeeze the jaws tightly against the blade. In the butt type contact, only one side of the blade's contact surface is held in tension against the conducting path. Electromagnetic forces due to high current faults tend to force the contacts apart, causing them to burn severely. Consequently, the knife blade and jaw type construction is the preferred type for use on all heavy duty switches. The action of the blades moving in and out of the jaws aids in cleaning the contact surfaces. All current-carrying parts of these switches are plated to reduce heating by keeping oxidation at a minimum. Switch blades and jaws are made of copper for high conductivity. Spring-clamped blade hinges are another feature that help assure good contact surfaces and cool operations. "Visible blades" are utilized to provide visual evidence that the circuit has been opened.

WARNING!   Before changing fuses or performing maintenance on any safety switch, always visibly check the switch blades and jaws to ensure that they are in the OFF position.

6.1.2   Fuse Clips

Fuse clips are plated to control corrosion and to keep heating to a minimum. All fuse clips on heavy duty switches have steel reinforcing springs for increased mechanical strength and firmer contact pressure. See Fig. 19.

6.1.3   Terminal Lugs

Most heavy duty switches have front removable, screw-type terminal lugs. Most switch lugs are suitable for copper or aluminum wire except NEMA TYPES 4, 4X, 5 stainless and TYPES 12 and 12K switches which have all copper current-carrying parts and lugs designated for use with copper wire only. Heavy duty switches are suitable for the wire sizes and number of wires per pole as listed in Fig. 20.

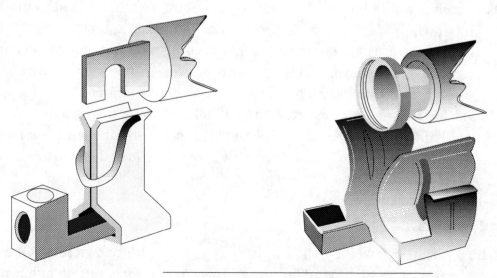

Figure 19. Detail of safety switch fuse clips.

TYPE 1 AND 3R HEAVY DUTY TERMINAL LUG DATA

Ampere Rating	Conductors Per Phase	Wire Range Wire Bending Space Per NEC Table 373-6	Lug Wire Range
30	1	#12-6 AWG (Al) or #14-6 AWG (Cu)	#12-2 AWG (Al) or #14-2 AWG (Cu)
60	1	#12-3 AWG (Al) or #14-3 AWG (Cu)	#12-2 AWG (Al) or #14-2 AWG (Cu)
100	1	#12-1/0 AWG (Al) or #14-1/0 AWG (Cu)	#12-1/0 AWG (Al) or #14-1/0 AWG (Cu)
200	1	#6 AWG-250 MCM (Al/Cu)	#6 AWG-300 MCM (Al/Cu)
400	1 or 2	#3/0 AWG-750 MCM (Al/Cu) or #6 AWG-300 MCM (Al/Cu)	#3/0 AWG-750 MCM (Al/Cu) and #6 AWG-300 MCM (Al/Cu)
600	2	#3/0 AWG-500 MCM (Al/Cu)	#3/0 AWG-500 MCM (Al/Cu)
800	3	#3/0 AWG-750 MCM (Al/Cu)	#3/0 AWG-750 MCM (Al/Cu)
1200	4	#3/0 AWG-750 MCM (Al/Cu)	#3/0 AWG-750 MCM (Al/Cu)

Type 4, 4X, 5 Stainless, and Type 12 and 12K Heavy Duty Terminal Lug Data

Ampere Rating	Conductors Per Phase	Wire Range Wire Bending Space Per NEC Table 373-6	Lug Wire Range
30	1	#14-6 AWG (Cu)	#14-2 AWG (Cu)
60	1	#14-4 AWG (Cu)	#12-2 AWG (Cu)
100	1	#14-1 AWG (Cu)	#14-1 AWG (Cu)
200	1	#6 AWG-250 MCM(Cu)	#6 AWG-250 MCM(Cu)
400	1 or 2	#1/0 AWG-600 MCM(Cu) or #6 AWG-250 MCM(Cu)	#1/0 AWG-600 MCM(Cu) and #6 AWG-250 MCM(Cu)
600	2	#4 AWG-350 MCM(Cu)	#4 AWG-350 MCM(Cu)

Figure 20. Safety switch lug specifications.

6.1.4 Insulating Material

As the voltage rating of switches is increased, arc suppression becomes more difficult and the choice of insulation material becomes more critical. Arc suppressors are usually made of insulation material and magnetic suppressor plates when required. All arc suppressor materials must provide proper control and extinguishing of arcs.

6.1.5 Operating Mechanism and Cover Latching

Most heavy duty safety switches have a spring driven quick-make, quick-break mechanism. A quick-breaking action is necessary if the switch is to be safely switched OFF under a heavy load.

The spring action, in addition to making the operation quick-make, quick-break, firmly holds the switch blades in the ON or OFF position. The operating handle is an integral part of the switching mechanism and is in direct control of the switch blades under normal conditions.

A one-piece cross bar, connected to all switch blades, should be provided which adds to the overall stability and integrity of the switching assembly by promoting proper alignment and uniform switch blade operation.

Dual cover interlocks are standard on most heavy duty switches where the NEMA enclosure permits. However, NEMA Types 7 and 9 have bolted covers and obviously cannot contain dual cover interlocks. The purpose of dual interlock is to prevent the enclosure door from being opened when the switch handle is in the ON position and prevents the switch from being turned ON while the door is open. A means of bypassing the interlock is provided to allow the switch to be inspected in the ON position by qualified personnel. However, this practice should be avoided if at all possible. Heavy duty switches can be padlocked in the OFF position with up to three padlocks.

6.1.6 Enclosures

Heavy duty switches are available in a variety of enclosures which have been designed to conform to specific industry requirements based upon the intended use. Sheet metal enclosures (that is, NEMA Type 1) are constructed from cold-rolled steel which is usually phosphatized and finished with an electrode deposited enamel paint. The Type 3R rainproof and Type 12 and 12K dusttight enclosures are manufactured from galvannealed sheet steel and painted to provide better weather protection. The Type 4, 4X and 5 enclosures are made of corrosion resistant Type 304 stainless steel and requires no painting. Type 7 and 9 enclosures are cast from copper-free aluminum and finished with an enamel paint. Type 1 switches are general purpose and designed for use indoors to protect the enclosed equipment from falling dirt and personnel from live parts. Switches rated through 200 amperes are provided with ample knockouts. 400 through 1200 ampere switches are provided without knockouts.

The following are the NEMA enclosure Types that will be encountered most often. Always make certain that the proper enclosure is chosen for the application.

Type 3R switches are designated "rainproof" and are designed for use outdoors. See Fig. 21.

Type 3R enclosures for switches rated through 200 amperes have provisions for interchangeable bolt-on hubs at the top endwall. Type 3R switches rated higher than 200 amperes have blank top endwalls. Knockouts are provided (below live parts only) on enclosures for 200 ampere and smaller Type 3R switches. Type 3R switches are available in ratings through 1200 amperes.

Figure 21. NEMA Type 3R enclosure.

Type 4, 4X, 5 stainless steel switches are designated dusttight, watertight and corrosion resistant and designed for indoor and outdoor use. Common applications include commercial type kitchens, dairies, canneries, and other types of food processing facilities, as well as areas where mildly corrosive liquids are present. All Type 4, 4X, and 5 stainless steel enclosures are provided without knockouts. Use of watertight hubs is required. Available switch ratings are 30 through 600 amperes. See Fig. 22.

Type 12 and Type 12K switches are designated dusttight (except at knockout locations on Type 12K) and are designed for indoor use. In addition, NEMA Type 12 safety switches are designated as raintight for outdoor use when the supplied drain plug is removed. Common applications include heavy industries where the switch must be protected from such materials as dust, lint, flyings, oil seepage, etc. Type 12K switches have knockouts in the bottom and top endwalls only. Available switch ratings are 30 through 600 amperes in Type 12 and 30 through 200 amperes in Type 12K.

NEMA Type 4, 4X, 5, Type 12 and 12K switch enclosures have positive sealing to provide a dusttight and raintight (watertight with stainless steel) seal. Enclosure doors are supplied with oil resistant gaskets. Switches rated 30 through 200 amperes incorporate spring loaded, quick-release latches. 400 and 600 ampere switches feature single-stroke sealing by operation of a cover mounted handle. 30, 60, and 100 ampere switches in these enclosures are provided with factory installed fuse pullers.

Figure 22. Stainless enclosure.

6.1.7 Interlocked Receptacles

Heavy-duty, 60 ampere Type 1 and Type 12 switches within interlocked receptacle are also available. This receptacle provides a means for connecting and disconnecting loads directly to the switch. A non-defeating interlock prevents the insertion or removal of the receptacle plug while the switch is in the ON position. It also prevents operation of the switch if an incorrect plug is used. See Fig. 23.

6.1.8 Accessories

Accessories available for field installation include Class R fuse kits, fuse pullers, insulated neutrals with grounding provisions, equipment grounding kits, watertight hubs for use with Type 4, 4X, 5 stainless or Type 12 switches, and interchangeable bolt-on hubs for Type 3R switches.

Figure 23. Switch with receptacle.

Electrical interlock consists of auxiliary contacts for use where control or monitoring circuits need to be switched in conjunction with the safety switch operation. Kits can be either factory or field installed, and they contain either one normally open and one normally closed contact or two normally open and two normally closed contacts. The electrical interlock is actuated by a pivot arm which operates directly from the switch mechanism. The electrical interlock is designed so that its contacts disengage before the blades of the safety switch open and engage after the safety switch blades close.

6.2.0 GENERAL DUTY SWITCHES

General duty switches for residential and light commercial applications are used where operation and handling are moderate and where the available fault current is 10,000 RMS symmetrical amperes or less. Some general duty safety switches, however, exceed this specification in that they are UL listed for application on systems having up to 100,000 RMS symmetrical amperes of available fault current when Class R fuses and Class R fuse kits are used. Class T fusible switches are also available in 400, 600 and 800 ampere ratings. These switches accept 300 VAC Class T fuses only. Some examples of general duty switch application include residential, farm, and small business services entrances, and light duty branch circuit disconnects.

General duty switches are rated up to 600 amperes at 240 volts AC in general purpose (Type 1) and rainproof (Type 3R) enclosures. Some general-duty switches are horsepower rated and capable of opening a circuit up to six times the rated current of the switch; others are not. Always check the switch's specifications before using under a horsepower-rated condition.

6.2.1   Switch Blades and Jaws

All current carrying parts of general duty switches are plated to minimize oxidation and reduce heating. Switch jaws and blades are made of copper for high conductivity. Where required, a steel reinforcing spring increases the mechanical strength of the jaws and contact pressure between the blade and jaw. Good pressure contact maintains the blade-to-jaw resistance at a minimum, which in turn, promotes cool operation. All general duty switch blades feature visible blade construction. With the door open, there is visually no doubt when the switch is OFF.

6.2.2   Fuse Clips

Fuse clips are normally plated to control corrosion and keep heating to a minimum. Where required, steel reinforcing springs are provided to increase the mechanical strength of the fuse clip. The result is a firmer, cooler connection to the fuses as well as superior fuse retention.

6.2.3   Terminal Lugs

Most general duty safety switches are furnished with mechanical set screw lugs which are suitable for aluminum or copper conductors.

6.2.4   Insulating Material

Switch and fuse bases are made of a strong, non-combustible, moisture-resistant material which provides the required phase-to-phase and phase-to-ground insulation for applications on 240VAC systems.

6.2.5   Operating Mechanism and Cover Latching

Although not required by either the UL or NEMA standards, some general duty switches have spring-driven quick-make, quick-break operating mechanisms. Operating handles are an integral part of the operating mechanism and are not mounted on the enclosure cover. The handle provides indication of the status of the switch. When the handle is up, the switch is ON. When the handle is down, the switch is OFF. A padlocking bracket is provided which allows the switch handle to be locked in the OFF position. Another bracket is provided which allows the enclosure to be padlocked closed.

6.2.6   Enclosures

General duty safety switches are available in either Type 1 for general purpose, indoor applications, or Type 3R for rainproof, outdoor applications. See Section 6.1.5 for information on enclosure types.

6.3.0   DOUBLE-THROW SAFETY SWITCHES

Double-throw switches are used as manual transfer switches and are not intended for use as motor circuit switches; thus, horsepower ratings are generally unavailable.

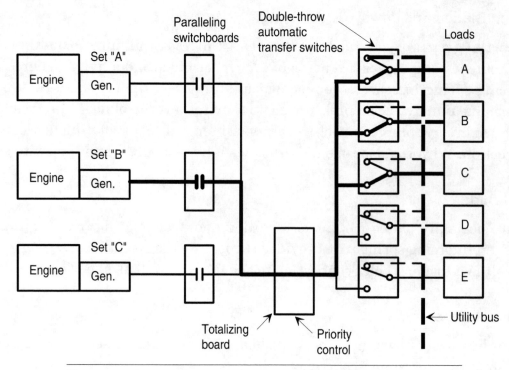

Figure 24. Practical application of a double-throw safety switch.

Double-throw switches are available as either fused or nonfusible devices and two general types of switch operation are available:

● Quick-make, quick-break
● Slow-make, slow-break

Figure 24 shows a practical application of a double-throw safety switch used as a transfer switch in conjunction with a stand-by emergency generator system.

6.4.0   NEC SAFETY-SWITCH REQUIREMENTS

Safety switches, in both fusible and nonfusible types, are used as a disconnecting means for services, feeders, and branch circuits. Installation requirements involving safety switches are found in several places throughout the NEC, but mainly in the following articles and sections:

● NEC Article 373
● NEC Article 380
● NEC Article 430-H
● NEC Article 440-B
● NEC Section 450-8(c)

When used as a service disconnecting means, the major installation requirements are listed in Fig. 25.

Application	Requirement	NEC Section
General	A means must be provided to disconnect all conductors serving the premises from service-entrance conductors. Each disconnecting means must be identified. Disconnecting means must be installed at a readily accessible location near point of entrance of service-entrance conductors. In multiple-occupancy buildings, access to the disconnecting means for each occupant must be provided.	230-70
Number of disconnects	Service disconnecting means can consist of no more than six switches for each service.	230-71
Working space	Requirements for any electrical equipment apply; that is, the minimum headroom of working spaces about service equipment must be 6.5 feet or more. The dimensions of working space in the direction of access to live parts operating at 600 volts, nominal, or less to ground and likely to require examination, adjustment, servicing, or maintenance while energized must not be less than indicated in NEC Table 110-16(a).	110-16
Type of disconnect	A manually- or power-operated safety switch meets NEC requirements.	230-76
Connections to terminals	Service conductors must be connected to the disconnecting means by pressure connectors, clamps, or other approved means. Soldered connections are forbidden. Each building must have an individual disconnecting means and each disconnecting means must be suitable for use as service equipment. A second service drop is permitted where the existing service capacity is over 2000 amperes.	230-81 and 230-2, Ex. 4
More than one building on same premises	In industrial establishments, the disconnecting means for several buildings may be conveniently located if conditions are met.	225-8(b), Ex. 1

Figure 25. NEC installation requirements governing switches used for service disconnects.

Summary

A *device*—by NEC definition—is a unit of an electrical system that is intended to carry, but not utilize, electric energy. This covers a wide assortment of system components that include, but are not limited to, the following:

- Switches
- Relays
- Contactors
- Receptacles
- Conductors

The purpose of a switch is to make and break an electrical circuit, safely and conveniently. NEC Article 380 covers most of the installation requirements for switches.

A receptacle is a contact device installed at the outlet for the connection of a single attachment plug. A single receptacle is a single contact device with no other contact device on the same yoke. A multiple receptacle is a single device containing two or more receptacles—the most common being the *duplex receptacle*. NEC sections dealing mainly with receptacles are as follows:

- Spacing: 210-52(a)
- Countertops: 210-52(c)
- Bathrooms: 210-52(d)
- Outdoors: 210-52(e)
- Basements: 310-52(g)
- Garages: 210-52(g)
- Hallways: 210-52(h)
- Guest rooms: 210-60
- Show windows: 210-62
- Small appliances: 220-4(b) and 210-52(b)
- Laundry: 220-4(c), 210-50(c), and 210-52(f)

References

For a more advanced study of topics covered in this Task Module, the following works are suggested:

American Electricians Handbook, Latest Edition, Croft, McGraw-Hill, New York, NY

National Electrical Code Handbook, Latest Edition, NFPA, Quincy, MA

ELECTRICAL TRAINEE TASK MODULE 20305

SELF-CHECK REVIEW/PRACTICE QUESTIONS

1. Which of the following statements is true concerning duplex receptacles?

   a. CU/AL markings indicate the use of both copper and aluminum wire
   b. CU/AL markings indicate the use of aluminum or copper-clad wire
   c. CO/ALR markings indicate the use of both copper and aluminum wire
   d. CO/ALR markings indicate the use of copper wire only

2. Which of the following slots in a duplex receptacle is for connection to a grounded conductor?

   a. The long slot
   b. The short slot
   c. The grounding slot
   d. The slot connected to the green screw

3. Which of the following wire sizes may usually be used in quick-connect wire clips in 15-ampere receptacles and switches?

   a. #12 AWG
   b. #6 AWG
   c. #4 AWG
   d. #2 AWG

4. What the is maximum distance that outlet boxes may be set back from the finished surface in concrete walls?

   a. $\frac{1}{16}''$
   b. $\frac{1}{8}''$
   c. $\frac{1}{4}''$
   d. $\frac{1}{2}''$

5. What the is maximum distance that outlet boxes may be set back from the finished surface in wood-paneled walls?

   a. $\frac{1}{16}''$
   b. $\frac{1}{8}''$
   c. $\frac{1}{4}''$
   d. $0''$

6. Which of the following best describes the intent of NEC Section 210-52?

    a. To prevent extension cords longer than 6 feet from being used
    b. To ensure amble amperage for cord-and-plug appliances
    c. To prevent the use of more appliances than the circuit can handle
    d. To enable extension cords longer than 6 feet to be used

7. Within what distance of a kitchen or bar sink must receptacles be protected with a GFCI?

    a. 2 feet
    b. 3 feet
    c. 4 feet
    d. 6 feet

8. The NEC requires at least one receptacles for a certain number of linear feet of show window in a store building. What is the number of linear feet?

    a. 3 feet
    b. 6 feet
    c. 12 feet
    d. 18 feet

9. Which of the following switches should be used to break all ungrounded conductors in a 240-volt, 2-wire, single-phase circuit?

    a. Single-pole, single-throw
    b. Double-pole, single-throw
    c. Triple-pole, single-throw
    d. Single-pole, double-throw

10. In which of the following sections in a set of electrical specifications would you most likely find information concerning wall switches for lighting control?

    a. Section 15408
    b. Section 16500
    c. Section 16700
    d. Section 16800

PERFORMANCE/LABORATORY EXERCISE

1.   Mount and connect a 440-volt receptacle as supplied by the instructor.

2.   Identify different types of receptacles as supplied by the instructor .

3.   Identify different types of switches as supplied by the instructor.

Answers to Self-Check Questions

1. c

2. a

3. a

4. c

5. d

6. a

7. d

8. c

9. b

10. b

Distribution Equipment

Module 20306

Electrical Trainee Task Module 20306

DISTRIBUTION EQUIPMENT

Objectives

Upon completion of this module, the trainee will be able to:

1. List the voltage convention classifications used in the industry.
2. Describe the purpose of switchgear.
3. Describe the basic physical makeup of a switchboard.
4. List the four general classifications of circuit breakers.
5. List the major circuit breaker ratings.
6. Describe switchgear construction.
7. Describe switchgear metering layouts.
8. Describe switchgear wiring requirements.
9. Describe switchgear maintenance.
10. List National Electrical Code (NEC) requirements pertaining to switchgear
11. Describe the visual and mechanical inspections and electrical tests associated with low voltage and medium voltage cables, metal-enclosed busways, and metering and instrumentation.
12. Describe a ground fault.
13. Describe a ground fault relay system.
14. Describe how to test a ground fault system.
15. Describe a Square D HVL switch.
16. Describe a bolted pressure switch.
17. Describe bolted pressure switch maintenance requirements.
18. Describe a typical switchgear transformer.
19. Describe the maintenance and tests associated with switchgear transformers.
20. List the safety precautions associated with instrument transformers.
21. Describe instrument transformer maintenance requirements.

Prerequisites

Successful completion of the following Task Modules is required before beginning study of this Task Module: Common Core Curricula, Electrical Levels 1 and 2, Electrical Level 3, Modules 20301 through 20305.

Required Student Material

1. Student Module
2. Appropriate Personal Protective Equipment
3. Copy of the latest edition of the National Electrical Code

Course Map Information

This course map shows all of the *Wheels of Learning* task modules in the third level of the Electrical curricula. The suggested training order begins at the bottom and proceeds up. Skill levels increase as a trainee advances on the course map. The training order may be adjusted by the local Training Program Sponsor.

Course Map: Electrical, Level 3

LEVEL 3 COMPLETE

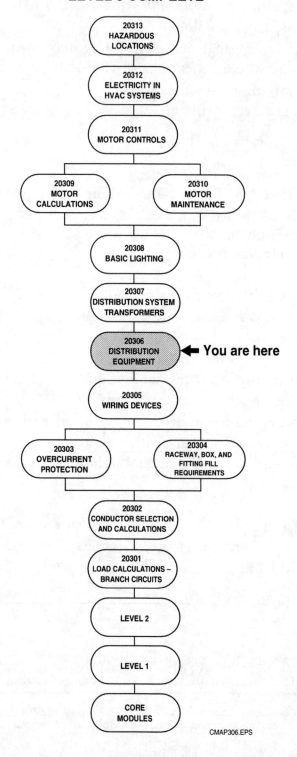

20313
HAZARDOUS
LOCATIONS

20312
ELECTRICITY IN
HVAC SYSTEMS

20311
MOTOR CONTROLS

20309
MOTOR
CALCULATIONS

20310
MOTOR
MAINTENANCE

20308
BASIC LIGHTING

20307
DISTRIBUTION SYSTEM
TRANSFORMERS

20306
DISTRIBUTION
EQUIPMENT ← **You are here**

20305
WIRING DEVICES

20303
OVERCURRENT
PROTECTION

20304
RACEWAY, BOX, AND
FITTING FILL
REQUIREMENTS

20302
CONDUCTOR SELECTION
AND CALCULATIONS

20301
LOAD CALCULATIONS –
BRANCH CIRCUITS

LEVEL 2

LEVEL 1

CORE
MODULES

CMAP306.EPS

TABLE OF CONTENTS

TABLE OF CONTENTS (Continued)

TABLE OF CONTENTS (Continued)

TABLE OF CONTENTS (Continued)

TABLE OF CONTENTS (Continued)

Trade Terms Introduced In This Module

Air circuit breaker: A circuit breaker in which the interruption occurs in air.

Ampacity: Current rating in amperes, as of a conductor.

Autotransformer: A transformer in which at least two windings have a common section.

AWG: Abbreviation for American Wire Gauge.

BIL: Abbreviation for basic impulse insulation levels, which are reference levels expressed in impulse-crest voltage.

Branch circuit: A set of conductors that extend beyond the last overcurrent device in the low-voltage system of a given building.

Bus: A conductor or group of conductors that serves as a common connection for two or more circuits in a switchgear assembly.

Bushing: An insulating structure including a through conductor, or providing a passageway for such a conductor, for the purpose of insulating the conductor from the barrier and conducting from one side of the barrier to the other.

Capacity: The rated load-carrying ability expressed in kilovolt-amperes or kilowatts of generating equipment or other electric apparatus.

Circuit breaker: A device that interrupts a circuit without injury to itself so that it can be reset and reused over again.

Contactor: An electric power switch, operated automatically, and designed for frequent operation.

Disconnecting switch: A mechanical switching device used for changing the connections in a circuit or for isolating a circuit or equipment for the source of power.

Distribution system equipment: Switchboard equipment that is downstream from the service entrance equipment.

Distribution transformer: A transformer for transferring electric energy from a primary distribution circuit to a secondary distribution circuit. Usually rated between 5 and 500kVA.

Effectively grounded: Grounded by means of a ground connection of sufficiently low impedance that fault grounds which may occur cannot build up voltages dangerous to connected equipment.

Feeder: A set of conductors originating at a main distribution center and supplying one or more secondary distribution centers, one or more branch circuit distribution centers, or any combination of these two types of load.

Fuse: An overcurrent protective device with a circuit-opening fusible part that is heated and severed by the passage of overcurrent through it.

Metal-enclosed switchgear: Switchgear that is primarily used in indoor applications and up to 600 volts.

Service entrance equipment: Equipment located at the service entrance of a given building that provides overcurrent protection to the feeder and service conductors and provides a means of disconnecting the feeders from energized service equipment.

Switchboard: A large, single panel, frame, or assembly of panels on which are mounted switches, fuses, buses, and usually instruments.

Switchgear: A general term covering switching or interrupting devices and any combination thereof, with associated control, instrumentation, metering, protective, and regulating devices.

1.0.0   INTRODUCTION

An electrical power system consists of several subsystems on both the utility (supply) side and the customer (user) side. Electricity generated in power plants is stepped up to transmission voltage and fed into a nationwide grid of transmission lines. This power is then bought, sold, and dispatched as needed. Local utility companies take power from the grid and reduce the voltage to levels suitable for subtransmission and distribution through various substations to the customer. This may range from the common 200-amp 240/120-volt residential service to hundreds of thousands of amps at voltages from 480 to 69kV in an industrial facility.

From the point of service, the customer must control, distribute, and manage the power to supply their electrical needs. This module will discuss how this is done using a typical industrial facility as an example. We will discuss the various components of the distribution system and their interdependence. An understanding of single line diagrams will allow analysis of a facilities distribution system.

Note   The voltage conventions used in this module are industry standards
              for distribution systems.

2.0.0   VOLTAGE CONVENTION

Electrical equipment is usually divided by voltage. The various voltage levels are classified as low, medium, high, extra high, and ultra high, and are discussed in the following paragraphs.

2.1.0   LOW VOLTAGE (LV)

Low voltage is considered to be 600V and below. It is typically used to supply nominal voltage directly to electrical loads. Common voltages in this category are 120V, 208V, 240V, and 480V.

2.2.0   MEDIUM VOLTAGE (MV)

Medium voltage is considered to range from 601V to 15kV. It is mainly used for distribution purposes and for supplying large electrical loads.

2.3.0   HIGH VOLTAGE (HV)

High voltage is considered to range from 15kV to 230kV. It is mainly used for transmission purposes.

2.4.0   EXTRA HIGH VOLTAGE (EHV) OR VERY HIGH VOLTAGE (VHV)

Extra high or very high voltage is considered to range from 230kV to 800kV. It is also used only for transmission purposes.

2.5.0   ULTRA HIGH VOLTAGE (UHV)

Ultra high voltage is considered to be voltage greater than 800kV. Presently, the common voltages in this category range between 1100kV and 1500kV.

3.0.0   SWITCHBOARDS

According to the National Electrical Code, the term **switchboard** may be defined as *a* large single panel, frame, or assembly of panels on which are mounted on the face or back or both, switches, overcurrent and other protective devices, buses, and usually instruments. Switchboards are generally accessible from the rear and from the front and are not intended to be installed in cubicles.

The design of switchboards makes it possible, convenient, and safe for electrical power delivery to the customer.

3.1.0   APPLICATIONS

Switchboards are used in the modern distribution system to subdivide large blocks of electrical power.

One location for switchboards is typically where the main power enters the building. In this location, the switchboard would be referred to as **service entrance equipment**. The other location common for switchboards is downstream from the service entrance equipment. In the downstream location, the switchboard is commonly referred to as **distribution system equipment**.

3.2.0   GENERAL DESCRIPTION

A switchboard consists of a stationary structure that includes one or more freestanding units of uniform height that are mechanically and electrically joined to make a single coordinated installation. These cubicles contain circuit interrupting devices. They take up less space in a plant, have more eye appeal, and eliminate the need to have a separate room to protect personnel from contact with lethal voltages.

The main portion of the switchboard is formed from heavy gauge steel welded with members across the top and bottom to provide a rigid enclosure. Most switchboard enclosures are divided into three sections: the front section, the **bus** section, and the cable section. These three sections are physically separated from one another by metal partitions. This confines any damage that may occur to any one section and keeps it from spreading to other sections.

Typical switchboard components would include:

- Circuit breakers
- Fuses
- Motor starters
- Ground fault system
- Instrument transformers
- Switchboard metering
- Control power transformers

Switchboards are used in modern distribution systems to subdivide large blocks of electrical power.

Electrical ratings include 3 Ø, three-wire and 3 Ø, four-wire systems with a large voltage rating up to 600 volts and current rating up to 4,000 amps.

The switchboard enclosure is described as a dead front panel. Inside, however, it contains energized breakers. Bus bars can be standard size or customized. Standard sizes include silver or tin-plated copper or tin-plated aluminum. Conventional bus sizing is .25" x 2" up through .375" x 7". Bussing also is fabricated in customized sizes. When using copper, it is good for 1,000 amps per each square inch of cross-sectional area. When using aluminum, the amperage value is 750.

When two bus bars are bolted together using grade S hardware with proper torque, the ampacity of the connection is 200 amps per square inch of the lapped portion for aluminum or copper bussing.

Bussing joints shall be bolted together to the specified torque and shall include belleville washers or keps nuts. Aluminum bus bars must be tin-plated and copper bus bars over 600 amps shall be plated with tin or silver.

3.3.0 SWITCHBOARD FRAME HEATING

The following should be observed in order to keep heat losses in the iron switchboard frame members to a safe minimum. The following dimensions are recommended values and should be adhered to whenever possible. There are some standards that deviate from these values, but these have been tested and cannot be applied to custom bussing.

Amperes	Min. Distance from Phase Bus to Closest Steel Member	Min. Distance from Neutral Bus to Closest Steel Member
3000	4"	2"
4000	6"	3"
5,000 and over	12"	see below
5,000 to 6,000	An aluminum or non-magnetic material should be used in place of steel frame sections. You must maintain, wherever possible, 12" to steel members, and 6" to aluminum or non-magnetic members. Neutral spacing can be 6" and 3", respectively. If the main bus is tapered, it is permissible (at 4,000A and below) to use steel frames for those sections containing the tapered bus.	
6,000 and over	You must use an aluminum or non-magnetic material for frame sections, and maintain 12" to steel members and 6" to aluminum or non-magnetic menbers. Neutral spacing can be 6" and 3" respectively. The use of any steel frame members is to be discouraged. If the main bus is tapered, it is permissible (at 4,000A and below) to use steel frames for those sections containing the tapered bus.	
NOTE: For amperages above 8,000A, the neutral spacing must be 12" wherever possible.		

3.4.0 LOW VOLTAGE SPACING REQUIREMENTS

To minimize tracking or arcing from energized parts to ground, switchboard construction includes spacing requirements. These spacing requirements are measured between live parts of opposite polarity and between live parts and grounded metal parts. *Figure 1* illustrates switchboard spacing requirements.

VOLTAGE INVOLVED		MINIMUM SPACING BETWEEN LIVE PARTS OF OPPOSITE POLARITY		MINIMUM SPACING THRU AIR AND OVER SURFACE BETWEEN LIVE PARTS AND GROUNDED METAL PARTS
GREATER THAN	MAX.	THRU AIR INCH	OVER SURFACE INCH	INCH
0 --	125	1/2"	3/4"	1/2"
125 --	250	3/4"	1-1/4"	1/2"
250 --	600	1"	2"	1" °

° A THROUGH AIR SPACING OF NOT LESS THAN 1/2 INCH IS ACCEPTABLE (1) AT A MOLDED-CASE CIRCUIT BREAKER OR A SWITCH, OTHER THAN A SNAP SWITCH, (2) BETWEEN UNINSULATED LIVE PARTS OF A METER MOUNTING OR GROUNDED DEAD METAL, AND (3) BETWEEN GROUNDED DEAD METAL AND THE NEUTRAL OF A 480Y/277-V, 3-PHASE, 4-WIRE SWITCHGEAR SECTION.

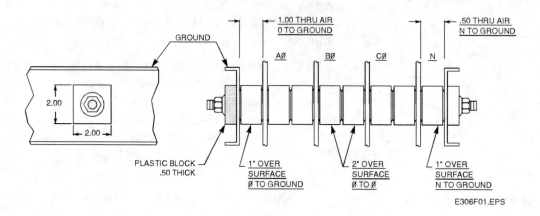

Figure 1.   Spacing Requirements

An isolated dead metal part, such as a screwhead or washer, interposed between uninsulated live parts of opposite polarity, or between an uninsulated live part and grounded dead metal, is to be considered to reduce the spacing by an amount equal to the dimension of the interposed part along the path of measurement.

When measuring over-surface spacing, any slot, groove, and the like that is 0.013 inch (0.33 mm) wide or less and in the contour of the insulating material is to be disregarded.

In measuring spacing, an air space of 0.013 inch or less between a live part and an insulating surface is to be disregarded, and the live part is to be considered in contact with the insulating material. A pressure wire connector shall be prevented from turning that would result in less than the minimum acceptable spacings. The means for turn prevention shall be reliable, such as a shoulder or boss. A lock-washer alone is not acceptable.

Exception: Means to prevent turning need not be provided if spacings are not less than the minimum accepted values:

a. When the connector, and any connector of opposite polarity, have each been turned 30 degrees toward the other.
b. When the connector has been turned 30 degrees toward other opposite-polarity live parts and toward grounded dead metal parts.

3.5.0   CABLE BRACING

All constructions employing conductors and having a short circuit current rating greater than 50,000 rms symmetrical amperes require a cable brace positioned as close to supply lugs as possible. The cable brace is intended to be mounted in the same area that is allotted for wire bending. It is not necessary to provide additional mounting height to accommodate the cable brace.

The cable brace requirement does not apply to load-side cables, main breakers, or switches. It only applies when cables are connected directly to an unprotected line side bus.

Line side bus – bus restrictions:

a. There can be no splice in edgewise mounting of 2,100 amps or less rated at 50,000 rms symmetrical amperes.
b. There can be no splice in flatwise mounting of 600 amps or less rated over 50,000 rms symmetrical amperes.

Line side bus – cable restrictions:

a. Bussing of 600 amps or less rated over 50,000 rms symmetrical amperes cannot use cables. It must be bus-connected.
b. If cabling is required, 800 amp minimum bussing must be used.

The no splice restriction does not apply to connections made from the through bus to a switch or **circuit breaker**.

Cable bracing requirements may be excluded if the bussing is able to fully withstand the total available short circuit current.

4.0.0   SWITCHGEAR

The term **switchgear** is a general term used to describe switching and interrupting devices, and assemblies of those devices with control, metering, protective, and regulatory equipment with the associated interconnections and supporting structures.

Switchgear performs two basic functions:

- Provides the means for switching or disconnecting power system apparatus.
- Provides power system protection by automatically isolating faulty components.

Switchgear can be classified as follows:

- Metal enclosed (low voltage)
- Metal clad (low and medium voltage)
- Metal enclosed interrupter
- Unit substation

The low and medium voltage switchgear assemblies are completely enclosed on all sides and top with sheet metal, except for ventilating openings and inspection windows. They contain primary power circuit switching or interrupting devices, buses, connections, and control and auxiliary devices. *Figure 2* shows a typical low voltage metal-clad switchgear.

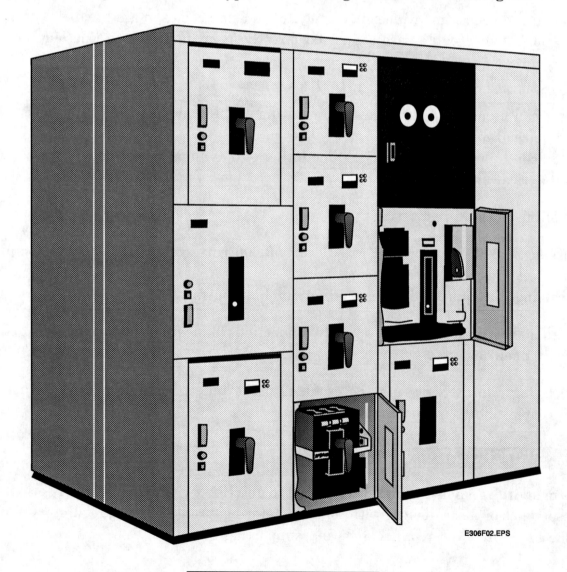

E306F02.EPS

Figure 2. Typical Switchgear

ELECTRICAL TRAINEE TASK MODULE 20306

The station-type cubicle switchgear consists of indoor and outdoor types with power circuit breakers rated from: 14.4 to 34.5kV, 1,200 to 5,000A, and 1,500 to 2,500mVA interrupting capacity.

4.1.0 DESCRIPTION OF SWITCHGEAR

Switchgear consists of a stationary structure that includes one or more free-standing units of uniform height that are mechanically and electrically joined to make a single coordinated installation. These units, commonly referred to as *cubicles*, contain circuit interrupting devices such as circuit breakers. They take up less space in a plant or installation, have more eye appeal, and eliminate the need to have a separate room to protect personnel from contact with lethal voltages.

The main portion of switchgear is formed from heavy gauge sheet steel welded or bolted together. Structural members across the top, sides, and bottom provide a rigid enclosure. Most switchgear enclosures are divided into three sections: the front section, the bus section, and the cable or termination section.

These three sections are physically separated from one another by metal partitions. This confines any damage that may occur to any one section and keeps it from spreading to other sections. It also separates power between the sections for ease and safety of maintenance.

The rigid enclosure provides primary structural strength of the switchgear assembly and the means by which the switchgear is fastened to its foundation. The strength of the enclosure and its mounting system will vary depending on its intended use. For example, switchgear used in a nuclear application must meet certain seismic qualifications.

The enclosure also provides the required supports and mounts for items to be located in the switchgear and provides for the necessary interconnections between the switchgear and other plant systems.

The number of sections and physical makeup of switchgear will vary depending on the voltage and current ratings, plant specifications, and the specific manufacturer.

4.2.0 SWITCHGEAR CONSTRUCTION

The Square D 5-15kV two-high, metal-clad switchgear is designed for use on electrical distribution systems rated from 2,400 volts to 13,800 volts nominal, see *Figures 3, 4* and *5*.

Ratings include:

- 2.4 to 13.8kV
- 1,200 to 3,000 amps (3,500A with cooling fans)
- 250 to 750mVA interrupting capacity
- 60 and 95kV BIL

- Stored energy operating mechanism
- Drawout construction
- Insulated bussing
- Indoor NEMA Size 1

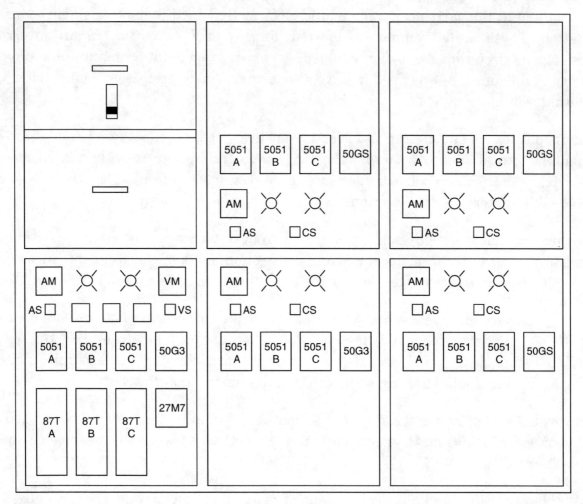

E306F03.EPS

Figure 3. Square D Two-High Metal-Clad Switchgear

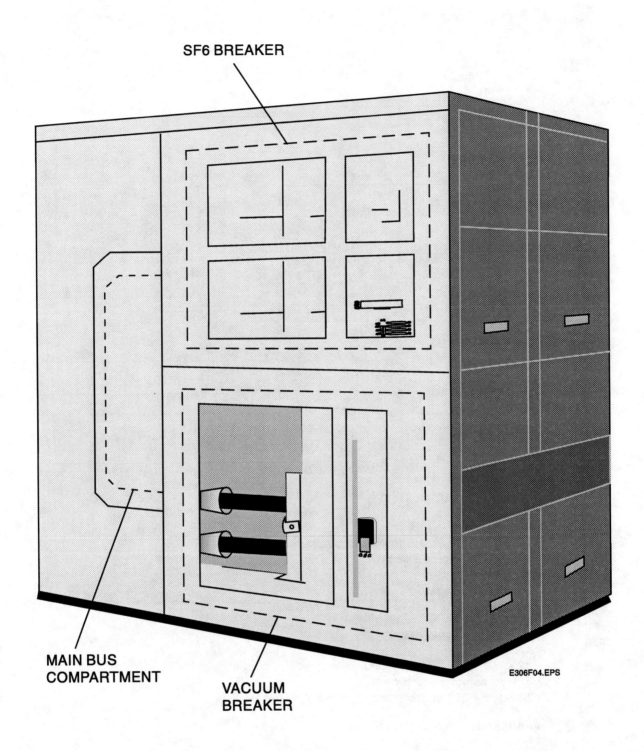

SF6 BREAKER

MAIN BUS COMPARTMENT

VACUUM BREAKER

E306F04.EPS

Figure 4.   Square D Two-High Metal-Clad Switchgear (Side View)

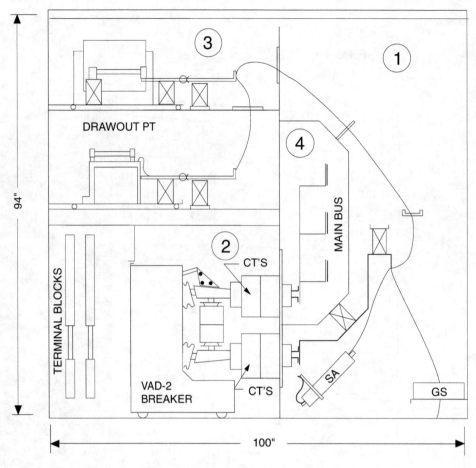

AVAILABLE OPTIONS

① **CABLE COMPARTMENT**

- Bottom or top cable entry
- Ground sensor current transformer (zero sequence)
- Space for stress cone termination
- 1/C or 3/C pothead
- Surge arresters, if required
- Fixed mounted CPT, with no cable entry at the bottom

② **BREAKER COMPARTMENT**

- 1200/2000 SF6 or Vacuum circuit breaker in two high construction
- Side space available for terminal block connections
- Maximum 4 CT's per phase (two on either side of breaker pole)

③ **AUXILIARY COMPARTMENT**

- Fused drawout PT's or single phase CPT (15 kVA maximum)
- Drawout fuses

④ **MAIN BUS COMPARTMENT**

- 1200/2000 ampere insulated aluminum (copper optional) main bus

---

Figure 5.   Square D Two-High Metal-Clad Switchgear (Internal View)

4.3.0 PRINCIPAL COMPONENTS

4.3.1 Hinged Front Door

Relays, instruments, and meters are mounted on the door space in a standardized arrangement.

4.3.2 Horizontal Drawout Circuit Breaker

Circuit breakers are horizontal drawout design. Disconnect, test, and connect positions are provided with the door in the closed position.

4.3.3 Main Bus Barriers

Main bus barriers (not shown) between bays are track-resistant, flame-retardant, glass polyester, with porcelain inserts at 8.25kV and 15kV ratings.

4.3.4 Current Transformers

Space is available for four front accessible bushing-type current transformers per phase. Two of these current transformers may be placed on the line side and two on the load side of the breaker on each phase.

4.3.5 Cable Space

Top or bottom cable entry with adequate space for 4,750 MCM cables per phase, potheads, cable supports, and surge arresters is provided.

4.3.6 Automatic Shutters

When the circuit breaker is withdrawn from the connected position, the breaker forces the steel shutter to rotate automatically into a position which covers the energized components.

4.3.7 Main Bus And Supports

An insulated bus is provided with fluidized bed epoxy insulation, and with a special design combination of porcelain and main bus supports. Porcelain or glass polyester bus supports and insulators as specified are provided in the cable and auxiliary compartments.

4.3.8 Insulating Tubes

Porcelain insulating tubes are used to insulate the primary stationary contacts and breaker runbacks at both 5kV and 15kV ratings.

4.3.9 Compartment Barriers

All main compartments are separated by grounded metal barriers.

4.3.10 Racking Mechanism

The gear-driven racking system incorporates safety interlocks.

4.3.11 Frame And Housing

Steel frames provide a strong rigid structure. The structure is engineered for flexibility to allow modifications and future addition of equipment.

4.3.12 Voltage Transformers

Front accessible drawer-mounted voltage transformers can be completely withdrawn through the use of rollers and drawer-mounted cantilever rails. For safety, the voltage transformers are grounded when moved to the withdrawn position.

4.3.13 Control Power Transformers

Control power transformers rated up to 15kVA are drawer-mounted and can be completely withdrawn from the front for ease of accessibility.

4.3.14 Circuit Breakers

FG-2 and VAD-2 circuit breakers using SF_6 and vacuum interrupters offer application flexibility.

The breakers are designed, manufactured, and tested per ANSI standards, and are of compact size and weight (as shown).

4.3.15 Breaker Lift Truck

A portable lifter device is required to install the breaker in the upper tier breaker compartment. This device uses a height adjustable platform, and is latched to the cell for safety. All breakers located in the bottom compartments can be rolled directly on the floor, and the use of a breaker lifting truck is not required.

4.3.16 Breaker Compartment

The stationary primary disconnect contacts are automatically covered by bright orange steel shutters when the breaker is withdrawn. Accidental contact with primary voltage live parts is prevented.

The shutters should not be forced into the open position or removed for inspection or maintenance when the unit is energized.

4.3.17 Doors

Relays, instruments, indicating lights, control switches, and other control devices may be mounted on the front door of the breaker and auxiliary components.

The racking mechanism is an internal part of the breaker. It consists of a self-engaging, self-aligning screw device that becomes spin-free in the connected position at the end of the breaker travel. This prevents possible damage due to overtightening.

Safety interlocks, as required by ANSI, are provided to prevent electrical or mechanical operation of the breaker during the racking procedure.

4.4.0 CONTROL AND METERING WIRE STANDARDS

Switchboard control and meter wiring standards shall meet the requirements of *Machine Tool Wires and Cables Safety Standard UL 1063*, *Switchboard Safety Standard UL 891*, and *Service Equipment Safety Standard UL 869*, with the following highlights:

1. Stranded copper conductor with thermoplastic insulation, UL and CSA approved.
2. Insulation thickness for #16 - #10 AWG size conductor is (2/64) .030 inch thick.
3. 600 volts, 90°C minimum rating for dry locations, 60°C minimum rating when exposed to oil and moisture.
4. Conductors no smaller than #14 gauge for control wiring.

4.5.0 WIRING SYSTEM

The NEC requires wiring to be supported mechanically to keep the wiring in place. Wire harnessing generally is used within the switchboard with the following restrictions:

1. Each bundle or cable of wires to run in a vertical or horizontal direction, securing the harness by means of plastic cable ties or cable clips.
2. Plastic wire cable clamps shall be placed at strategic locations along the harnessing to hold the harness firmly in place to prevent interference with control components' required electrical, mechanical, and arcing clearances.
3. Apply wire ties to the harnessed wiring every 3 to 4 inches with self-adhesive cable ties at 12-inch spacing.
4. Precautions to be observed when wiring the switchboard electrical components:
   a. Keep control wires at least 1/2-inch from moving parts.
   b. Avoid running wires across sharp metal edges. To protect the wiring from mechanical damage, use the following approved cable protectors:

- Nylon clip cable guard
- Wire guard for edge protection
- Special edge protection molding

c. Wires shall not touch exposed bare electrical parts of opposite polarity.

d. Wires shall not interfere with adjustment or replacement of components.

e. Wires should be as straight and as short as possible.

f. Wires shall not be spliced.

g. To eliminate possible strain on the control wire, a certain amount of slack should be given to the individual or harnessed conductor terminated at a component connection.

h. Equipment ground bus bar shall not be used as a portion of the control or metering circuits.

i. Do not use pliers for bending control wiring. Use your hands or an approved wire bending device.

4.5.1 Door-Mounted Wiring Restrictions

No incoming wiring connections shall be made directly to the door-mounted devices. Wires from the door-mounted equipment to the panel terminal block should be a minimum of 19 strand wire.

Wires from the door shall be neatly cabled so that the door can be opened easily without placing excessive strain on the wire terminal connections. In some cases, the cable must be separated into two bundles to accomplish this. Insulated sleeving, tubing, or vinyl tape shall be used to bundle and protect the flexible wires.

4.5.2 Terminal Connections

All control or metering wiring entering or leaving the switchboard should terminate at terminal blocks, leaving one side of the terminal block free for user's connections. No factory connections are allowed on the user's terminal connection point. For factory wiring, allow a maximum of two control wires on the same side of a terminal block. No more than three connections are allowed on terminals of control transformers, meters, meter selector switches, and metering equipment.

Since bolted pressure switches or any 100% current rated molded case circuit breaker's line and load power terminals are allowed a higher maximum operating temperature than the recommended insulated conductor's operating temperature, the control wires cannot be placed directly on the 100% rated disconnect device's line and load connections.

In all cases, control wires cannot touch any exposed part of opposite electrical polarity.

4.6.0 METERING CURRENT AND POTENTIAL TRANSFORMERS

Ground connections on potential or current transformer's secondary terminal must be connected to the ground bus. Current transformer (CT) secondary terminals must be shorted, if no metering equipment is connected to the current transformer.

Potential transformers (PTs) are required to have primary and secondary fusing. If protective circuits such as ground fault or phase failure protective systems are placed in the secondary circuit of the potential transformer, no secondary fusing is required.

Metering circuit connections made directly to the incoming bus are to be provided with current limiting fuses, equal in rating to the available interrupting capacity.

Note   CTs and PTs will be discussed later.

4.7.0 FUSES

Fuses are to be easily accessible, mounted in a vertical position, and the incoming wires connected to the top terminals of the fuse blocks.

4.8.0 SWITCHGEAR HANDLING, STORAGE AND INSTALLATION

The following is a basic guide for the handling of switchgear. It is important to emphasize when working with this type of equipment that these guides supplement the manufacturer's instruction. Manufacturers include instruction books and drawings with their equipment. It is necessary for personnel to review these documents before handling equipment.

4.8.1 Switchgear Handling

Immediately upon receipt of switchgear, an inspection for damage during transit should be performed. If damage is noted, the transportation company should be notified immediately.

4.8.2 Switchgear Rigging

Instructions for switchgear should be found in the manufacturer's instruction books and drawings. Verify that the rigging is suitable for the size and weight of the equipment.

4.8.3 Switchgear Storage

Indoor switchgear that is not being installed should be stored in a clean, dry location. The equipment should be level and protected from the environment if construction is proceeding. The longer equipment is in storage, the more care is required for protection of the equipment. If a temporary cover is used to protect the equipment, this cover should not prevent air circulation. If the building is not heated or temperature controlled, heaters should be used to prevent moisture/condensation buildup.

Outdoor switchgear that cannot be installed immediately must be provided with temporary power. This power will allow operation of the space heaters provided with the equipment.

4.8.4 Bus Connections

The main bus that is usually removed during shipping should be reconnected. Ensure the contact surfaces are clean and that pressure is applied in the correct manner. The conductivity of the joints is dependent on the applied pressure at the contact points. The manufacturer's torque instructions should be referenced.

4.8.5 Cable Connections

When making cable connections, verify the phasing of each cable. This procedure is done in accordance with the connection diagrams and the cables tagged. When forming and mounting cables, ensure the cables are tightened per the manufacturer's instructions.

4.8.6 Grounding

Sections of ground bus previously disconnected at shipping sections should be reconnected when the units are installed. All secondary wiring should be connected to the switchgear ground bus. The ground bus should be connected to the system ground with as direct a connection as possible. If it is to be run in metal conduit, adequate bonding to the circuit is required. The ground connection is necessary for all switchgear and should be of sufficient **ampacity** to handle any abnormal condition.

5.0.0 SWITCHBOARD MAINTENANCE

To be able to work on switchboards or any piece of electrical equipment, you must always follow safety procedures. If you are not sure if the procedures you are following are safe, contact your supervisor. It is better to check the procedure than to go ahead and follow unsafe work practices.

Always follow procedures established for grounding and locking out a circuit, and to ensure you have control over the switchboard voltage source(s).

5.1.0 GENERAL GUIDELINES

5.1.1 Visual

1. Check exterior for proper fit of doors and covers, paint, etc.
2. Check interior, particularly current-carrying parts.
   a. Busbars dirty, corroded, or overheated.
   b. Infrared or thermographic test: If required/needed, note discoloration. This is a poor bus joint.

c. Check busbar supports for cracks.
d. Check electrical spacing.
e. Check torque of all joints.

5.1.2 Cleaning

1. Vacuum interior (do not use compressed air).
2. Wipe down with rags. Use nonconductive, non-residue solution such as contact cleaner or denatured alcohol.

5.1.3 Operation

1. Manually open and close circuit breakers and switches.
2. Electrically operate all components such as ground fault, sure trip metering, current transformers, test blocks, ground lights, blown main fuse detector, and phase failure.

5.1.4 Megger Test

1. Isolate bus by opening all circuit breakers and switches.
2. Disconnect any devices such as relays and transformers that may be connected to the busbars.
3. Make sure all personnel are clear of switchboard.
4. Use 1,000-volt megger to check resistance phase-to-phase and phase-to-ground.
5. Megger readings should be 10 megohms minimum.
6. If low readings are found and moisture is thought to be a problem, add heat to switchboard (200 watts per section). Retest in 8 hours. Readings should improve substantially.

5.2.0 SPECIFIC GUIDELINES

5.2.1 Thermographic Survey

A thermographic survey involves checking switches, busway, open buses, switchgear, cables and bus connections, circuit breakers, rotating equipment, and load tap changer.

Infrared surveys should be performed during periods of maximum possible loading but not less than 40% of the rated load of the electrical equipment being inspected.

Test Results:

1. Temperature gradients of 1°C to 3°C indicate possible deficiency and warrant investigation.
2. Temperature gradients of 4°C to 15°C indicate deficiency. Repair as time permits.
3. Temperature gradients of 16°C and above indicate major deficiency. Secure power and repair as soon as possible.

DISTRIBUTION EQUIPMENT

5.2.2 Metal-Enclosed Switchgear And Switchboard

Visual and Mechanical Inspection:

1. Inspect for physical, electrical, and mechanical condition.
2. Compare equipment nameplate information with latest single-line diagram and report discrepancies.
3. Check for proper anchorage, required area clearances, physical damage, and proper alignment.
4. Inspect all doors, panels, and sections for missing paint, dents, scratches, fit, and missing hardware.
5. Inspect all bus connections for high resistance. Use low resistance ohmmeter, or check tightness of bolted bus joints by calibrated torque wrench method.
6. Test all electrical and mechanical interlock systems for proper operation and sequencing.
   - Closure attempt shall be made on locked open devices. Opening attempt shall be made on locked closed devices.
   - Key exchange shall be made with devices operated in off-normal positions.
7. Clean entire switchgear using manufacturer's approved methods and materials.
8. Inspect insulators for evidence of physical damage or contaminated surfaces.
9. Lubrication:
   - Verify appropriate contact lubricant on moving current-carrying parts.
   - Verify appropriate lubrication of moving and sliding surfaces.
   - Exercise all active components.
   - Inspect all indicating devices for proper operation.

Electrical Tests:

1. Perform ratio and polarity tests on all current and voltage transformers.
2. Perform ground resistance tests.
3. Perform insulation resistance tests on each bus section, phase-to-phase and phase-to-ground for one minute. Refer to *Table 1*.

MAXIMUM VOLTAGE RATING OF EQUIPMENT (VOLTS)	MINIMUM TEST VOLTAGE (VDC)	RECOMMENDED MINIMUM INSULATION RESISTANCE (OHMS)
250	500	25
600	1,000	100
5,000	2,500	1,000
8,000	2,500	2,000
15,000	2,500	5,000
25,000	5,000	20,000
35,000	15,000	100,000+
46,000	15,000	100,000+
69,000	15,000	100,000+

Table 1. Insulation Resistance Tests On Electrical Apparatus And Systems

Nominal System (Line) Voltage* (kV)	Insulation Class	Minimum AC Factory Test (kV)	Minimum Field Applied AC Test (kV)	Field Applied DC Test (kV)
1.2	1.2	10	6	8.5
2.4	2.5	15	9	12.7
4.8	5	19	11.4	16.1
8.3	8.7	26	15.6	22.1
14.4	15	34	20.4	28.8
18	18	40	24	33.9
25	25	50	30	42.4
34.5	35	70	42	59.4
46	46	95	57	80.6
69	69	140	84	118.8

* Intermediate voltage ratings are placed in the next higher insulation class.

Table 2. Overpotential Test Voltage For Electrical Apparatus Other Than Inductive Equipment

4. Perform an overpotential test on each bus section (phase-to-ground) for five minutes. Refer to *Table 2*.
5. Perform insulation resistance test on control wiring. (Do not perform this test on wiring connected to solid-state components.)
6. Calibrate all meters at mid-scale. Calibrate watt-hour meters to one-half percent (0.5%). Verify multipliers.
7. Perform phasing check on double-ended switchgear to ensure proper bus phasing from each source.

Test Values:

Insulation resistance test to be performed in accordance with *Table 1*. Values of insulation resistance less than this table or manufacturer's minimum should be investigated. Overpotential tests should not proceed until insulation resistance levels are raised above minimum values.

Overpotential test voltages shall be applied in accordance with *Table 2*.

• Test results are evaluated on a go/no-go basis by slowly raising the test voltage to the required value. The final test voltage shall be applied for five minutes for DC test potentials and one minute for AC test potentials.

5.2.3   Cables – Low Voltage – 600V Maximum

Visual and Mechanical Inspection:

1. Inspect cables for physical damage and proper connection in accordance with single-line diagram.
2. Test cable mechanical connections to manufacturer's recommended values with a calibrated torque wrench.
3. Check cable color coding with applicable engineer's specifications and NEC standards.

Electrical Test:

1. Perform insulation resistance test on each conductor with respect to ground and adjacent conductors. Applied potential to be 1,000 VDC for one minute.
2. Perform continuity test to ensure proper cable connection.

Test Values:

Minimum insulation resistance values shall not be less than 2 megohms.

5.2.4   Cables – Medium Voltage – 15kV Maximum

Visual and Mechanical Inspection:

1. Inspect exposed sections for physical damage.
2. Inspect for shield grounding, cable support, and termination.
3. Inspect for proper fireproofing in common cable areas.
4. If cables are terminated through window-type CTs, make an inspection to verify that neutrals and grounds are properly terminated for normal operation of protective devices.
5. Visually inspect jacket and insulation condition.
6. Inspect for proper phase identification and arrangement.

5.2.5   Metal Enclosed Busways

Visual and Mechanical Inspection:

1. Inspect bus for physical damage.
2. Inspect for proper bracing, suspension, alignment, and enclosure.
3. Check tightness of bolted joints using the calibrated torque wrench method.
4. Check for proper physical orientation per manufacturer's labels to ensure proper cooling. Perform continuity tests on each conductor to verify that proper phase relationships exist.
5. Check outdoor busway for removal of *weep-hole* plugs, if applicable, and the proper installation of joint shield.

Electrical Tests:

1. Perform an insulation resistance test. Measure insulation resistance on each bus run (phase-to-phase and phase-to-ground) for one minute.
2. Perform AC or DC overpotential tests on each bus run, phase-to-phase and phase-to-ground.
3. Perform contact resistance test on each connection point of uninsulated bus. On insulated bus, measure resistance of bus section and compare values with adjacent phases.
4. Insulation resistance test voltages and resistance values to be in accordance with manufacturer's specifications or *Table 1*.
5. Apply overpotential test voltages in accordance with *Table 2*.

5.2.6 Metering And Instrumentation

Visual and Mechanical Inspection:

1. Examine all devices for broken parts, indication of shipping damage, and wire connection tightness.
2. Verify that meter connections are in accordance with appropriate diagrams.

Electrical Test:

1. Check calibration of meters at all cardinal points.
2. Calibrate watt-hour meters to one-half of one percent (0.5%).
3. Verify all instrument multipliers.

6.0.0 NATIONAL ELECTRICAL CODE

This section is designed to provide a brief description of the NEC articles that are applicable to switchboards, their installation, and accessories.

6.1.0 REQUIREMENTS FOR ELECTRICAL INSTALLATIONS

6.1.1 Interrupting Rating (Article 110-9)

Interrupting rating is the maximum current a device can interrupt without damage to itself. The NEC defines equipment interrupting rating as *sufficient to interrupt the available fault current at the line side terminals of the equipment.*

6.1.2 Deteriorating Agents (Article 110-11)

Provides for protection of equipment and conductors from environments which could cause deterioration (such as gases, vapors, liquids, or moisture) unless specifically designed for such environments.

6.1.3 Mechanical Execution Of Work (Article 110-12)

Electrical equipment is to be installed in a neat and professional manner. Any openings provided by the equipment manufacturer or at the time of installation that are not being used will be sealed equivalent to the structure wall. This section also forbids the use of electrical equipment with damaged parts that may affect the safe operation or mechanical strength of the equipment.

6.1.4 Mounting And Cooling (Article 110-13)

Electrical equipment shall be securely fastened to its mounting surface by mechanical fasteners excluding wooden plugs driven into concrete or masonry. Equipment shall be located so as not to restrict airflow required for convection or forced-air cooling.

6.1.5 Electrical Connections (Article 110-14)

Due to the resistive oxidation created when dissimilar metals are connected, splicing devices and pressure connectors will be identified for the conductor material they are intended for use with. Dissimilar metal conductors will not be mixed in terminations. Fluxes, solders, and anti-oxidation compounds will be suitable for use and shall not adversely affect conductors. Terminals for use with more than one conductor or aluminum will be identified as such.

6.1.6 Working Space (Article 110-16)

Suitable access and working space shall be maintained around electrical equipment to permit safe operation and maintenance. A minimum clearance of 3 feet is required in front of all electrical enclosures; in all cases, space will be adequate to allow doors or hinged parts to open to a 90-degree angle. In special case installations, clearances in Table 110-16(a) will be adhered to. Storage of any kind is not permitted within the clearance area. At least one entrance of ample size will be provided to access the work area. In cases of services over 1,200 amperes and over 6 feet wide, two entrances are required. The work space shall be adequately illuminated.

6.1.7 Markings (Article 110-21)

Manufacturer's trademark or logo, as well as system ratings, including voltage, current, wattage, etc., shall be permanently attached to equipment.

6.1.8 Disconnect Identification (Article 110-22)

Each disconnecting means (circuit breaker, fused switch, **feeder**, unfused disconnects) shall be clearly marked as to its purpose at its point of origin unless located in such a manner that its purpose is evident.

6.2.0 REQUIREMENTS FOR CONDUCTORS

6.2.1 Neutrals (Article 200-6)

Grounded conductors (neutrals) size #6 and smaller are color coded with a solid white or gray marking for the entire length of the conductor. Conductors size #6 and larger are color coded with a solid white or gray marking tape at termination points at the time of installation. Where different electrical systems are run together, each system's grounded conductor must be distinctively identified.

6.2.2 Protection (Article 240-3)

Branch circuit conductors shall be protected by overcurrent devices as specified in Article 240-3.

6.2.3 Loading (Article 220-3)

Protective device calculations for continuous duty circuits are calculated at 125% of continuous load. This equates to an 80% loading factor on the branch circuit.

6.2.4 Tap Rules (Article 240-21B)

Tap conductors are conductors which are tapped onto the line side bus of the switchboard to feed control circuits, control power transformers, metering devices, etc. Overcurrent devices, typically fuses, are connected where the conductor to be protected receives its supply. Tap conductors do not require protection if the following conditions are met:

1. The length of the conductor is not over 10 feet.
2. The ampacity of the conductor is not less than the combined loads supplied by the conductor.
3. The conductors do not extend beyond the switchboard.
4. The conductors are enclosed in a raceway except at the point of connection to the bus.
5. For field installations where the tap conductors leave the enclosure or vault in which the tap is made, the rating of the overcurrent device on the line side of the tap conductors shall not exceed 1,000% of the tap conductor's ampacity.

6.2.5 Markings (Article 310-11)

All conductors and cable shall be permanently marked to indicate manufacturer, voltage, **AWG** size, and insulation type (Article 310-12, Article 384-3).

Grounded conductors (neutrals) size #6 and smaller shall have a continuous marking of white or gray for the entire length of the conductor. Larger conductors will be marked at each termination with marking tape.

Grounding conductors (ground wires) shall be permitted to be bare wire. In cases of insulated grounding conductors, the conductor will have a continuous marking of green for the entire length of the conductor. Larger conductors will be marked at each termination with marking tape.

Ungrounded conductors (phase wires) shall be distinguishable from grounded or grounding conductors with colors other than white, gray, or green. Typical ungrounded conductor identification colors are black, red, blue, brown, orange, and yellow. Conductors size #6 or less shall have continuous marking. Larger cables shall be marked at each termination.

In switchboards fed by a four-wire delta system in which one phase is grounded at its midpoint, the phase having the higher voltage to ground shall be marked with an orange marker.

6.2.6   Ampacities (Article 310-15)

Ampacities of cable shall be determined by cable size and insulation grade.

6.3.0   GROUNDING

6.3.1   Grounding Requirements (Article 250-5)

AC systems between 50 and 1,000 volts shall be grounded when any of the following conditions are met:

1. Where the system can be grounded in such a way that maximum voltage from phase-to-ground does not exceed 150 volts.
2. When the system is three-phase, four-wire, wye-connected, and the neutral is used as a circuit conductor.
3. When the system is three-phase, four-wire, delta-connected, and the midpoint of a phase is used as a conductor (developed neutral).

6.3.2   Ground Electrode Conductor (Article 250-53;94)

A grounding electrode (ground rod) conductor of the proper size shall be used to connect the equipment grounding conductors (ground bus) and the service equipment enclosure.

For grounded systems, delta or wye, an unspliced main bonding jumper in the service equipment shall be used to connect the grounding conductor to the equipment grounding conductor and enclosure.

Note   Some systems are ungrounded, which will not blow fuses.

6.3.3 Made Electrodes (Article 250-83;84)

Made electrodes such as rod or pipe shall extend a minimum of 8 feet into soil. The electrode shall be no less than 3/4-inch in diameter and shall be galvanized or 5/8-inch for copper-coated to resist corrosion. Underground structures such as water piping systems may be used as the made electrode. Underground gas piping systems shall not be used. Aluminum electrodes are not permitted. Made electrodes shall maintain a resistance of no more than 25 ohms to ground. If the resistance is above 25 ohms, an additional electrode(s) is required to maintain minimum resistance.

6.3.4 Grounding Of Ground Wire Conduits (Article 250-92)

A grounding conductor or its enclosure shall be securely mounted to the surface which it runs along. In cases where the conductor is enclosed, the enclosure shall be electrically continuous and shall be firmly grounded.

6.3.5 Ground Connection Surfaces (Article 250-118)

Nonconducting coatings such as paint, enamel, or insulating materials shall be thoroughly removed at any point where a grounding connection is made.

6.4.0 SWITCHBOARDS AND PANELBOARDS

6.4.1 Dedicated Air Space (Article 384-4)

Panelboards and switchboards will be installed in spaces specifically designed for such purposes. No other piping, ducts, or devices will be allowed to be installed or pass through such areas, except equipment that is necessary to the operation of the electrical equipment.

6.4.2 Inductive Heating (Article 384-3)

Busbars and conductors shall be arranged so as to avoid overheating due to inductive forces.

6.4.3 Phasing (Article 384-3)

Phasing in switchboards shall be arranged A, B, C from front to back, top to bottom, and left to right, when facing the front of the switchboard. In systems containing a high leg B-phase shall be the phase conductor having a higher voltage to ground.

6.4.4 Wire Bending Space (Article 384-35)

Wire bending space shall be in accordance with Table 373-6.

6.4.5   Minimum Spacing (Article 384-36)

Spacing between bare metal parts and conductors shall be as specified in Table 384-36.

6.4.6   Conductor Insulation (Article 384-9)

Insulated conductors within switchboards shall be listed flame-retardant and shall be rated not less than the voltage applied to them or any adjacent conductors they may come in contact with.

7.0.0   GROUND FAULTS

Ground faults exist when an unintended current path is established between an ungrounded conductor and ground. These faults can occur due to deteriorated insulation, moisture, dirt, rodents, foreign objects such as tools, and careless installation.

Ground faults are usually high arcing and low level in nature, which conventional breakers will not detect. Ground fault protection is used to protect equipment and cables against these low level faults.

Ground fault protection is required per the NEC on solidly grounded wye services of more than 150 volts to ground but not exceeding 600 volts phase-to-phase with each service disconnecting means of 1,000 amps or more.

7.1.0   GROUND FAULT SYSTEMS

Generic types of ground fault systems are:

*   Ground strap
*   Residual
*   Zero sequence

The ground-powered ground fault relay is an all solid-state device designed for industrial environments. Ground-powered ground fault relays are intended for use only on power systems which include a grounded conductor (neutral or ungrounded phase). The grounded conductor must be grounded at the service equipment, but the neutral may or may not be used in the feeder or branch circuits.

7.2.0   SYSTEM OPERATION

When circuit conditions are normal, the currents from all the phase conductors and neutral (if used) add up to zero and the sensor current transformer produces no signal. When any ground fault occurs, the currents add up to equal the ground fault current and the sensor produces a signal proportional to the ground fault. This signal provides power to the ground fault relay, which trips the circuit breaker.

A ground fault lasting for less than the time delay period will not pick up the ground trip coil, thus eliminating nuisance tripping of self-clearing faults.

The ground fault relay is a high reliability device due to its all solid-state construction. The use of redundant, self-protecting, and high reliability components further improves the performance.

Self-protection against failure is through an internal fuse which will blow and result in a tripping function if the solid-state circuitry failed during a ground fault situation.

7.3.0   SENSOR MOUNTING

The sensor current transformer (sensor) should be mounted so that all phase and neutral (if used) conductors pass through the core window once. The ground conductor (if used) must not pass through the core window. The neutral conductors must be free of all grounds after passing through the core window (see *Figure 6*).

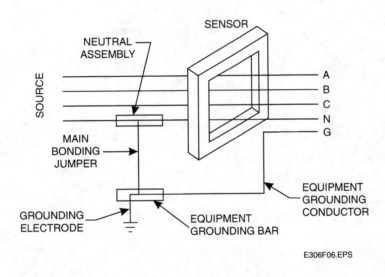

Figure 6.   Sensor Location

When so specified by the system design engineer, the sensor may be mounted so that only the conductor connecting the neutral to ground at the service equipment passes through the core window. In such cases, the sensor must provide power to the particular ground fault relay which is associated with the main circuit breaker.

Maintain at least two inches clearance from the iron core of the sensor to the nearest bus bar or cable to avoid false tripping. Cable conductors should be bundled securely and braced to hold them at the center of the core window.

The sensor should be mounted within an enclosure and protected from mechanical damage.

7.4.0 RELAY MOUNTING

The ground fault relay should be mounted in a vertical position within an enclosure with the terminal block at the lower end. The location of the relay should be such that the trip setting knob is accessible without exposing the operator to contact with live parts or arcing from disconnect operation.

7.5.0 CONNECTIONS

Connections for standard application should be made according to the wiring diagram in *Figure 7*. Wires from the sensor to the ground fault relay should be no longer than 25 feet and no smaller than #14 wire. Wires from the ground fault relay to the trip coil should be no longer than 50 feet and no smaller than #14 wire. All these wires should be protected from arcing fault and physical damage by barriers, conduit, armor, or location in equipment enclosure. Do not disconnect or short circuit wires to the circuit breaker trip coil at any time when the power is turned on.

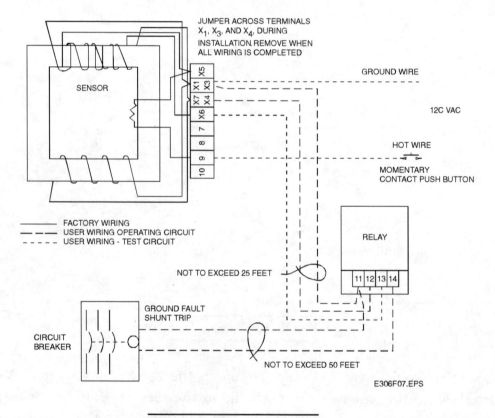

Figure 7. Wiring Diagram

7.6.0 RELAY SETTINGS

The ground fault relay has an adjustable trip setting. The amount of time delay is factory set and is available in nominal time delays of 0.1, 0.2, 0.3, and 0.5 seconds. When ground fault protection is used in downstream steps, the feeder should have the next lower time delay curve than the main and the branch the next lower curve than the feeder, etc.

High trip settings on main and feeder circuits are desirable to avoid nuisance tripping. High settings usually do not reduce the effectiveness of the protection if the ground path impedance is reasonably low. Ground faults usually quickly reach a value of 40 percent or more of the available short circuit current in the ground path circuit.

7.6.1 Coordination With Downstream Circuit Breakers

It is recommended that the magnetic trips of any downstream circuit breakers which are *not* equipped with ground fault protection be set as low as possible. Likewise, the ground fault relay trip settings for main or feeder circuits should be higher than the magnetic trip settings for unprotected downstream breakers where possible. This will minimize nuisance tripping of the main or feeder breaker for ground faults occurring on downstream circuits.

7.6.2 Instantaneous Trip Feature

Standard ground-powered ground fault relays have a built-in *instantaneous trip feature*. This instantaneous trip has a fixed time delay of approximately 1-1/2 cycles and the fixed trip setting is higher than found on most feeder or branch breakers to avoid nuisance tripping. Its purpose is to interrupt very high current ground faults on main disconnects as quickly as possible and to protect the ground fault relay components.

7.7.0 GENERIC NATIONAL ELECTRICAL TESTING ASSOCIATION GROUND FAULT SYSTEM TEST

7.7.1 Procedures

Visual Inspection:

1. Inspect components for physical damage.
2. Determine if ground sensor was located properly around appropriate conductor(s).
   a. Zero sequence sensing requires all phases and the neutral to be encircled by the sensor(s).
   b. Ground return sensing requires the sensor to encircle the main bonding jumper.
3. Inspect main bonding jumper to assure:
   a. Proper size.
   b. Termination on line side of neutral disconnect link.
   c. Termination on line side of sensor on zero sequence systems.
4. Inspect grounding electrode conductor to assure:
   a. Proper size.
   b. Correct switchboard termination.
5. Inspect ground fault control power transformer for proper installation and size. When control transformer is supplied from line side of ground fault protection circuit interrupting device, overcurrent protection and a circuit disconnecting means must be provided.

6. Visually inspect switchboard neutral bus downstream of neutral disconnect line to verify absence of ground connections.

Electrical Tests:

1. Ground fault system performance including correct response of the circuit interrupting device was confirmed by (primary/secondary) ground sensor current injection.
   a. Relay pickup current was measured.
   b. Relay time delay was measured at two values above pickup.
2. Test system operation at 57 percent rated voltage.
3. Functionally check operation of ground fault monitor panel for:
   a. Trip test.
   b. No-trip test.
   c. Nonautomatic reset.
4. Verify proper sensor polarity on phase and neutral sensors for residual systems.
5. Measure system neutral insulation resistance downstream of neutral disconnect link to verify absence of grounds.
6. Test systems (zone interlock/time coordinates) by simultaneous ground sensor current injection and monitoring proper response.

Test Result Evaluation:

1. System neutral insulation resistance should be above 100 ohms, and preferably 1 megohm or greater.
2. The maximum pickup setting of the ground fault protection shall be 1,200 amperes, and the maximum time delay shall be one second for ground fault currents equal to or greater than 3,000 amperes (NEC Article 230-95).
3. The relay pickup current should be within 10 percent of the manufacturer's calibration marks or fixed setting.
4. Relay timing should be in accordance with the manufacturer's published time-current characteristics.

8.0.0 SQUARE D HVL SWITCH

Figure 8 shows the general appearance of an HVL switch. The HVL switch is a switching device for primary circuits up to the full interrupting current of the switch. The switches are single-throw devices designed for use on 2.4 to 34.5 kilovolt systems.

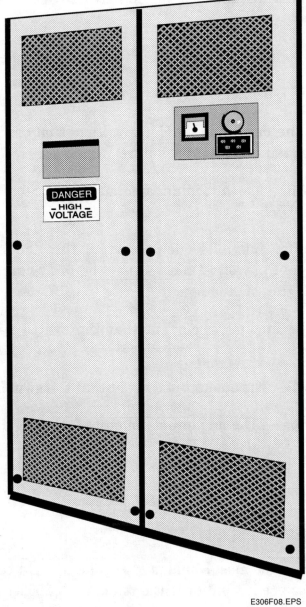

Figure 8. HVL Switch

HVL switches function as a prime component of the system and may provide both switching and overcurrent protection. HVL switches are commonly used as a service disconnect in unit substations and for sectionalizing medium voltage feeder systems. The HVL switch is designed to conform to ANSI standards as **metal-enclosed switchgear**.

8.1.0   RATINGS

Switch kV – The design voltage for the switch. Of course, nominal system voltage is the normal application method. Thus, the 5kV switch may be used for nominal system voltages of 2.4kV or 4.16kV, etc.

BIL (kV) – This is the maximum voltage pulse that the equipment will withstand.

Frequency (Hertz) – All HVL switches may be used in either 50 or 60 Hertz power systems.

Withstand (kV) – This is the maximum 60 Hertz voltage that can be applied to the switch for one minute without causing insulation failure.

Capacitor Switching (KVAR) – The maximum amount of capacitors expressed in KVAR that can be switched with the HVL.

Fault Close – The maximum, fully offset fault current that the switch can be closed into without sustaining damage. Fully offset means that the fault current will have a decaying DC component as well as the AC component.

Short Time Current – The amount of current that the switch will carry for ten seconds without damage.

Continuous Current (Amps) – The amount of current that the switch will carry continuously.

Interrupting Current (Amps) – The maximum amount of current that the switch will safely interrupt.

8.2.0   VARIATIONS

8.2.1   Upright/Inverted

The upright switch design is the most common type. The upright construction of the service entry, jaws, and arc chutes are located near the top of the cubicle. The hinge point is *below* the jaws and arc chutes.

The inverted switch design has the terminals, jaws, and arc chutes located near the *bottom* of the cubicle. The hinge point is *above* the jaw and arc chute. This type of switch is used primarily as a main switch to a lineup of other switches. Its handle operation is identical to that of an upright switch: to close the switch, the handle is moved up, and to open it, the handle is moved down.

8.2.2 Fused/Unfused

HVL switches are available in both fused and unfused models. If equipped with fuses, the entire HVL switch has the fault interrupting capacity of the fuse. Thus, the fused-type switch provides fault protection.

Either current limiting or boric acid fuses may be used in the HVL switch.

8.2.3 Duplex

A duplex switch is actually two switches, each in its own bay. The bays are mechanically connected and the switches are electrically connected on the load side. This switch may be used to supply power to a single load from two different sources.

8.2.4 Selector

A selector allows an HVL switch to have double throw characteristics. The selector switch is a single switch with a load connected to the moving or switch mechanism. Throwing the switch to one side connects the load to one source, while throwing it the other way connects it to a second source.

The selector switch will be interlocked with another switch to prevent the selector switch from interrupting current flow. The selector serves a purpose similar to the duplex switch. However, the selector switch is not an interrupter. It is a disconnect.

8.2.5 Motor-Operated Switches

This type of switch is most commonly used as the major component in an automatic transfer scheme. It can also be used when open and close functions are to be initiated from remote locations.

8.3.0 OPENING OPERATION

In the closed position, the main switch blade is engaged on the stationary interrupting contacts. The circuit current flows through the main blades.

As the switch operating handle is moved towards the open position, the stored energy springs are charged. After the springs become fully charged, they toggle over the dead center position, discharging force to the switch operating mechanism.

The action of the switch operating mechanism forces the movable main blade off the stationary main contacts, while the interrupting contacts are held closed, momentarily carrying all the current without arcing. Once the main contacts have separated well beyond the striking distance, the interrupting blade contact, held captive, has charged the interrupter blade hub spring and the interrupter blade is suddenly forced free and flips open.

The resulting arc, drawn between the stationary and movable interrupting contacts, is elongated and cooled as the plastic arc chute absorbs heat and generates an arc extinguishing gas to break up and blow out the arc. The combination of arc stretching, arc cooling, and extinguishing gas causes a quick interruption with only minor erosion of the contacts and arc chutes.

The movable main and interrupting contacts continue to the fully open position and are maintained there by spring pressure.

8.4.0   CLOSING

When the switch operating handle is moved towards the closed position, the stored energy springs are being charged and the main blades begin to move. As the main and interrupter blades approach the arc chute, the stored energy springs become fully charged and toggle over the dead center position.

When the main and movable blades approach the main stationary contacts, a high voltage arc leaps across the diminishing air gap attempting to complete the circuit. The arc occurs between the tip of the stationary main contacts and a remote corner of the movable main blades. This arc is short and brief since the fast closing blades minimize the arcing time.

Spring pressure and the momentum of the fast-moving main blades completely close the contacts.

The force is great enough to cause the contacts to close even against repelling short circuit magnetic forces if a fault exists. At the same time, the interrupter blade tip is driven through the twin stationary interrupting contacts, definitely latching and preparing them for an interrupting operation when the switch is opened.

8.5.0   MAINTENANCE

The HVL switch should be operated several times. Observe the mechanism and check for binding.

Inspect the interrupting and main blades every 100 operations for excessive wear or damage. Replace as necessary. Additionally, inspect the arc chutes for damage.

Clean the switch and its compartment thoroughly.

Use clean cloths and avoid solvents. Next, lubricate the switch. The pivot points on the switch should be greased. The switch contacts should also be lubricated with a light film of grease after being cleaned.

Final maintenance aspects include phase-to-ground and phase-to-phase meggers. If meggers are satisfactory, then a DC high potential test shall be performed.

8.6.0 SLUGGISH OPERATION

A switch that is operating sluggishly hesitates on the opening cycle. This contrasts with the normal *snapping* action. Observing the interrupter blade during the opening operation is the proper way to determine sluggish operation. Sluggishness must be repaired to prevent the switch from locking up completely. Perform the following operations:

1. Tease the switch closed and then open again while watching the interrupter blades closely. Sluggishness on close will be noted by the main blade's being engaged behind the contacts of the arc chute. On opening, the interrupter blades may hesitate momentarily.
2. Disconnect the links from the operating shaft. Never operate the switch with the links off as this may break the handle crank casting. This is because the main spring energy is absorbed by the handle crank rather than the main blades.
3. Rotate the handle approximately 45 degrees and hold it in this position while trying to operate the switch by hand. Excessive binding will prevent rotation of the shaft.
4. Check the contact adjustment at the jaw and hinge.
5. Check for binding between the interrupter blade and the arc chute.
6. Remove the front panel over the operating mechanism and disconnect the spring yoke from the cam. Check for binding between the spring pivot and the sides of the operator. Check the spring for breaks.

9.0.0 BOLTED PRESSURE SWITCHES

Bolted pressure switches (*Figures 9* and *10)* are used frequently on service entrance feeders. They are used in lieu of circuit breakers because they are inexpensive. Bolted pressure switches can be manually-operated or motor-operated. However, unlike a circuit breaker, it can only be automatically tripped by three events: ground fault, phase failure, or a blown main fuse detector.

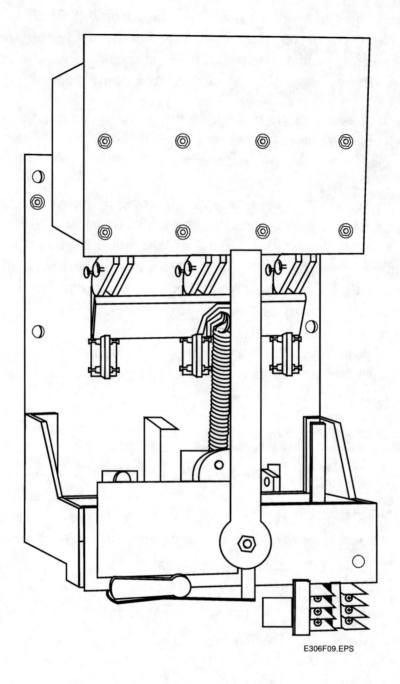

E306F09.EPS

Figure 9. Bolted Pressure Switch

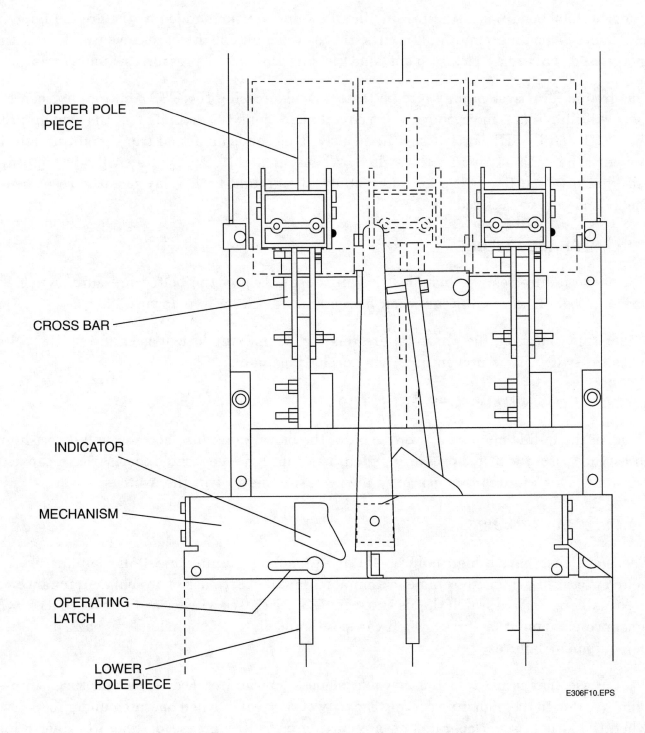

UPPER POLE
PIECE

CROSS BAR

INDICATOR

MECHANISM

OPERATING
LATCH

LOWER
POLE PIECE

E306F10.EPS

Figure 10. Bolted Pressure Switch (Front View)

9.1.0   GROUND FAULT

Under normal conditions, the currents in all conductors surrounded by the ground fault CT equal zero. When a ground fault occurs, this sensed current increases, eventually reaching the ground fault relay pickup point, and causing the bolted pressure switch to trip.

The ground fault system may also be tested. By depressing the TEST button, a green test light will illuminate, indicating correct circuit operations. To actually test the switch, press the TEST and RESET buttons simultaneously. This sends an actual trip signal through the current sensor, thus tripping the switch. Whenever a bolted pressure switch is tripped, a red light or a red flag will trip. Additionally, the ground fault relay must be reset before the switch can be reclosed.

9.2.0   PHASE FAILURE

If a phase failure relay is installed, it will cause a trip of the bolted pressure switch if a phase is lost. This could occur when a tree limb knocks a line down.

Under this condition, the phase failure relay will sense the lost phase and trip the bolted pressure switch, thus preventing a single phasing condition.

9.3.0   BLOWN MAIN FUSE DETECTOR

If one of the in-line main fuses were to blow, the blown main fuse detector will detect it and cause the trip of the bolted pressure switch. The trip signal generated comes from a capacitor trip unit. This ensures that power is always available to trip the switch.

9.4.0   MAINTENANCE

These switches have a high failure rate due to lack of maintenance. All manufacturers of bolted pressure switches recommend annual maintenance. A lack of annual maintenance will eventually result in a switch that is stuck shut. Since these switches are often used as service entrance equipment, a stuck switch can pose immediate personnel safety hazards as well as equipment failures.

The grease that is used on the movable blades deteriorates over time and turns into an adhesive due to the high interrupting capacity of the switch when operated under load. Even when the switch is not operated on a recurring basis, the grease deteriorates due to high temperatures associated with the current drawn by the phase.

The deterioration has been shown to cause the switch to stick shut. The grease must be cleaned off yearly with denatured alcohol and replaced. Be careful not to use any grease not specifically recommended by the appropriate manufacturer. Regular electrical grease cannot be used. Molilith AW-1 is the specific type of grease prescribed by most bolted pressure switch manufacturers.

Additionally, infrared scanning of in-service bolted pressure switches revealed a marked heating concern in switches. The use of a digital low resistance ohmmeter (DLRO) is used to ensure that all three phases carry similar current loads. DLRO readings should never be greater than 75 microhms and there should not be more than a 5 percent difference between the phases.

Typical annual maintenance includes:

1. Deenergizing the switch and a preliminary operational check.
2. Pre-maintenance DLRO readings recorded.
3. With the switch open, disassemble the cross bar to free all three phases.
4. Clean off all old grease with denatured alcohol or similar solvent.
5. Inspect arc tips and arc chutes for damage.
6. Adjust all pivotal connections on each blade to within manufacturer's recommended tolerances.
7. Apply new grease (Molilith AW-1) to movable blades and the area where the blades come in contact with the stationary assembly.
8. Check pullout torque on each individual blade prior to cross bar reassembly. It should be in accordance with manufacturer's prescribed limits. Too much torque will result in a switch that will be unable to open under load.
9. Record DLRO readings.
10. Reassemble cross bar assembly.
11. Close and open the switch manually several times. Ensure that no phases hang up on arc chute assembly.
12. Megger the switch.
13. Energize and test accessories (ground fault, phase failure, blown main fuse detector).

Remember, if the switch is physically stuck shut, deenergize the switch from the incoming power supply and take extra precautions when trying to unstick the switch. It is acceptable to pry the blades open with a large screwdriver, but beware of the excessive outward force that will result from a charged opening spring. To alleviate this, discharge the opening spring before commencing any work on the switch.

10.0.0 TRANSFORMERS

Transformers are used to step voltage up and down in the power transmission and distribution system.

The reason for such high transmission voltages is twofold. First, as a transformer increases transmission voltage, the required current decreases in the same proportion; therefore, larger amounts of power can be transmitted and line losses reduced. Second, to send large amounts of power over long distances at a high current and a low voltage requires a very large diameter wire. Reduction in current reduces the conductor size requirement, which results in cost reduction.

A transformer is an electrical device which, by electromagnetic induction, can change the levels of voltage and current in an AC circuit without changing the frequency and with very little loss of power.

10.1.0 TRANSFORMER THEORY

As current flows through a conductor, a magnetic field is produced around the conductor. This magnetic field begins to form at the instant current begins to flow and expands outward from the conductor as the current increases in magnitude.

When the current reaches its peak value, the magnetic field is also at its peak value. When the current decreases, the magnetic field also decreases.

Alternating current (AC) changes direction twice per cycle. These changes in direction or alternation create an expanding and collapsing magnetic field around the conductor.

If the conductor is wound into a coil, the magnetic field expanding from each turn of the coil cuts across other turns of the coil. When the source current starts to reverse direction, the magnetic field collapses, and again the field cuts across the other turns of the coil.

The result in both cases is the same as if a conductor is passed through a magnetic field. An electromotive force (emf) is induced in the conductor. This emf is called a *self-induced emf* because it is induced in the conductor carrying the current.

The direction of this induced emf is always opposite the direction of the emf which caused the current to flow initially. This principle is known as Lenz's law:

* An induced emf always has such a direction as to oppose the action that produced it.
* For this reason, the emf induced is also known as a *counter-electromotive force* (cemf).

The counter-electromotive force reaches a value nearly equal to the applied voltage, thus the primary current is limited when the secondary is open-circuited.

10.1.1 No-Load Operation

The operation of a transformer is based on the principle that electrical energy can be transferred efficiently by mutual induction from one winding to another. When the primary winding is energized from an AC source, an alternating magnetic flux is established in the transformer core. This flux links the turns of the primary with the secondary, thereby inducing a voltage in them. Since the same flux cuts both windings, the same voltage is induced in each turn of both windings. Whenever the secondary of a transformer is left disconnected (or open), there is no current drawn by the secondary winding. The primary winding draws the amount of current required to supply the magnetomotive force, which produces the transformer core flux. This current is called the *exciting* current or *magnetizing* current.

ELECTRICAL TRAINEE TASK MODULE 20306

The exciting current is limited by the counter emf of the primary and a small amount of resistance which cannot be avoided in any current-carrying conductor.

10.1.2 Load Operation

When a load is connected to the secondary of a transformer, the secondary current flowing through the secondary turns produces a counter-magnetomotive force. According to Lenz's law, this magnetomotive force is in a direction that opposes the flux which produced it. This opposition tends to reduce the transformer flux, and is accompanied by a reduction in the cemf in the primary. Since the primary current is limited by the internal impedance of the primary winding and the counter emf in the winding, whenever the counter emf is reduced, the primary current continues to increase until the original transformer flux reaches a state of equilibrium.

10.2.0 TRANSFORMER TYPES

Transformers can be divided into two main categories: power transformers and **distribution transformers**. Power transformers handle large amounts of power and are generally used at transmission level voltages. Distribution transformers are designed to handle larger currents at lower voltage levels. Distribution transformer have smaller kVA ratings and physically are a lot smaller than power transformers. Power transformers often have auxiliary means of cooling, such as fans and radiators. Distribution transformers are usually self-cooled with no fans or other cooling methods. Where distribution transformers may be pole-mounted or pad-mounted, power transformers are always pad-mounted.

Although there is some overlap between power and distribution transformers, a transformer that is rated at more than 500kVA and/or 34.5kV is generally a power transformer. A transformer rated below these values can be considered a distribution transformer. Remember, there is overlap in kVA capacity and voltage, depending on the system and power requirements.

10.3.0 DRY TRANSFORMERS (AIR-COOLED)

Many transformers do not use an insulating liquid to immerse the core and windings. Dry or air-cooled transformers are used for many jobs where small, low kVA transformers are required. Distribution transformers of large size are usually oil-filled for better cooling and insulating. However, for installations in buildings and other locations where the oil in oil-filled transformers would be a serious fire hazard, dry transformers are used. These transformers are generally of the core form. The core and coils are similar to those of other transformers. A three-phase, dry-type transformer is shown in *Figure 11*. The enclosing side plates on the high-voltage side have been removed to show the baffles which control the direction of air circulation.

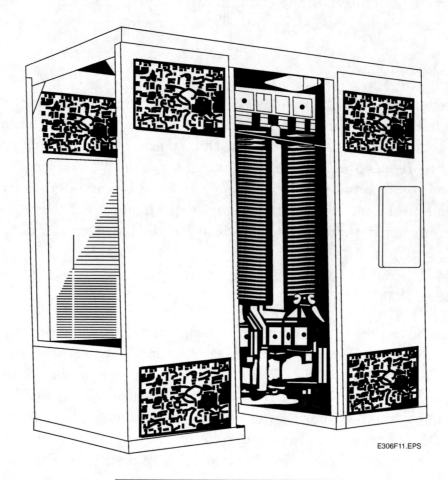

Figure 11. Dry-Type Transformer

The case is made of sheet metal and provided with ventilating louvers for the circulation of cooling air. To increase the output, fans can be installed to draw cooling air through the coils at a faster rate than is possible with natural circulation.

Either Class B or Class H insulation is used for the windings. Class B insulation may be operated safely at a hot-spot temperature of 130°C. Class H insulation may be operated safely at a hot-spot temperature of 180°C. The use of these materials makes possible smaller transformers. Both Class B and Class H insulation consist of mica, asbestos, fiberglass, and similar inorganic material. High-temperature-resistant organic varnishes are used as the binder for Class B insulation. Silicone or fluorine compounds or similar materials are used as the binder for Class H insulation. Such transformers use high-temperature insulation only in locations where the high temperature requires such insulation.

10.4.0   SEALED DRY TRANSFORMERS

Hermetically sealed dry transformers are constructed for voltages above 15kV and in large sizes. They are used for installations in buildings and other locations where oil-filled transformers would be a serious fire hazard, but may also be used for lower voltages and kVA ratings, and for water-submersible transformers in locations subject to floods.

Nitrogen and sulfur hexafluoride (SF_6) are typical gases that are used for the insulation and cooling of sealed dry transformers.

10.5.0   TRANSFORMER NAMEPLATE DATA

10.5.1   Electrical Ratings

The information relating to the transformer electrical parameters can be found on the nameplate.

10.5.2   Voltage Ratings

The voltage rating identifies the nominal RMS voltage value at which the transformer is designed to operate. A transformer can operate within a ±5 percent range of its rated primary voltage. If the primary voltage is increased more than 5 percent, the windings of the transformer can overheat. Operation of the transformer at more than 5% below its rated voltage decreases its power output proportional to the percent voltage reduction. Transformer windings are rated as follows:

Phase-to-phase and phase-to-neutral for wye windings.

> Example: 480Y/277 VAC

Phase-to-phase for delta windings.

> Example: 480 VAC

Dual voltage windings.

> Example: 4160 VAC x 17,470 VAC

When transformers are equipped with a tap changer, the voltage ratings in the nameplate indicate the nominal voltages.

10.5.3   Basic Impulse Level (BIL)

Basic impulse level (BIL) identifies the maximum impulse voltage the winding insulation can withstand without failure.

10.5.4 Phase

The phase information indicates the number of phase windings contained in a transformer tank.

10.5.5 Frequency

The frequency rating of a transformer is the normal operating system frequency. When a transformer is operated at a lower frequency than rated, the reactance of the primary winding decreases. This causes a higher than normal exciting current and an increase in flux density. In addition, there is an increase in core loss, which results in overall heating.

10.5.6 Class

Transformers are classified by the type of cooling they employ.

10.5.7 Temperature Rise

The temperature rise rating is the maximum elevation above ambient temperature that can be tolerated without causing some insulation damage.

10.5.8 Capacity

The transformer's **capacity** to transfer energy is related to its ability to dissipate the heat produced in the windings. The capacity rating is the product of the rated voltage and the current that can be carried at that voltage without exceeding the temperature rise limitation.

10.5.9 Impedance

Impedance identifies a transformer's opposition to the passage of short circuit current.

10.5.10 Phasor Diagrams

Phasor diagrams show phase and polarity relationships of the high and low windings. They can be used with the schematic connection diagram to provide test connection points and to provide proper external system connections.

10.6.0 TRANSFORMER CASE INSPECTIONS

When inspecting the inside of a transformer case, look for the following:

- Bent, broken, or loose parts
- Debris on floor or coils
- Corrosion of any part
- Worn or frayed insulation

- Shifted core members
- Damaged tap changer mounts or mechanisms
- Misaligned core spacers, loose coil elements
- Broken or loose blocking

Upon the completion of the inspection, replace the covers and bolt securely. All information should be recorded on appropriate inspection sheets.

10.7.0 TRANSFORMER TESTS

The following list of tests are the recommended minimum tests that should be included as part of a maintenance program. These tests are conducted to determine and evaluate the present condition of the transformer. From the results of these tests, a determination is made as to whether the transformer is suitable for service.

10.7.1 Continuity And Winding Resistance Test

There should be a continuity check of all windings. If possible, measure the winding resistance and compare to the factory test values. An increase of more than 10 percent could indicate loose internal connections.

10.7.2 Insulation Resistance Test

To ensure that no grounding of the windings exist, a 1,000-volt insulation resistance test should be made.

10.7.3 Ratio Test

A turns ratio test should be made to ensure proper transformer ratios and that all connections were made. If equipped with a tap changer, all positions should be checked.

10.7.4 Core Ground

This test is performed in the same way as the insulation resistance test, except the measurement is made from the core to the frame and ground bus. Remove the core ground strap before the test.

10.7.5 Heat Scanning

After the transformer is energized, a heat scan test should be done to detect loose connections. This test is performed using an infrared scanning device which shows or indicates hot spots.

11.0.0   INSTRUMENT TRANSFORMERS

For all practical purposes, the voltages and currents used in the primary circuits of substations are much too large to be used to provide operating quantities to relaying or metering circuits. In order to reduce voltage and currents to usable levels, instrument transformers are employed.

Instrument transformers are used to:

1. Protect personnel and equipment from the high voltages and/or currents used in electric power transmission and distribution.
2. Provide reasonable use of insulation levels and current-carrying capacity in relay and metering systems and other control devices.
3. Provide a means to combine voltages and/or currents phasorially to simplify relaying or metering.

There are two major classifications of instrument transformers. They are: potential (voltage) transformers and current transformers.

A potential transformer is designed to reduce high primary system voltages down to usable levels.

Potential transformers are used whenever the system primary voltage exceeds 600 volts, and on many 240-volt and 480-volt systems.

The standard secondary circuit voltage level for a potential transformer circuit is 120 volts for circuits below 25kV and 115 volts for circuits above 25kV at the potential transformer's rated primary voltage. These voltages correspond to typical transformation ratios of standard transmission voltages. The current flowing in the secondary of the potential transformer circuit is very low under normal operating conditions, typically less than one ampere.

Potential transformers are constructed as lightly loaded distribution transformers with the design emphasis on winding ratio accuracy rather than thermal ratings. Potential transformer construction can be of the air insulated dry type, case epoxy insulated, oil filled, or SF_6 insulated, depending upon the primary circuit voltage level.

Instrument transformers are manufactured with a multitude of different ratios to provide standard output for the many different system primary voltage levels and load currents.

A current transformer is designed to reduce high primary system currents down to usable levels.

Current transformers are used whenever system primary voltage isolation is required.

The standard secondary circuit current for a current transformer circuit is 5 amperes with full rated current flowing in the primary circuit. The voltage level in the secondary circuit is typically very low under normal operating conditions, However, the voltage level across a current transformer's secondary terminals can rise to a very dangerous level if the secondary circuit becomes opened while the primary circuit is energized.

Current transformers are constructed in four basic types: oil filled, window, bushing, and bar type. The primary considerations in current transformer design are current-carrying capability and saturation characteristics. Insulation systems are of the same generic types as potential transformers; however, SF_6 insulation is used infrequently in current transformer construction.

11.1.0 POTENTIAL TRANSFORMERS

Potential transformers are basically lightly loaded distribution transformers where the rating is based on ratio accuracy and not thermal capabilities. The standard output voltage of potential transformers is either 120 or 69.3 volts, depending on whether the primary connections are from phase-to-phase or phase-to-neutral. Understanding the operation of a potential transformer is simplified by the inspection of its equivalent circuit.

Potential transformers MUST have their secondary circuits grounded for safety reasons in the event that a short circuit might develop between the primary and secondary windings and to negate the effects of parasitic capacitance between the primary and secondary. *Figure 12* shows the connection of an *ideal* potential transformer circuit.

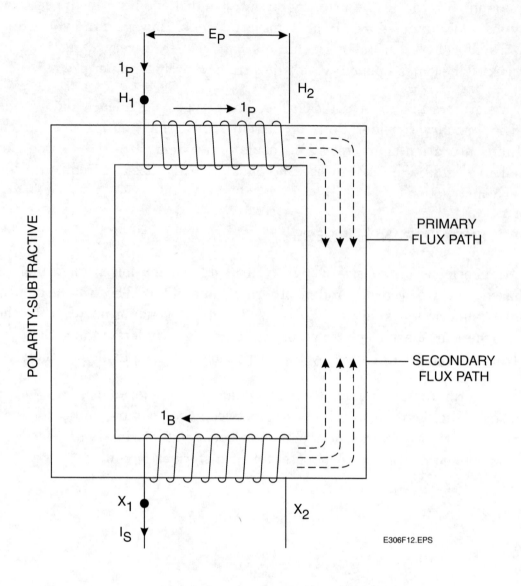

Figure 12. Potential Transformer Construction

11.2.0 CURRENT TRANSFORMERS

Current transformers are manufactured in four basic types: oil-filled, bar, window, and **bushing** type. The bushing types are normally applied on circuit breakers or power transformers. The other types are used for the remaining indoor and outdoor installations. *Figure 13* illustrates some common current transformer constructions.

The major criterion for the selection of the current transformer for relaying is its primary current rating, maximum burden, and saturation characteristics. Saturation is particularly important in relaying due to the fact that many relays are called upon to operate only under fault conditions.

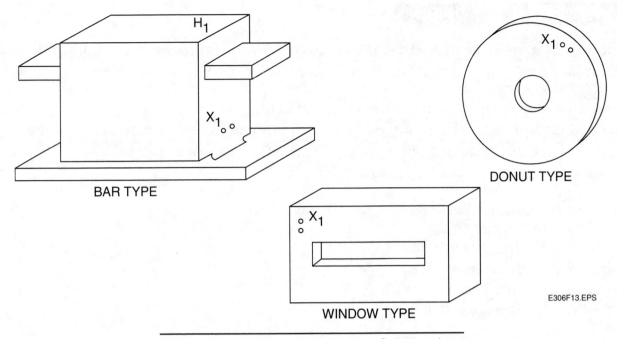

BAR TYPE

DONUT TYPE

WINDOW TYPE

E306F13.EPS

Figure 13.   Current Transformer Constructions

Current circuits operate at a very low voltage. Connected loads (burdens) range from 0.2 to 2 ohms. These small impedances, together with maximum continuous current of up to 5 amperes, keeps these circuits at low potentials. The voltage can become high momentarily during faults when large secondary currents flow. This voltage is a function of the current, burden, and transformer VA capability.

As with potential transformers, current transformers MUST also have their secondary windings grounded in the event of an insulation breakdown between the primary and secondary, and to negate the effects of parasitic capacitance.

11.3.0   INSTRUMENT TRANSFORMER MAINTENANCE

11.3.1   Visual And Mechanical Inspection

This inspection should cover the following points:

1. Inspect for physical damage and nameplate information for compliance with instructions and specification requirements.
2. Verify proper connection of transformers with system requirements.
3. Verify tightness of all bolted connections and ensure adequate clearances exist between primary circuits to secondary circuit wiring.
4. Verify that all required grounding and shorting connections provide good contact.
5. Test proper operation of transformer withdrawal mechanism (trip out) and grounding operation when applicable.

12.0.0   CIRCUIT BREAKERS

Circuit breakers are the only circuit interrupting devices which combine full fault current interruption rating and the ability to be manually or automatically opened or closed.

A circuit breaker is defined as a mechanical switching device capable of making, carrying, and breaking currents under normal circuit conditions and also making, carrying (for a specified time), and breaking currents under specified abnormal circuit conditions such as those of a short circuit (according to IEEE).

The four general classifications of circuit breakers are:

1.   Air Circuit Breakers (ACBs)
2.   Oil Circuit Breakers (OCBs)
3.   Vacuum Circuit Breakers (VCBs)
4.   Gas Circuit Breakers (GCBs)

Circuit breakers may conveniently be divided into low, medium, and high voltage classes. Although there is considerable overlap among these classes, each one has certain characteristic features.

12.1.0   CIRCUIT BREAKER RATINGS

Circuit breaker ratings are given on the breaker nameplates. The information from these nameplates should be taken and reviewed in considering any breaker selection problem. The same rating information should be included in any documentation about breaker applications. The rating information includes some of the following items:

Rated Voltage: Rated voltage is the maximum voltage for which the circuit breaker is designed.

Rated Current: This is the continuous current that a circuit breaker can carry without exceeding a standard temperature rise (usually 55°C).

Interrupting Rating: This is the maximum value of current at rated voltage that a circuit breaker is required to successfully interrupt for a limited number of operations under specified conditions. The term is usually applied to abnormal or emergency conditions.

13.0.0   ELECTRICAL DRAWING IDENTIFICATION

Before looking at actual plant diagrams, it is necessary to understand the symbology used to condense electrical drawings. The designer uses symbols and abbreviations as a type of shorthand. This section will present the standard symbols, abbreviations, and device numbers which make up the designer's shorthand.

13.1.0 ELECTRICAL DIAGRAM SYMBOLOGY

It is imperative that every line, symbol, figure, and letter is a diagram have a specific purpose for being present, and the information being presented is in its most concise form. For example, when the rating of a current transformer is given, an abbreviation such as "CT" is not needed; the information is implied by the symbol itself. Writing the unit of measure (amp) in this case is also unnecessary, since a current transformer is always rated in amperes. Thus, the numerical rating and the transformer symbol are sufficient. The key to reading and interpreting electrical diagrams is to understand and use the electrical legend. The legend shows the symbols used in the diagrams, and also contains general notes and other important information. Most electrical legends are very similar; however, there are some variations between legends developed by different companies. Only the legend specifically designed for a given set of drawings should be used for those drawings.

The legend prevents the necessity to memorize all the symbols presented on a diagram and can be used as a reference for unfamiliar symbols. Typically, the legend will be found in the bottom right corner of a print or on a separate drawing. In addition to symbols, abbreviations are an important part of the designer's shorthand. For example, a circle can be used to symbolize a meter, relay, motor, or indicating light. A circle's application can generally be distinguished by its location in the circuit; however, the designer uses a set of standard abbreviations to make the distinction clear. The following abbreviations are used to distinguish what type of meter a given circle represents.

A - Ammeter	PH - Phase meter
AH - Ampere-hour meter	SYN - Synchroscope
CRO - Oscilloscope	TD - Transducer
DM - Demand meter	V - Voltmeter
F - Frequency meter	VA - Volt-ammeter
GD - Ground detector	VAR - VAR meter
OHM - Ohmmeter	VARH - VAR hour meter
OSC - Oscillograph	W - Wattmeter
PF - Power factor meter	WH - Watt-hour meter

As mentioned earlier, indicating lamps may also be represented by circles. The following abbreviations are used to distinguish indicating lamps:

A - Amber	G - Green
B - Blue	R - Red
C - Clear	W - White

Another component commonly represented by circles is relays. The following abbreviations are used for relays:

CC - Closing coil	TD - Time delay relay
CR - Closing/control relay	TDE - Time delay energize
TC - Trip coil	TDD - Time delay deenergize
TR - Trip relay	X - Auxiliary relay

Still another component commonly represented by a circle is the motor. Motors usually have the horsepower rating in or near the circle representing them. The abbreviation for horsepower is HP. Any other piece of equipment represented by a circle will be identified on the diagram, in the legend, notes, or spelled out on the diagram.

Contacts and switches are also identified by standard abbreviations. The following is a list of these abbreviations:

A - Breaker "A" contact	PB - Push button
B - Breaker "B" contact	PS - Pressure switch
BAS - Bell alarm switch	PSD - Differential pressure switch
BLPB - Backlighted push-button	TDO - Time delay open
CS - Control Switch	TDC - Time delay closed
FS - Flow switch	TS - Temperature switch
LS - Limit switch	XSH - Auxiliary switch

The following figures illustrate examples of the previously discussed abbreviations and symbols.

Figure 14 shows "A" and "B" contacts in their normally deenergized state. If relay CR is deenergized, the "A" contact is open and the "B" contact is shut. When relay CR is energized, the "A" contact is shut and the "B" contact is open.

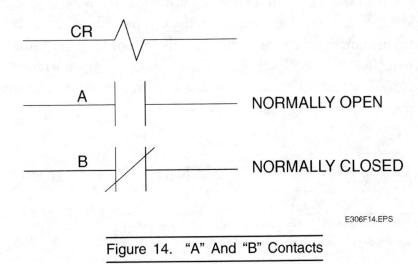

E306F14.EPS

Figure 14. "A" And "B" Contacts

Figure 15 illustrates a control switch and its associated contacts. Contacts 1 through 4 open and close as a result of the operation of control switch #1 (CS1).

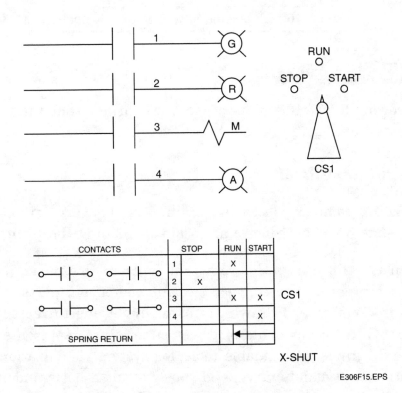

E306F15.EPS

Figure 15. Switch Development

In the stop position, contact 2 is shut and the red indicating lamp is lit. In the start position, contacts 3 and 4 are shut, energizing the M coil and the amber indicating lamp, respectively. When the switch handle is released, the spring returns to the run position, and contact 4 opens, deenergizing the amber lamp and closing contact 1 to energize the green lamp. There are many abbreviations used on electrical drawings. The designer makes an effort to use standard abbreviations; however, the technician will encounter nonstandard abbreviations. Nonstandard abbreviations will typically be defined in the diagram notes or legend. *Figure 16* defines abbreviations commonly used in writing prints and specifications. The symbols further illustrate descriptions of the abbreviations.

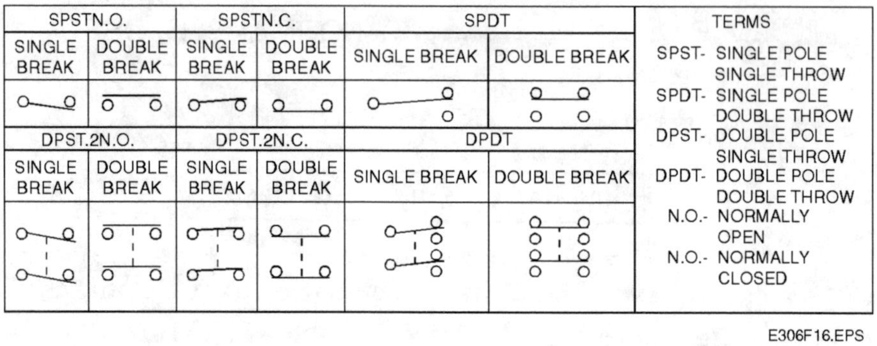

E306F16.EPS

Figure 16.   Supplementary Contact Symbols

14.0.0   ELECTRICAL PRINTS

This section will cover the specific types of electrical prints that the trainee needs to be familiar with to install and maintain electrical systems.

14.1.0   SINGLE-LINE DIAGRAMS

Analyzing and reading complex electrical circuits can be very difficult. Diagrams are simplified to single-line (one-line) diagrams to aid in reading the prints.

A one-line diagram is defined as a diagram that indicates by means of single lines and standard symbology the paths, interconnections, and component parts of an electric circuit or system of circuits. This type of drawing uses a single line to represent all conductors (phases) of the system. All components of power circuits are represented by symbols and notations. One-line diagrams are valuable tools for system visualization during planning, installation, operation, and maintenance, and they provide a basic understanding of how a portion of the electrical system functions in terms of physical components of the circuit.

There are two types of single-line diagrams: the overall plant single-line diagram; and the project single-line diagram. Overall plant single-line diagrams show the electrical power distribution system, in simplified form, from the utility or generated supply to the load side of substation protection devices. The overall plant single-line diagrams do not include distribution panels, motor control centers, motors, or similar electrical equipment located on the load side of the substation. Project single-line diagrams show the electrical power distribution and utilization for a particular project or local plant area. These diagrams are the continuation of overall plant single-line diagrams and indicate the power distribution from the load side of the substation protective device to the final point of utilization on a branch circuit. This diagram will generally be of more use than the overall plant diagram. An example of a single-line diagram is shown in *Figure 17*.

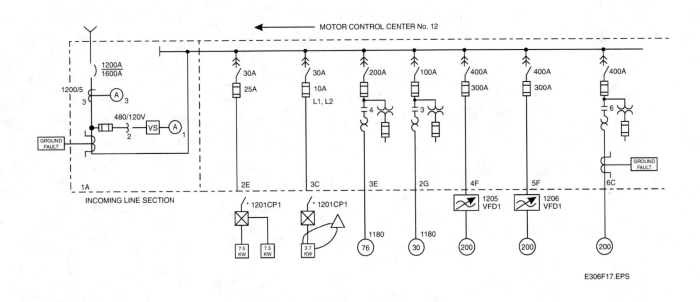

Figure 17. One-Line Diagram

14.2.0   ELEMENTARY DIAGRAMS

An elementary diagram is a drawing that falls between one-line diagrams and schematics in complexity. An elementary diagram is a wiring diagram showing how each individual conductor is connected.

This section will discuss elementary diagrams, interconnection, and connection diagrams. All of these diagrams fall under the previous definition of complexity and illustrate individual conductors. Elementary diagrams are used to show wiring of instrument and electrical control devices in elementary ladder or schematic form. The elementary diagrams reflect the control wiring required to achieve the operation and sequence of operations described in logic diagrams. *Figure 18* is an example of an elementary diagram.

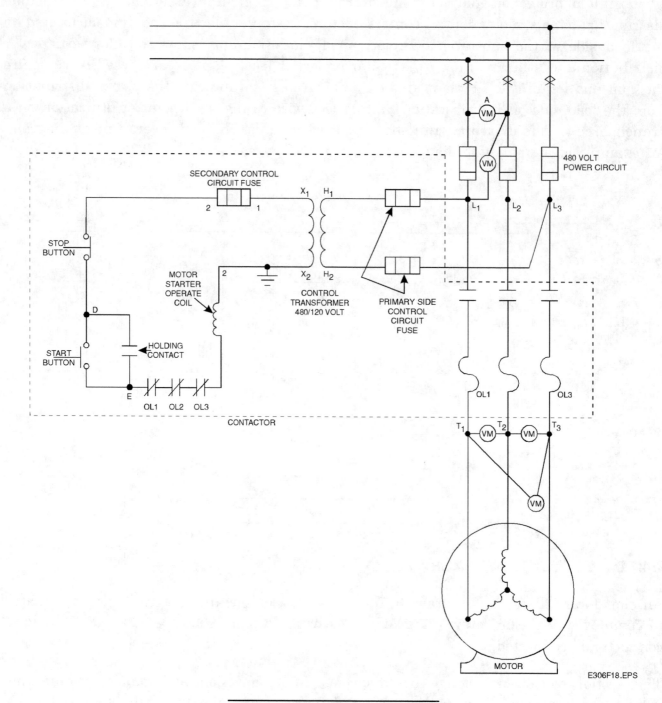

Figure 18. Elementary Diagram

When presented in the ladder form the vertical lines in each ladder diagram represent the hot and neutral wires of a 120VAC circuit. If a number of schemes are connected to the same 120VAC circuit, the vertical lines are continuous from the top to the bottom of the ladder. The hot wire is always shown at the left side of the ladder and the neutral at the right. A ground symbol is not shown on the neutral wire.

The hot and neutral wire numbers are shown at the top of the vertical lines on each ladder diagram. The circuit identification number and source of the 120VAC circuit is also shown at the top center of each ladder. If two or more 120VAC circuits are represented in a single ladder, the vertical lines are broken and the wire numbers and circuit identification entered at the top of each ladder segment. Each horizontal line in a ladder diagram represents a circuit path. All devices shown on a single horizontal line represent a series circuit path; parallel circuit paths are shown on two or more horizontal lines.

14.3.0   INTERCONNECTION DIAGRAMS

When troubleshooting electrical circuits, the trainee may utilize an elementary circuit diagram to determine the cause of a failures; however, since elementary diagrams are drawn without regard to physical locations, connection diagrams should be used to aid the technician in locating faulty components. Interconnection and connection diagrams are structured such that they present all the wires which were shown in the elementary drawing in their actual locations. These drawings show all electrical connections within an enclosure, with each wire labeled such that it indicates where each end of the wire is terminated.

The interconnection diagram is made to show the actual wiring connections between unit assemblies or equipment. Internal wiring connections within unit assemblies or equipment are usually omitted. The interconnection diagrams will appear adjacent to the schematic diagram or on a separate drawing, depending upon the format chosen when making the schematic diagram. The development of the interconnection diagram is integrated with that of the schematic diagram and only the equipment, terminal blocks, and wiring pertinent to the accompanying schematic diagram appears in the interconnection diagram.

A typical interconnection diagram will contain the following information:

- The outline of the equipment involved in its relative physical location.
- Terminal blocks in the equipment that are concerned with the wiring illustrated on the schematic.
- Wire numbers, cable sizes, cable numbers, cable routing, and cable tray identification. This information should not be repeated on the interconnection diagram except where it is necessary to simplify the reading of the diagram.
- Wiring between equipment usually is shown as individual cables. However, on complex drawings, cables may be identified and combined as they leave and enter the equipment and be shown as a single line to simplify the drawing.
- Equipment identification information.

14.4.0   CONNECTION DIAGRAMS

The connection diagram shows the wiring connections for an installation, component devices, or any related parts. A connection diagram may cover internal connections, external connections, or both, and will contain as much detail as necessary to make or trace any electrical connections involved.

A connection diagram generally shows the physical arrangement of component electrical connections, like an interconnection diagram. The connection diagram differs from the interconnection diagram by typically not including external connections between two or more unit assemblies or equipment.

Using the previously discussed examples, it would seem the connection diagram would be more useful to the trainee than an interconnection diagram. This is typically the case; an interconnection diagram will rarely be of any assistance in troubleshooting.

The schematic diagram will show internal wiring and functions, while the wiring diagram will allow the technician to easily locate terminals and wires. The schematic, like the elementary drawing, is not laid out with respect to physical locations, while the wiring diagram is. A wiring diagram in conjunction with a schematic greatly aids in troubleshooting a given piece of equipment. Connection diagrams can be shown in various forms. The following section illustrates different types of connection diagrams the technician may see when working in the field.

14.4.1   Point-To-Point Method

This form is used for the simpler diagrams where sufficient space is available to show each individual wire without sacrificing clarity of the diagram. *Figure 19* is a point-to-point connection diagram.

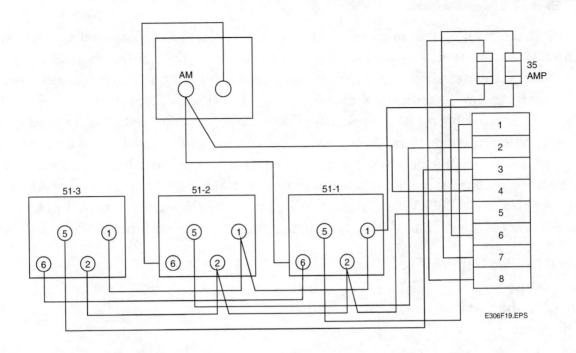

Figure 19. Point-To-Point Connection Diagram

14.4.2 Cable Method

For the more complex diagram, individual wires are "cabled" so as to conserve drawing space. *Figure 20* is a cable method connection diagram.

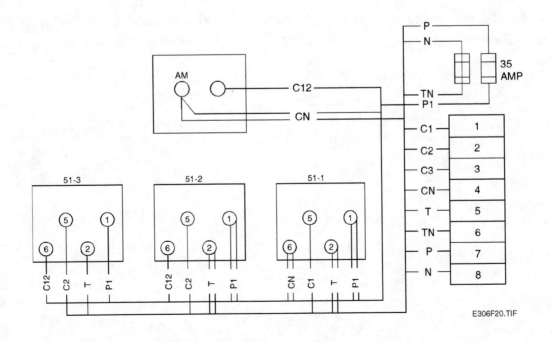

Figure 20. Cable Method Connection Diagram

15.0.0 INTERPRETING ONE-LINE DIAGRAMS

The one-line diagram shown in *Figures 21* and *22* is typical of those used to show workers how an electrical system is to be installed. In general, a one-line diagram is never drawn to scale. Such drawings show the major components in an electrical system and then utilize only one drawing line to indicate the connections between these components. Even though only one line is used between components, this single line may indicate a raceway of two, three, four, or more conductors. Notes, symbols, tables, and detailed drawings are used to supplement and clarify a one-line diagram. Referring again to *Figures 21* and *22*, these drawings were prepared by an electrical manufacturing company to give workers at the job site an overview of a 2,000kVA substation utilizing a 13.8kV primary and a 4.16kV, 3-phase, 3-wire, 60Hz secondary. Note that this drawing sheet is divided into the following sections:

- Service order numbers
- Unit numbers
- One-line diagram
- Title block
- Revision notes

Service order numbers are arranged at the top of the drawing sheet in a time-sequence, bar-chart type arrangement. For example, S.O. #58454 deals with the primary side of the 2,000kVa transformer, including the transformer itself. This section includes a high-voltage switchgear, with an indoor-outdoor enclosure. The switchgear itself consists of two HLP-C interrupter switches, each rated at 15kV, 600 amperes, and 150E current-limiting fuses (CLF).

Service order #58455 deals with the wiring and related components on the secondary side of the transformer and begins with a low-voltage switchgear with an indoor-outdoor enclosure.

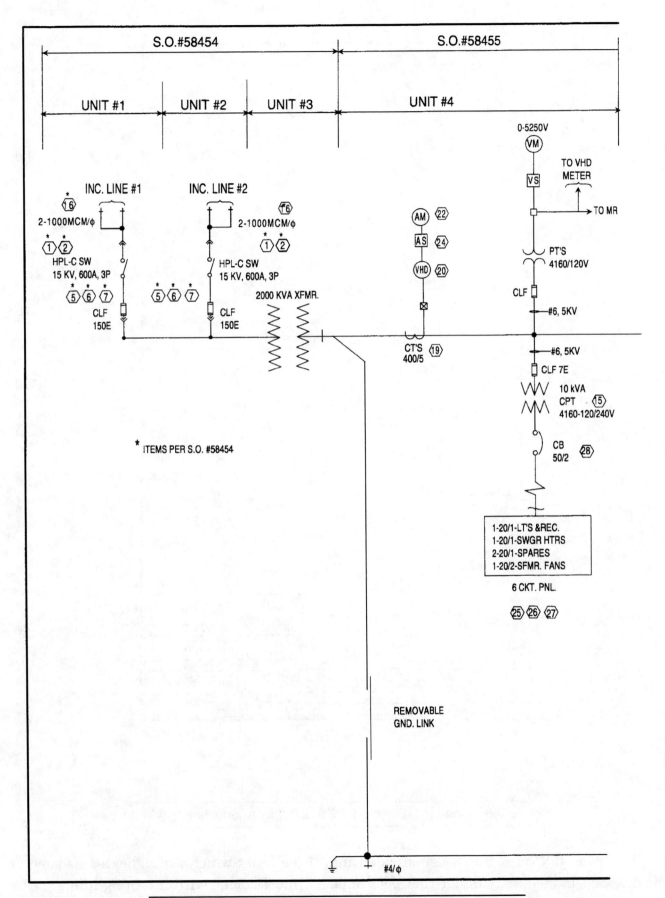

Figure 21.   One-Line Diagram Of A 2,000kVA Substation

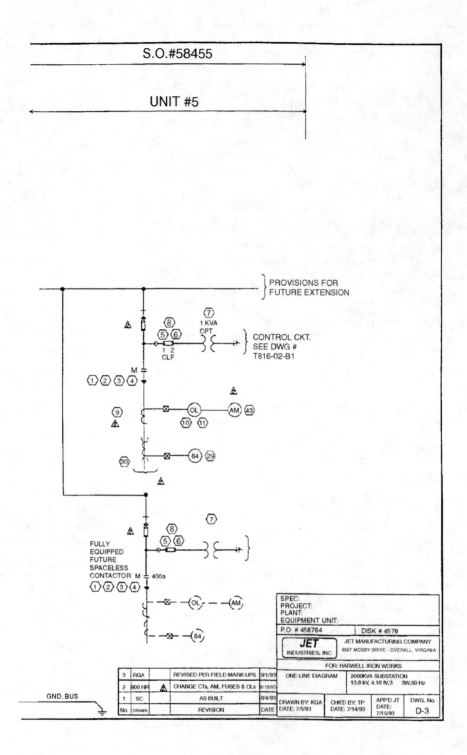

Figure 22.   One-Line Diagram Of A 2,000kVA Substation (Cont.)

Service order #58454 is further subdivided into three units which are indicated as such on the drawing immediately under the S.O. number. Unit #1 deals with incoming line #1, Unit #2 deals with incoming line #2, and Unit #3 covers the 2,000kVA transformer and its related connections and components.

Service order #58455 is further subdivided into two units: Unit #4 and Unit #5. Basically, Unit #4 covers grounding, the installation of current transformers, various meters, potential transformers, a 10kVA 4,160/240V transformer, and a 6-circuit panel, all derived from a 600-ampere, 4,160-volt, 3-wire, 60Hz main bus.

Unit #5 continues with the main bus and covers the installation and connection of a complete motor control center, along with another fully-equipped future space, less contactors.

The one-line diagram takes up most of the drawing sheet and gives an overview of the entire installation. Let's begin at the left side of the drawing where incoming line #1 is indicated. This section of the drawing is shown in *Figure 23*.

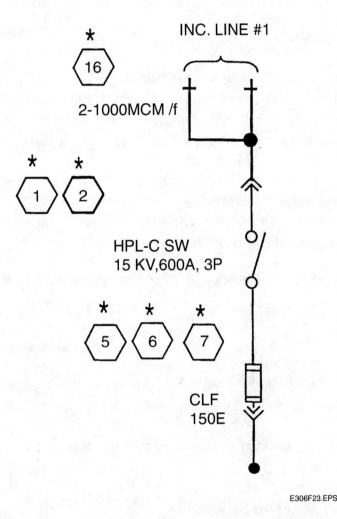

Figure 23. Incoming High-Voltage Line

Incoming line #1 (partially abbreviated on the drawings as "INC. LINE #1") consists of two, 1,000 MCM (Kemil) conductors per phase as indicated by note "2-1000MCM/f." In other words, parallel 1,000 kemilconductors. Since this is a three-phase system, a total of six 1,000 kemilconductors are utilized.

The single-line continues to engage separable connectors at the single-throw 15kV, 600-ampere, 3-pole switch. Overcurrent protection is provided by current-limiting fuses as indicated by the fuse symbol combined with a note. The single-line continues to the high-voltage bus (*Figure 21*) which connects to the primary side of the 2,000kVA transformer. Incoming line #2 (partially abbreviated on the drawings as "INC. LINE #2") has identical components as line #1. This line also connects to the high-voltage bus, which connects to the primary side of the 2,000kVA transformer. Notice the numerals, each enclosed by a hexagon, placed near various components in these two high-voltage primaries. Note also that an asterisk is placed above each of these marks. A note on the drawing indicates the following:

*ITEMS PER S.O. #58454

These marks appear in a supplemental schedule known as the "Bill of Materials", which describes the marked items, lists the number required, manufacturer, catalog number, and a brief description of each. Such schedules are extremely useful to estimators, job superintendents, and workers on the job to ensure that each required item is accounted for and installed.

Every electrical drawing should have a title block, and it is normally located in the lower right-hand corner of the drawing sheet; the size of the block varies with the size of the drawing and also with the information required.

In general, the title block for an electrical drawing should contain the following:

- Name of the project
- Address of the project
- Name of the owner or client
- Name of the person or firm who prepared the drawing
- Date the drawing was made
- Scale(s), if any
- Initials of the drafter, checker, designer, and engineer, with dates under each
- Job number
- Drawing sheet number
- General description of the drawing

The title block for the project in question is shown in *Figure 24*.

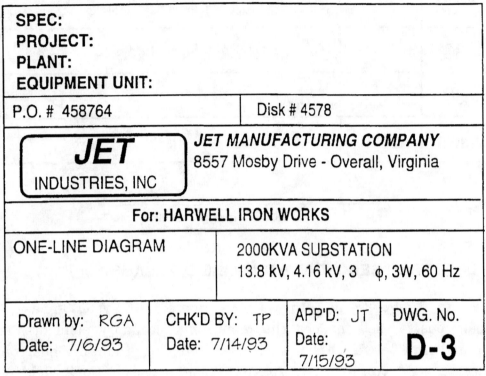

SPEC:	
PROJECT:	
PLANT:	
EQUIPMENT UNIT:	
P.O. # 458764	Disk # 4578

JET INDUSTRIES, INC

JET MANUFACTURING COMPANY
8557 Mosby Drive - Overall, Virginia

For: **HARWELL IRON WORKS**

ONE-LINE DIAGRAM	2000KVA SUBSTATION 13.8 kV, 4.16 kV, 3 φ, 3W, 60 Hz

Drawn by: RGA Date: 7/6/93	CHK'D BY: TP Date: 7/14/93	APP'D: JT Date: 7/15/93	DWG. No. **D-3**

E306F24.EPS

Figure 24.   Typical Drawing Title Block

Sometimes electrical drawings will have to be partially redrawn or modified during the planning or construction of a project. It is extremely important that such modifications are noted and dated on the drawings to ensure that the workers have an up-to-date set of drawings to work from. In some situation, sufficient space is left near the title block for dates and description of revisions, as shown in *Figure 25*.

CAUTION: When a set of electrical drawings has been revised, always make certain that the most up-to-date set is used for all future layout work. Either destroy the obsolete set of drawings, or else clearly mark on the affected drawing sheets, "Obsolete Drawing - Do Not Use." Also, when working with a set of working drawings and written specifications for the first time, thoroughly check each page to see if any revisions or modifications have been made to the originals. Doing so can save much time and expense to all concerned with the project.

No.	Drawn	App'd	Revision	Date
3	RGA		Revised per field mark-ups	9/1/93
2	SC		Change CTs, AM, fuses & OLs	8/18/93
1	SC		As built	8/4/93

E306F27.EPS

Figure 25.   Typical Drawing Revision Block

15.1.0   INTERPRETING SECONDARY ONE-LINE DIAGRAMS

Referring again to *Figure 21*, note that a 600-ampere, 4,160-volt, 3-phase, 3-wire, 60Hz aluminum main bus is used to feed the remaining secondary elements. Also note the removable ground link between the transformer and the grounding bus connection. The drawing shows this conductor to be No. 4/0 AWG. Looking back at the 2,000kVA transformer, note that the main bus continues in a horizontal line to the right of the transformer symbol. The first group of equipment encountered is the metering section. Note the current transformers (CTs) which are designated by both symbol and note. The "400/5" note indicates that the CTs have a ratio of 400 to 5; that is, if 400 amperes are flowing in the main bus, only 5 amperes will flow to the meters. Again, numerals enclosed in hexagons are placed at each component in this section. Referring to the "Bill of Materials" schedule in *Figure 26*, we see a description of Item #19 to be "CTs 400/5 Type JAF-0." Two are required; the catalog number is 750X10G304 and the manufacturer is GE (General Electric). Continuing from the CTs down to Item #20, the schedule describes this as a three-phase, three-wire, watt-hour meter with a 15-minute demand and is designed to register with CTs with a ratio of 44/5 and PTs with primary/secondary at 4,160/240 volts. Locate the remaining numerals in this group and find their description in the schedule in *Figure 26*.

Mark	Req'd	Cat. No.	Mfg.	Description
1	2	IC2957B103C	GE	Disc. Handle & Elec. Interlock ASM. (400A) (CAT#116C9928G1)
2	2	IC2957B108E	GE	Vert. Bus (CAT#195B4010G1)(400A) Shutter ASM. (CAT#116C9927G1) (400A)
3	2	1C2957B10BF	GE	Coil Finger ASM. (CAT#194A6949G1) (400A) Safety Catch (CAT#194A6994G1) (400A) Stab Fingers (CAT#232A6635G) (400A)
4	1		Toshiba	5kV, 300A, 3P, Vacuum Contactor 120VAC Rectified Control Type CV461J-GAT2
5	4	2033A73G03	W	5kV Fuse MTG (2/CPT)
6	4	677C592G09	W	5kV, CLF, 2E Fuses Type CLE-PT
7	2	HN1K0EG15	Micron	1kVA, 4160-120 CPT
8	3	9F60LJD809	GE	CLF Size 9R (170A) Type EJ-2 (600HP)
9	3	615X3	GE	CT'S 150/5A Type JCH-0
10	1	CR224C610A	GE	200 Line Block O.L. Rly. 3 Elements Ambient Compensated W/INC. Contact
11	0	CR123C3.56A	GE	O.L. HTR (600HP)
11A	3	CR123C3.26A	GE	O.L. HTR. (2.79A) (700HP)
12	1.	7022AB	AG	Off Delay R.Y .5-5 SEC.
13	0	CR2810A14A	GE	Machine Tool RLY. INO&INC 120VAC (MR)
14	1	CR294OUM301	GE	Emergency Stop PB (Push to Stop Pull to Reset) W/NP
15	1	9T28Y5611	GE	10kVA CPT. 4160-120/240V
16	2	643X92	GE	PT'S 4160/120V Type JVM-3/2FU
17	2	9F60CED007	GE	CLF 7E, 4.8kV Type EJ-1
18	2	9F61BNW451	GE	Fuse Clips Size C

E306F28.TIF

Figure 26. Bill Of Materials Schedule

Mark	Req'd	Cat. No.	Mfg.	Description
19	2	750X10G304	GE	CT'S 400/5 Type JAF-0
20	1	700X64G885	GE	DWH-Meter 3φ, 3W, 60HZ, Type DSM-63 W/15MIN. Demand Register CT'S Ratio 400/5 & PT'S 4160-120V
21	1	50-103021P	GE	VM Scale 0-5250V Type AB-40
22	1	50-103131L	GE	AM Scale 0-400A Type AB-40
23	1	10AA004	GE	VS Type SBM
24	1	10AA012	GE	AS Type SBM
25	1	TL612FL	GE	6 CKT. PNL.
26	4	TQL1120	GE	20/1 C/B Type TQL.
27	1	TQL2120	GE	20/2 C/B Type TQL.
28	1	TEB12050WL	GE	50/2 C/B Type TEB
29	1	3512C12H02	W	Type GR Groundgard RLY. Solid State
30	1	3512C13H03	W	GRD. Sensor
31	2	H	Smout Hollman	1/2 LT. REC.
32	2	7604-1	GE	LT. SW. & Receptacle
33	2	4D846G20	GE	120VAC, 250W HTR
34	1		Econo	Econo Lift for Contactor
35	11	Lot	Cook	NP/Schedule DWG. 58455-A1
36	3	Hold	T & B	Lug
37	0	50250440LSPK	GE	AM Scale 0-100A PNL. Type 2% ACC. Type 250 4-1/2 Case
38	1	NON10	Bus	10A, 250V Fuse
39	1	CP232	AH	2P, 250V Pull-Apart Fuse Block
40	1		Cook	SWGR NP S.O.#58455

E306F28B.EPS

Figure 26. Bill Of Materials Schedule (Cont.)

The two taps from the main bus in the drawing in question are for feeding two motor control centers (MCC); one to be put into use immediately while the other is a fully-equipped MCC, less contactors, for future use. Let's look at the complete MCC first. An enlarged view of this section is shown in *Figure 27* for clarification. This feeder is provided with overcurrent protection by means of current-limiting fuses (CLF9R), which are fuse type EJ-2, rated at 170 amperes. Immediately beneath this device, note that a tap is taken from the main line, fused with 5kV MTG fuses and also 5kV, CLF, 2E fuses before terminating at a 1kVA, 4,160/120-volt CPT transformer. This transformer is provided to accommodate the 120-volt control circuit shown in *Figure 27*. Since motor controls and motor control circuits are covered in Electrical Task Module 20311, the circuit in *Figure 27* will not be explained in depth in this module. It is presented here to give the trainee visual knowledge of a motor control circuit, and not necessarily how it functions. Now let's backtrack to the main feeder and continue downward to a contactor before another group of current transformers (CTs) are installed in the circuit. These CTs are accompanied by notes and Mark #9. Referring to the schedule in *Figure 26* for a description of Mark #9, we see that these three CTs have a ratio of 150/5; that is, when the circuit is drawing 150 amperes, the metering devices will receive only 5 amperes, but the meter itself will indicate 50 amperes. This circuit continues to a 200 line block overload relay with three elements, and then on to an ammeter with a range of 0 to 150 amperes.

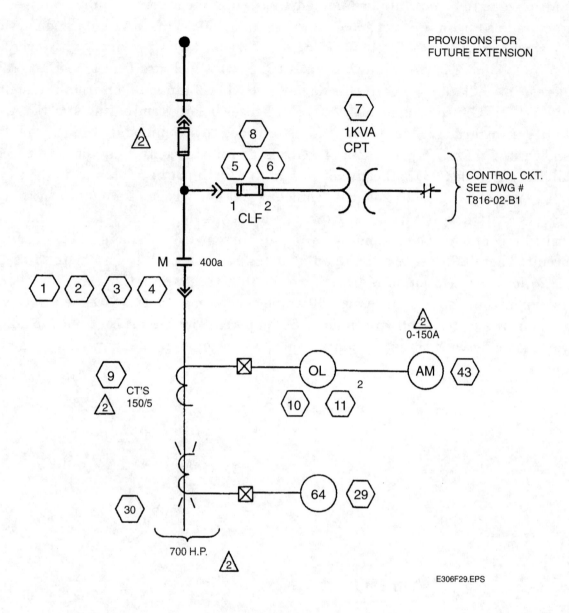

PROVISIONS FOR
FUTURE EXTENSION

1KVA
CPT

CONTROL CKT.
SEE DWG #
T816-02-B1

CLF

M   400a

0-150A

OL   AM   43

CT'S
150/5

700 H.P.

E306F29.EPS

Figure 27.   Enlarged View Of The MCC Feeder

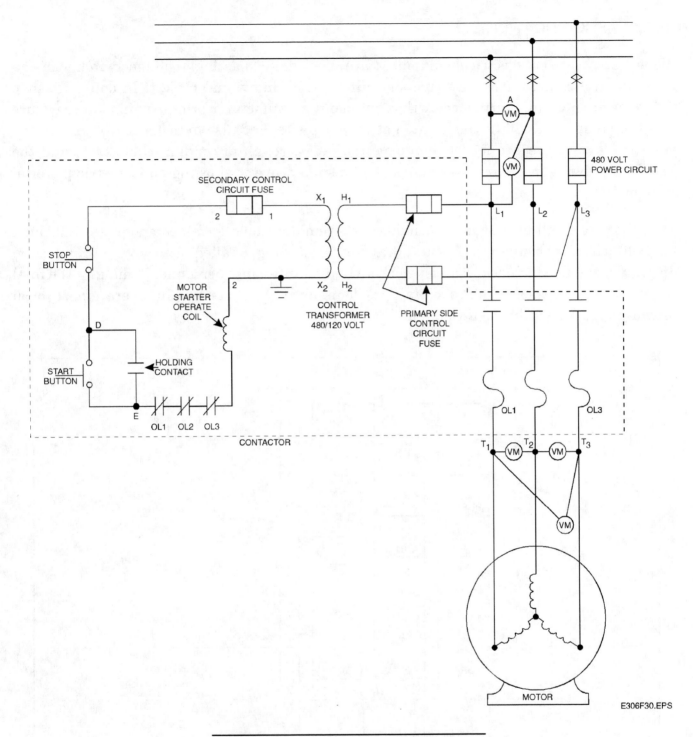

Figure 28. Motor Control Circuit Diagram

The next item on this main vertical feeder is a ground sensor which is connected to a solid-state ground guard relay. The feeder then enters, and connects to the busbars, in a motor control center (MCC) enclosure as shown in *Figure 28*. The remaining feeder in the one-line wiring diagram under consideration is for future use and is similar to the circuit just described.

15.2.0 SHOP DRAWINGS

When large pieces of electrical equipment are needed, such as high-voltage switchgear and motor control centers, most are custom built for each individual project. In doing so, shop drawings are normally furnished by the equipment manufacturer prior to shipment to ensure that the equipment will fit the location at the shop site, and also to instruct workers on the job as to how to prepare for the equipment; that is, rough in conduit, cable tray, and the like. The drawing in *Figure 29* is one page of a shop drawing showing an isometric pictorial view of the enclosure.

Shop drawings will also usually include connection diagrams for all components that must be "field wired" or connected. As-built drawings, including detailed factory wired connection diagrams, are also included to assist workers and maintenance personnel in making the final connections, and then troubleshooting problems once the system is in operation. Typical drawings are shown in *Figures 30* and *31*.

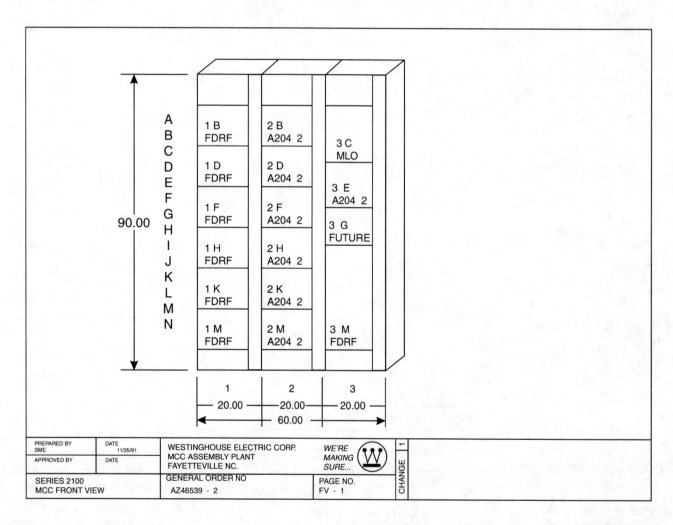

E306F31.EPS

Figure 29. Isometric View Of A Motor Control Center

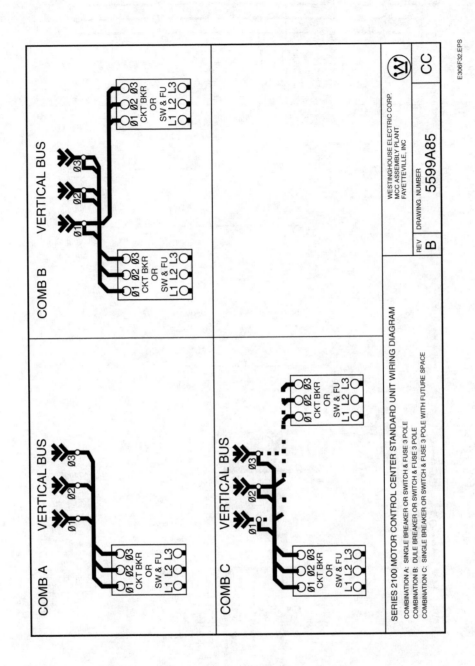

Figure 30. Motor Control Center Standard Unit Wiring Diagram

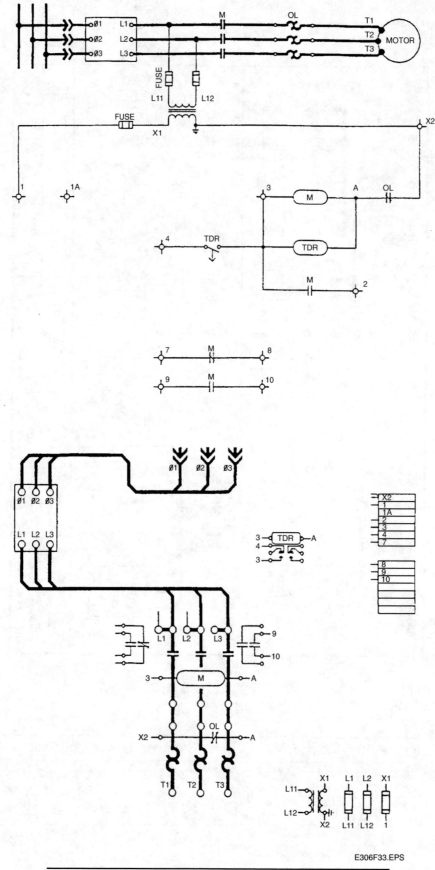

E306F33.EPS

Figure 31. Unit Diagrams For Motor Control Center

16.1.0   PANELBOARDS

Circuit control and overcurrent protection must be provided for all circuits and the power-consuming devices connected to these circuits. Lighting and power panels located throughout large buildings being supplied with electrical energy provide this control and protection. Fifteen panelboards are provided in a typical industrial building to feed electrical energy to the various circuits (*Figure 32*).

PANEL NO.	LOCATION	MAINS	VOTAGE RATING	NO. OF CIRCUITS	BREAKER RATINGS	POLES	PURPOSE
P-1	BASEMENT N. CORRIDOR	BREAKER 100 A	208/120 V 3 ø, 4 W	19 2 5	20 A 20 A 20 A	1 2 1	LIGHTING AND RECEPTACLES SPARES
P-2	BASEMENT N. CORRIDOR	BREAKER 100 A	208/120 V 3 ø, 4 W	24 2 0	20 A 20 A	1 2	LIGHTING AND RECEPTACLES SPARES
P-3	2nd FLOOR N. CORRIDOR	BREAKER 100 A	208/120 V 3 o, 4 W	24 2 0	20 A 20 A	1 2	LIGHTING AND RECEPTACLES SPARES
P-4	BASEMENT S. CORRIDOR	BREAKER 100 A	208/120 V 3 o, 4 W	24 2 0	20 A 20 A	1 2 1	LIGHTING AND RECEPTACLES SPARES
P-5	1st FLOOR S. CORRIDOR	BREAKER 100 A	208/120 V 3 o, 4 W	23 2 1	20 A 20 A 20 A	1 2 1	LIGHTING AND RECEPTACLES SPARES
P-6	2nd FLOOR S. CORRIDOR	BREAKER 100 A	208/120 V 3 o, 4 W	22 2 2	20 A 20 A 20 A	1 1 1	LIGHTING AND RECEPTACLES SPARES
P-7	Mfg. AREA S. WALL E.	BREAKER 100 A	208/120 V 3 o, 4 W	5 7 2	20 A 20 A 20 A	1 1 1	LIGHTING AND RECEPTACLES SPARES
P-8	Mfg. AREA S. WALL W.	BREAKER 100 A	208/120 V 3 o, 4 W	5 7 2	20 A 20 A 20 A	1 1 1	LIGHTING AND RECEPTACLES SPARES
P-9	Mfg. AREA S. WALL E.	BREAKER 100 A	208/120 V 3 o, 4 W	5 7 2	50 A 20 A 20 A	1 1 1	LIGHTING AND RECEPTACLES SPARES
P-10	Mfg. AREA S. WALL W.	BREAKER 100 A	208/120 V 3 o, 4 W	5 7 2	50 A 20 A 20 A	1 1 1	LIGHTING AND RECEPTACLES SPARES
P-11	Mfg. AREA EAST WALL	LUGS ONLY 225 A	208/120 V 3 o, 4 W	6	20 A	3	BLOWERS AND VENTILATORS
P-12	BOILER ROOM	BREAKER 100 A	208/120 V 3 o, 4 W	10 4	20 A 20 A	1 1	LIGHTING AND RECEPTACLES SPARES
P-13	BOILER ROOM	LUGS ONLY 225 A	208/120 V 3 o, 4 W	6	20 A	3	OIL BURNERS AND PUMPS
P-14	Mfg. AREA EAST WALL	LUGS ONLY 400 A	208/120 V 3 o, 4 W	3 2 1	175 A 70 A 40 A	3 3 3	CHILLERS FAN COIL UNITS FAN COIL UNITS
P-15	Mfg. AREA WEST WALL	LUGS ONLY 600 A	208/120 V 3 o, 4 W	5	100 A	3	TROLLEY BUSWAY AND ELEVATOR

E306F34.EPS

Figure 32.   Schedule Of Electric Panelboards For The Industrial Building

16.2.0 PANELBOARD CONSTRUCTION

In general, panelboards are constructed so that the main feed bus bars run the height of the panelboard. The buses to the branch circuit protective devices are connected to the alternate main buses. In an arrangement of this type, the connections directly across from each other are on the same phase and the adjacent connections on each side are on different phases. As a result, multiple protective devices can be installed to serve the 208-volt equipment. Examples of panelboards are shown in *Figures 33, 34,* and *35.*

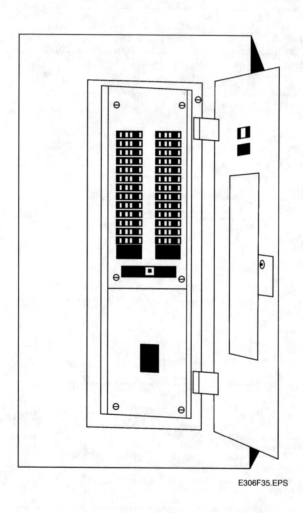

E306F35.EPS

Figure 33. Lighting And Appliance Panelboard With Main Breaker

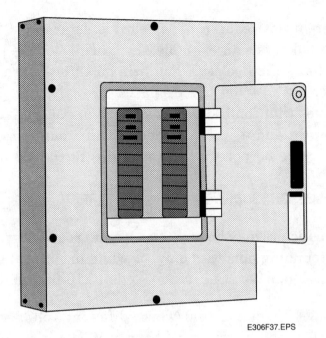

E306F37.EPS

Figure 34. Panelboard

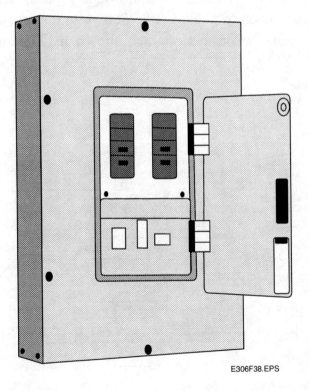

E306F38.EPS

Figure 35. Panelboard

16.2.1 Number Of Circuits

The number of overcurrent devices in a panelboard is determined by the needs of the area being served. Using the bakery as an example, there are 13 single-pole circuits and 5 three-pole circuits. This is a total of 28 poles. When using a three-phase supply, the incremental number is 6 (a pole for each of the three phases on both sides of the panelboard). The minimum number of poles that could be specified for the bakery is 30. This would limit the power available for growth, and it would not permit the addition of a three-pole lead. The reasonable choice is to go to 36 poles, which provides flexibility for growth loads.

16.2.2 Identification Of Conductors

The hot ungrounded conductors may be any color *except* green (or green with a yellow stripe), which is reserved for grounding purposes only, or white or natural gray, which are reserved for the grounded circuit conductor. See NEC Section 210-6(a) and (b).

Section 210-4(d) of the NEC requires that where different voltages exist in a building, the ungrounded conductors for each system must be identified and posted at each branch circuit panelboard. The Fine Print Note states that for the purposes of identification, "color coding, marking, tape, tagging, or other equally effective means is acceptable."

This situation occurs where, for instance, the building is served with 480/277 volts, and where step-down transformers are used to provide 208/120 volts for lighting and receptacle outlets. Examples of panelboard wiring connections are shown in *Figures 36, 37, 38,* and *39.*

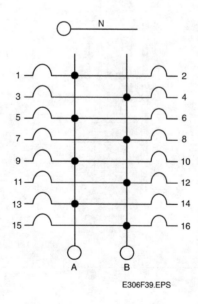

E306F39.EPS

Figure 36. Lighting And Appliance Branch Circuit Panelboard – Single-Phase, Three-Wire Connections

ELECTRICAL TRAINEE TASK MODULE 20306

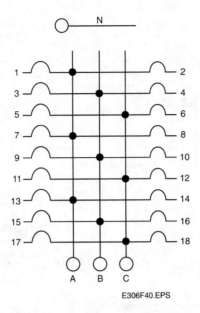

E306F40.EPS

Figure 37. Lighting And Appliance Branch Circuit Panelboard – Three-Phase, Four-Wire Connections

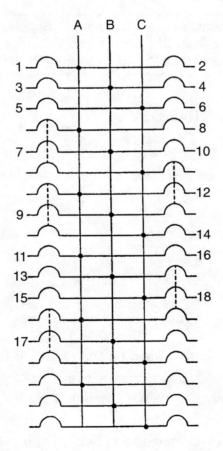

E306F41.EPS

Figure 38. Bakery Panelboard Circuit Numbering Scheme

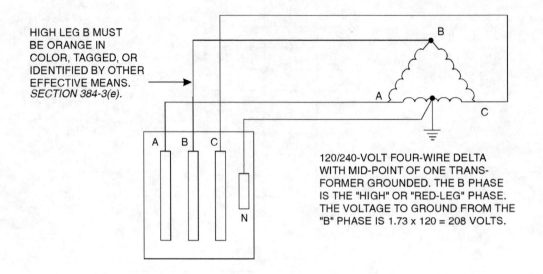

HIGH LEG B MUST
BE ORANGE IN
COLOR, TAGGED, OR
IDENTIFIED BY OTHER
EFFECTIVE MEANS.
SECTION 384-3(e).

120/240-VOLT FOUR-WIRE DELTA
WITH MID-POINT OF ONE TRANS-
FORMER GROUNDED. THE B PHASE
IS THE "HIGH" OR "RED-LEG" PHASE.
THE VOLTAGE TO GROUND FROM THE
"B" PHASE IS 1.73 x 120 = 208 VOLTS.

E306F43.EPS

Figure 39. Panelboards And Switchboards Supplied By Four-Wire,
Delta-Connected Systems Shall Have The "B" Phase Connected To
The Phase Having The Higher Voltage To Ground, NEC Section 384-3(e)

16.3.0 PANELBOARD PROTECTIVE DEVICE

The main for a panelboard may be either a fuse or a circuit breaker. This section concentrates on the use of circuit breakers. The selection of the circuit breaker should be based on the necessity to:

- Provide the proper overload protection
- Ensure a suitable voltage rating
- Provide a sufficient interrupting current rating
- Provide short circuit protection
- Coordinate the breaker(s) with other protective devices

The choice of the overload protection is based on the rating of the panelboard. The trip rating of the circuit breaker cannot exceed the amperage capacity of the bus bars in the panelboard. The number of branch circuit breakers generally is not a factor in the selection of the main protective device except in a practical sense. It is a common practice to have the total amperage of the branch breakers exceed the rating of the main breaker by several times; however, it makes little sense for a single branch circuit breaker to be the same size as or larger than the main breaker.

The voltage rating of the breaker must be higher than that of the system. Breakers are usually rated at 250 to 600 volts.

The importance of interrupting rating is covered in detail as follows. The trainee should recall that if there is any question as to the exact value of the short circuit current available at a point, a circuit breaker with a high interrupting rating is to be installed.

Many circuit breakers used as the main protective device are provided with an adjustable magnetic trip, *Figure 40*. Adjustments of this trip determine the degree of protection provided by the circuit breaker if a short circuit occurs. The manufacturer of this device provides exact information about the adjustments to be made. In general, a low setting may be ten or twelve times the overload trip rating. Two rules should be followed whenever the magnetic trip is set:

• The lower setting provides the greater protection.
• The setting should be lower than the value of the short circuit current available at that point.

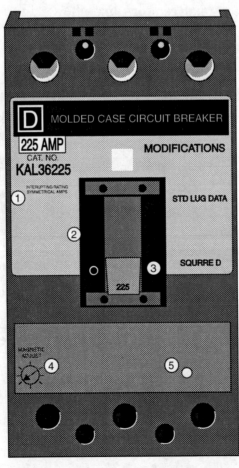

E306F44.EPS

Figure 40. Circuit Breaker With Adjustable Magnetic Trip

If subfeed lugs are used, the electrician must ensure that the lugs are suitable for making multiple breaker connections, as required by NEC Section 110-14(a). In general, this means that a separate lug is to be provided for each conductor being connected.

If taps are made to the subfeeder, they can be reduced in size according to NEC Section 240-21. This specification is very useful in cases such as that of panel P-12. For this panel, a 100-ampere main breaker is fed by a 350 MCM conductor. Within the distances given in the section, a conductor with a 100-ampere rating may be tapped to the subfeeder and connected to the 100-ampere main breaker in the panel.

The temperature rating of conductors must be selected and coordinated so as not to exceed the lowest temperature rating of any connected termination, conductor, or device [Section 110-14(c)].

16.4.0   BRANCH CIRCUIT PROTECTIVE DEVICES

The schedule of panelboards for the industrial building *(Figure 32)*, shows that lighting panels P-1 through P-6 have 20-ampere circuit breakers, including two double-pole breakers (to supply special receptacle outlets). A double-pole breaker requires the same installation space as two single-pole breakers. Breakers are shown in *Figure 41*.

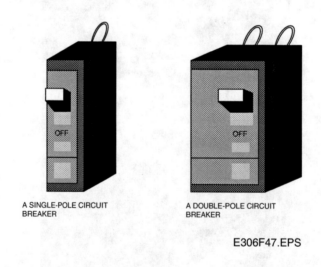

A SINGLE-POLE CIRCUIT BREAKER

A DOUBLE-POLE CIRCUIT BREAKER

E306F47.EPS

Figure 41.   Branch Circuit Protective Devices

Summary

An electrical drawing consists of lines, symbols, dimensions, and notations to accurately convey an engineer's design to workers who install the electrical system on the job. The trainee should keep in mind that the workers must be able to take a complete set of electrical drawings, related written specifications, and supplemental drawings, and without further instruction, install or produce the electrical system as the engineer or designer intended it to be accomplished. An electrical drawing, therefore, is an abbreviated language for conveying a large amount of exact, detailed information, which would otherwise take many pages of manuscript or hours of verbal instruction to convey. In every branch of electrical work, there is often occasion to interpret an electrical drawing. Electricians, for example, who are responsible for installing the electrical system in a new installation, usually consult a set of electrical drawings and specifications to locate the incoming electric service, switchgear, main distribution panels, subpanels, motor control centers, routing of raceways and circuits, and similar details. Electrical estimators must refer to electrical drawings in order to determine the quantity of materials needed in preparing a bid. Electricians in industrial plants consult schematic diagrams when wiring electrical controls for motor applications. Plant maintenance personnel use electrical drawings in troubleshooting problems that occur. Circuits may be tested and checked against the original drawings to help locate any faulty points in the installation.

References

For advanced study of topics covered in this Task Module, the following works are suggested:

American Electricians Handbook, Latest Edition, Croft, McGraw-Hill, New York, NY.
National Electrical Code Handbook, Latest Edition, NFPA, Quincy, MA.

SELF CHECK REVIEW / PRACTICE QUESTIONS

1. Which of the following changes the voltage level of an electrical system or circuit?
   a. Transducer
   b. Transformer
   c. Capacitor
   d. Rectifier

2. The term ampacity refers to the:
   a. starting current of a motor.
   b. maximum full load current of a motor.
   c. current in amperes a conductor can carry continuously under conditions of use without exceeding its ratings.
   d. trip setting of a circuit breaker.

3. The term interrupting rating refers to the:
   a. trip setting of a circuit breaker.
   b. voltage rating of a fuse.
   c. highest voltage level a device can withstand.
   d. maximum current a device will safely interrupt at rated voltage.

4. A device that is designed to protect equipment from ground faults is a:
   a. molded case circuit breaker.
   b. dual element fuse.
   c. ground fault relay.
   d. ground fault circuit interrupter.

5. Which of the following is *not* a requirement for mounting a fuse in switchgear?
   a. Mounted horizontally
   b. Mounted vertically
   c. Accessible
   d. Incoming wires connected to top terminals on the fuse blocks

6. The key to reading and interpreting electrical diagrams is to understand and use the:
   a. legend.
   b. index.
   c. glossary.
   d. symbol chart.

7. What diagram shows the actual wiring connections between unit assemblies or equipment?
   a. Schematic diagram
   b. Front panel diagram
   c. Interconnection diagram
   d. Block diagram

8. Which of the following is *not* a general classification of circuit breakers?
   a. Fuse circuit breaker
   b. Air circuit breaker
   c. Oil circuit breaker
   d. Gas circuit breaker

9. Transformers can be divided into two main categories: _____ and _____.

10. What gas is typically used to cool transformers? _____

11. If a device number "51" is indicated on an electrical print, it would be a:
    a. circuit breaker.
    b. reverse power relay.
    c. field circuit breaker.
    d. AC time overcurrent relay.

12. Choose from among the following the reason for using instrument transformers which is *not* correct.
    a. Instrument transformers have uniform secondary quantity ratings.
    b. Instrument transformers make it possible to use reasonable insulation ratings in relays.
    c. Instrument transformers provide isolation between the power system and the relays.
    d. Instrument transformers are more accurate than direct monitoring of power system voltages and currents.

13. Which of the following is *not* a necessary element in a basic transformer?
    a. A core
    b. A primary winding
    c. A secondary winding
    d. A magnetic shield

14. The maximum voltage which a piece of equipment can withstand is:
    a. interrupting capacity.
    b. (BIL) basic impulse level.
    c. current limit.
    d. frequency.

15. When an unintended path is established between an ungrounded conductor and ground, it is called a:
    a. phase fault.
    b. open circuit.
    c. ground fault.
    d. short circuit.

Appendix Standard Device Numbers

To simplify electrical diagrams, many switchgear devices are not labeled with reference to their function. Standard numbers are commonly used instead of standard abbreviations. These standard numbers, like abbreviations, allow the designer to produce an uncluttered drawing by minimizing the amount of writing.

The following is a list of standard numbers for labeling switchgear devices. This list can be used for quick reference; the technician does not need to memorize the numbers. Numbers commonly used will become as familiar to the technician as common abbreviations. Standard numbers for switchgear devices are as follows:

1 - Master element
2 - Time delay closing relay
3 - Interlocking relay
4 - Master contactor
5 - Stopping device
6 - Starting circuit breaker
7 - Anode circuit breaker
8 - Control power disconnect device
9 - Reversing device
10 - Unit sequence switch
11 - Reserved for future application
12 - Overspeed device
13 - Synchronous-speed device
14 - Underspeed device
15 - Speed or frequency matching device
16 - Reserved for future application
17 - Shunting or discharge switch
18 - Accelerating or decelerating device
19 - Starting-to-running transition device
20 - Electrically operated valve
21 - Distance relay
22 - Equalizer circuit breaker
23 - Temperature control device
24 - Reserved for future application
25 - Synchronizing check device
26 - Apparatus thermal device
27 - Undervoltage relay
28 - Flame detector
29 - Isolating contactor

30	-	Annunciator relay
31	-	Separate excitation device
32	-	Directional power relay
33	-	Position switch
34	-	Sequence device
35	-	Brush operating device
36	-	Polarity device
37	-	Undercurrent/underpower device
38	-	Bearing protective device
39	-	Mechanical condition monitor
40	-	Field relay
41	-	Field circuit breaker
42	-	Running circuit breaker
43	-	Manual transfer device
44	-	Unit sequence starting relay
45	-	Atmospheric condition monitor
46	-	Reverse phase relay
47	-	Phase sequence voltage relay
48	-	Incomplete sequence relay
49	-	Thermal relay
50	-	Instantaneous overcurrent relay
51	-	AC time overcurrent relay
52	-	AC circuit breaker
53	-	Exciter or DC generator relay
54	-	High speed DC circuit breaker
55	-	Power factor relay
56	-	Field application relay
57	-	Short-circuiting or grounding device
58	-	Rectifier failure relay
59	-	Overvoltage relay
60	-	Voltage balance relay
61	-	Current balance relay
62	-	Time delay relay
63	-	Pressure switch
64	-	Ground protective relay
65	-	Governor
66	-	Jogging device
67	-	AC directional overcurrent delay
68	-	Blocking delay
69	-	Permissive control device
70	-	Electrically operated rheostat
71	-	Level switch

72	-	DC circuit breaker
73	-	Load resistor contactor
74	-	Alarm relay
75	-	Position changing mechanism
76	-	DC overcurrent relay
77	-	Pulse transmitter
78	-	Phase angle relay
79	-	AC reclosing relay
80	-	Flow switch
81	-	Frequency relay
82	-	DC reclosing relay
83	-	Automatic transfer relay
84	-	Operating mechanism
85	-	Carrier receiver relay
86	-	Lockout relay
87	-	Differential relay
88	-	Auxiliary motor
89	-	Line switch
90	-	Regulating device
91	-	Voltage directional relay
92	-	Voltage/power directional relay
93	-	Field changing contactor
94	-	Tripping relay
95 to 99	-	Used for specific applications

The following gives a brief description of the function of each of the switchgear devices in the previous list.

Device Number	Function and Description
1	Master Element - The initiating device (control switch, voltage relay, float switch, etc.) that places equipment into or out of operation. This is done either directly or through a permissive device, such as a protective or time delay relay.
2	Time Delay Starting (or Closing) Relay - A device that provides a given amount of time delay before or after any operation in a switching sequence or protective relay system. This is true except as specifically provided by devices 62 and 79, described later.
3	Checking (or Interlocking) Relay - A relay that operates in response to the position of other devices, or to the predetermined conditions in equipment. This relay allows an operating sequence to continue, stops the sequence, or provides a check of the position of the devices or the predetermined conditions for any purpose.
4	Master Contactor - A device that makes or breaks the necessary control circuits to place equipment into or out of service when the required conditions exist. This device is generally controlled by a master element (device 1), and the necessary permissive and protective devices.
5	Stopping Device - A device whose primary function is to place and hold equipment out of service.
6	Starting Circuit Breaker - A device that is used in the anode circuits of a power rectifier to interrupt the rectifier current if an arc-back occurs.
7	Anode Circuit Breaker - A device that is used in the anode circuits of a power rectifier to interrupt the rectifier current if an arc-back occurs.
8	Control Power Disconnecting Device - A device (knife switch, circuit breaker, or pullout fuse block) that is used to connect or disconnect the control power to and from the control bus or equipment. Control power includes auxiliary power that supplies small motors and heaters.
9	Reversing Device - A device that is used to reverse a machine's field or to perform any other reversing function.
10	Unit Sequence Switch - A switch that changes the sequence in which units may be placed into or out of service in a multiunit system.
11	Reserved for future application.

12	Over-Speed Device - A direct-connected speed switch that functions when a machine overspeeds.
13	Synchronous-Speed Device - A device (centrifugal speed switch, slip-frequency relay, voltage relay, undercurrent relay, or any other type of device) that operates at approximately the synchronous speed of a machine.
14	Under-Speed Device - A device that functions when the speed of a machine falls below a predetermined value.
15	Speed (or Frequency) Matching Device - A device that matches and holds the speed (frequency) of a machine or a system equal (or approximately equal) to that of another machine, source, or system.
16	Reserved for future application.
17	Shunting (Discharge) Switch - A switch that opens or closes a shunting circuit around any piece of apparatus (except a resistor), such as: a machine field, a machine armature, a capacitor, or a reactor. This excludes devices that perform the shunting operations that are necessary when a machine is started by devices 6 or 42. Device 73, the switching of a load resistor, is also excluded.
18	Accelerating (Decelerating) Device - A device that closes circuits used to increase or decrease the speed of a machine.
19	Start-To-Running Transition Contactor - A device that causes the automatic transfer of a machine from the starting to the running power connection.
20	Electrically Operated Valve - A motor operated valve that is used in vacuum, air, gas, oil, water, or similar lines. The function of the valve may be indicated by the insertion of a descriptive word, such as *brake* in the function name (i.e., electrically operated brake valve).
21	Distance Relay - A relay that functions when the circuit impedance, or reactance increases or decreases beyond predetermined limits.
22	Equalizer Circuit Breaker - A breaker that controls the equalizer or current balancing connections for a machine field, or for regulating equipment, in a multiunit system
23	Temperature Control Device - A device that raises or lowers the temperature of a machine or other apparatus (or of any medium), when its temperature falls below or rises above a predetermined value. An example is a thermostat that switches on a space heater in a switchgear assembly when the temperature falls below the predetermined value. This is different than a device that provides automatic temperature regulation between close limits (Device 90).
24	Reserved for future application.

25	Synchronizing (Synchronism Check) Device - A device that permits or causes the paralleling of two AC sources when they are within the desired limits of frequency, phase angle, and voltage.
26	Apparatus Thermal Device - A device that functions when the temperature of the field of a machine, a load limiting or shifting resistor, a liquid, or any other medium exceeds a predetermined limit. It also functions if the temperature of the protected apparatus, such as a power rectifier, decreases below a predetermined limit.
27	Undervoltage Relay - A relay that functions on a given value of undervoltage.
28	Flame Detector - A device that monitors the presence of the pilot or main flame in apparatus such as a gas turbine or a steam boiler.
29	Isolating Contactor - A contactor that is used expressly to disconnect one circuit from another to perform emergency operations, maintenance, or tests.
30	Annunciator Relay - A non-automatic reset device that gives a number of visual indications upon the functioning of a protective device. It may also be arranged to perform a lockout function.
31	Separate Excitation Device - A device that connects a circuit, such as the shunt field of a synchronous converter, to a source of separate excitation during the starting sequence. This device also energizes the excitation and ignition circuits of a power rectifier.
32	Directional Power Relay - A device that functions on a desired value of power flow in a given direction, or upon reverse power resulting from arc-back in the anode or cathode circuits of a power rectifier.
33	Position Switch - A switch that makes or breaks contact when the main device or piece of apparatus, which has no device function number, reaches a given position.
34	Sequence Device - A multi-contact switch that fixes the operating sequence of the major devices during starting, stopping, or other sequential switching operations.
35	Brush Operating (Slip-Ring Short Circuiting) Device - A device that raises, lowers, or shifts the position of the brushes in a machine, or that short circuits its slip-rings. This device also engages or disengages the contacts of a mechanical rectifier.
36	Polarity Device - A device that operates or permits the operation of another device on a predetermined polarity only.
37	Undercurrent (Underpower) Relay - A relay that functions when the current or power flow decreases below a predetermined value.

38	Bearing Protective Device - A device that functions on excessive bearing temperature or on other abnormal mechanical conditions, such as undue wear, which may eventually result in excessive bearing temperature.
39	Mechanical Condition Monitor - A device that functions upon the occurrence of an abnormal mechanical condition (except that associated with bearings as covered under Device 38). Examples of abnormal conditions are: excessive vibration, eccentricity, expansion, shock, tilting, or seal failure.
40	Field Relay - A relay that functions on a given low value, or failure, of machine field current. This relay also functions on an excessive value of the reactive component of armature current in an AC machine, which indicates abnormally low field excitation.
41	Field Circuit Breaker - A device that applies or removes the field excitation of a machine.
42	Running Circuit Breaker - A device that connects a machine to its source of running voltage, after the machine has been brought up to the desired speed on the starting connection (motor starter).
43	Manual Transfer (Selector) Device - A manually operated device that transfers the control circuits so as to modify the plan of operation of the switching equipment or of some of the devices.
44	Unit Sequence Starting Relay - A relay that functions to start the next available unit, in a multiunit system, on the failure or the nonavailability of the normally preceding unit.
45	Atmospheric Condition Monitor - A device that functions upon the occurrence of an abnormal atmospheric condition, such as fumes, explosive mixtures, smoke, and fire.
46	Reverse-Phase (Phase-Balance) Current Relay - A relay that functions when the polyphase currents are reverse-phase sequenced, unbalanced, or contain negative phase sequence components above a given amount.
47	Phase-Sequence Voltage Relay - A relay that functions on a predetermined value of polyphase voltage in the desired phase sequence.
48	Incomplete Sequence Relay - A relay that returns the equipment to the normal or off position and locks it out. This relay functions if the normal starting, operating, or stopping sequence is not properly completed within a predetermined time.
49	Machine (Transformer) Thermal Relay - A relay that functions when the temperature of a machine armature or other load-carrying winding (element), a power rectifier, or a power transformer (including a power rectifier transformer) exceeds a predetermined value.

50	Instantaneous Overcurrent (Rate-Of-Rise) Relay - A relay that functions instantaneously on an excessive value of current, or on an excessive rate of current rise. This relay is used to indicate a fault in the apparatus or circuit being protected.
51	AC Time Overcurrent Relay - A relay with either a definite or inverse time characteristic that functions when the current in an AC circuit exceeds a predetermined value.
52	AC Circuit Breaker - A device that closes or interrupts an AC power circuit under normal conditions. The breaker also interrupts the circuit under fault or emergency conditions.
53	Exciter (DC Generator) Relay - A relay that forces a DC machine's field excitation to build up during starting or that functions when the machine's voltage has built up to a given value.
54	High Speed DC Circuit Breaker - A circuit breaker that functions to reduce the current in the main circuit in 0.01 seconds or less. The breaker functions after the occurrence of a DC overcurrent or excessive rate of current rise.
55	Power Factor Relay - A relay that operates when the power factor in an AC circuit rises above or falls below a predetermined value.
56	Field Application Relay - A relay that automatically controls the application of field excitation to an AC motor at some predetermined point in the slip cycle.
57	Short Circuiting (Grounding) Device - A power or stored energy device that short circuits (or grounds) a circuit in response to automatic or manual means.
58	Power Rectifier Misfire Relay - A relay that functions if one or more of the power rectifier anodes fails to fire.
59	Overvoltage Relay - A relay that functions on a given value of overvoltage.
60	Voltage Balance Relay - A relay that operates on a given difference in voltage between two circuits.
61	Current Balance Relay - A relay that operates on a given difference in the current input or output of two circuits.
62	Time-Delay Stopping (Opening) Relay - A time-delay relay that serves in conjunction with the device that initiates the shutdown, stopping, or opening operation in an automatic sequence.
63	Liquid or Gas Pressure, Level, or Flow Relay - A relay that operates on given values of liquid pressure, gas pressure, flow, or level, or on a given rate of change of these values. This relay is normally an auxiliary relay. Also see Devices 71 and 80.

64	Ground Protective Relay - A relay that functions on the failure of the insulation of a machine, a transformer, or any other apparatus to ground. This relay also functions on a flash over to ground in a DC machine. This function is only assigned to a relay that detects current flow from the frame of a machine (enclosing case or the structure of an apparatus) to ground. This relay also detects grounds on normally ungrounded windings or circuits. This function is not applied to a device connected in the secondary circuit or neutral of a current transformer, or current transformers connected in the power circuit of a normally grounded system.
65	Governor - The equipment that controls the gate or valve opening of a prime mover.
66	Notching (Jogging) Device - A device that functions to allow only a specified number of operations of a given device or equipment, or a specified number of successive operations within a given time of each other. It is also a device that functions to energize a circuit periodically, or that is used to permit intermittent acceleration (jogging) of a machine at low speed for mechanical positioning.
67	AC Directional Overcurrent Relay - A relay that functions on a desired value of AC overcurrent flowing in a predetermined direction.
68	Blocking Relay - A relay that initiates a pilot signal for blocking a trip on external faults in a transmission line, or in any other apparatus, under predetermined conditions. It also cooperates with other devices to trips or reclosings on an out-of-step condition.
69	Permissive Control Device - Generally, a two-position, manually operated switch that permits the closing of a circuit breaker, or the placing of equipment into operation in one position, and prevents the circuit beaker or the equipment from being operated in the other position.
70	Electrically Operated Rheostat - A rheostat that is used to vary the resistance of a circuit in response to some means of electrical control.
71	Level Switch - A switch that operates on given values, or on a given rate of change of level.
72	DC Circuit Breaker - A circuit breaker that closes or interrupts a DC circuit under normal conditions. The breaker also interrupts the circuit under fault or emergency conditions.
73	Load Resistor Contractor - A contactor that is used to shunt or to insert a step of load limiting, shifting, or indicating resistance in a power circuit. This device is also used to switch a space heater in a circuit, and to switch a light on the regenerative load resistor of a power rectifier, or other machine into or out of a circuit.

74	Alarm Relay - A relay, other than an annunciator (Device 30), which is used to operate, or operate in conjunction with, a visual or audible alarm.
75	Position Changing Mechanism - The mechanism that is used to move a removable circuit breaker unit to and from the connected, disconnected, and test position.
76	DC Overcurrent Relay - A relay that functions when the current in a DC circuit exceeds a given value.
77	Pulse Transmitter - Generates and transmits pulses over a telemetering or pilot-wire circuit to the remote indicating or receiving device.
78	Phase Angle Measuring (Out-Of-Step Protective) Relay - A relay that functions at a predetermined phase angle between two voltages, two currents, or a voltage and current.
79	AC Reclosing Relay - A relay that controls the automatic reclosing and locking out of an AC circuit interrupter.
80	Flow Switch - A switch that operates on a given value, or on a given rate of change of flow.
81	Frequency Relay - A relay that functions on a predetermined value of frequency (either above or below normal system frequency), or rate of change of frequency.
82	DC Reclosing Relay - A relay that controls the closing and reclosing of a DC circuit interrupter, generally in response to load circuit conditions.
83	Automatic Selective Control (Transfer) Relay - A relay that operates to automatically select between certain sources or conditions in equipment. The relay can also perform automatic transfer operations.
84	Operating Mechanism - The complete electrical mechanism (or servo-mechanism), including the operating motor, solenoids, position switches, etc., for a tap change or any piece of apparatus that has no device number.
85	Carrier (Pilot-Wire) Receiver Relay - A relay that is operated or restrained by a signal used in conjunction with carrier-current or DC pilot-wire fault directional relaying.
86	Lock-Out Relay - An electrically operated (hand or electrically reset) device that functions to shut down and hold equipment out of service on the occurrence of an abnormal condition.
87	Differential Protective Relay - A protective relay that functions on a percentage, a phase angle, or other quantitative difference between two currents, or some other electrical quantities.

88	Auxiliary Motor (Motor Generator) - A motor used to operate auxiliary equipment, such as pumps, blowers, exciters, rotating magnetic amplifiers, etc.
89	Line Switch - A switch used as a disconnecting or isolating switch in an AC or DC power circuit, when the device is electrically operated or has electrical accessories such as an auxiliary switch, a magnetic lock, etc.
90	Regulating Device - A device that controls a quantity (or quantities), such as voltage, current, power, frequency, temperature, and load at a given value or between certain limits, for machines, tie lines, or other apparatus.
91	Voltage Directional Relay - A relay that operates when the voltage across an open circuit breaker or contactor exceeds a given value in a given direction.
92	Voltage and Power Directional Relay - A relay that permits or causes the connection of two circuits when the voltage difference between them exceeds a given value in a predetermined direction, and causes these two circuits to be disconnected from each other when the power flowing between them exceeds a given value in the opposite direction.
93	Field Changing Contactor - A contactor that functions to increase or decrease, in one step, the value of field excitation on a machine.
94	Tripping (Trip-Free) Relay - A relay that functions to trip a circuit breaker, a contactor or equipment, or to permit immediate tripping by another device. This device may also prevent the immediate reclosing of a circuit interrupter, should it open automatically, even though its closing circuit is maintained closed.
95 to 99	Used only for specific applications on individual installations where none of the assigned numbered functions from 1 to 94 are suitable. When the standard numbers are used on an electrical diagram, they are sometimes preceded (or followed) by an additional number or letter. This is used for more precise identification. For example, there may be differential relay protection on two different buses. The designation of a differential relay is 87. However, if this number is used for both buses, confusion could result. If the two buses are the 6,900-volt bus and the 4,160-volt bus, the 6,900-volt bus relay could be 687 and the 4,160-volt bus relay could be 487.

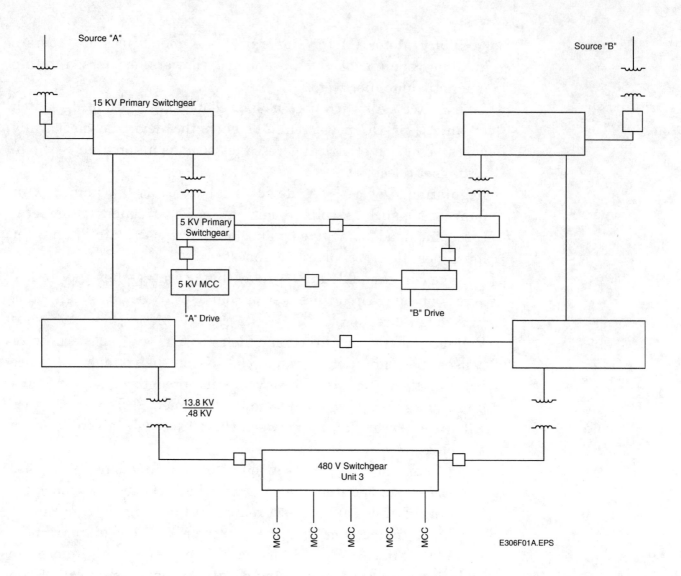

Source "A"

Source "B"

15 KV Primary Switchgear

5 KV Primary
Switchgear

5 KV MCC

"A" Drive

"B" Drive

13.8 KV
.48 KV

480 V Switchgear
Unit 3

MCC MCC MCC MCC MCC

E306F01A.EPS

ELECTRICAL TRAINEE TASK MODULE 20306

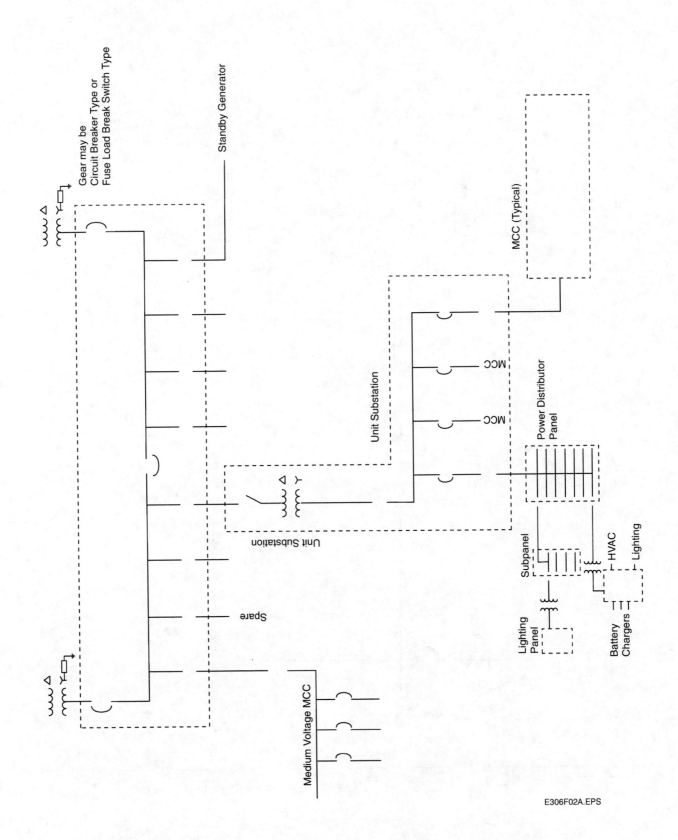

Gear may be
Circuit Breaker Type or
Fuse Load Break Switch Type

Standby Generator

MCC (Typical)

Unit Substation

MCC

MCC

Power Distributor
Panel

Unit Substation

Spare

Subpanel

HVAC

Lighting

Lighting
Panel

Battery
Chargers

Medium Voltage MCC

E306F02A.EPS

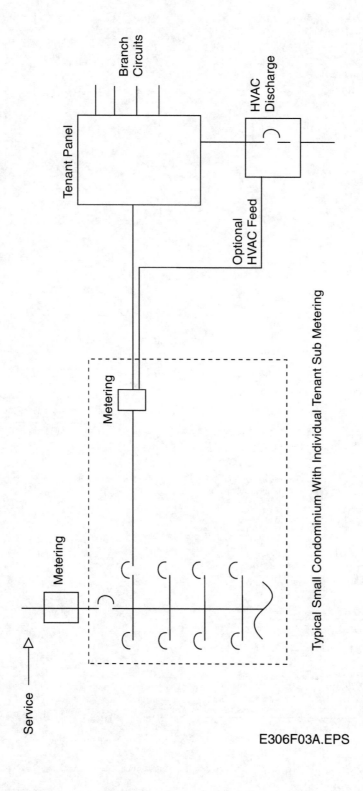

Typical Small Condominium With Individual Tenant Sub Metering

E306F03A.EPS

Distribution System Transformers

Module 20307

Electrical Trainee Task Module 20307

DISTRIBUTION SYSTEM TRANSFORMERS

Objectives

Upon completion of this module, the trainee will be able to:

1. Describe how transformers operate.
2. Explain the principle of mutual inductance.
3. Describe the operating characteristics of the various types of transformers.
4. Connect a multi-tap transformer for the required secondary voltage.
5. Explain NEC requirements governing the installation of transformers.
6. Compute transformer sizes for various applications.
7. Explain types and purposes of grounding transformers.
8. Connect a control transformer for a given application.
9. Size the maximum load allowed on open-delta systems.
10. Describe how current transformers are used in conjunction with watt-hour meters.
11. Apply capacitors and rectifiers to practical applications.
12. Calculate the power factor of any given electrical circuit.

Prerequisites

Successful completion of the following Task Module is required before beginning study of this Task Module: Core Curricula, Electrical Levels 1 and 2, Electrical Level 3, Modules 20301 through 20306.

Required Student Materials

1. Trainee Task Module
2. Copy of the latest edition of the National Electrical Code

COURSE MAP

This course map shows all of the *Wheels of Learning* task modules in the third level of the Electrical curricula. The suggested training order begins at the bottom and proceeds up. Skill levels increase as a trainee advances on the course map. The training order may be adjusted by the local Training Program Sponsor.

Course Map: Electrical, Level 3

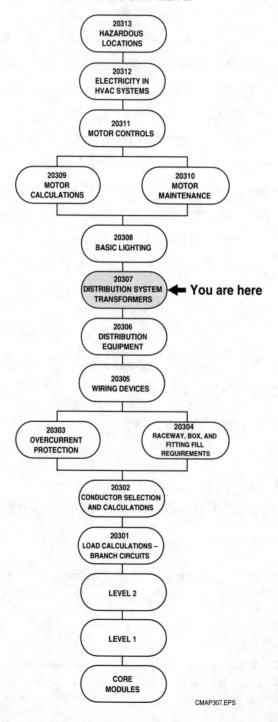

LEVEL 3 COMPLETE

20313 HAZARDOUS LOCATIONS

20312 ELECTRICITY IN HVAC SYSTEMS

20311 MOTOR CONTROLS

20309 MOTOR CALCULATIONS

20310 MOTOR MAINTENANCE

20308 BASIC LIGHTING

20307 DISTRIBUTION SYSTEM TRANSFORMERS ← **You are here**

20306 DISTRIBUTION EQUIPMENT

20305 WIRING DEVICES

20303 OVERCURRENT PROTECTION

20304 RACEWAY, BOX, AND FITTING FILL REQUIREMENTS

20302 CONDUCTOR SELECTION AND CALCULATIONS

20301 LOAD CALCULATIONS – BRANCH CIRCUITS

LEVEL 2

LEVEL 1

CORE MODULES

CMAP307.EPS

TABLE OF CONTENTS

Trade Terms Introduced in This Module

ampere-turn: The product of amperes times the number of turns in a coil.

autotransformer: Any transformer where primary and secondary connections are made to a single cell. The application of an autotransformer is a good choice for some users where a 480Y/277- or 208Y/120-volt, three-phase, four-wire distribution system is utilized.

capacitance: The storage of electricity in a capacitor; the opposition to voltage change; the unit of measurement is the farad (f) or microfarad (mf).

current transformer: A single-phase instrument transformer connected in series in a line that carries the full-load current. The turns ratio is designed to produce a reduced current in the secondary suitable for the current coil of standard measuring instruments and in proportion to the load current.

flux: 1) The rate of flow of energy across or through a surface. 2) A substance used to promote or facilitate soldering or welding by removing surface oxides.

hertz: The derived SI unit for frequency; one hertz equals one cycle per second.

hum: Interference from AC power, normally of low frequency and audible.

induction: The production of magnetization or electrification in a body by the mere proximity of magnetized or electrified bodies, or of an electric current in a conductor by the variation of the magnetic field in its vicinity.

induction coil: Essentially a transformer with open magnetic current, in which an alternating current of high voltage is induced in the secondary by means of a pulsating direct current in the primary.

inductor: A device having winding(s) with or without a magnetic core for creating inductance in a circuit.

kVA: Kilovolts times amperes. Also referenced as 1000 volt-amperes.

labeled: Items to which a label, trademark, or other identifying mark of nationally recognized testing labs has been attached to identify the items as meeting appropriate standards.

loss: Power expended without doing useful work.

magnetic coil: The winding of an electromagnet. A coil of wire wound in one direction, producing a dense magnetic field capable of attracting iron or steel when carrying a current of electricity.

magnetic density: The number of lines of force or induction per unit area taken perpendicular to the induction. In free space, flux density and field intensity are the same numerically, but within magnetic material, the two are quite different.

magnetic field: The area around a magnet in which the effect of the magnet can be felt.

magnetic induction: The number of magnetic lines or the magnetic flux per unit of cross-sectional area perpendicular to the direction of the flux.

magnetism: The ability of two pieces of iron to be attracted to each other by physical means or electrical means.

mutual inductance: The condition of voltage in a second conductor because of current in another conductor.

potential transformer: Designed for use in measuring high voltage; normally the secondary voltage is 120 volts.

power transformer: Designed to transfer electrical power from the primary circuit to the secondary circuit(s) to 1) step-up the secondary voltage at less current or 2) step-down the secondary voltage at more current; with the voltage-current product being constant for either primary or secondary.

reactance: The imaginary part of impedance. The opposition to AC due to capacitance and/or inductance.

rectifiers: Devices used to change alternating current to direct current.

resonance: A condition reached in an electrical circuit when the inductive reactance neutralizes the capacitance reactance leaving ohmic resistance as the only opposition to the flow of current.

safety-isolation transformer: Inserted to provide a nongrounded power supply such that a grounded person accidentally coming in contact with the secondary circuit will not be electrocuted.

self-inductance: Magnetic field induced in the conductor carrying the current.

transformer: A static device consisting of winding(s) with or without a tap(s) with or without a magnetic core for introducing mutual coupling by induction between circuits.

transformer-rectifier: Combination transformer and rectifier in which input in AC may be varied and then rectified into DC.

turn: The basic coil element that forms a single conducting loop comprised of one insulated conductor.

turn ratio: The ratio between the number of turns between windings in a transformer; normally primary to secondary, except for current transformers it is the ratio of the secondary to the primary.

vault-type transformers: Suitable for occasional submerged operation in water.

1.0.0 INTRODUCTION TO TRANSFORMERS

The electric power produced by alternators in a generating station is transmitted to locations where it is utilized and distributed to users. Many different types of transformers play an important role in the distribution of electricity. Power transformers are located at generating stations to step up the voltage for more economical transmission. Substations with additional power transformers and distribution equipment are installed along the transmission line. Finally, distribution transformers are used to step down the voltage to a level suitable for utilization.

Transformers are also used quite extensively in all types of control work, to raise and lower AC voltage on control circuits. They are also used in 480Y/277 systems to reduce the voltage for operating 208Y/120-volt lighting and other electrically-operated equipment. Buck-and-boost transformers are used for maintaining appropriate voltage levels in certain electrical systems.

It is important for anyone working with electricity to become familiar with transformer operation; that is, how they work, how they are connected into circuits, their practical applications and precautions to take during the installation or while working on them. This module is designed to cover these items as well as overcurrent protection and grounding. Other subjects include correcting power factor with capacitors, and the application of rectifiers.

2.0.0 TRANSFORMER BASICS

A very basic transformer consists of two coils or windings formed on a single magnetic core as shown in Fig. 1. Such an arrangement will allow transforming a large alternating current at

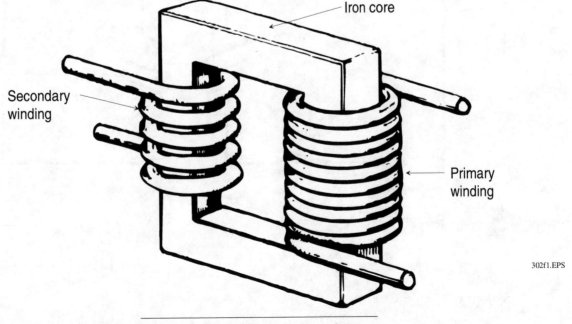

Figure 1. Basic parts of a transformer.

low voltage into a small alternating current at high voltage, or vice versa. But let's start at the beginning. What makes a transformer work?

2.1.0   MUTUAL INDUCTION

The term *mutual induction* refers to the condition in which two circuits are sharing the energy of one of the circuits. It means that energy is being transferred from one circuit to another.

Consider the diagram in Fig. 2. Coil A is the primary circuit which obtains energy from the battery. When the switch is closed, the current starts to flow and a magnetic field expands out of coil A. Coil A then changes electrical energy of the battery into the magnetic energy of a magnetic field. When the field of coil A is expanding, it cuts across coil B, the secondary circuit, inducing a voltage in coil B. The indicator (a galvanometer) in the secondary circuit is deflected, and shows that a current, developed by the induced voltage, is flowing in the circuit.

The induced voltage may be generated by moving coil B through the flux of coil A. However, this voltage is induced without moving coil B. When the switch in the primary circuit is open, coil A has no current and no field. As soon as the switch is closed, current passes through the coil and the magnetic field is generated. This expanding field moves or "cuts" across the wires of coil B, thus inducing a voltage without the movement of coil B.

The magnetic field expands to its maximum strength and remains constant as long as full current flows. Flux lines stop their cutting action across the turns of coil B because expansion of the field has ceased. At this point the indicator needle on the meter reads zero because no

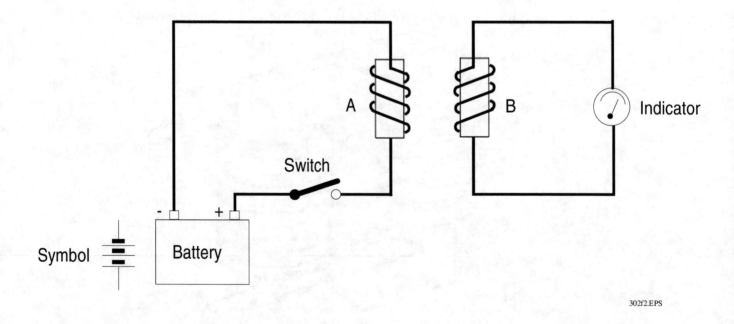

302f2.EPS

Figure 2. Mutual induction circuits.

ELECTRICAL TRAINEE TASK MODULE 20307

induced voltage exists anymore. If the switch is opened, the field collapses back to the wires of coil A. As it does so, the changing flux cuts across the wires of coil B, but in the opposite direction. The current present in the coil causes the indicator needle to deflect, showing this new direction. The indicator, then, shows current flow only when the field is changing, either building up or collapsing. In effect, the changing field produces an induced voltage exactly as does a magnetic field moving across a conductor. This principle of inducing voltage by holding the coils steady and forcing the field to change is used in innumerable applications. The transformer is particularly suitable for operation by mutual induction. Transformers are perfect components for transferring and changing AC voltages as needed.

Transformers are generally composed of two coils placed close to each other but not connected together. Refer once more to Fig. 1. The coil that receives energy from the line voltage source, etc., is called the "primary" and the coil that delivers energy to a load is called the "secondary." Even though the coils are not physically connected together they manage to convert and transfer energy as required by a process known as mutual induction.

Transformers, therefore, enable changing or converting power from one voltage to another. For example, generators that produce moderately large alternating currents at moderately high voltages utilize transformers to convert the power to very high voltage and proportionately small current in transmission lines, permitting the use of smaller cable and providing less power loss.

When alternating current (AC) flows through a coil, an alternating magnetic field is generated around the coil. This alternating magnetic field expands outward from the center of the coil and collapses into the coil as the AC through the coil varies from zero to a maximum and back to zero again, as discussed in an earlier module. Since the alternating magnetic field must cut through the turns of the coil, a self-inducing voltage occurs in the coil which opposes the change in current flow.

If the alternating magnetic field generated by one coil cuts through the turns of a second coil, voltage will be generated in this second coil just as voltage is induced in a coil which is cut by its own magnetic field. The induced voltage in the second coil is called the "voltage of mutual induction," and the action of generating this voltage is called "transformer action." In transformer action, electrical energy is transferred from one coil (called the primary) to another (the secondary) by means of a varying magnetic field.

2.2.0 INDUCTION IN TRANSFORMERS

A simple transformer consists of two coils located very close together and electrically insulated from each other. The coil to which the AC is applied is called the *primary*. It generates a magnetic field which cuts through the turns of the other coil, called the *secondary*, and generates a voltage in it. The coils are not physically connected to each other. They are, however, magnetically coupled to each other. Consequently, a transformer transfers electrical power from one coil to another by means of an alternating magnetic field.

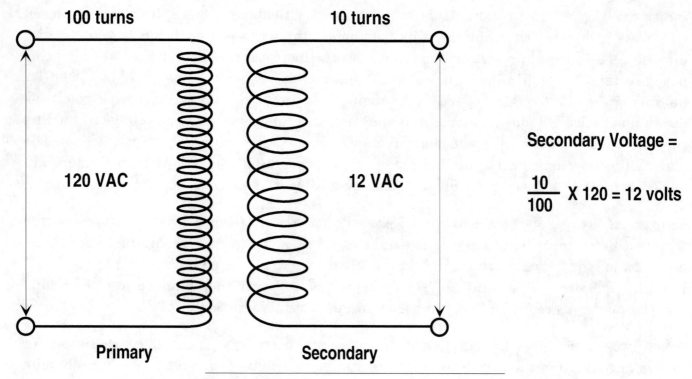

Figure 3. Step-down transformer with 10:1 ratio.

Assuming that all the magnetic lines of force from the primary cut through all the turns of the secondary, the voltage induced in the secondary will depend on the ratio of the number of turns in the primary to the number of turns in the secondary. For example, if there are 100 turns in the primary and only 10 turns in the secondary, the voltage in the primary will be 10 times the voltage in the secondary. Since there are more turns in the primary than there are in the secondary, the transformer is called a "step-down transformer." Transformers are rated in kVA because it is independent of power factor. Figure 3 shows a diagram of a step-down transformer with a ratio of 100:10 or 10:1.

$$\frac{10\ turns}{100\ turns} = 0.10 = .10 \times 120 = 12\ volts$$

Assuming that all the primary magnetic lines of force cut through all the turns of the secondary, the amount of induced voltage in the secondary will vary with the ratio of the number of turns in the secondary to the number of turns in the primary.

If there are more turns on the secondary than on the primary winding, the secondary voltage will be higher than that in the primary and by the same proportion as the number of turns in the winding. The secondary current, in turn, will be proportionately smaller than the primary current. With fewer turns on the secondary than on the primary, the secondary voltage will be proportionately lower than that in the primary, and the secondary current will be that much larger. Since alternating current continually increases and decreases in value, every change in the primary winding of the transformer produces a similar change of flux in the core. Every change of flux in the core and every corresponding movement of magnetic field around the core

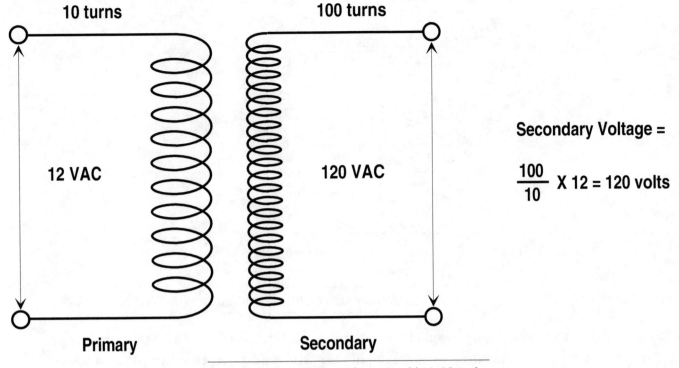

10 turns

100 turns

12 VAC

120 VAC

Secondary Voltage =

$$\frac{100}{10} \times 12 = 120 \text{ volts}$$

Primary

Secondary

Figure 4. Step-up transformer with 1:10 ratio.

produce a similarly changing voltage in the secondary winding, causing an alternating current to flow in the circuit that is connected to the secondary.

For example, if there are, say, 100 turns in the secondary and only 10 turns in the primary, the voltage induced in the secondary will be 10 times the voltage applied to the primary. See Fig. 4. Since there are more turns in the secondary than in the primary, the transformer is called a "step-up transformer."

$$\frac{100}{10} = 10 \times 12 = +120 \text{ volts}$$

Note   A transformer does not generate electrical power. It simply transfers electric power from one coil to another by magnetic induction. Transformers are rated in either volt-amperes (VA) or kilo-volt-amperes (kVA).

2.3.0   MAGNETIC FLUX IN TRANSFORMERS

Figure 5 shows a cross-section of what is known as a high-leakage flux transformer. In applications, if no load were connected to the secondary or output winding, a voltmeter would indicate a specific voltage reading across the secondary terminals. If a load were applied, the voltage would drop, and if the terminals were shorted, the voltage would drop to zero. During these circuit changes, the flux in the core of the transformer would also change — it is forced out of the transformer core and is known as *leakage flux*. Leaking flux can actually be

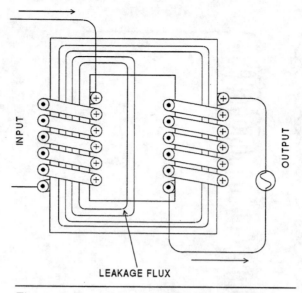

Figure 5. Transformer with high-leakage flux.

demonstrated with iron filings placed close to the transformer core. As the changes take place, the filings will shift their position — clearly showing the change in the flux pattern.

What actually happens is that as the current flows in the secondary, it tries to create its own magnetic field which is in opposition to the original flux field. This action, like a valve in a

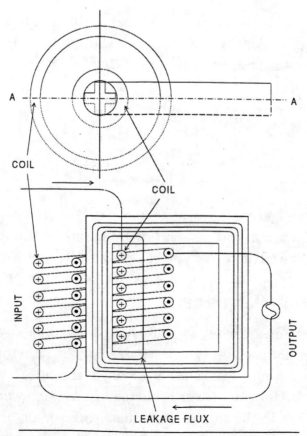

Figure 6. Transformer of low-leakage design.

water system, restricts the flux flow which forces the excess flux to find another path—either through air or in adjacent structural steel as in transformer housings or supporting clamps.

Note that the coils in Fig. 5 are wrapped on the same iron core, but separated from each other, while the transformer in Fig. 6 has its coils wrapped around each which results in a low-leakage transformer design.

3.0.0 TRANSFORMER CONSTRUCTION

Transformers designed to operate on low frequencies have their coils, called "windings," wound on iron cores. Since iron offers little resistance to magnetic lines, nearly all the magnetic field of the primary flows through the iron core and cuts the secondary.

Iron cores of transformers are constructed in three basic types — the open core, the closed core and the shell type. See Fig. 7. The open core is the least expensive to manufacture as the primary and secondary are wound on one cylindrical core. The magnetic path, as shown in Fig. 7, is partially through the core and partially through the surrounding air. The air path opposes the magnetic field, so that the magnetic interaction or "linkage" is weakened. The open core transformer, therefore, is highly inefficient.

The closed core improves the transformer efficiency by offering more iron paths and less air path for the magnetic field. The shell type core further increases the magnetic coupling and

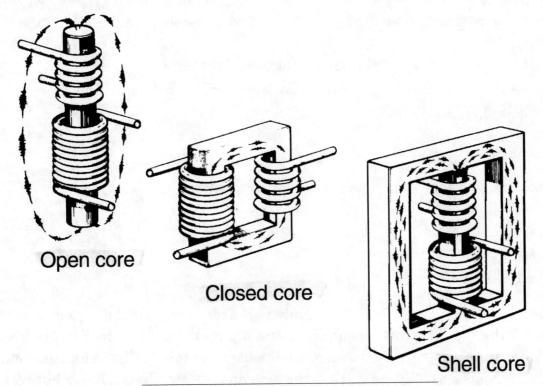

Open core

Closed core

Shell core

Figure 7. Three types of iron-core transformers.

therefore the transformer efficiency is greater due to two parallel magnetic paths for the magnetic field—providing maximum coupling between the primary and secondary.

3.1.0 CORES

Special core steel is used to provide a controlled path for the flow of magnetic flux generated in a transformer. In most practical applications, the transformer core is not a solid bar of steel, but is constructed of many layers of thin sheet steel called laminations.

While the specifications of the core steel are primarily of interest to the transformer design engineer, the electrical worker should at least have a conversational knowledge of the materials used.

Steel used for transformer core laminations will vary with the manufacturer, but a popular size is .014 inch thick and is called 29-gauge steel. It is processed from silicon iron alloys containing approximately $3\frac{1}{4}\%$ silicon. The addition of silicon to the iron increases its ability to be magnetized and also renders it essentially non-aging.

The most important characteristic of electrical steel is core loss. It is measured in watts per pound at a specified frequency and flux density. The core loss is responsible for the heating in the transformer and it also contributes to the heating of the windings. Much of the core loss is a result of eddy currents which are induced in the laminations when the core is energized. To hold this loss to a minimum, adjacent laminations are coated with an inorganic varnish.

Cores may be of either the "core type" as shown in Fig. 8 or the "shell type" as shown in Fig. 9. Of the two, the core type is favored for dry-type transformers because:

- Only three core legs require stacking; thus, reducing cost.
- Steel does not encircle the two outer coils; this provides better cooling.
- Floor space is reduced.

3.2.0 TYPES OF CORES

Transformer cores normally full under three distinct types:

- Butt
- Wound
- Mitered cores

The butt-and-lap core is shown in Fig. 10. Only two sizes of core steel are needed in this type of core due to the lap construction shown at the top and right side in Fig. 10. For ease of understanding, the core strips are shown much thicker than the .014 inch thickness mentioned earlier. Each strip is carefully cut so that the air gap indicated in the lower left corner is as small as possible. The permeability of steel to the passage of flux is about 10,000 times as

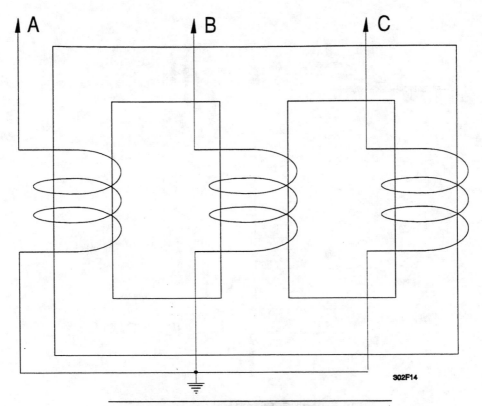

Figure 8. Core type transformer construction.

effective as air, hence the air gap must be held to the barest minimum to reduce the ampere turns necessary to achieve adequate flux density. Also, the amount of sound that emanates from a transformer due to magneto-striction is a function of the flux density and this poses an interesting difference between this construction and other types.

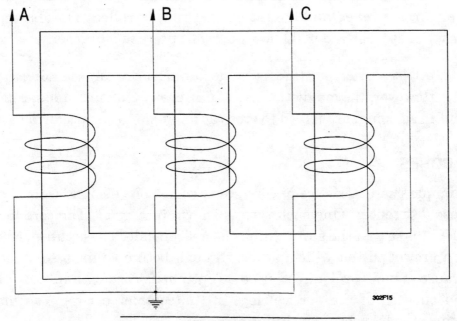

Figure 9. Shell type transformer core.

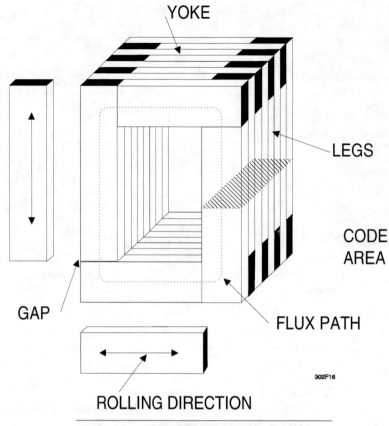

YOKE

LEGS

CODE
AREA

FLUX PATH

GAP

ROLLING DIRECTION

302F16

Figure 10. Butt-and-lap transformer core.

Another phenomenon in core steel is that the flux flows more easily in the direction in which the steel was rolled. Even this characteristic is different in hot-rolled versus cold-rolled steel. For example, the core loss due to flux passing at right angles to the rolling direction is almost 1½ times as great in hot rolled and 2½ times as great in cold-rolled when compared with the core loss in the direction of rolling. The difference in exciting current is more dramatic, with ratios of two to one in hot-rolled and almost 40 to 1 in cold-rolled. These are primarily the designer's concern, but at least you will know now that there is a difference.

Eddy currents are restricted from passage from one lamination to another due to the inorganic insulating coating. However, the magnetic lines of flux easily transfer at adjacent laminations in the lap area but in so doing, are forced to cross at an angle to the preferred direction.

3.3.0 WOUND CORES

Because of the unique characteristics of core steel, some core designs were made that took advantage of these differences. One such type is shown in Fig. 11. The core loops are cut to pre-determined lengths so that the gap locations do not coincide. These cuts permit assembling the core around a prewound coil which passes through both openings. Another design, now discontinued because of unfavorable cost, used a continuous core with no cuts. Separate coils had to be wound on each of the vertical legs of the completed core. You may encounter transformers of this type in existing installations.

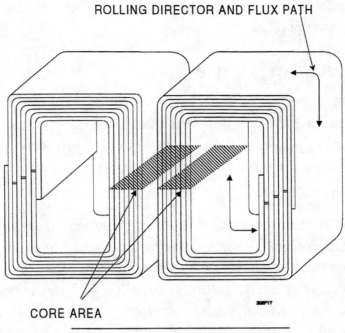

ROLLING DIRECTOR AND FLUX PATH

CORE AREA

Figure 11. Wound transformer coil.

3.4.0   MITERED CORES

Figure 12 shows a mitered-core design. In effect, it is a butt-lap core with the joints made at 45-degree angles. There are two benefits derived from this type of joint:

- It eliminates all cross grain flux thereby improves the core loss and exciting current values.
- It reduces the flux density in the air gap—resulting in lower sound levels.

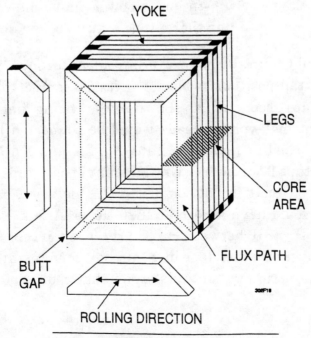

YOKE

LEGS

CORE
AREA

FLUX PATH

BUTT
GAP

ROLLING DIRECTION

Figure 12. Mitered transformer core.

This type of core is normally used only with cold-rolled, grain oriented steel and permits this steel to be used to its fullest capability.

3.5.0   TRANSFORMER CHARACTERISTICS

In a well-designed transformer, there is very little magnetic leakage. The effect of the leakage is to cause a decrease of secondary voltage when the transformer is loaded. When a current flows through the secondary in phase with the secondary voltage, a corresponding current flows through the primary in addition to the magnetizing current. The magnetizing effects of the two currents are equal and opposite.

In a perfect transformer, that is, one having no eddy-current losses, no resistance in its windings, and no magnetic leakage, the magnetizing effects of the primary load current and the secondary current neutralize each other, leaving only the constant primary magnetizing current effective in setting up the constant flux. If supplied with a constant primary pressure, such a transformer would maintain constant secondary pressure at all loads. Obviously, the perfect transformer has yet to be built; the closest to perfect has very small eddy-current loss where the drop in pressure in the secondary windings is not more than 1 to 3 percent, depending on the size of the transformer.

4.0.0   TRANSFORMER TAPS

If the exact rated voltage could be delivered at every transformer location, transformer taps would be unnecessary. However, this is not possible, so taps are provided on the secondary windings to provide a means of either increasing or decreasing the secondary voltage.

Generally, if a load is very close to a substation or power plant, the voltage will consistently be above normal. Near the end of the line the voltage may be below normal.

In large transformers, it would naturally be very inconvenient to move the thick, well-insulated primary leads to different tap positions when changes in source-voltage levels make this necessary. Therefore, taps are used, such as shown in the wiring diagram in Fig. 13. In this transformer, the permanent high-voltage leads would be connected to H_1 and H_2, and the secondary leads, in their normal fashion, to X_1 and X_2, X_3, and X_4. Note, however, the tap arrangements available at taps 2 through 7. Until a pair of these taps is interconnected with a jumper wire, the primary circuit is not completed. If this were, say, a typical 7200-volt primary, the transformer would have a normal 1620 turns. Assume 810 of these turns are between H_1 and H_6 and another 810 between H_3 and H_2. Then, if taps 6 and 3 were connected together with a flexible jumper on which lugs have already been installed, the primary circuit is completed, and we have a normal ratio transformer that could deliver 120/240 volts from the secondary.

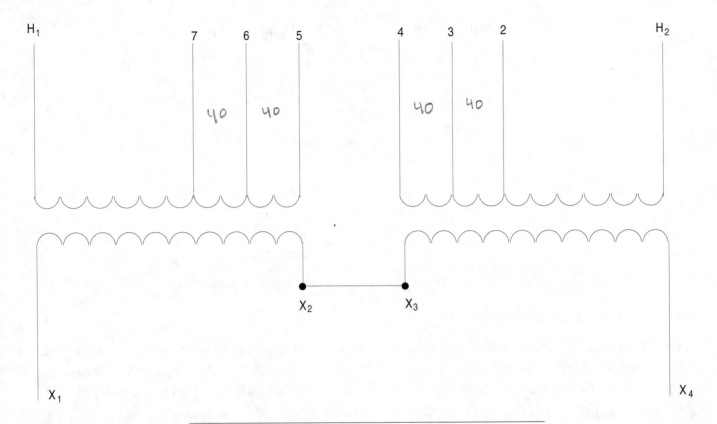

Figure 13. Transformer taps to adjust secondary voltage.

Between taps 6 and either 5 or 7, 40 turns of wire exist. Similarly, between taps 3 and either 2 or 4, 40 turns are present. Changing the jumper from 3 to 6 to 3 to 7 removes 40 turns from the left half of the primary. The same condition would apply on the right half of the winding if the jumper were between taps 6 and 2. Either connection would boost secondary voltage by 2½ percent. Had taps 2 and 7 been connected, 80 turns would have been omitted and a 5 percent boost would result. Placing the jumper between taps 6 and 4 or 3 and 5 would reduce the output voltage by 5 percent.

5.0.0   TRANSFORMER CONNECTIONS — BASIC

Transformer connections are many, and space does not permit the description of all of them here. However, an understanding of a few will give the basic requirements and make it possible to use manufacturer's data for others should the need arise.

5.1.0   SINGLE-PHASE LIGHT AND POWER

The diagram in Fig. 14 is a transformer connection used quite extensively for residential and small commercial applications. It is the most common single-phase distribution system in use today. It is known as the three-wire, 240/120-volt single-phase system and is used where 120 and 240 volts are used simultaneously.

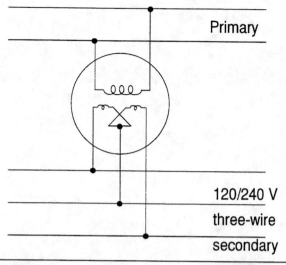

Figure 14. 120/240-volt, three-wire, single-phase transformer connection.

5.2.0 Y-Y FOR LIGHT AND POWER

The primaries of the transformer connection in Fig. 15 are connected in wye—sometimes called *star* connection. When the primary system is 2400/4160Y volts, a 4160-volt transformer is required when the system is connected in delta-Y. However, with a Y-Y system, a 2400-volt transformer can be used, offering a saving in transformer cost. It is necessary that a primary

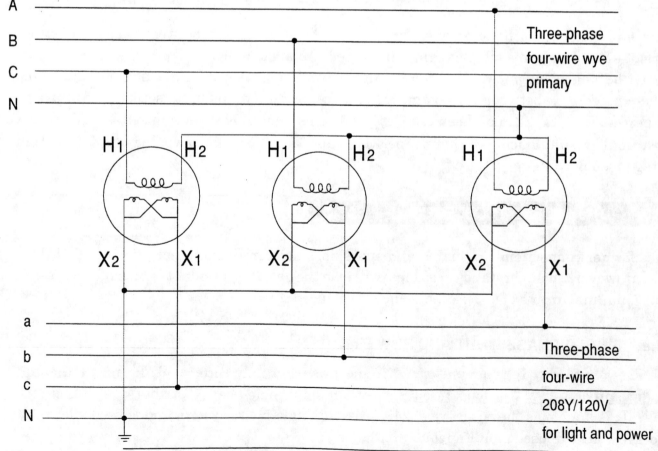

Figure 15. Three-phase, 4-wire, Y-Y connected transformer system.

neutral be available when this connection is used, and the neutrals of the primary system and the transformer bank are tied together as shown in the diagram. If the three-phase load is unbalanced, part of the load current flows in the primary neutral. For these reasons, it is essential that the neutrals be tied together as shown. If this tie were omitted, the line-to-neutral voltages on the secondary would be very unstable. That is, if the load on one phase were heavier than on the other two, the voltage on this phase would drop excessively and the voltage on the other two phases would rise. Also, varying voltages would appear between lines and neutral, both in the transformers and in the secondary system, in addition to the 60-hertz component of voltage. This means that for a given value of rms voltage, the peak voltage would be much higher than for a pure 60-hertz voltage. This overstresses the insulation both in the transformers and in all apparatus connected to the secondaries.

5.3.0 DELTA-CONNECTED TRANSFORMERS

The delta-connected system in Fig. 16 operates a little differently from the previously described wye-wye system. While the wye-connected system is formed by connecting one terminal from three equal voltage transformer windings together to make a common terminal, the delta-connected system has its windings connected in series, forming a triangle or the Greek delta symbol Δ. Note in Fig. 17 that a center-tap terminal is used on one winding to ground the

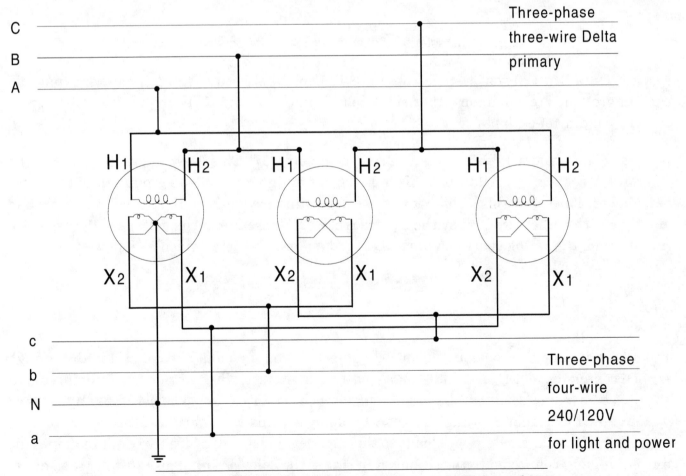

Figure 16. Three-phase, four-wire, delta-connected secondary.

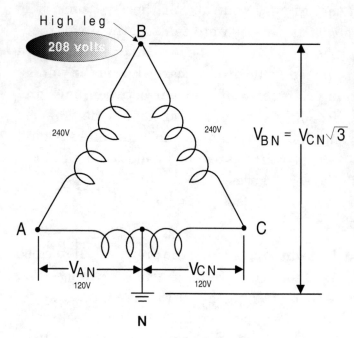

High leg
208 volts

B

240V   240V

$V_{BN} = V_{CN}\sqrt{3}$

A   C

V_{AN} V_{CN}
120V   120V

N

On a 3-phase, 4-wire 120/240V delta-connected system, the midpoint of one phase winding is grounded to provide 120V between phase A and ground; also between phase C and ground. Between phase B and ground, however, the voltage is higher and may be calculated by multiplying the voltage between C and ground (120V) by the square root of 3 or 1.73. Consequently, the voltage between phase B and ground is approximately 208 volts. Thus, the name "high leg."

The NEC requires that conductors connected to the high leg of a 4-wire delta system be color coded with orange insulation or tape.

Figure 17. Characteristics of a center-tap, delta-connected system.

system. On a 240/120-volt system, there are 120 volts between the center-tap terminal and each ungrounded terminal on either side; that is, phases A and C. There are 240 volts across the full winding of each phase.

Refer to Fig. 17 and note that a high leg results at point "B". This is known in the trade as the "high leg," "red leg," or "wild leg." This high leg has a higher voltage to ground than the other two phases. The voltage of the high leg can be determined by multiplying the voltage to ground of either of the other two legs by the square root of 3. Therefore, if the voltage between phase A to ground, the voltage between phase B to ground may be determined as follows:

$$120 \times \sqrt{3} \quad = \quad 207.84 = 208 \; volts$$

From this, it should be obvious that no single-pole breakers should be connected to the high leg of a center-tapped, 4-wire delta-connected system. In fact, NEC Section 215-8 requires that the phase busbar or conductor having the higher voltage to ground to be permanently marked by an outer finish that is orange in color. By doing so, this will prevent future workers from connecting 120-volt single-phase loads to this high leg which will probably result in damaging any equipment connected to the circuit. Remember the color *orange*; no 120-volt loads are to be connected to this phase.

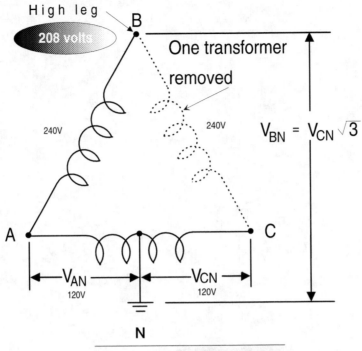

High leg

208 volts

One transformer
removed

B

240V

240V

$V_{BN} = V_{CN} \sqrt{3}$

A

V_{AN}
120V

V_{CN}
120V

C

N

Figure 18. Open delta system.

WARNING!   Always use caution when working on a center-tapped, 4-wire, delta-connected system. Phase B has a higher voltage to ground than phases A and C. Never connect 120-volt circuits to the high leg. Doing so will result in damage to the circuits and equipment.

5.3.1 Open Delta

Three-phase, delta-connected systems may be connected so that only two transformers are used; this arrangement is known as *open delta* as shown in Fig. 18. This arrangement is frequently used on a delta system when one of the three transformers becomes damaged. The damaged transformer is disconnected from the circuit and the remaining two transformers carry the load. In doing so, the three-phase load carried by the open delta bank is only 86.6% of the combined rating of the remaining two equal sized units. It is only 57.7% of the normal full-load capability of a full bank of transformers. In an emergency, however, this capability permits single- and three-phase power at a location where one unit burned out and a replacement was not readily available. The total load must be curtailed to avoid another burnout.

5.4.0 TEE-CONNECTED TRANSFORMERS

When a delta-wye transformer is used, we would usually expect to find three primary and three secondary coils. However, in a tee-connected three-phase transformer, only two primary and two secondary windings are used as shown in Fig. 19. If an equilateral triangle is drawn as

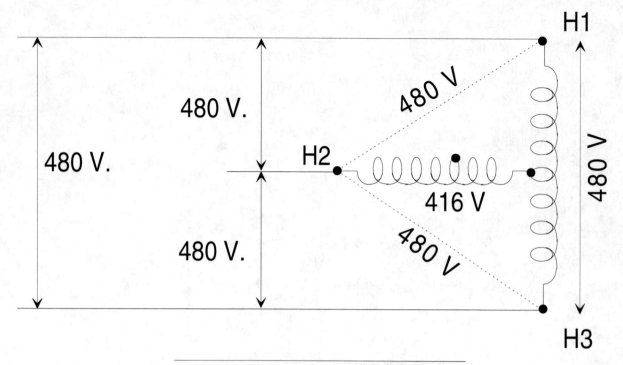

Figure 19. Typical Tee-connected transformer.

indicated by the dotted lines in Fig. 19 so that the distance between H_1 and H_3 is 4.8 inches, you would find that the distance between H_2 to the midpoint of H_1 - H_3 measures 4.16 inches. Therefore, if the voltage between outside phases is 480 volts, the voltage between H_2 to the midpoint of H_1 - H_3 will equal 480 volts × .866 = 415.68 or 416 volts. Also, if you were to place an imaginary dot exactly in the center of this triangle it would lay on the horizontal winding— the one containing 416 volts. If you measured the distance from this dot to H_2, you would find it to be twice as long as the distance between the dot and the midpoint of H_1 to H_3. The measured

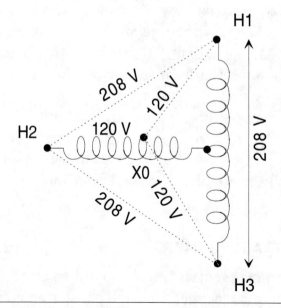

Figure 20. Secondary voltage on tee-connected system.

ELECTRICAL TRAINEE TASK MODULE 20307

distances would be 2.77 inches and 1.385 inches or the equivalent of 277 volts and 138½ volts respectively.

Now, let's look at the secondary winding in Fig. 20. By placing a neutral tap X_0 so that ⅓ the number of turns exist between it and the midpoint of X_1 and X_3, as exist between it and X_2, we then can establish X_0 as a neutral point which may be grounded. This provides 120 volts between X_0 and any of the three secondary terminals and the three phase voltage between X_1, X_2, and X_3, will be 208 volts.

5.5.0   PARALLEL OPERATION OF TRANSFORMERS

Transformers will operate satisfactorily in parallel on a single-phase, three-wire system if the terminals with the same relative polarity are connected together. However, the practice is not very economical because the individual cost and losses of the smaller transformers are greater than one larger unit giving the same output. Therefore, paralleling of smaller transformers is usually done only in an emergency. In large transformers, however, it is often practical to operate units in parallel as a regular practice. See Fig. 21.

In connecting large transformers in parallel, especially when one of the windings is for a comparatively low voltage, the resistance of the joints and interconnecting leads must not vary materially for the different transformers, or it will cause an unequal division of load.

Two three-phase transformers may also be connected in parallel provided they have the same winding arrangement, are connected with the same polarity, and have the same phase rotation. If two transformers—or two banks of transformers—have the same voltage ratings, the same turn ratios, the same impedances, and the same ratios of reactance to resistance, they will divide the load current in proportion to their kVA ratings, with no phase difference between

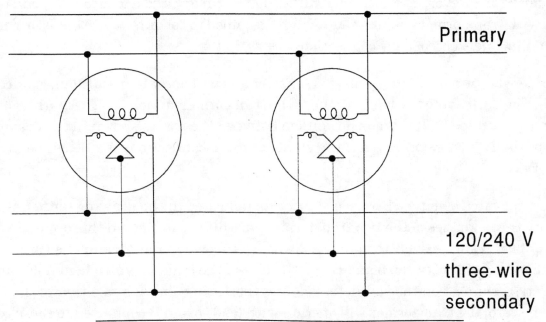

Figure 21. Parallel operation of single-phase transformers.

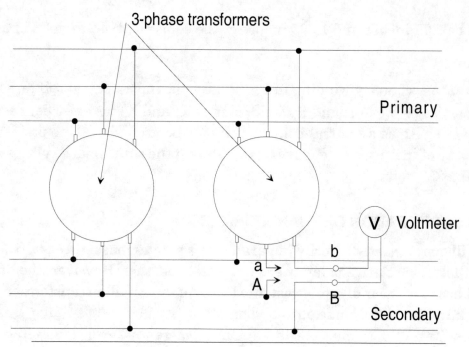

Figure 22. Testing three-phase transformers for parallel operation.

the currents in the two transformers. However, if any of the preceding conditions are not met, then it is possible for the load current to divide between the two transformers in proportion to their kVA ratings. There may also be a phase difference between currents in the two transformers or banks of transformers.

Some three-phase transformers cannot be operated properly in parallel. For example, a transformer having both its primary and secondary windings connected in delta cannot be connected in parallel with another transformer that is connected either with a primary delta or a secondary Y. However, a transformer with a delta primary and a Y secondary can be made to parallel with transformers having their windings joined in certain ways; that is, a Y primary connection and a delta secondary connection.

To determine whether or not three-phase transformers will operate in parallel, connect them as shown in Fig. 22, leaving two leads on one of the transformers unjoined. Test with a voltmeter across the unjoined leads. If there is no voltage between the points shown in the drawing, the polarities of the two transformers are the same, and the connections may then be made and put into service.

If a reading indicates a voltage between the points indicated in the drawing (either one of the two or both), the polarities of the two transformers are different. Should this occur, disconnect transformer lead A successively to mains 1, 2, and 3 as shown in Fig. 22 and at each connection test with the voltmeter between b and B and the legs of the main to which lead A is connected. If with any trial connection the voltmeter readings between b and B and either of the two legs is found to be zero, the transformer will operate with leads b and B connected to those two legs.

If no system of connections can be discovered that will satisfy this condition, the transformer will not operate in parallel without changes in its internal connections, and there is a possibility that it will not operate in parallel at all.

In parallel operation, the primaries of the two or more transformers involved are connected together, and the secondaries are also connected together. With the primaries so connected, the voltages in both primaries and secondaries will be in certain directions. It is necessary that the secondaries be so connected that the voltage from one secondary line to the other will be in the same direction through both transformers. Proper connections to obtain this condition for single-phase transformers of various polarities are shown in Fig. 23. In Fig. 23(a), both transformers A and B have additive polarity; in Fig. 23(b), both transformers have subtractive polarity; in Fig. 23(c), transformer A has additive polarity and B has subtractive polarity.

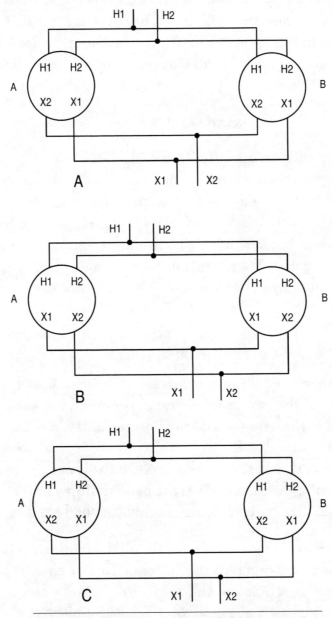

Figure 23. Transformers connected in parallel.

Transformers, even when properly connected, will not operate satisfactorily in parallel unless their transformation ratios are very close to being equal and their impedance voltage drops are also approximately equal. A difference in transformation ratios will cause a circulating current to flow, even at no load, in each winding of both transformers. In a loaded parallel bank of two transformers of equal capacities, for example, if there is a difference in the transformation ratios, the load circuit will be superimposed on the circulating current. The result in such a case is that in one transformer the total circulating current will be added to the load current, whereas in the other transformer the actual current will be the difference between the load current and the circulating current. This may lead to unsatisfactory operation. Therefore, the transformation ratios of transformers for parallel operation must be definitely known.

When two transformers are connected in parallel, the circulating current caused by the difference in the ratios of the two is equal to the difference in open-circuit voltage divided by the sum of the transformer impedances, because the current is circulated through the windings of both transformers due to this voltage difference. To illustrate, let I represent the amount of circulating current—in percent of full-load current—and the equation will be

$$I = \frac{Percent\ voltage\ difference \times 100}{Sum\ of\ percent\ impedances}$$

Let's assume an open-circuit voltage difference of 3% between two transformers connected in parallel. If each transformer has an impedance of 5%, the circulating current, in percent of full-load current, is I = (3 × 100)/5 + 5) = 30%. A current equal to 30% full-load current therefore circulates in both the high-voltage and low-voltage windings. This current adds to the load current in the transformer having the higher induced voltage and subtracts from the load current of the other transformer. Therefore, one transformer will be overloaded, while the other may or may not be—depending on the phase-angle difference between the circulating current and the load current.

5.5.1 Impedance in Parallel-Operated Transformers

Impedance plays an important role in the successful operation of transformers connected in parallel. The impedance of the two or more transformers must be such that the voltage drop from no load to full load is the same in all transformer units in both magnitude and phase. In most applications, you will find that the total resistance drop is relatively small when compared with the reactance drop and that the total percent impedance drop can be taken as approximately equal to the percent reactance drop. If the percent impedances of the given transformers at full load are the same, they will, of course, divide the load equally.

The following equation may be used to obtain the division of loads between two transformer banks operating in parallel on single-phase systems. In this equation, it can be assumed that the ratio of resistance to reactance is the same in all units since the error introduced by differences in this ratio is usually so small as to be negligible:

$$power = \frac{(kVA\text{-}1)/(z\text{-}1)}{[(kVA\text{-}1)/(z\text{-}1)] + [(kVA\text{-}2)/(z\text{-}2)]} \times total\ kVA\ load$$

where

kVA - 1 = kVA rating of transformer 1

kVA - 2 = kVA rating of transformer 2

Z - 1 = percent impedance of transformer 1

Z - 2 = percent impedance of transformer 2

The preceding equation may also be applied to more than two transformers operated in parallel by adding, to the denominator of the fraction, the kVA of each additional transformer divided by its percent impedance.

5.5.2 Parallel Operation of Three-Phase Transformers

Three-phase transformers, or banks of single-phase transformers, may be connected in parallel provided each of the three primary leads in one three-phase transformer is connected in parallel with a corresponding primary lead of the other transformer. The secondaries are then connected in the same way. The corresponding leads are the leads which have the same potential at all times and the same polarity. Furthermore, the transformers must have the same voltage ratio and the same impedance voltage drop.

When three-phase transformer banks operate in parallel and the three units in each bank are similar, the division of the load can be determined by the same method previously described for single-phase transformers connected in parallel on a single-phase system.

In addition to the requirements of polarity, ratio, and impedance, paralleling of three-phase transformers also requires that the angular displacement between the voltages in the windings be taken into consideration when they are connected together.

Phasor diagrams of three-phase transformers that are to be paralleled greatly simplify matters. With these, all that is required is to compare the two diagrams to make sure they consist of phasors that can be made to coincide; then connect together terminals corresponding to coinciding voltage phasors. If the diagram phasors can be made to coincide, leads that are connected together will have the same potential at all times. This is one of the fundamental requirements for paralleling. Phasor diagrams are covered later in this module. Also review Electrical Task Module 01201 — Alternating Current for a basic understanding of how phasor diagrams are developed.

An autotransformer is a transformer whose primary and secondary circuits have part of a winding in common and therefore the two circuits are not isolated from each other. See Fig. 24. The application of an autotransformer is a good choice for some users where a 480Y/277- or 208Y/120-volt, three-phase, four-wire distribution system is utilized. Some of the advantages are as follows:

- Lower purchase price
- Lower operating cost due to lower losses
- Smaller size; easier to install
- Better voltage regulation
- Lower sound levels

For example, when the ratio of transformation from the primary to secondary voltage is small, the most economical way of stepping down the voltage is by using autotransformers as shown in Fig. 25. For this application, it is necessary that the neutral of the autotransformer bank be connected to the system neutral.

An autotransformer, however, cannot be used on a 480- or 240-volt, three-phase, three-wire delta system. A grounded neutral phase conductor must be available in accordance with NEC Article 210-9, which states:

NEC Section 210-9: Circuits Derived from Autotransformers. Branch circuits shall not be supplied by autotransformers.

Exception No. 1: Where the system supplied has a grounded conductor that is electrically connected to a grounded conductor of the system supplying the autotransformer.

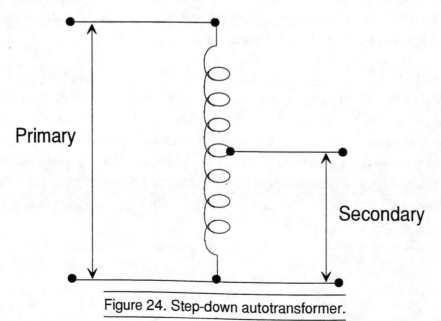

Figure 24. Step-down autotransformer.

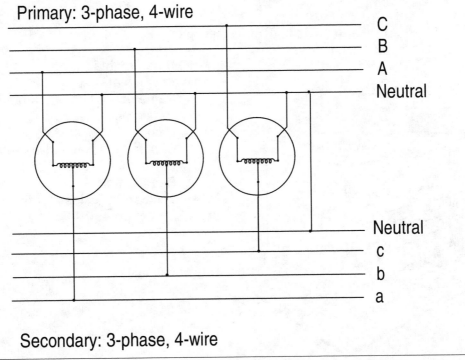

Primary: 3-phase, 4-wire

C
B
A
Neutral

Neutral
c
b
a

Secondary: 3-phase, 4-wire

Figure 25. Autotransformers supplying power from a 3-phase, 4-wire system.

Exception No. 2: An autotransformer used to extend or add an individual branch circuit in an existing installation for an equipment load without the connection to a similar grounded conductor when transforming from a nominal 208 volts to a nominal 240 volt supply or similarly from 240 volts to 208 volts.

7.0.0   TRANSFORMER CONNECTIONS — DRY TYPE

Electricians performing work on commercial and industrial installations will more often be concerned with the installation and connections of dry-type transformers as opposed to oil-filled ones. Dry-type transformers are available in both single- and three-phase with a wide range of sizes from the small control transformers to those rated at 500 kVA or more. Such transformers have wide application in electrical systems of all types.

NEC Section 450-11 requires that each transformer must be provided with a nameplate giving the manufacturer; rated kVA; frequency; primary and secondary voltage; impedance of transformers 25 kVA and larger; required clearances for transformers with ventilating openings; and the amount and kind of insulating liquid where used. In addition, the nameplate of each dry-type transformer must include the temperature class for the insulation system. See Fig. 26.

In addition, most manufacturers include a wiring diagram and a connection chart as shown in Fig. 27 for a 480-volt delta primary to 208Y/120-volt secondary. It is recommended that all transformers be connected as shown on the manufacturer's nameplate.

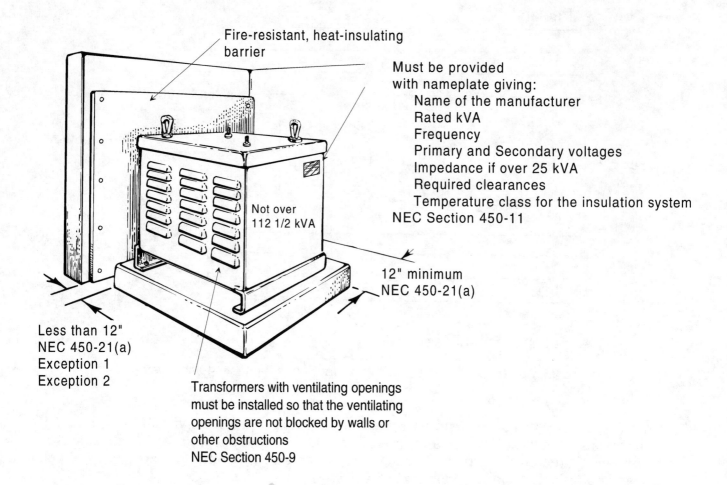

Fire-resistant, heat-insulating barrier

Must be provided
with nameplate giving:
    Name of the manufacturer
    Rated kVA
    Frequency
    Primary and Secondary voltages
    Impedance if over 25 kVA
    Required clearances
    Temperature class for the insulation system
NEC Section 450-11

Not over
112 1/2 kVA

12" minimum
NEC 450-21(a)

Less than 12"
NEC 450-21(a)
Exception 1
Exception 2

Transformers with ventilating openings
must be installed so that the ventilating
openings are not blocked by walls or
other obstructions
NEC Section 450-9

Figure 26. Dry-type transformer installed indoors.

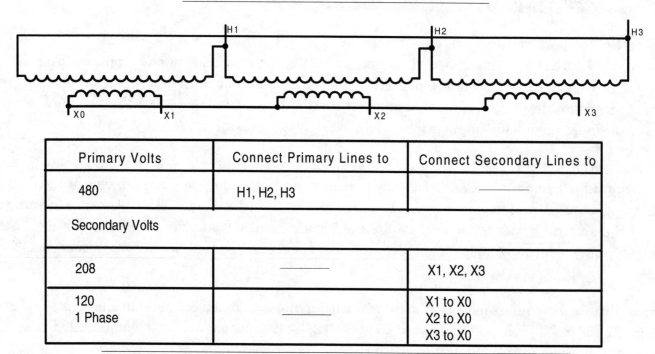

Primary Volts	Connect Primary Lines to	Connect Secondary Lines to
480	H1, H2, H3	————
Secondary Volts		
208	————	X1, X2, X3
120 1 Phase	————	X1 to X0 X2 to X0 X3 to X0

Figure 27. Typical transformer manufacturer's wiring diagram — delta-wye.

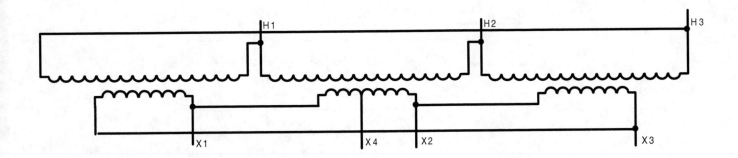

Primary Volts	Connect Primary Lines to	Connect Secondary Lines to
480	H1, H2, H3	————
Secondary Volts		
240	————	X1, X2, X3
120	————	X1, X4 or X2, X4

Figure 28. 480V delta to 240V delta transformer connections.

In general, this wiring diagram and accompanying table indicates that the 480-volt, 3-phase, 3-wire primary conductors are connected to terminals H₁, H₂, and H₃, respectively — regardless of the desired voltage on the primary. A neutral conductor, if required, is carried from the primary through the transformer to the secondary. Two variations are possible on the secondary side of this transformer: 208-volt, 3-phase, 3- or 4-wire or 120-volt, 1-phase, 2-wire. To connect the secondary side of the transformer as a 208-volt, 3-phase, 3-wire system, the secondary conductors are connected to terminals X₁, X₂, and X₃; the neutral is carried through with conductors usually terminating at a solid-neutral bus in the transformer.

Another popular dry-type transformer connection is the 480-volt primary to 240-volts delta/120 volts secondary. This configuration is shown in Fig. 28. Again, the primary conductors are connected to transformer terminals H₁, H₂, and H₃. The secondary connections for the desired voltages are made as indicated in the table.

7.1.0 ZIG-ZAG CONNECTIONS

There are many occasions where it is desirable to upgrade a building's lighting system from 120-volt fixtures to 277-volt fluorescent lighting fixtures. Oftentimes these buildings have a 480/240-volt, three-phase, four-wire delta system. One way to obtain 277 volts from a 480/240-volt system is to connect 480/240-volt transformers in a zig-zag fashion as shown in Fig. 29. In doing so, the secondary of one phase is connected in series with the primary of another phase, thus changing the phase angle.

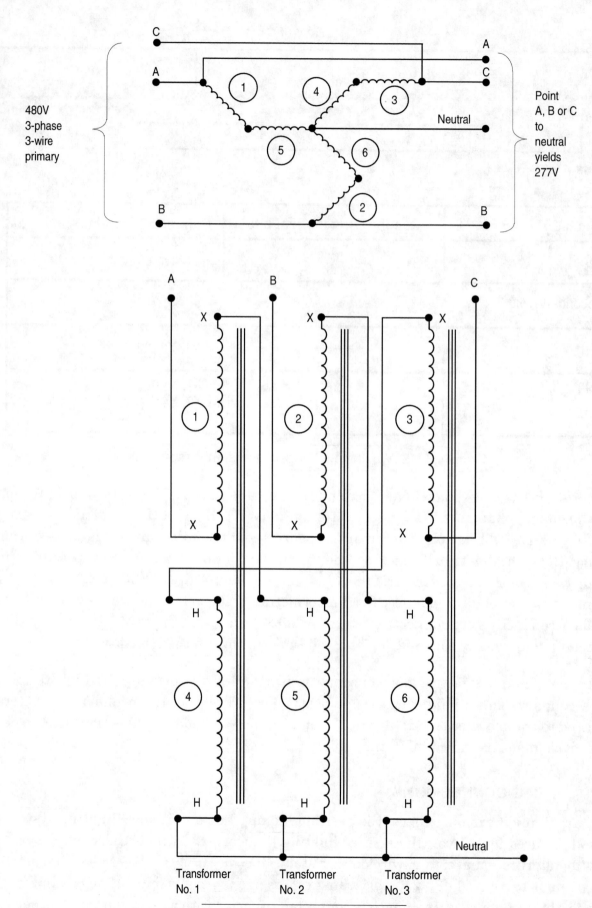

480V
3-phase
3-wire
primary

Neutral

Point
A, B or C
to
neutral
yields
277V

Transformer
No. 1

Transformer
No. 2

Transformer
No. 3

Neutral

Figure 29A. Zig-zag connection.

The zig-zag connection may also be used as a grounding transformer where its function is to obtain a neutral point from an ungrounded system. With a neutral being available, the system may then be grounded. When the system is grounded through the zig-zag transformer, its sole function is to pass ground current. A zig-zag transformer is essentially six impedances connected in a zig-zag configuration.

The operation of a zig-zag transformer is slightly different from that of the conventional transformer. We will consider current rather than voltage. While a voltage rating is necessary for the connection to function, this is actually line voltage and is not transformed. It provides only exciting current for the core. The dynamic portion of the zig-zag grounding system is the fault current. To understand its function, the system must also be viewed backward; that is, the fault current will flow into the transformer through the neutral as shown in Fig. 29B.

The zero sequence currents are all in phase in each line; that is, they all hit the peak at the same time. In reviewing Fig. 29B, we see that the current leaves the motor, goes to ground, flows up the neutral, and splits three ways. It then flows back down the line to the motor through the fuses which then open — shutting down the motor.

The neutral conductor will carry full fault current and must be sized accordingly. It is also time rated (0-60 seconds) and can therefore be reduced in size. This should be coordinated with the manufacturer's time/current curves for the fuse. See Electrical Task Module 20303 — Overcurrent Protection.

To determine the size of a zig-zag grounding transformer, proceed as follows:

Step 1.   Calculate the system line-to-ground asymmetrical fault current.

Step 2.   If relaying is present, consider reducing the fault current by installing a resistor in the neutral.

Step 3.   If fuses or circuit breakers are the protective device, you may need all the fault current to quickly open the overcurrent protective devices.

Step 4.   Obtain time/current curves of relay, fuses, or circuit breakers.

Step 5.   Select zig-zag transformer for:

        a. Fault current — the line-to-ground
        b. Line-to-line voltage
        c. Duration of fault (determined from time/current curves)
        d. Impedance per phase at 100%; for any other, contact manufacturer

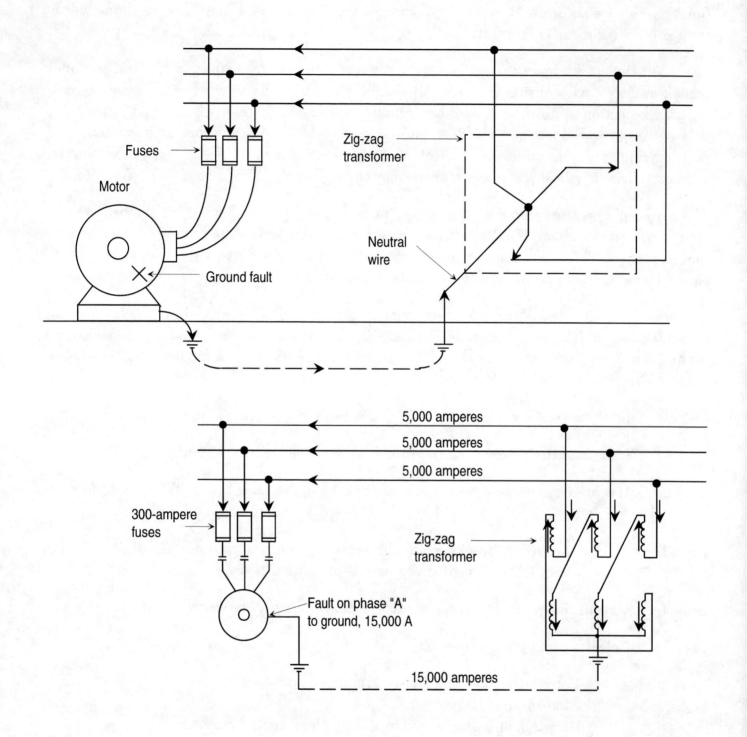

Figure 29B. Fault-current paths for three-phase system.

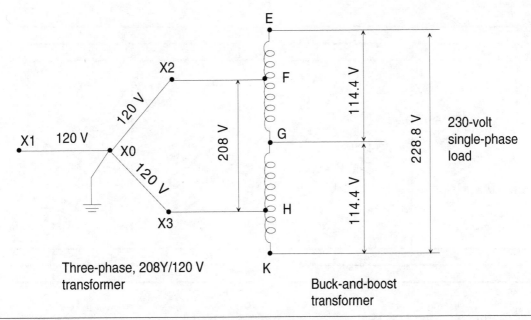

Figure 30. Buck-and-boost transformer connected to 208-volt system to obtain 230 volts.

7.2.0 BUCK-AND-BOOST TRANSFORMERS

The buck-and-boost transformer is a very versatile unit for which a multitude of applications exist. Buck-and-boost transformers, as the name implies, is designed to raise (boost) or lower (buck) the voltage in an electrical system or circuit. In their simplest form, these insulated units will deliver 12 or 24 volts when the primaries are energized at 120 or 240 volts respectively. Their prime use and value, however, lies in the fact that the primaries and the secondaries can be interconnected—permitting their use as an autotransformer.

Let's assume that an installation is supplied with 208Y/120V service, but one piece of equipment in the installation is rated for 230 volts. A buck-and-boost transformer may be used on the 230-volt circuit to increase the voltage from 208 volts to 230 volts. See Fig. 30. With this connection, the transformer is in the "boost" mode and delivers 228.8 volts at the load. This is close enough to 230 volts that the load equipment will function properly.

If the connections were reversed, this would also reverse the polarity of the secondary with the result that a voltage would be 208 volts minus 20.8 volts = 187.2 volts. The transformer is now operating in the "buck" mode.

It is important to know how to calculate sizes of buck-and-boost transformers for any given application. However, due to the amount of basic material covered in this module, advanced sizing and application techniques for buck-and-boost transformers are presented in Year-Four, Specialty Transformers. Still, the trainee should be familiar with the basic buck-and-boost wiring diagrams at this time. Consequently, transformer connections for typical three-phase buck-and-boost open-delta transformers are shown in Fig. 31. The connections shown are in the "boost" mode; to convert to "buck" mode, reverse the input and output.

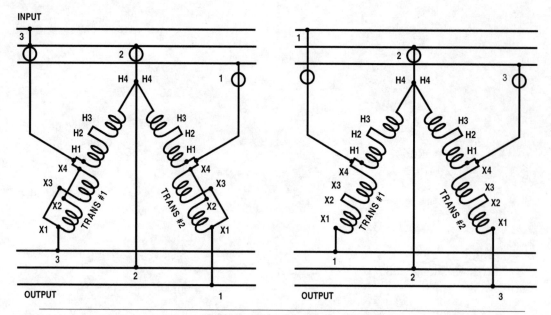

Figure 31. Open delta, three-phase, buck-and-boost transformer connections.

Another three-phase buck-and-boost transformer connection is shown in Fig. 32; this time wye-connected. While the open-delta transformers (Fig. 31) can be converted from buck to boost or vice-versa by reversing the input/output connections, this is not the case with the three-phase, wye-connected transformer. The connection shown (Fig. 32) is for the boost mode only.

Several typical single-phase buck-and-boost transformer connections are shown in Fig. 33. Other diagrams may be found on the transformer's nameplate or with packing instructions that come with each new transformer.

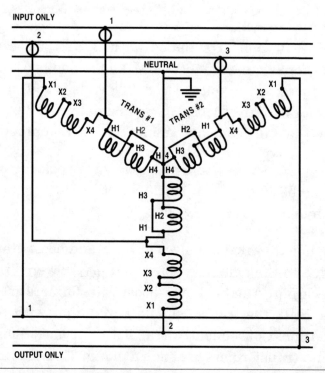

Figure 32. Three-phase, wye-connected buck-and-boost transformer in the "boost" mode.

ELECTRICAL TRAINEE TASK MODULE 20307

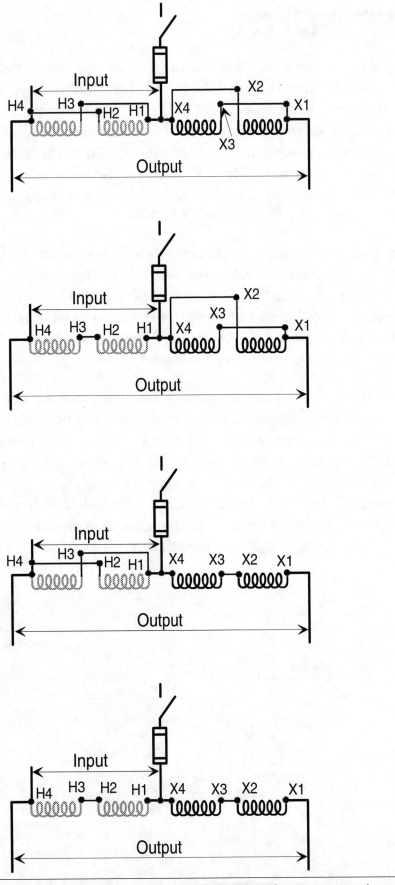

Figure 33. Typical single-phase buck-and-boost transformer connections.

8.0.0 CONTROL TRANSFORMERS

Control transformers are available in numerous types, but most control transformers are dry-type step-down units with the secondary control circuit isolated from the primary line circuit to assure maximum safety. See Fig. 34. These transformers and other components are usually mounted within an enclosed control box or control panel, which has a pushbutton station or stations independently grounded as recommended by the NE Code. Industrial control transformers are especially designed to accommodate the momentary current inrush caused when electromagnetic components are energized, without sacrificing secondary voltage stability beyond practical limits. See NEC Section 470-32.

Other types of control transformers, sometimes referred to as control and signal transformers, normally do not have the required industrial control transformer regulation characteristics. Rather, they are constant-potential, self-air-cooled transformers used for the purpose of supplying the proper reduced voltage for control circuits of electrically operated switches or other equipment and, of course, for signal circuits. Some are of the open type with no protective casing over the winding, while others are enclosed with a metal casing over the winding.

In seeking control transformers for any application, the loads must be calculated and completely analyzed before the proper transformer selection can be made. This analysis involves every electrically energized component in the control circuit. To select an appropriate control transformer, first determine the voltage and frequency of the supply circuit. Then determine the total inrush volt-amperes (watts) of the control circuit. In doing so, do not neglect the current requirements of indicating lights and timing devices that do not have inrush volt-amperes, but are energized at the same time as the other components in the circuit. Their total volt-amperes should be added to the total inrush volt-amperes.

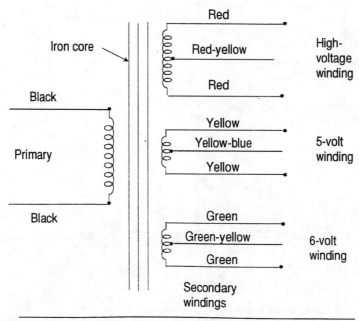

Figure 34. Typical control-transformer wiring diagram.

Again, control transformers will be covered in more detail in Year-Four, as will potential and current transformers. The material presented in this module is designed to introduce the trainee to these devices; more knowledge of practical applications is forthcoming.

9.0.0 POTENTIAL AND CURRENT TRANSFORMERS

In general, a potential transformer is used to supply voltage to instruments such as voltmeters, frequency meters, power-factor meters, and watt-hour meters. The voltage is proportional to the primary voltage, but it is small enough to be safe for the test instrument. The secondary of a potential transformer may be designed for several different voltages, but most are designed for 120 volts. The potential transformer is primarily a distribution transformer especially designed for good voltage regulation so that the secondary voltage under all conditions will be as nearly as possible a definite percentage of the primary voltage.

9.1.0 CURRENT TRANSFORMERS

A current transformer (Fig. 35) is used to supply current to an instrument connected to its secondary, the current being proportional to the primary current, but small enough to be safe for the instrument. The secondary of a current transformer is usually designed for a rated current of 5 amperes.

A current transformer operates in the same way as any other transformer in that the same relation exists between the primary and the secondary current and voltage. A current transformer is connected in series with the power lines to which it is applied so that line current

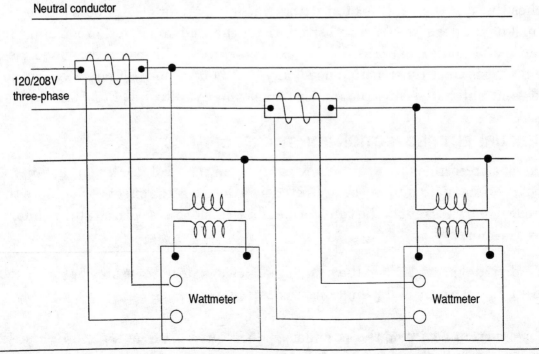

Figure 35. Current and potential transformers used in conjunction with watt-hour meter.

flows in its primary winding. The secondary of the current transformer is connected to current devices such as ammeters, wattmeters, watt-hour meters, power-factor meters, some forms of relays, and the trip coils of some types of circuit breakers.

When no instruments or other devices are connected to the secondary of the current transformer, a short-circuit device or shunt is placed across the secondary to prevent the secondary circuit from being opened while the primary winding is carrying current, if the secondary circuit is open there will be no secondary ampere turns to balance the primary ampere turns, so the total primary current becomes exciting current and magnetizes the core to a high flux density. This produces a high voltage across both primary and secondary windings and endangers the life of anyone coming in contact with the meters or leads. A CT is the only transformer which may be short circuited on the secondary while energized.

10.0.0   NEC REQUIREMENTS

Transformers must normally be accessible for inspection except for dry-type transformers under certain specified conditions. Certain types of transformers with a high voltage or kVA rating are required to be enclosed in transformer rooms or vaults when installed indoors. The construction of these vaults is covered in NE Code Sections 450-41 through 450-48.

In general, the NE Code specifies that the walls and roofs of vaults must be constructed of materials that have adequate structural strength for the conditions with a minimum fire resistance of 3 hours. However, where transformers are protected with an automatic sprinkler system, water spray, carbon dioxide, or halon, the fire resistance construction may be lowered to only 1 hour. The floors of vaults in contact with the earth must be of concrete and not less than 4 inches thick. If the vault is built with a vacant space or other floors (stories) below it, the floor must have adequate structural strength for the load imposed thereon and a minimum fire resistance of 3 hours. Again, if the fire extinguishing facilities are provided, as outlined above, the fire resistance construction need only be 1 hour. The NE Code does not permit the use of studs and wall board construction for transformer vaults. See Fig. 36.

10.1.0 OVERCURRENT PROTECTION FOR TRANSFORMERS

The overcurrent protection for transformers is based on their rated current, not on the load to be served. The primary circuit may be protected by a device rated or set at not more than 125% of their rated primary current of the transformer for transformers with a rated primary current of 9 amperes or more.

Instead of individual protection on the primary side, the transformer may be protected only on the secondary side if all the following conditions are met.

- The overcurrent device on the secondary side is rated or set at not more than 125% of the rated secondary current.

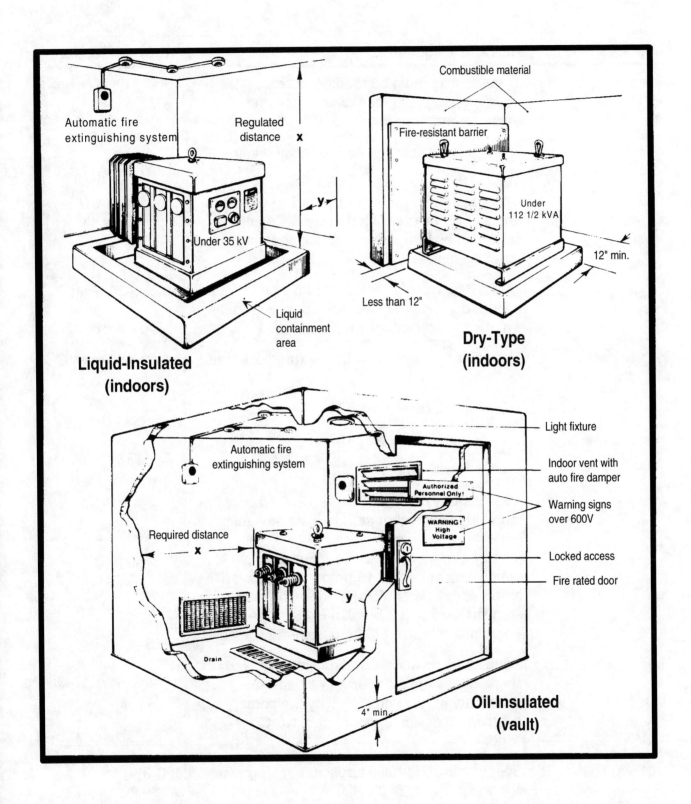

Figure 36. Summary of NEC transformer installation requirements.

Application	NE Code Regulation	NE Code Section
Location	Transformers must be readily accessible to qualified personnel for maintenance and replacement.	450-13
	Dry-type transformers may be located in the open.	
	Dry-type transformers not exceeding 600 volts and 50 kVA are permitted in fire-resistant hollow spaces of buildings under conditions as specified in the NE Code.	
	Liquid-filled transformers must be installed as specified in the NE Code and usually in vaults when installed indoors.	Article 450, Part B
Overcurrent protection	The primary protection must be rated or set as follows: 9 amperes or more, 125% Less than 9 amperes, 167% Less than 2 amperes, 300% If the primary current (line side) is 9 amperes or more, the next higher standard size overcurrent protective device greater than 125% of the primary current may be used. For example, if the primary current is 15 amperes, 125% of 15 amperes equals 18.75 amperes. The next standard size circuit breaker is 20 amperes. Therefore, this size (20 amperes) may be used. Conductors on the secondary side of a single-phase transformer with a two-wire secondary may be protected by the primary overcurrent device under certain NE Code conditions.	Article 450, part C
Over 600 volts	Special NE Code rules apply to transformers operating at over 600 volts.	450-3(a)

Figure 36. Summary of NEC transformer installation requirements. *(Cont.)*

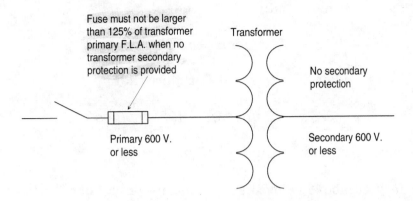

Fuse must not be larger than 125% of transformer primary F.L.A. when no transformer secondary protection is provided

Transformer

No secondary protection

Primary 600 V. or less

Secondary 600 V. or less

Primary Current	Primary Fuse Rating
9 amps or more	125% or next higher standard rating if 125% does not correspond to a standard fuse size
2 amps to 9 amps	167% maximum
Less than 2 amps	300% maximum

Figure 37. Transformer circuit with primary fuse only.

- The primary feeder overcurrent device is rated or set at not more than 250% of the rated primary current.

For example, if a 12 kVA transformer has a primary current rating of:

12,000 watts/480 volts = 25 amperes

and a secondary current rated at

12,000 watts/120 volts = 100 amperes

the individual primary protection must be set at

1.25 x 25 amperes = 31.25 amperes

In this case, a standard 30-ampere cartridge fuse rated at 600 volts could be used, as could a circuit breaker approved for use on 480 volts. However, if certain conditions are met, individual primary protection for the transformer is not necessary in this case if the feeder overcurrent-protective device is rated at not more than

$$2.5 \times 25 \text{ amperes} = 62.5 \text{ amperes}$$

and the protection of the secondary side is set at not more than

$$1.25 \times 100 \text{ amperes} = 125 \text{ amperes}$$

A standard 125 ampere circuit breaker could be used.

Note   The example cited above is for the transformer only; not the secondary conductors. The secondary conductors must be provided with overcurrent protection as outlined in NEC Section 210-20.

The requirements of NEC Section 450-3 cover only transformer protection; in practice, other components must be considered in applying circuit overcurrent protection. For circuits with transformers, requirements for conductor protection per NEC Articles 240 and 310 and for panelboards per NEC Article 384 must be observed. Refer to NEC Sections 240-3, Exceptions 2 and 5; 240-21, Exceptions 2 and 8; 384-16d.

Primary Fuse Protection Only (NEC Section 450-3b1): If secondary fuse protection is not provided, then the primary fuses must not be sized larger than 125% of the transformer primary full-load amperes except if the transformer primary F.L.A. is that shown in NEC Section 450-3b1 (See Fig. 37).

Individual transformer primary fuses are not necessary where the primary circuit fuse provides this protection.

Primary and Secondary Protection: In unsupervised locations, with primary over 600 volts, the primary fuse can be sized at a maximum of 300%. If the secondary is also over 600 volts, the secondary fuses can be sized at a maximum of 250% for transformers with impedances not greater than 6%; 225% for transformers with impedances greater than 6% and not more than 10%. If the secondary is 600 volts or below, the secondary fuses can be sized at a maximum of 125%. Where these settings do not correspond to a standard fuse size, the next higher standard size is permitted.

In supervised locations, the maximum settings are as shown in Fig. 38 except for secondary voltages of 600 volts or below, where the secondary fuses can be sized at a maximum of 250%.

Primary Protection Only: In supervised locations, the primary fuses can be sized at a maximum of 250%, or the next larger standard size if 250% does not correspond to a standard fuse size.

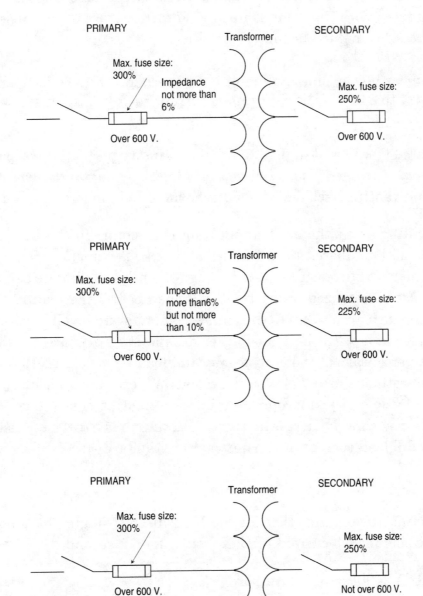

Figure 38. Minimum overcurrent protection for transformers in supervised locations.

Note: The use of "Primary Protection Only" does not remove the requirements for compliance with NEC Articles 240 and 384. See (FPN) in NEC Section 450-3 which references NEC Sections 240-3 and 240-100 for proper protection for secondary conductors.

10.1.1 Overcurrent Protection for Small Power Transformers

Low amperage, E-Rated medium voltage fuses are general purpose current limiting fuses. The E rating defines the melting-time-current characteristic of the fuse and permits electrical interchangeability of fuses with the same E Rating. For a general purpose fuse to have an E Rating the following condition must be met:

- The current responsive element shall melt in 300 seconds at an rms current within the range of 200% to 240% of the continuous current rating of the fuse, fuse refill, or link. (ANSI C37.46).

Low amperage, E-Rated fuses are designed to provide primary protection for potential, small service, and control transformers. These fuses offer a high level of fault current interruption in a self-contained non-venting package which can be mounted indoors or in an enclosure.

As for all current-limiting fuses, the basic application rules found in the NEC and manufacturer's literature should be adhered to. In addition, potential transformer fuses must have sufficient inrush capacity to successfully pass through the magnetizing inrush current of the transformer. If the fuse is not sized properly, it will open before the load is energized. The maximum magnetizing inrush currents to the transformer at system voltage and the duration of this inrush current varies with the transformer design. Magnetizing inrush currents are usually denoted as a percentage of the transformer full load current, i.e., 10X, 12X, 15X, etc. The inrush current duration is usually given in seconds. Where this information is available, an easy check can be made on the appropriate minimum-melting curve to verify proper fuse selection. In lieu of transformer inrush data, the rule of thumb is to select a fuse size rated at 300% of the primary full load current or the next larger standard size.

Example

The transformer manufacturer states that an 800 VA 240 Volt, single phase potential transformer has a magnetizing inrush current of 12 X lasting for 0.1 second.

A. I_{FL} = 800VA/2400V = 0.333 Ampere

Inrush Current = 12×0.333 = 4 Amperes

Since the voltage is 2400 volts we can use either a JCW or a JCD fuse. The proper fuse would be a JCW-1E, or JCD-1E.

B. Using the rule of thumb—300% of .333 Ampere is .999 ampere.

Therefore we would choose a JCW-1E or JCD-1E.

Typical Potential Transformer Connections: The typical potential transformer connections encountered in industry can be grouped into two categories:

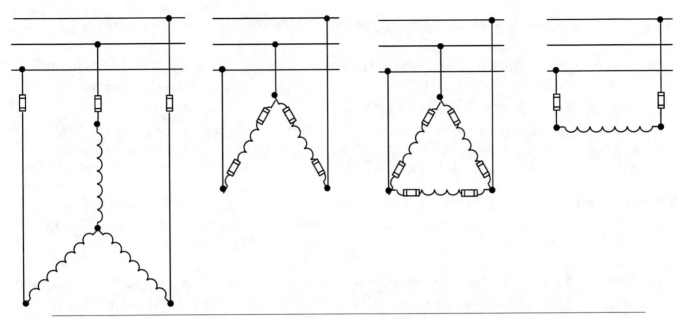

Figure 39. Connections requiring fuses to pass only the magnetizing inrush of one transformer.

- Those connections that require the fuse to pass only the magnetizing inrush of one potential transformer. See Fig. 39.

- Those connections that must pass the magnetizing inrush of more than one potential transformer. See Fig. 40.

Fuses for Medium Voltage Transformers and Feeders: E-rated, medium-voltage fuses are general purpose current limiting fuses. The fuses carry either an "E" or an "X" rating which defines the melting-time-current characteristic of the fuse. The ratings are used to allow electrical interchangeability among different manufacturers' fuses.

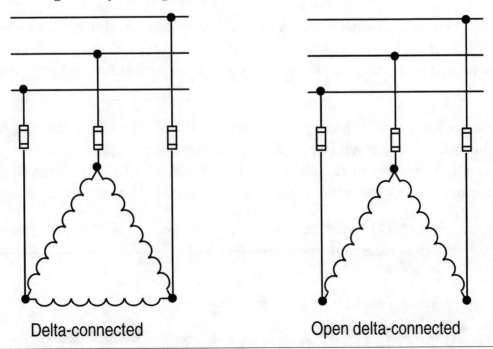

Delta-connected   Open delta-connected

Figure 40. Connections requiring fuses to pass the magnetizing inrush of more than one transformer.

For a general purpose fuse to have an E rating, the following conditions must be met:

- The current responsive element with ratings 100 amperes or below shall melt in 300 seconds at an rms current within the range of 200% to 240% of the continuous current rating of the fuse unit. (ANSI C37.46).

- The current responsive element with ratings above 100 amperes shall melt in 600 seconds at an rms current within the range of 220% to 264% of the continuous current rating of the fuse unit. (ANSI C37.46).

A fuse with an "X" rating does not meet the electrical interchangeability for an "E" rated fuse but offers the user other ratings that may provide better protection for his particular application.

Transformer protection is the most popular application of E-Rated fuses. The fuse is applied to the primary of the transformer and is solely used to prevent rupture of the transformer due to short circuits. It is important, therefore, to size the fuse so that it does not clear on system inrush or permissible overload currents. Magnetizing inrush must also be considered with sizing a fuse. In general, power transformers have a magnetizing inrush current of 12X the full load rating for a duration of $\frac{1}{10}$ second.

10.2.0 TRANSFORMER GROUNDING

Grounding is necessary to remove static electricity and also as a precautionary measure in case the transformer windings accidentally come in contact with the core or enclosure. All should be grounded and bonded to meet NE Code requirements and also local codes, where applicable.

The tank of every power transformer should be grounded to eliminate the possibility of obtaining static shocks from it or being injured by accidental grounding of the winding to the case. A grounding lug is provided on the base of most transformers for the purpose of grounding the case and fittings.

The NE Code specifically states the requirements of grounding and should be followed in every respect. Furthermore, certain advisory rules recommended by manufacturers provide additional protection beyond that of the NE Code. In general, the code requires that separately derived alternating current systems be grounded as stated in Article 250-26.

Figure 41 summarizes NEC regulations governing the grounding of transformers to provide for fault current to trip overcurrent protective devices.

The noncurrent-carrying metal
parts of transformers must be
effectively bonded together
NEC Section 250-71(a)

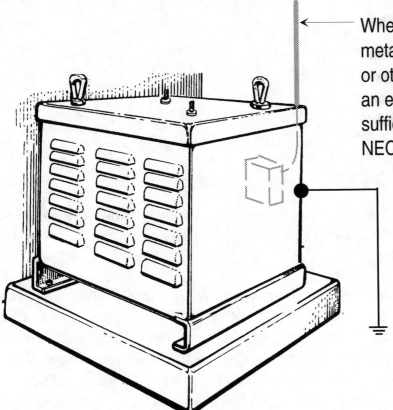

Where supplied by a metal-clad,
metal-sheathed, metal-raceway,
or other wiring method that provides
an equipment ground qualifies as
sufficiently grounding the transformer
NEC Section 250-42

Exposed noncurrent-carrying
metal parts of transformer
installations, including fences,
guards, etc., must be grounded
where required unded the
conditions and in the manner
specified for electrical equipment
and other exposed metal parts.
NEC Section 450-10

The path to ground must be
permanent and continuous;
have capacity to conduct safely
any fault current likely to be
imposed on it; and have
sufficiently low impedance to
limit the voltage to ground
and to facilitate the operation
of the overcurrent protective
devices
NEC Section 250-51

The main and equipment
bonding jumpers must be
of copper or other corrosion-
resistant material
NEC Section 250-79(a)

Figure 41. Summary of NEC requirements for transformer grounding.

Power factor was covered in the Year-Two Module, Alternating Current, so a brief review of the subject should suffice here. The equation for power factor is as follows:

$$PF = \frac{Kw}{Kva}$$

where:

PF = Power Factor

KW = Kilowatts

Kva = Kilovolt-amperes

Calculating the power factor of an electrical system requires that we determine the true power, inductive reactance, and capacitive reactance of the system. An analogy will enhance understanding these terms.

Imagine a farm wagon to which three horses are hitched as shown in Fig. 42. The horse in the middle (#1) is pulling straight ahead; we'll call this horse "True Power," because all his effort is in the direction that the work should be done. Horse #2 wants to nibble at the grass growing along side the road. This horse doesn't contribute an ounce of pull in the desired direction, but causes havoc by pulling the wagon toward the ditch. We'll call horse #2 "Inductance." The third horse has about the same strength as horse #2 and enjoys the grass on the opposite side of the road which causes this horse to pull in the exact opposite direction of horse #2. Horse #3 also contributes nothing to the forward motion. This horse is called "Capacitance."

Let's forget about horse #3 for the moment. If only horse #1 and #2 were pulling, the wagon would go in the direction of the dotted line "A". Notice that the length of that line is greater than the line to #1, so the horse named "Inductance" has an effect on the final result. The direction and length of A might well be called "apparent power" and happens to be the hypotenuse (or diagonal) of a right angle triangle.

If the #2 horse were unhitched, then horses #1 and #3 would cause the wagon to move in the direction of "B" and the length of that line would also be "apparent power." If all three horses were pulling, horses #2 and #3 would cancel one another, and the only useful animal is reliable horse #1—True Power.

In an AC circuit, there are always three forces working in varying lengths. Inductance (horse #2) is present in every magnetic circuit and always works in a 90-degree angle with True Power. Therefore, we have a power factor of less than 100%, because the wagon does not move straight ahead, but travels towards the right due to the pull of horse #2. However, we can improve the

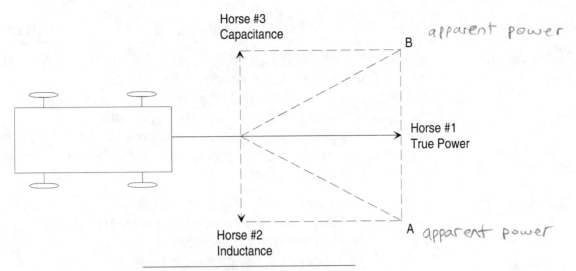

Horse #3
Capacitance

B   *apparent power*

Horse #1
True Power

Horse #2
Inductance

A   *apparent power*

Figure 42. Depicting power factor.

power factor by adding horse #3, Capacitance, which tends to cancel out the pull of inductance, enabling the wagon to travel straight ahead.

11.1.0   CAPACITORS

NEC Article 460 states specific rules for the installation and protection of capacitors other than surge capacitors or capacitors that are part of another apparatus. The chief use of capacitors is to improve the power factor of an electrical installation or an individual piece of electrically-operated equipment. This efficiency, in general, lowers the cost of power.

The NE Code requirements for capacitors operating under 600 volts are summarized in Fig. 43.

Since capacitors may store an electrical charge and hold a voltage that is present even when a capacitor is disconnected from a circuit, capacitors must be enclosed, guarded, or located so that persons cannot accidentally contact the terminals. In most installations, capacitors are installed out of reach or are placed in an enclosure accessible only to qualified persons. The stored charge of a capacitor must be drained by a discharge circuit either permanently connected to the capacitor or automatically connected when the line volt age of the capacitor circuit is removed. The windings of a motor or a circuit consisting of resistors and reactors will serve to drain the capacitor charge.

Capacitor circuit conductors must have an ampacity of not less than 135% of the rated current of the capacitor. This current is determined from the kVA rating of the capacitor as for any load. A 100 kVA (100,000 watts) three-phase capacitor operating at 480 volts has a rated current of

$$100,000 \text{ kVA}/1.73 \times 480 \text{ volts} = 120.3 \text{ amperes}$$

The minimum conductor ampacity is then:

Application	NE Code Regulation	NE Code Section
Enclosing and guarding	Capacitors must be enclosed, located, or guarded so that persons cannot come into accidental contact or bring conducting materials into accidental contact with exposed energized parts, terminals, or buses associated with them. However, no additional guarding is required for enclosures accessible only to authorized and qualified persons.	460-2(b)
Stored charge	Capacitors must be provided with a means of draining the stored charge. The discharge circuit must be either permanently connected to the terminals of the capacitor or capacitor bank, or provided with automatic means of connecting it to the terminals of the capacitor bank on removal of voltage from the line. Manual means of switching or connecting the discharge circuit shall not be used.	460-6
Capacitors on circuits over 600 volts	Special NE Code regulations apply to capacitors operating at over 600 volts.	Article 460
Conductor ampacity	The ampacity of capacitor circuit conductors must not be less than 135% of the rated current of the capacitor.	460-8(a)
Capacitors on motor circuits	The ampacity of conductors that connect a capacitor to the terminals of a motor or to motor circuit conductors shall not be less than one third the ampacity of the motor circuit conductors and in no case less than 135% of the rated current of the capacitor.	460-8(a)
Overcurrent protection	Overcurrent protection is required in each ungrounded conductor unless the capacitor is connected on the load side of a motor-running overcurrent device. The setting must be as low as practicable.	460-8(b)
Disconnecting means	A disconnecting means is required for a capacitor unless it is connected to the load side of a motor-running overcurrent device. The rating must not be less than 135% of the rated current of the capacitor.	460-8(c)

Figure 43. NEC requirements for capacitor installations.

Application	NE Code Regulation	NE Code Section
Improved power factor	The total kilovar rating of capacitors connected to the load side of a motor controller must not exceed the value required to raise the no-load power factor to 1.	460-7
Overcurrent protection for improved power factor	If the power factor is improved, the motor-running overcurrent device must be selected based on the reduced current drawn; not the full-load current of the motor.	460-9
Grounding	Capacitor cases must be grounded except when the system is designed to operate at other than ground potential.	460-10

Figure 43. NEC requirements for capacitor installations. *(Cont.)*

1.35 x 120.3 amperes = 162.4 amperes

When a capacitor is switched into a circuit, a large inrush current results to charge the capacitor to the circuit voltage. Therefore, an overcurrent protective device for the capacitor must be rated or set high enough to allow the capacitor to charge. Although the exact setting is not specified in the NE Code, typical settings vary between 150% and 250% of the rated capacitor current.

In addition to overcurrent protection, a capacitor must have a disconnecting means rated at not less than 135% of the rated current of the capacitor unless the capacitor is connected to the load side of the motor running overcurrent device. In this case, the motor disconnecting means would serve to disconnect the capacitor and the motor.

A capacitor connected to a motor circuit serves to increase the power factor and reduce the total kVA required by the motor-capacitor circuit. The power factor is defined as the true power in kilowatts divided by the total kVA or

$$pf = \frac{kW}{kVA}$$

where the power factor is a number between .0 and 1.0. A power factor less than one represents a lagging current for motors and inductive devices. The capacitor introduces a leading current that reduces the total kVA and raises the power factor to a value closer to unity. If the inductive load of the motor is completely balanced by the capacitor, a maximum power factor of unity results and all of the input energy serves to perform useful work.

The capacitor circuit conductors for a power factor correction capacitor must have an ampacity of not less than 135% of the rated current of the capacitor. In addition, the ampacity must not be less than one-third the ampacity of the motor circuit conductors.

The connection of a capacitor reduces current in the feeder up to the point of connection. If the capacitor is connected on the load side of the motor-running overcurrent device, the current through this device is reduced and its rating must be based on the actual current, not on the full-load current of the motor.

11.2.0 RESISTORS AND REACTORS

NEC Article 470 covers the installation of separate resistors and reactors on electric circuits. However, this Article does not cover such devices that are component parts of other machines and equipment.

In general, the NEC requires resistors and reactors to be placed where they will not be exposed to physical damage. Therefore, such devices are normally installed in a protective enclosure such as a controller housing or other type of cabinet. When these enclosures are constructed of metal, they must be grounded as specified in NEC Article 250. Furthermore, a thermal barrier must be provided between resistors and/or reactors and any combustible material that is less than 12 inches away. A space of 12 inches of more between the devices and combustible material is considered sufficient distance so as not to require a thermal barrier.

Insulated conductors used for connections between resistors and motor controllers must be rated at not less than 90°C (194°F) except for motor-starting service. In this latter case, other conductor insulation is permitted provided other Sections of the NEC is not violated.

11.3.0 DIODES AND RECTIFIERS

The diode and the rectifier are the simplest form of electronic components. The only difference between the two is their size; that is, a component that is rated less than 1 ampere is called a *diode*, while a similar component rated above 1 ampere is called a *rectifier*. The main purpose of either device is to convert or rectify alternating current to direct current.

Diodes and rectifiers are composed of two material types: P and N—one has free electrons while the other has a shortage of electrons. When the two types of material are bonded together, a solid-state component is produced that will allow electrons to flow in one direction and act as an insulator when voltage is reversed.

Diodes and rectifiers are used extensively in control circuits. For example, Fig. 44 shows an AC voltage supplying a control transformer which must supply a DC electronic controller. Consequently, two rectifiers are installed on the secondary side of the transformer to change AC to DC. The rectifiers allow current to flow in one direction, but will not allow the normal alternating current reversal—simulating direct current.

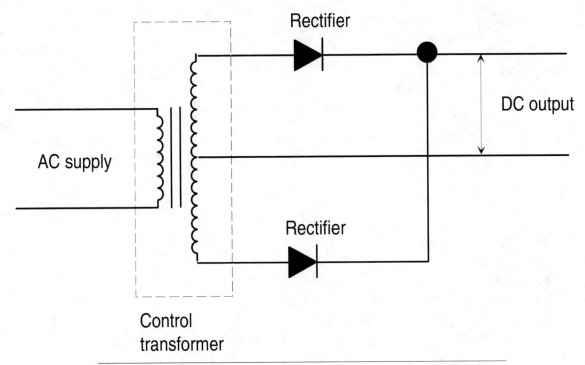

Figure 44. Rectifiers used in a control circuit to change AC to DC.

Theoretical study of conditions in a transformer includes the use of phasor diagrams (vectors) that represent graphically voltages and currents in transformer windings.

A vector or phasor diagram is a line with direction and length. Vectors are like reading a road map and not much more difficult, just a bit more refined. For example, road directions that instruct you to go east 40 miles, then south 30 miles to get to your destination, are simple to understand. In other words, you had to drive 70 miles to get there. However, had you been able to drive as the crow flies, the distance would have been shorter—only 50 miles. See Fig. 45A. If the miles were converted into electrical terms, this same triangle would be the classical 3, 4, 5 triangle or the 80% power factor relationship as shown in Fig. 45B.

Referring to Fig. 45B, if AB is 4 amperes and BC is 3 amperes, then the diagonal line AC (hypotenuse of a right angle triangle) will be 5 amperes. This value may be proven by drawing lines AB and BC to scale and then connecting line AC and measuring it. However, the same results may be obtained mathematically using the right angle triangle rule which states:

● The hypotenuse is equal to the square root of the sum of the squares of the other two sides; that is:

$$AC = \sqrt{AB^2 + BC^2}$$

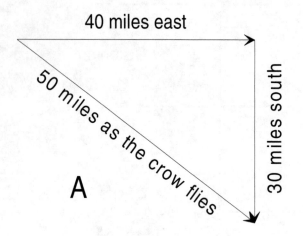

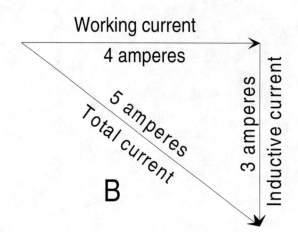

Figure 45. Typical vectors.

$$AC = \sqrt{4^2 + 3^2} = \sqrt{16 + 9} = \sqrt{25}$$

$$AC = 5$$

In this example, the working current is related to kilowatts (kW) and the total current, related to kVA, are the ratio of 4/5 = 0.80. Since power factor $= \dfrac{kW}{kVA}$, the power factor of this circuit is 80%.

12.1.0   PRACTICAL APPLICATIONS OF PHASOR DIAGRAMS

When three single-phase transformers are used as a three-phase bank, the direction of the voltage in each of the six phase windings may be represented by a voltage phasor. A voltage phasor diagram of the six voltages involved provides a convenient way to study the relative direction and amounts of the primary and the secondary voltages.

The same is true for one three-phase transformer, which also has six voltages to be considered, because its phase windings on the high-voltage side and the low-voltage side are connected together in the same way as the phase windings of three single-phase transformers.

To show how a phasor diagram is drawn, consider the three-phase transformer shown in Fig. 46. Here we have a Y-delta-connected three-phase transformer with three legs and each carrying a high- and low-voltage winding. The high-voltage windings are connected in Y (with a common neutral point at N) with the leads to the high-voltage terminals designated H$_1$, H$_2$,

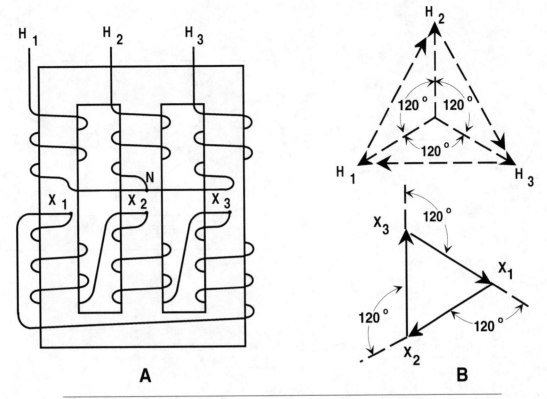

Figure 46. Windings and phasor diagrams of three-phase transformer.

and H₃. The three low-voltage windings are connected in delta. The junction points of the three windings serve as low-voltage terminals X₁, X₂, and X₃.

The voltages in the low-voltage windings are assumed to be equal to each other in amount but are displaced from each other 120°.

When drawing the phasor diagram for the low-voltage windings, phasor X_1X_2 is drawn first in any selected direction and to any convenient scale. The arrowhead indicates the instantaneous direction of the alternating voltage in winding X_1X_2, and the length of the phasor represents the amount of voltage in the winding. The broken lines that extend past the arrowheads represent reference lines for phase angles. Since winding X_2X_3 is physically connected to the end of X_2 in the X_1X_2 winding, phasor X_2X_3 will be started at point X_2, 120— out of phase with phasor X_1X_2 in the clockwise direction. The length of phasor X_2X_3 is equal to the length of phasor X_1X_2. Winding X_3X_1 is drawing in a similar manner.

The high-voltage phasor comes next. Since the low-voltage winding X_1X_2 is wound on the same leg as the high-voltage winding NH_1, the voltages in these two windings are in phase and are represented by parallel voltage phasors. Therefore, the phasor NH_1 is drawn from a selected point N parallel to X_1X_2. Note that all three high-voltage windings are physically connected to a common point N. Therefore, the phasors representing the voltages NH_1, NH_2, and NH_3 will all start at the common; point N in the phasor diagram. Again, the high-voltage phasors are all of the same length but displaced from each other by 120°as shown.

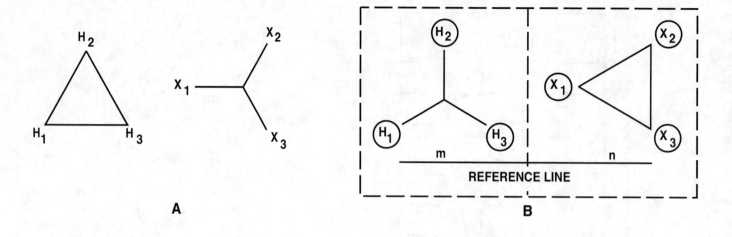

Figure 47. Phasor diagrams for three-phase transformer banks in parallel.

The high-voltage phasors have the same length as the low-voltage phasors. Each phase of the high voltage is, however, proportional to the low voltage in the same phase according to the turns ratio, making the voltage in each phase of the high-voltage windings are connected in Y, each line voltage is 1.732 times higher than a phase voltage in any one winding. It can be geometrically proved, for example, that $H_1H_2 \times NH_1 = 1.732 \times NH_2$ and so forth for each line voltage.

In comparing three-phase transformers for possible operation in parallel, draw the voltage phasor diagram for transformer bank A and mark the terminals as shown in Fig. 47A. Next draw the phasor diagram for bank B on drafting or other transparent paper; draw a heavy reference line m-n as shown in the drawing in Fig. 47B. Cut the transparent diagram into two parts as indicated by the dashed lines in the drawing. The two parts of the reference line are now marked m and n. Place diagram m on the high-voltage diagram of (a) so that the terminals which are desired to be connected together coincide. Place diagram n on the low-voltage diagram of (a) so that the heavy reference line of n is parallel to the heavy reference line of m. If, under these condition;s, the terminals of n can be made to coincide with the low-voltage terminals of (a), the terminals which coincide can be connected together for parallel operation. If the low-voltage terminals cannot be made to coincide, parallel operation is not possible with the assumed high-voltage connection.

12.2.0 VOLTAGE DROP

Figure 48 shows a simple 100 percent power factor circuit in which a 1 ohm resistance appears between the source of power and the load. The current in this series circuit is 10 amperes and the voltage at the source is 120. Because Ohm's law reminds us that E=IR, we must have a

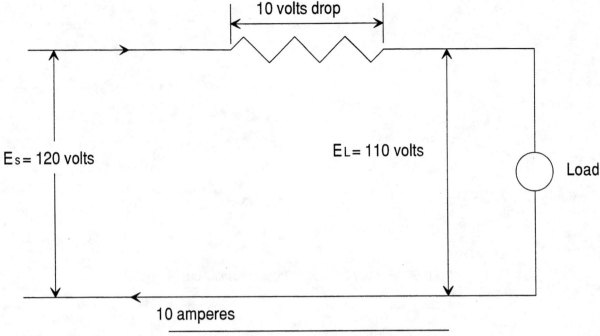

Figure 48. Circuit containing resistance only.

voltage drop across the resistance equal to 10 times 1, or 10 volts. With voltage and current in phase, only 110 volts will be available at the load because we must subtract the resistance drop from the source voltage. Voltage drop is covered more thoroughly in Electrical Task Modules 01201 and 20302.

Voltage drop, however, becomes a little more complicated when inductance is introduced into the circuit as shown in Fig. 49. In this case, the load voltage will be equal to the source voltage minus the voltage drops through R and X, but they cannot be added arithmetically! The vector

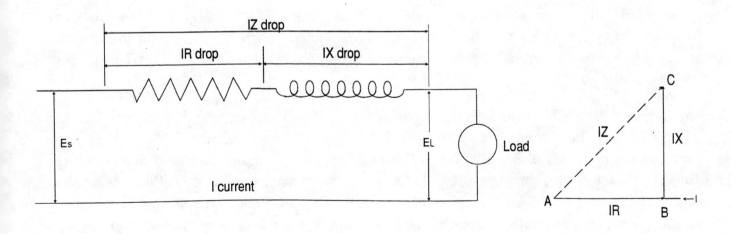

Figure 49. Electrical circuit with both resistance and inductance.

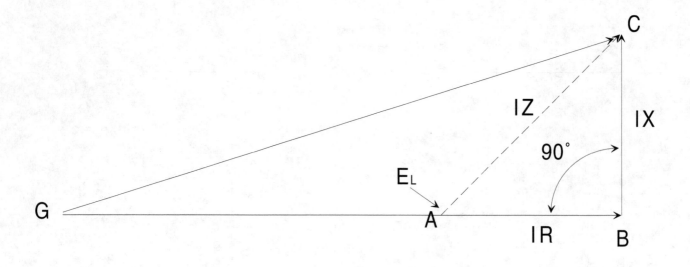

Figure 50. Effect of voltage drop in an AC circuit.

diagram in Fig. 50 shows that the dotted line, AC, is the combination of the two voltages across R and X and represents the voltage drop in the line only.

Figure 50 shows that line GA, which is the load voltage D_L, is less than the source voltage GC, due to the voltage drop in the line AC. Because of the effect of the reactance in the line, this voltage drop cannot be subtracted arithmetically from the source voltage to obtain the load voltage; vectors are the way to go.

Calculations of impedance can be simplified by using equivalent circuits. Reactance voltage drop is governed by the leakage flux, and the voltage regulation depends on the power factor of the load. To determine transformer efficiency at various loads, it is necessary to first calculate the core loss, hysteresis loss, eddy-current loss, and load loss.

However, for all practical purposes in electrical construction applications, a transformer's efficiency may be considered to be 100%. Therefore, for our purposes, a transformer may be defined as a device that transfers power from its primary circuit to the secondary circuit without any significant loss.

Since power (W) equals voltage (E) times current (I), if E_PI_P represents the primary power and E_SI_S represents the secondary power, then $E_PI_P = E_SI_S$ (the subscript P = primary, and the subscript S = secondary). See Fig. 50. If the primary and secondary voltages are equal, the primary and secondary currents must also be equal. Let's assume that E_P is twice as large as E_S. For E_PI_P to equal E_SI_S, I_P must be one-half of I_S as shown in Fig. 51. Therefore, a transformer that steps voltage down always steps current up. Conversely, a transformer that steps voltage up always steps current down. However, transformers are classified as step-up or step-down only in relation to their effect on voltage.

ELECTRICAL TRAINEE TASK MODULE 20307

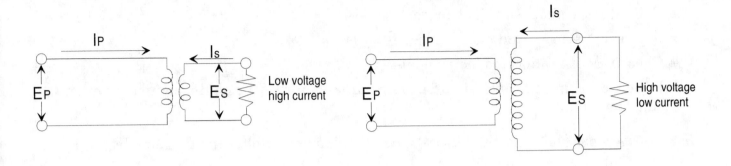

Figure 51. When voltage is stepped down, current is stepped up, or vice-versa.

13.0.0 TROUBLESHOOTING

Since transformers are an essential part of every electrical installation, electricians must know how to test and locate troubles that develop in transformers—especially in the smaller power supply or control transformers. The procedure for accomplishing this is commonly known as "troubleshooting."

The term, "troubleshooting" as used in this module covers the investigation, analysis and corrective action required to eliminate faults in electrical systems, including circuits, components, and equipment. Most troubles are simple and easily corrected; examples are an open circuit, a ground fault or short circuit, or a change in resistance.

The many troubleshooting charts available are very useful when troubleshooting. To use troubleshooting charts, most list the complaint on the left side of the charts, then the possible cause; followed by the proper corrective action.

Think before acting: Study the problem thoroughly, then ask yourself these questions:

- What were the warning signs preceding the trouble?
- What previous repair and maintenance work has been done?
- Has similar trouble occurred before?
- If the circuit, component, or piece of equipment still operates, is it safe to continue operation before further testing?

The answers to these questions can usually be obtained by:

- Questioning the owner of the equipment.
- Taking time to think the problem through.

- Looking for additional symptoms.
- Consulting troubleshooting charts.
- Checking the simplest things first.
- Referring to repair and maintenance records.
- Checking with calibrated instruments.
- Double checking all conclusions before beginning any repair on the equipment or circuit components.

Note   Always check the easiest and obvious things first; following this simple rule will save time and trouble.

13.1.0   DOUBLE-CHECK BEFORE BEGINNING

The source of many problems can be traced not to one part alone but to the relationship of one part with another. For instance, a tripped circuit breaker may be reset to restart a piece of equipment, but what caused the breaker to trip in the first place? It could have been caused by a vibrating "hot" conductor momentarily coming into contact with a ground; or a loose connection could eventually cause overheating; and any number of other causes.

Too often, electrically-operated equipment are completely disassembled in search of the cause of a certain complaint and all evidence is destroyed during disassembly operations. Check again to be certain an easy solution to the problem has not been overlooked.

13.2.0   FIND AND CORRECT BASIC CAUSE OF TROUBLE

After an electrical failure has been corrected in any type of electrical circuit or piece of equipment, be sure to locate and correct the cause so the same failure will not be repeated. Further investigation may reveal other faulty components.

Also be aware that although troubleshooting charts and procedures greatly help in diagnosing malfunctions, they can never be complete. There are too many variations and solutions for a given problem.

To solve electrical problems consistently, you must first understand the basic parts of electrical circuits, how they function, and for what purpose. If you know that a particular part is not performing its job, then the cause of the malfunction must be within this part or series of parts.

13.3.0   TROUBLESHOOTING TRANSFORMERS

Open Circuit: Should one of the windings in a transformer develop a break or "open" condition, no current can flow and therefore, the transformer will not deliver any output. The symptoms of an open circuited transformer is that the circuits which derive power from the transformer are de-energized or "dead." Use an AC voltmeter to check across the transformer output terminals as shown in Fig. 52. A reading of zero volts indicates an open circuit.

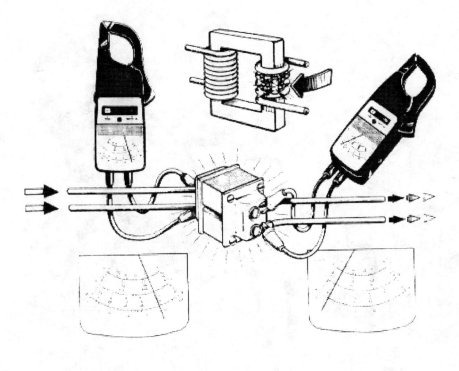

Figure 52. Checking transformer for an open winding.

Then take a voltage reading across the input terminals. If a voltage reading is present, then the conclusion is that one of the windings in the transformer is open. However, if no voltage reading is on the input terminals either, then the conclusion is that the open is elsewhere on the line side of the circuit; perhaps a disconnect switch is open.

WARNING!   Make absolutely certain that your testing instruments are designed for the job and are calibrated for the correct voltage. Never test the primary side of any transformer over 600 volts unless you are qualified, have the correct high-voltage testing instruments, and the test is made under the proper supervision.

However, if voltage is present on the line or primary side and none on the secondary or load side, open the switch to de-energize the circuit, and place a warning tag (tag-out and lock) on this switch so that it is not inadvertently closed again while someone is working on the circuit. Disconnect all of the transformer primary and secondary leads, check each winding in the transformer for continuity (a continuous circuit), as indicated by a resistance reading taken with an ohmmeter as shown in Fig. 53.

Continuity is indicated by a relatively low resistance reading on control transformers, while an open winding will be indicated by an infinite resistance reading on the ohmmeter. In most

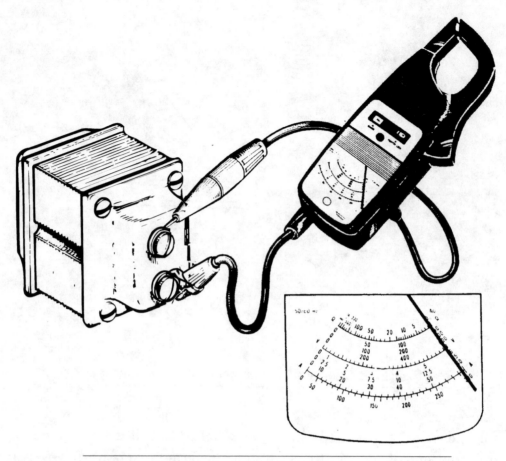

Figure 53. Checking for open winding with a continuity test.

cases, such small transformers will have to be replaced, unless of course the break is accessible and can be repaired.

Shorted Turns: Sometimes a few turns in the secondary winding of a transformer will acquire a partial short, which in turn will cause a voltage drop across the secondary. The symptom of this condition is usually overheating of the transformer caused by large circulating currents flowing in the shorted windings. The most accurate way to check this condition is with a transformer turns ratio tester (TTR). However, another way to check this condition is with a voltmeter set at the proper voltage scale (Fig. 54). Take a reading on the line or primary side of the transformer first to make certain normal voltage is present. Then take a reading on the secondary side. If the transformer has a partial short or ground fault, the voltage reading should be lower than normal.

Replace the faulty transformer with a new one and again take a reading on the secondary. If the voltage reading is now normal and the circuit operates satisfactorily, leave the replacement transformer in the circuit, and either discard or repair the original transformer.

A highly sensitive ohmmeter may also be used to test this condition when the system is de-energized and the leads are disconnected; a lower reading on the ohmmeter than normal

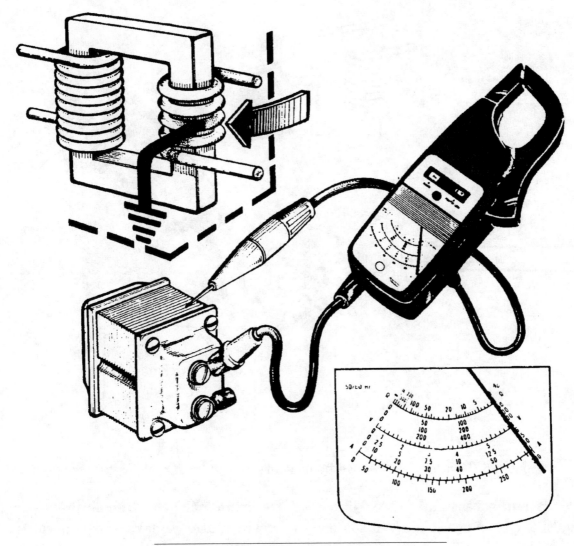

Figure 54. Testing transformer for ground fault.

indicates this condition. However, the reading will usually be so slight that the average ohmmeter is not sensitive to detect the difference. Therefore, the recommended way is to use the voltmeter test.

Complete Short: Occasionally a transformer winding will become completely shorted. In most cases, this will activate the overload protective device and de-energize the circuit, but in other instances, the transformer may continue trying to operate with excessive overheating—due to the very large circulating current. This heat will often melt the wax or insulation inside the transformer, which is easily detected by the odor. Also, there will be no voltage output across the shorted winding and the circuit across the winding will be dead.

The short may be in the external secondary circuit or it may be in the transformer's winding. To determine its location, disconnect the external secondary circuit from the winding and take a reading with a voltmeter. If the voltage is normal with the external circuit disconnected, then the problem lies within the external circuit. However, if the voltage reading is still zero across the secondary leads, the transformer is shorted and will have to be replaced.

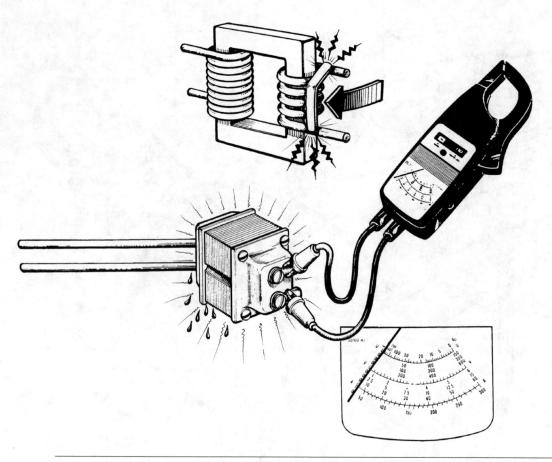

Figure 55. Transformers that overheat usually have a partial short in the windings.

Grounded Windings: Insulation breakdown is quite common in older transformers—especially those that have been overloaded. At some point, the insulation breaks or deteriorates and the wire becomes exposed. The exposed wire often comes into contact with the transformer housing and grounds the winding.

If a winding develops a ground, and a point in the external circuit connected to this winding is also grounded, part of the winding will be shorted out. The symptoms will be overheating, which is usually detected by feel or smell, and a low voltage reading as indicated on a voltmeter reading as shown in Fig. 55. In most cases, transformers with this condition will have to be replaced.

A megohmmeter is the best test instrument to check for this condition. Disconnect the leads from both the primary and secondary windings. Tests can then be performed on either winding by connecting the megger negative test lead to an associated ground and the positive test lead to the winding to be measured.

Insulation resistance should then be measured between the windings themselves. This is accomplished by connecting one test lead to the primary and the second test lead to the secondary. All such tests should be recorded on a record card under proper identifying labels.

Summary

When the AC voltage needed for an application is lower or higher than the voltage available from the source, a transformer is used. The essential parts of a transformer are the primary winding, which is connected to the source, and the secondary winding, which is connected to the load, both wound on an iron core. The two windings are not physically connected. The alternating voltage in the primary winding induces an alternating voltage in the secondary winding. The ratio of the primary and secondary voltages is equal to the ratio of the number of turns in the primary and secondary windings. Transformers may step up the voltage applied to the primary winding and have a higher voltage at the secondary terminals, or they may step down the voltage applied to the primary winding and have a lower voltage available at the secondary terminals. Transformers are applied in AC systems only, single-phase and polyphase, and would not work in DC systems since the induction of voltage depends on the rate of change of current.

A transformer is constructed as a single-phase or a three-phase apparatus. A three-phase transformer has three primary and three secondary windings which may be connected in delta (Δ) or star (Y). Combinations such as Δ-Δ, Δ-Y, Y-Δ, and Y-Y are possible connections of the primary and secondary windings. The first symbol indicates the connection of the primary winding, and the second, that of the secondary winding. A bank of three single-phase transformers can serve the same purpose as one three-phase transformer.

References

For more advanced study of topics covered in this Task Module, the following works are suggested:

National Electrical Code Handbook, Latest Edition, NFPA, Quincy, MA
American Electricians Handbook, Latest Edition, Croft, McGraw-Hill, New York, NY

SELF-CHECK REVIEW/PRACTICE QUESTIONS

1. What is the main purpose of a transformer?

    a. To change the current
    b. To improve the power factor
    c. To change the voltage
    d. To enter impedance into the circuit

2. When AC flows through a transformer coil, what type of field is generated around the coils?

    a. Magnetic field
    b. Non-magnetic field
    c. High-impedance field
    d. Rotating field

3. When the field from one coil cuts through the turns of a second coil, what will occur in the second coil?

    a. Nothing
    b. The coils will rotate
    c. Voltage will be generated
    d. No current will flow in the circuit

4. What causes voltage to be induced in a transformer?

    a. Transformer taps
    b. Mutual induction
    c. Reluctance
    d. Capacitance

5. What are the three parts of a very basic transformer?

    a. Housing, lifting hooks, and base
    b. Dry, oil-filled, and gas-filled
    c. Shell, open, and closed
    d. Primary winding, secondary winding, and core

6. In a transformer with a winding ratio of 5:1 (primary has 5 times the number of turns as the secondary), what will be the voltage on the secondary if the primary voltage is 120?

    a. 12 volts
    b. 24 volts
    c. 48 volts
    d. 60 volts

7. What are the three basic types of iron-core transformers?

    a. Dry, oil-filled, and auto
    b. Closed, open, and shell
    c. Control, power, and lighting
    d. Metal, nonmetallic, and high-temperature

8. What is one effect caused by magnetic leakage in transformers?

    a. Reactance voltage drop
    b. Low impedance
    c. Higher voltage
    d. Fewer amperes

9. Describe the symptoms of a transformer with an open circuit.

    a. High voltage on the secondary
    b. Excessive overheating
    c. No output on the secondary
    d. Voltage drop across the secondary

10. What is the main purpose of transformer taps?

    a. For mounting to concrete walls
    b. To vary the voltage level on the secondary
    c. To thread mounting holes in transformer housings
    d. To alter the primary voltage

PERFORMANCE/LABORATORY EXERCISE

1. Connect the secondary terminals of a multi-tap control transformer to operate a 24-volt control circuit, and the primary terminals for connection to a 120-volt power source.

2. Test the circuit in No. 1 above for operation. Use a voltmeter to measure the exact voltage on the secondary. If the resulting voltage is other than 24 volts, adjust the transformer taps to obtain the required voltage.

3. Calculate the maximum load that should be connected to a three-phase, open-delta system (one of the three transformers removed), using two 60 kVA transformers.

Answers to Self-Check Questions

1. c

2. a

3. c

4. b

5. d

6. b

7. b

8. d

9. c

10. b

Basic Lighting

Module 20308

Electrical Trainee Task Module 20308

BASIC LIGHTING

Objectives

Upon completion of this module, the trainee will be able to:

1. Explain how light enables us to see.
2. Identify the operating characteristics of incandescent, fluorescent and high intensity discharge lamps.
3. Explain how to design a residential lighting system.
4. Describe the various lighting controls and how each operates.
5. Explain how lighting fixtures are installed and connected.
6. Interpret lighting layouts on commercial and industrial blueprints.
7. Interpret lighting fixture schedules on commercial and industrial blueprints.
8. Describe the NEC requirements for fixtures.

Prerequisites

Successful completion of the following Task Modules is required before beginning study of this Task Module: Common Core Curricula, Electrical Levels 1 and 2, Electrical Level 3, Modules 20301 through 20307.

The trainee should also read NEC Article 410–Lighting Fixtures, Lampholders, Lamps and Receptacles.

Required Student Material

1. Trainee Task Module
2. Appropriate Personal Protective Equipment
3. Copy of the latest edition of the National Electrical Code

Course Map Information

This course map shows all of the *Wheels of Learning* task modules in the third level of the Electrical curricula. The suggested training order begins at the bottom and proceeds up. Skill levels increase as a trainee advances on the course map. The training order may be adjusted by the local Training Program Sponsor.

Course Map: Electrical, Level 3

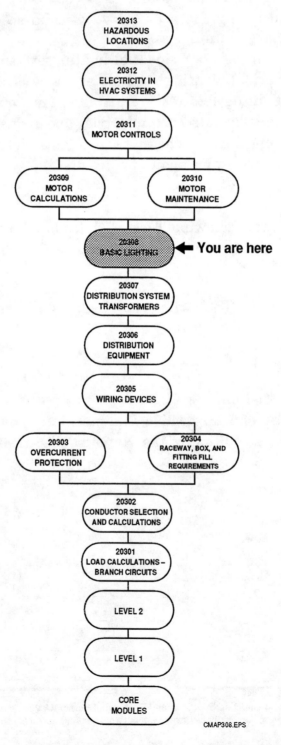

LEVEL 3 COMPLETE

20313
HAZARDOUS
LOCATIONS

20312
ELECTRICITY IN
HVAC SYSTEMS

20311
MOTOR CONTROLS

20309
MOTOR
CALCULATIONS

20310
MOTOR
MAINTENANCE

20308
BASIC LIGHTING ← **You are here**

20307
DISTRIBUTION SYSTEM
TRANSFORMERS

20306
DISTRIBUTION
EQUIPMENT

20305
WIRING DEVICES

20303
OVERCURRENT
PROTECTION

20304
RACEWAY, BOX, AND
FITTING FILL
REQUIREMENTS

20302
CONDUCTOR SELECTION
AND CALCULATIONS

20301
LOAD CALCULATIONS –
BRANCH CIRCUITS

LEVEL 2

LEVEL 1

CORE
MODULES

CMAP308.EPS

TABLE OF CONTENTS

TABLE OF CONTENTS (Continued)

TABLE OF CONTENTS (Continued)

TABLE OF CONTENTS (Continued)

Trade Terms Introduced In This Module

Average luminance (average brightness): The luminance value of a luminaire (at a specific angle of view) expressed in candles per square inch or footlamberts. It is the average value of the luminous subtended area and is obtained by dividing the candlepower at the specified angle by the luminous subtended area of the luminaire. In the case of luminaires, which depend upon secondary reflecting, surfaces may be taken into account under controlled conditions in the calculations. Average luminance values are calculated from related data as opposed to maximum luminance values which are measured.

Baffle: In luminaire designing, a blade of opaque or similar opaque material used primarily to control the light distribution or shield the lamp.

Ballast: Essentially an electromagnetic or electronic device used with electrical discharge lamps to control the light distribution for optimum lamp performance.

Black body: Theoretically a full radiator, one which absorbs all incident radiation, reflecting none. Such a body will emit radiation in which the energy contained in any frequency range is related with that frequency and with the temperature of the radiator.

Candela: Unit of luminous intensity, one candela is equal to 1/60 of one square centimeter of projected area of a black body radiator operating at the freezing point of platinum.

Candlepower: Luminous intensity expressed in candela. In the early days of photometry, luminous intensity was measured by rating sources in terms of ordinary candles. With the advent of more precise standards of luminous intensity, more precise measurements are possible through the use of the standard candle in terms of the luminous intensity of a black body radiator at the temperature of freezing platinum.

Cathode: An electrode where positive current leaves a device which employs electrical conduction other than through solids, such as electrical discharge lamps. A fluorescent lamp has an electrode at each end that serves as a cathode or anode, depending upon the current alternations.

Cold cathode lamp: An electrical discharge lamp designed to start without the need to preheat electrodes; that is, "Slimline" (instant start) lamps.

Diffuser: A device to redirect or scatter the light from a source. It may do so by diffuse reflection or diffuse transmission. In luminaire design, the device is primarily utilized for diffuse transmission.

Fluorescence: The process of emission of electromagnetic radiation by a substance as the result of absorption of energy from some other electromagnetic or particulate radiation, provided that the emission continues only as long as the stimulus that produces the emission is maintained.

Footcandle: The unit of illumination on a surface when the foot is the unit of length. It equals one lumen per square foot, provided that the luminous flux is distributed uniformly over the area.

Glare: A general term used to describe the sensation produced by luminances within the visual field that are sufficiently greater than the luminance to which the eyes are adapted; to cause annoyance, discomfort, or loss in visual performance or visibility.

Halogen: One of a family of elements such as iodine, bromine, fluorine and chlorine, which combine directly with metals to produce halides. The use of iodine in incandescent lamps to improve life characteristics has identified such lamps as halogen lamps.

Incandescence: The emission of visible radiant energy by a heated body which may be a solid, liquid or gas. The temperature of a body required to produce incandescence will vary depending upon the material. Excitation to produce incandescence can be achieved by several means. In lighting, the effect is produced by the passage of an electrical current as in an incandescent lamp.

Inverse square law: As applied to illumination, states that the illumination at a point on a surface varies directly with the candlepower and inversely as the square of the distance between the source and the point. This law holds true where the distance between source and point is at least five times maximum dimension of the source.

Lumen: The unit of luminous flux. It is equal to the flux through a unit solid angle from a uniform point source of one candela (candle).

Luminaire: Lighting fixture.

Luminance: The current accepted term for brightness.

Mounting height: The distance from the finished floor to the light center of the luminaire, or to the plane of the ceiling in recessed applications.

Opal glass: A white diffusing glass material.

Phosphor: Any substance which exhibits luminescence.

Preheat fluorescent lamps: A fluorescent lamp requiring the use of a starter circuit to preheat the filaments prior to striking the arc.

Quartz-iodine lamps: A tungsten filament lamp in a quartz envelope utilizing the halogen regenerative cycle to provide excellent lamp lumen maintenance together with small bulb size. It is classed as an incandescent lamp. Sometimes referred to as a halogen lamp.

Rapid start lamp: Fluorescent lamps which, in conjunction with appropriate ballasts, eliminate the need for auxiliary starting aids to affect lamp operation.

Electric light in the home greatly improves the appearance of the home as well as people and objects in the home. It also speeds up household chores, reduces eye strain, and makes it a pleasure for members of the family to work or play during evening hours. Electric lighting is not only cleaner, safer, and more convenient than any other form of artificial light, it is also inexpensive enough to be within the means of almost every family.

In industry, electric lighting speeds up production, reduces errors, increases safety, and generally improves the morale of employees.

In stores, hotels, and office buildings, electric illumination is used on a large scale to improve the efficiency of employees, to aid in the selling of merchandise, and to reduce eye strain.

The exteriors of some buildings are beautifully floodlighted and streets are lighted brightly with electric lamps. The lighting of outdoor sport areas enables us to view football, baseball, and other sports at night. Television would not be possible without electricity and artificial light.

The cases are endless, and almost everyone today realizes the value of better lighting. This field also provides some of the most fascinating and enjoyable work for the electrician.

1.1.0   THE EYE AND VISION

Since the effect of light upon the eye gives us the sensation of sight, any study of lighting must begin with a consideration of the eye and the seeing process. An understanding of the eye's mechanism will help you to understand the primary function of illumination—to provide light for the performance of visual tasks with a maximum of comfort and a minimum of strain and fatigue.

1.1.1   The Seeing Mechanism

The human eye is a fine precision instrument that is often compared to a camera (see *Figure 1*). Both the eye and camera have a covering or housing. Each has a lens that focuses an inverted image on a light-sensitive surface—the retina in the eye and the film in the camera. The camera shutter corresponds to the eyelid. In front of the lens in the camera is a diaphragm, which may be opened or closed to regulate the amount of light entering the camera. The iris of the eye performs the same function.

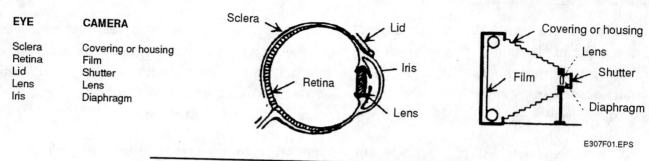

EYE	CAMERA
Sclera	Covering or housing
Retina	Film
Lid	Shutter
Lens	Lens
Iris	Diaphragm

E307F01.EPS

Figure 1. Comparison Of The Eye To The Camera

There are, however, some important differences between the eye and the camera—the most important being the fact that the eye is a living organ. Taking pictures in poor-quality light will do no harm to the camera. But using the eyes under light of poor quality will result in unnecessary fatigue and may lead to headaches and inflammation of the eyes. Consistent misuse of the eyes can cause permanent damage to them and may also contribute to the development of disorders in other parts of the body. You should now begin to realize the importance of proper lighting design for visual tasks.

When a beam of light passes through the transparent protective outer layers of the eye, it is bent or refracted. The amount of light coming through the eye is controlled automatically by the contraction or expansion of the iris. The light continues on through the lens, which focuses the rays on to the retina. From this point on, the process is electrochemical. Pulsations are set up and are carried to the optic nerve, which, in turn, transmits them to the brain where they are interpreted as light, or where they cause the sensation of sight. Thus, the brain and the eye working together transform radiant energy (light) into the sensation of sight.

1.1.2   Objective Factors In The Process Of Seeing

Investigation has shown that the quality of sight depends upon four primary conditions associated with the visual object in question:

- Size of object.
- Luminance (brightness) of object.
- The luminance contrast between the object and its immediate background.
- The time available for seeing the object.

Size: The size of the object is the most generally recognized and accepted factor in seeing.

Everyone is familiar with the conventional eye test chart that is used by schools, optometrists, tests for driving licenses, and others for testing visual defects. The larger the object in terms of *visual angle,* the better it can be seen.

Figure 2a shows a 30-foot telephone pole approximately 53 feet away from the person viewing it. The angle of sight formed from the eye to the object is 30°. If the viewer then backs away from the telephone pole another 53 feet, *(Figure 2b)*, the object has not become smaller; it is still a 30-foot telephone pole. However, since the angle from the eye to the object has become smaller, the object will appear smaller than it did in *Figure 2a* and cannot be seen as clearly.

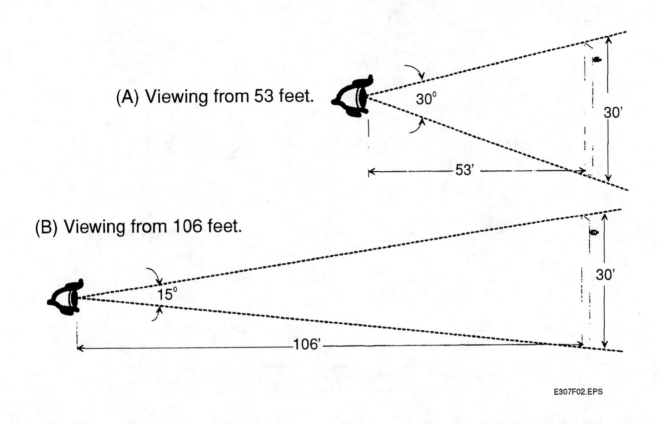

(A) Viewing from 53 feet.

30°

53'

30'

(B) Viewing from 106 feet.

15°

106'

30'

E307F02.EPS

Figure 2 (a&b). Viewing A 30-Foot Telephone Pole From Varying Distances

To further illustrate, if we take two objects, one 12 feet high and the other 1 foot high *(Figure 3a)* and place both objects at the same distance from the viewer, the larger of the two objects can be seen more clearly since the angle from the eye to the object will be greater from the larger object. However, if we move the smaller object closer to the viewer so that the angle becomes the same as the larger object *(Figure 3b)*, both objects can then be viewed with the same clarity.

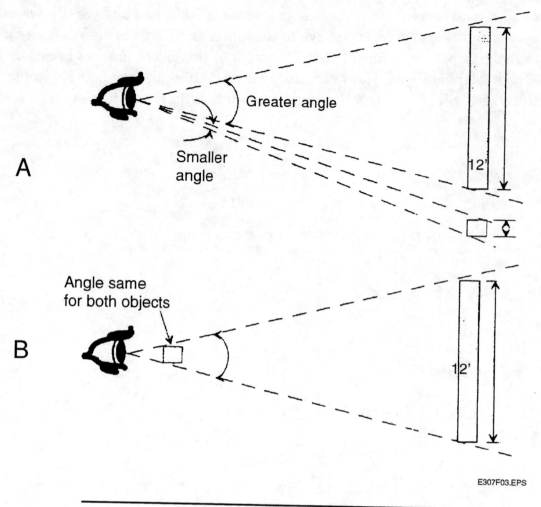

A — Greater angle — Smaller angle — 12'

B — Angle same for both objects — 12'

E307F03.EPS

Figure 3 (a&b). Viewing Two Objects Of Different Sizes

Luminance (brightness): The brightness of an object depends upon the amount of light striking it, and the proportion of that light reflected in the direction of the eye. A light surface will have a much higher brightness than a dark surface. However, by adding enough light to the dark surface, it is possible to make it as bright as the light object. Since the brighter object will be seen first, it will require a lesser amount of light than the darker object for good visibility. Thus, it would take a greater amount of general illumination to adequately light a room with walls painted a dark color, or a dark wood stain, than it would for one with the walls painted a light pastel color.

Contrast: The contrast in brightness or color between the visual object and its immediate background is as important for sight as the general brightness is. The difference in the visual effort required to read the two halves of the circle in *Figure 4* demonstrates this fact; that is, black on white is easier to read than black on gray.

Notice that the print is the same in this message, but due to different backgrounds, part of this message is difficult to read.

E307F04.EPS

Figure 4.   Contrast Between The Object And Its Immediate Background

Again, the higher levels of illumination will partly compensate for brightness contrast where such conditions cannot be avoided.

Time: Seeing is not an instantaneous process; it requires **time**. The eye can see very well under low levels of brightness if sufficient time is allowed. If the object must be seen quickly, then more light is required. In fact, high lighting levels actually make moving objects appear to move slower, and this greatly increases their visibility. It then stands to reason that it would take a greater amount of illumination to properly light a baseball field than it would a football field, since the baseball usually will be traveling at a higher rate of speed than the football will. Size, brightness, contrast, and time are mutually interrelated and interdependent. A deficiency in one can usually be corrected, within limits, by an adjustment in one or more of the others. Of these four conditions, brightness and contrast are usually under the direct control of the interior decorator. With proper control of brightness and contrast, unfavorable conditions, such as size of the object and time given for seeing this object, can be overcome.

1.1.3   Summary

- The purpose of lighting is to make vision possible.
- The mechanism of the human eye is similar to that of a camera.
- Proper illumination is necessary to protect the eyes.
- Good lighting can do much to relieve the eyestrain involved in the performance of difficult visual tasks.
- Research reveals that the advantages of high illumination are even more advantageous to older eyes than to young, normal eyes.
- Size, brightness, contrast, and time are the four basic conditions considered when evaluating the quality of sight.

2.0.0   CHARACTERISTICS OF LIGHT

Light may be defined as a radiant energy evaluated in terms of its capacity for producing the sensation of sight.

All light travels in a straight line unless it is modified or redirected by means of a reflecting, refracting, or diffusing condition.

2.1.0   LIGHT COLORS

The different colors of light are due to the different wave frequencies, which are considered to be of an electromagnetic nature, and are known to be of extremely high frequency and much shorter in length than the shortest television waves.

Ordinary sunlight, while it appears white, is actually made up of a number of colors. In 1666, Sir Isaac Newton passed a beam of light through a prism and discovered that it contained all colors of the rainbow. The three basic colors are red, yellow, and blue, but by continuously blending together, they also produce violet, orange, and green. See *Figure 5*.

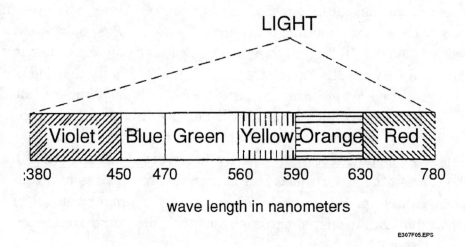

Figure 5.   The Visible Light Spectrum

ELECTRICAL TRAINEE TASK MODULE 20308

Artificial white or daylight is generally the most desirable form of light for illuminating purposes, but it must contain a certain number of the colors which compose sunlight. It is the reflection to our eyes of these various colors from the object they strike that enables us to see objects and to get an impression of their color. Certain surfaces and materials absorb light on one color and frequency and reflect that of another color; this gives us our color distinction in seeing different things.

White and light-colored surfaces reflect more light than dark surfaces do. (Remember *contrast* in the first part of this lesson.)

The ordinary tungsten-filament electric lamp (the ordinary light bulb) is a good example of nearly white artificial light that is excellent for most applications. The molecules of the tungsten wire are caused to vibrate rapidly and produce heat when an electric current is applied. When enough current is passed through the wire, it becomes incandescent (white light).

2.2.0   UNITS OF LIGHT MEASUREMENT

Before you undertake to perform actual lighting layouts or select lighting fixtures for certain applications, it is necessary to learn more about actual quantities of light, etc. An understanding of these interesting units and principles will enable you to better understand the nature of light.

When you purchase an incandescent lamp, you would normally refer to the rating of the lamp in terms of watts (a 60-watt lamp, 100-watt lamp, 150-watt lamp, etc.). While the rating in watts usually gives a general idea of the lamp size, it does not tell how much light a certain lamp can be expected to produce.

For example, one might expect a 100-watt incandescent lamp to produce more light than a 40-watt fluorescent lamp. However, the average inside frosted 100-watt incandescent lamp emits light at the rate of about 1490 lumens, while the flow of light from a 40-watt fluorescent lamp is about 3200 lumens—over twice the amount of light given off by the incandescent lamp.

The preceding example implies that the total amount of light actually given off by a light source is measured in terms of the unit *lumen*.

2.3.0   LUMEN

A lumen may be defined as the quantity of light that will strike a surface of one square foot, all points of which are one foot distance from a light source of one candlepower (one standard candle for our purposes).

A lumen of light, may be visualized by placing a standard candle inside a hollow sphere that has a radius of one foot or a diameter of two feet, and the inside of which is completely black to prevent any reflection of light. If a one foot square is cut out of the sphere as shown in *Figure 6*, the amount of light that will escape through this hole will be one lumen.

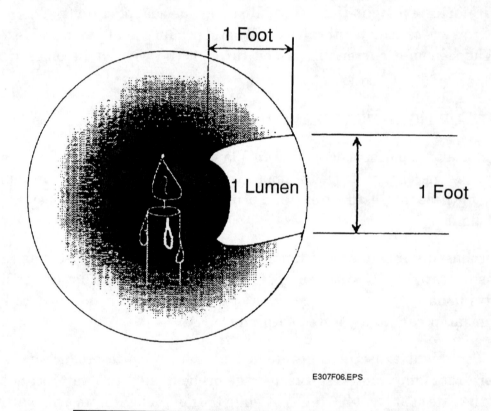

E307F06.EPS

Figure 6.   Using A Sphere To Illustrate One Lumen

If the area of the opening was 1/4 square foot, then the light emitted from the hole would be 1/4 lumen; if the hole was 1/2 square foot, the escaping light would be 1/2 lumen and so on. See *Figure 7*.

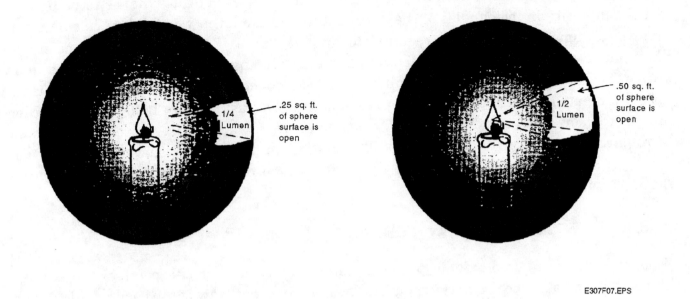

Figure 7. Using A Sphere To Illustrate A Portion Of A Lumen

A sphere with a one-foot radius has a total area of 12.57 square feet, so if the entire sphere was removed from the candle, the total lumens emitted by the standard candle would be 12.57. From this we find that we can determine the approximate amount of lumens given by any lamp by multiplying the average candlepower of the lamp by 12.57.

2.4.0 FOOTCANDLES

Electric lamps are a source of light, and the result of this light striking various surfaces is called *illumination*.

While the lumen serves as a unit to measure the total light obtained from any light source, we must also have a unit to measure the intensity of this light on a given surface. (Desk tops and work benches would be examples of such surfaces.) The unit used for this purpose is called the *footcandle*.

A footcandle is a unit of measure which represents the intensity of illumination that will be produced on a surface that is one foot away from a source of one candlepower, and at right angles to the light rays from the source. This means that if you should hold a sheet of 8 1/2 x 11 paper one foot from a candle, the edges of the paper must be bent towards the candle so that all surfaces are one foot away. The footcandle is the unit used in illumination calculations to determine the proper level of illumination on any working plane or surface. When one lumen of light is evenly distributed over a surface of one square foot, that same area is illuminated to an intensity of one footcandle.

The preceding paragraph shows that if we know the area of a surface that is to be lighted and the intensity in footcandles of the desired illumination level, we can then calculate the number of lumens that will be required to light the area. For example, if we desire to illuminate a surface of 100 square feet to an average intensity of 10 footcandles by a light source one foot from the surface, we multiply the desired footcandles by the area of the surface to be illuminated—10 x 100. Therefore, the source of light must produce 1000 lumens in order to light the surface to an intensity of 10 footcandles. In actual practice, more lumens would be required to produce 10 footcandles of illumination on this surface, since the efficiency of any light source is rarely 100%. More light will also be required as the distance between the light source and surface increased.

2.5.0   INVERSE SQUARE LAW FOR LIGHT

Footcandle units are used to indicate the illumination level at a specific point, or the average illumination on a service or working plane. The inverse square law is the basis of calculation in the point-by-point method of lighting design. The interior designer will have use for this method in many commercial applications. Therefore, a brief description of the inverse square law is warranted.

The inverse square law states that the illumination on a surface varies directly with the candlepower of the source of light, and inversely with the square of the distance from the source.

This really means that a small change in distance from a light source will make a great change in the illumination level on a surface. *Figure 8* illustrates the reasons for this.

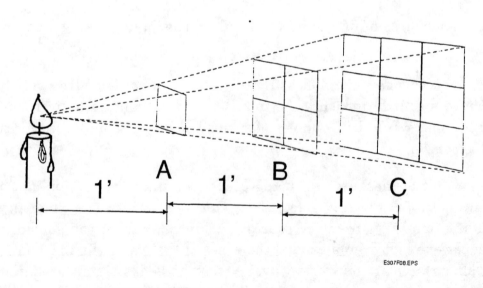

E307F08.EPS

Figure 8.   The Inverse Square Law Illustrated

In *Figure 8* we have a light source of one candlepower, and since the surface "A" is one foot from the candle, its illumination intensity will be one footcandle. If we move the surface or plane to "B," which is two feet from the source, the same number of light rays will have to spread over four times the area, as that area increases in all directions. Then the illumination intensity at twice the distance is only 1/4 the amount it was before, as the distance of two squared is four ($2^2 = 4$), and this is the number of times the illumination is reduced.

If we continue to move the surface to "C," which is three feet away from the light source, the rays are now spread over nine times the original area ($3^2 = 9$), and the intensity of illumination on the surface will now be only 1/9 of its former value.

Various models of convenient portable footcandle meters are available to measure the footcandle level on surfaces or planes. A discussion of these instruments and their uses is found later in this module.

2.6.0   LIGHT REFLECTION

Light can be reflected from certain light-colored or highly polished surfaces. This fact must also be considered when performing lighting calculations.

Some surfaces and materials are much better reflectors than others. Usually, the lighter colors reflect more light and absorb less light than darker colors.

The percentage of light will be reflected from many common materials are as follows:

White plaster	90% to 92%
Mirrored glass	80% to 90%
White paint	75% to 90%
Metallized plastic	75% to 85%
Polished aluminum	75% to 85%
Stainless steel	55% to 65%
Limestone	35% to 65%
Marble (white)	45%
Concrete	40%
Dark red glazed bricks	30%

The better classes of reflectors are used in directing light. The colors of walls, ceilings, and floors and their reflecting ability are also considered in interior lighting design—the darker areas, the more light required.

2.7.0 FIELD MEASUREMENTS

There are a number of large and elaborate devices used in laboratories for making exact tests and measurements on light and lighting fixtures. But for practical use in the field, a portable light or footcandle meter is quite satisfactory and relatively inexpensive.

To use a footcandle meter, first remove the cover. Hold the meter in a position so the cell is facing toward the light source and at the level of the work plane where tile illumination is required. The shadow of your body should not be allowed to fall on the meter cell during tests. A number of such tests at various points in a room will give the average illumination level in footcandles—indicating if additional lighting fixtures are required. Such tests, however, are normally done by lighting engineers, but the practice is a good way for the Trainee to observe the type of lighting system, and then find out the amount of light produced. Much can be learned from instructional material, but not nearly as much (or as quickly) as by actually putting your knowledge to practical use.

2.8.0 SUMMARY

- While ordinary sunlight appears white, it is actually made up of a number of colors: blue, green, yellow, orange, and red.
- Artificial white or daylight is generally the most desirable form of light for illuminating purposes.
- White and light-colored surfaces reflect more light than dark surfaces do.
- The total amount of light actually given off by a light source is measured in terms of the unit *lumen*.
- The level of illumination on any surface or work plane is measured in terms of the unit *footcandle*.
- The instrument used to measure the illumination level is called the *footcandle meter*.
- The farther any surface is from a light source, the less light it receives from that source.

3.0.0 LIGHT SOURCES

Now that we know something of the nature of light and the fundamentals of good illumination, we can discuss the three common sources of electric light:

- Incandescent lamps
- Gaseous discharge lamps
- Electroluminescent lamps

Despite continuous improvement, none of these light sources have a high overall efficiency. The very best light source converts only approximately 1/4 of its input energy into visible light. The remaining input energy is converted to heat or invisible light. *Figure 9* illustrates the energy distribution of a typical cool-white fluorescent lamp.

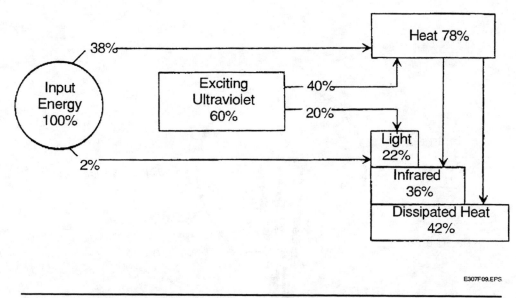

Figure 9.   Energy Distribution Of A Typical Cool-White Fluorescent Lamp

3.1.0   INCANDESCENT LAMPS

Incandescent lamps are made in thousands of different types and colors, from a fraction of a watt to over 10,000 watts each, and for practically any conceivable lighting application. Extremely small lamps are made for instrument panels, flashlights, etc., while large incandescent lamps, over twenty inches in diameter, have been used for spotlights and street lighting.

Regardless of the type or size, all incandescent filament lamps consist of a sealed glass envelope containing a filament, *Figure 10*. The incandescent filament lamp produces light by means of a filament heated to incandescence (white light) by its resistance to a flow of electric current. Most of these elements are capable of producing 11 to 22 lumens per watt, and some produce as high as 33 lumens per watt.

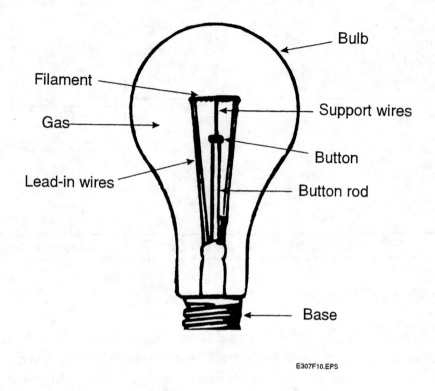

Figure 10.   Parts Of An Incandescent Lamp

The filaments of incandescent lamps were originally made of carbon. Now, tungsten is used for virtually all lamp filaments because of its higher melting point, better spectral characteristics, and strength—both hot and cold.

3.2.0   TUNGSTEN-HALOGEN LAMPS

The quartz-iodine tungsten-filament, lamp is basically an incandescent lamp, since light is produced from the incandescence of its coiled tungsten filament. However, the lamp envelope, made of quartz, is filled with an iodine vapor which prevents the evaporation of the tungsten filament. This evaporation is what normally occurs in conventional incandescent lamps. When the bulb begins to blacken, light output deteriorates, and eventually the filament burns out. While the quartz-iodine lamp has approximately the same efficiency as an equivalent conventional incandescent lamp, it has the advantages of double the normal life, low lumen depreciation, and a small bulb for a given wattage.

CAUTION: Never touch a quartz lamp with the bare hands; oil from the hands can adversely affect the lamp envelope.

3.2.1 Filaments

A filament designation consists of a prefix letter to indicate whether the wire is straight or coiled, and a number to indicate the arrangement of the filament on the supports. Prefix letters include: S (straight)—wire is straight or slightly corrugated; C (coiled)—wire is wound into a helical coil or it may be deeply fluted; CC (coiled coil)—wire is wound into a helical coil and this coiled wire again wound into a helical coil.

3.3.0 ELECTRIC DISCHARGE LAMPS   *HID—High intensity discharge*

The electric discharge lamp category includes the well-known fluorescent, neon, and mercury-vapor lamps as well as the newer metal halide and sodium lamps. In this group of lamps, light is produced by the passage of an electric current through a tungsten filament. The application of an electrical voltage ionizes the gas and permits current to flow between two electrodes located at opposite ends of the lamp. This arc discharge accelerates to tremendous speeds, and when the current collides with the atoms of the gas it temporarily alters the atomic structure. Light results from the energy given off by these altered atoms as they return to their normal state—resulting in a most efficient lamp output.

The incandescent lamp has certain characteristics that make it inherently inefficient as a source of light; maximum possible values for this type of lamp have probably already been approached. However, the electric discharge lamp produces light by an entirely different process and is capable of achieving a much higher efficiency of light output.

3.4.0 FLUORESCENT LAMPS

Of all the electric discharge light sources, the fluorescent lamp is the best known and most widely used. Since fluorescent lighting was introduced to the general public, during the 1933 Chicago Centennial Exposition, it has almost completely taken the place of the incandescent lamp in all branches of construction, except for specialty lighting and possibly for residential use.

One of the reasons for the popularity of fluorescent lighting is its high efficiency as compared to incandescent lamps. The average 40-watt incandescent lamp delivers approximately 470 initial lumens, while a 40-watt fluorescent lamp delivers about 3150 lumens. This power efficiency not only saves on the cost of power consumed, but it lessens the heat and reduces air conditioning loads. Fluorescent lighting allows more comfort for those working under bright lights during warm weather.

Fluorescent lamps are made with long glass tubes sealed at both ends and contain an inert gas, generally argon, and low pressure mercury vapor. Built into each end is a cathode which supplies the electrons to start and maintain the gaseous discharge.

The inside of the lamp tubing is coated with a thin layer of materials called *phosphors*. A phosphor is a substance that becomes luminous or which glows with visible light when struck by streams of electrons, that are caused to pass through the space between filaments inside the lamp. When the phosphors are thus made luminous, the action is called fluorescent—giving the lamp its name. The main components of a fluorescent lamp are shown in *Figure 11*, and different types of fluorescent lamps are shown in *Figure 12*.

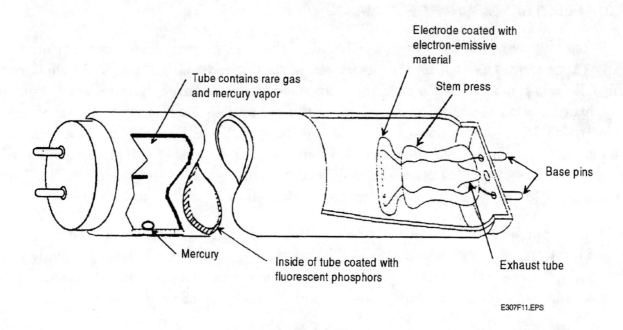

E307F11.EPS

Figure 11.   Main Components Of A Fluorescent Lamp

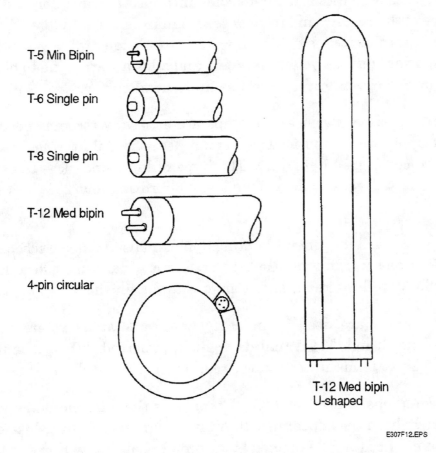

Figure 12. Fluorescent Lamps Come In Many Different Configurations: Straight, U-Shaped, And Circular

3.4.1 Fluorescent Colors

Color is achieved in fluorescent lamps by mixing various phosphors, and with these mixtures, a wide range of visible light colors is possible.

Cool-white: This lamp is often selected for offices, factories, and commercial areas where a psychologically cool working atmosphere is desirable. This is the most popular of all fluorescent lamp colors since it gives a natural outdoor lighting effect and is one of the most efficient fluorescent lamps manufactured today.

Deluxe cool-white: This lamp is used for the same general applications as the cool-white, but contains more red which emphasizes pink tones. Deluxe cool-white is used in food display because it gives a good appearance to lean meat, keeps fats looking white, and emphasizes the fresh, crisp appearance of green vegetables. This type of lamp is generally chosen wherever very uniform color rendition is desired, although it is less efficient than cool-white.

Warm-white: Warm-white lamps are used whenever a warm social atmosphere is desirable in areas that are not color critical. It approaches incandescent in color and is suggested whenever a mixture of fluorescent and incandescent lamps is used. Yellow, orange, and tan interior finishes are emphasized by this lamp, and its beige tint gives a bright warm appearance to reds, brings out the yellow in green, and adds a warm tone to blue. It imparts a yellowish white or yellowish gray appearance to neutral surfaces.

Deluxe warm-white: Deluxe warm-white lamps are generally recommended for home or social environment applications and for commercial use where flattering effects on people and merchandise are considered important. This type of lamp enhances the appearance of poultry, cheese, and baked goods. These lamps are approximately 25% less efficient that warm-white lamps.

White: White lamps are used for general lighting applications in offices, schools, stores, and homes where either a cool working atmosphere or warm social atmosphere is not critical. They emphasize yellow, yellow-green, and orange interior finishes.

Daylight: Daylight lamps are used in industry and work areas where the blue color associated with the "north light" of actual daylight is preferred. While it makes blue and green bright and clear, it tends to tone down red, orange, and yellow.

In general, the designations "warm" and "cool" represent the differences between artificial light and natural daylight in the appearance they give to an area. Their deluxe counter parts have a greater amount of red light, supplied by a second phosphor within the tube. The red light shows colors more naturally, but at a sacrifice in efficiency.

Other colors of fluorescent lamps are available in sizes that are interchangeable with white lamps. These colored lamps are best used for flooding large areas with colored light; where a colored light of small area must be projected at a distant object, incandescent lamps using colored filters are best.

3.4.2   Ballast

The length of an incandescent lamp's filament limits the amount of electrical current passing through the lamp, regulating its light output. However, the fluorescent lamp, with its arc replacing the filament, needs an additional electrical device for starting and power regulation. This device—necessary to fluorescent lamp operation is called a *ballast*.

Fluorescent lamps sold in the U.S. today are available in a wide variety of shapes and sizes. They range from miniature versions rated at 4 watts, six inches in length, with a diameter of 5/8 inches—to models rated at 215 watts, eight feet or more in length, with diameters exceeding two inches.

Voltage required to start the lamps is dependent on lamp length and diameter, with larger lamps requiring higher voltages. Each fluorescent lamp must be operated by a ballast that is specifically designed to provide the proper starting and operating voltage required by that particular lamp.

In *all* fluorescent lighting systems, the ballast performs three basic tasks:

- Provides the proper voltage to establish an arc between the two electrodes.
- Regulates the electric current flowing through the lamp to stabilize light output.
- Supplies the correct voltage required for proper lamp operation and compensates for voltage variations in the electrical current.

In rapid start fluorescent systems, the ballast may perform an additional task, providing continuous voltage to maintain heat in the lamp electrodes—at a level recommended by the lamp manufacturer—while the lamp operates. If the electrode filaments of a rapid start lamp are not continuously heated, they may deteriorate prematurely, shortening lamp life.

All indoor fluorescent fixtures must incorporate ballast thermal protection according to the new National Electrical Code. (Fixtures employing a simple reactive type ballast are excepted.)

Underwriters' Laboratories, Inc. has developed a new standard including the ballast protection required by the code. Ballasts meeting the standard are designated "Class P."

Class P ballasts have the protector within the ballast case to prevent physical damage and tampering. Operation of this thermally actuated automatic reclosing protective device will disconnect the ballast from the power line in the event of overtemperature.

Class P ballasts also help protect against excessive voltage supply, internal ballast short circuiting, inadequate lamp maintenance and improper fixture application. They also eliminate the need for individual fixture fusing.

When replacing Class P or non-Class P ballasts be sure to use an exact equivalent. (For instance, nuisance tripping may occur if a non-Class P is replaced with a Class P ballast.)

3.4.3   Non-Resetting Fuse Protection

The wide use of Class P ballasts has reduced the need for fusing of ballasts. Individual fusing is sometimes considered when many fixtures operate on a single circuit and where it is desirable to isolate an inoperative fixture quickly. This helps avoid complete circuit outage.

If used, fuses should be of the slow-blow type and should accommodate inrush current and abnormal starting cycle currents of the ballast.

Some fuse link ballasts incorporate a built-in temperature sensitive non-resetting protector. It is sensitive to heat, and designed to open when the ballast is subject to high temperature from any internal or external cause.

The fuse link ballasts protect against ballast leakage and end-of-life hazards. When a condition occurs to cause either link to open, the ballast must be replaced.

3.4.4   Ballast Sound

The slight hum present in fluorescent lighting installation originates from the inherent magnetic action in the core and coil assembly of the ballasts. There are three possible ways this sound may be amplified:

1. Method of mounting the ballast in the fixture. It is recommended that all ballast holes be used to mount the ballast securely to the fixture.
2. Loose parts in the fixture.
3. Off of ceilings, walls, floors, and furniture.

The choice of fluorescent lamp ballasts should be made on the basis of selecting the one rated quietest for a specific location. Ballasts are sound rated by a letter code, and some have a more discernible hum due to basic construction features and electrical ratings.

Ambient noise level of the interior is an important consideration also. It is obvious that consideration of ballast sound is more important in a radio station than in a busy store. Sound charts are available for assistance in selecting the proper sound rated ballast.

3.5.0   PREHEAT CIRCUIT

Preheat fluorescent lamps require the lamp electrodes to be preheated before lighting. A manual or automatic starter switch must be placed in series with the filaments, thereby shunting the lamp.

As power is impressed, current passes through the ballast, electrodes of the lamp, and starting switch. Once the lamp becomes lit, the current heats the electrodes until they emit electrons that travel the length of the lamp.

During the starting cycle, the ballast limits the current flow to a calibrated value for preheating the electrodes. In a few seconds the electrodes attain the proper temperature, at which time the starting switch automatically opens. The opening of the starting switch breaks the shunting wire path for the flow of current, leaving the gas in the lamp as the only other path to travel.

With assistance from the power supplied by the fluorescent lamp ballast, the current follows the path of the gas in the lamp, igniting the lamp.

3.6.0 RAPID START CIRCUIT

Rapid start lamps use short, low voltage cathodes which are automatically preheated by the fluorescent lamp ballast, also eliminating the need for a starter. The rapid start ballast preheats the cathodes by means of a heater winding unit built into the ballast. This heater winding continues to produce current to the lamp after ignition.

Because of the continuously heated electrodes, less voltage is required for the initial surge to light than with Slimline instant start. Rapid start lamps light immediately at low brightness and are fully lighted in about two seconds.

3.7.0 SLIMLINE INSTANT START

To overcome the starting delay of preheat fluorescent lamps, a Slimline instant start lamp was introduced. With a Slimline instant start circuit, lamps are started by impressing a high initial voltage between the lamp electrodes without the assistance of a starter. This high initial voltage requires a larger autotransformer as an integral part of the ballast.

A larger "choke coil" or "reactor" must also be included to reduce the starting voltage produced and the exclusion of the lamp starters.

This ballast provides the necessary starting voltage to ignite each lamp independently of the other. The lamps are wired in parallel.

To reduce the size, weight, and cost of the Slimline instant start lead lag ballast, a series-sequence ballast was introduced. In this ballast circuit two lamps are operated in series with the lamps starting in sequence. This circuit was pioneered by ADVANCE®.

The series-sequence circuit differs from all others in that each of the circuits performs a separate function. The starting section supplies sufficient voltage and current to light one lamp with the remaining lamp igniting in sequence from the same voltage and current. Because the lamps are in series, the operating circuit is not required to supply individual lamp currents.

The reduction in power requirements makes it possible to produce a fluorescent lamp ballast for operating Slimline instant start lamps which is lighter, smaller and provides higher efficiency.

3.8.0 VENTILATION

A fluorescent lamp ballast, like other electrical equipment, generates heat during normal operation. Underwriters' Laboratories stipulates the temperature limitation of this ballast using Class A insulation at normal operation should have a maximum ballast coil temperature of 105°C (221°F) and a maximum ballast case temperature of 90°C (194°F) at its hottest spot. Ballast life will be reduced if it is operated at temperatures above these limits.

Where more than one ballast is installed in an enclosure, the ballasts should be positioned far enough apart to provide for the combined normal heating effects. To assist in limiting the temperature rise of ballasts, the following procedures are recommended:

- Mount ballast with maximum number of sides in direct contact with metal channel of fixture. Radiators are an excellent means of dissipating heat.
- Provide fixture ventilation.
- Paint the unpainted fixture channels with a nonmetallic finish to increase radiation.
- Place the ballast in a cooler location outside the fixture.
- Place fixture to attain maximum dissipation of heat by conduction, convection or radiation.

3.9.0   COLD WEATHER OPERATION

Lumen ratings of fluorescent lamps apply for operation in still air that has a temperature of 77°F. While many fluorescent lamps and fluorescent lamp ballasts are designed to give their best performance at 77°F, they will provide reasonable good light output down to 50°F. Further decreases in ambient temperature will result in decreased light output. There are ballasts for 0° to -20°F.

Such variables as humidity, line voltage, fixture design and variations within the particular design of the lamp and the fluorescent lamp ballast play an important part in determining the low temperature starting limit.

The 800 mA and 1000 mA, VHO and power groove lamps with their higher wall bulb temperatures are recommended for most efficient cold weather operation. Even with high output fluorescent lamps, satisfactory operation of the lamps depends upon adequate shielding to permit them to reach recommended operating temperatures. Care must be exercised in fixture designs for the prevention of overheating of the fluorescent lamp ballast in summertime operation.

3.10.0   INOPERATIVE FIXTURE

Often when a fixture becomes inoperative, the cause is not attributable to the ballast. It is therefore important to examine all components of the fixture before removing the ballast for replacement. The following procedure is recommended:

1. Change or check all lamps to insure satisfactory operation.
2. As lamps are removed, examine all sockets to assure proper and positive contact with lamp pins.
3. If starters are used, each starter should be checked and replaced wherever necessary.
4. Examine all connections within the fixture to insure their conformance with the wiring instructions appearing on the ballast.
5. Examine and test ballast.

3.10.1 Testing Preheat Installations

An ammeter, with a scale from zero to one ampere and a voltmeter, with a scale from zero to 300 volts are required to test preheat ballasts. To measure starting current and operating current the ammeter must be connected between the colored high voltage secondary lead of the ballast and the lamp. To determine starting voltage, remove lamp and connect voltmeter between respective primary and secondary leads of each lamp. The following chart shows examples of current and voltage requirements for various fluorescent lamp types.

LAMP TYPE	OPERATING CURRENT (Ampere)	STARTING CURRENT (Ampere)	STARTING VOLTAGE (Minimum) (Open Circuit)
F4WT5	.12 to .13	.13 to .18	108
F6WT5	.14 to .15	.16 to .25	108
F8WT5	.16 to .18	.18 to .27	108
F14WT12	.37 to .40	.44 to .65	108
F15WT8	.29 to .32	.44 to .65	108
F15WT12	.32 to .35	.44 to .65	108
F20WT12	.34 to .37	.44 to .65	108
F22WT9	.37 to .41	.50 to .70	108
F25WT12	.49 to .54	.65 to .90	108
F30WT8	.33 to .36	.44 to .65	176*
F32WT10	.41 to .45	.55 to .75	132*
F40WT12	.39 to .43	.44 to .75	176*
F90WT17	1.50 to 1.70	1.35 to 2.20	132*

*For single lamp, measure voltage between red & white leads. For two lamps, measure voltage between red & white and blue and white leads.

3.10.2 Troubleshooting Preheat Installations

One of the major causes of trouble with a preheat circuit is the incorrect wiring of the fluorescent ballast. This condition can be noted by short lamp or starter life, non-starting of the lamp, or premature failure of the ballast. For example, with a two-lamp ballast, the starter leads from the two pairs of lamp holders may be criss-crossed. If both starters open at the same time, the lamp will start. However, if one lamp starts before the other, the non-starting lamp may blink ON and OFF for a long time before starting, if it will start at all.

To determine if the ballast is wired correctly, short the terminals of a fluorescent with a fine bar wire. Remove all starters from the fixture but leave the lamps in. Insert the shorter starter in one starter slot. If the fixture is wired properly, both ends of the same lamp will glow. If wired incorrectly, one end of each lamp will glow signifying cross-wiring.

Another common complaint is when a preheat ballast produced prior to 1954 is replaced with a new ballast. Ballasts produced prior to 1954 had the low voltage end of the lamps on a two-lamp ballast connected to the white lead of the ballast. After 1954, the lamps were connected to the black leads of the ballast. Thus, when the old ballast is replaced and the new unit is wired in the same way as before, the lamps will not light.

There have been many installations of preheat fluorescent lighting wherein two-lamp ballasts are operating with one lamp on and one lamp out or with shorted starters. These conditions will cause premature ballast failures due to the ballast coils being operated above the 105°C coil temperature limitation. Thus, it is advisable that all inoperative lamps be immediately replaced and, at the same time, a new starter be put in the fixture if the old one is not of the reset design.

This will greatly reduce maintenance. For example, as a lamp reaches end of life, one fluorescent lamp will become deactivated and the lamp will not properly ignite, the starter will keep cycling, trying to start the lamp. If the lamp is not replaced within a reasonable time, damage will be done to the non-resetting starter. Thus, it will fail in a short period of time.

Other causes of difficulty could be (1) low or high circuit voltage, (2) improper lampholder contact, (3) pinched wires or (4) improper lamps.

3.10.3   Rapid Start

For testing 430 mA, 800 mA, and 1000 mA. Power groove and VHO, a voltmeter with a zero to 1000 volts scale is required. To measure starting voltage, connect voltmeter between the highest reading red lead and blue lead. To measure filament voltage on a single lamp unit, read voltage between red-red and blue-blue leads. For two-lamp units, read voltage between red-red, blue-blue, and yellow-yellow leads. The following chart lists starting voltage for various rapid start lamps.

RAPID START 430 mA LAMP TYPE	STARTING VOLTAGE (Minimum @ 50°F)		FILAMENT VOLTAGE
	SINGLE LAMP	TWO LAMP	
F14T12	108	162	7.5 – 9.0
F15T8	108	162	7.5 – 9.0
F15T12	108	162	7.5 – 9.0
F20T12	108	162	7.5 – 9.0
F22T9	185	-	3.4 – 3.9
F30T12/RS	150	215	3.4 – 3.9
F32T10	205		3.4 – 3.9
F40T12/RS	205	256	3.4 – 3.9
22–32 (Circline Comb.)		235	3.4 – 3.9
32–40 (Circline Comb.)		295	3.4 – 3.9

RAPID START 800 & 1000 mA LAMP TYPE	STARTING VOLTAGE (Minimum)					
	Single Lamp			Two Lamp		
	50°F	0°F	20°F	50°F	0°F	20°F
F24T12/HO	85	110	140	145	195	225
F36T12/HO	115	155	190	195	235	260
F48T12/HO	165	215	255	256	290	310
F60T12/HO	210	240	290	325	350	365
F64T12/HO	240	270	330	345	370	388
F72T12/HO	275	300	360	395	410	420
F84T12/HO	285	325	370	430	445	455
F96T12/HO	280	330	360	465	480	490

NOTE: Filament Voltage 3.4–4.3

RAPID START 1500 mA LAMP TYPE	STARTING VOLTAGE (Minimum)					
	Single Lamp			Two Lamp		
	50°F	0°F	20°F	50°F	0°F	20°F
F48PG, VHO, SHO	160	205	240	250	265	300
F72PG, VHO, SHO	225	270	310	350	360	400
F96PG, VHO, SHO	300	355	400	470	470	500

NOTE: Filament Voltage 3.4–4.3

3.10.4   Troubleshooting Rapid Start Installations

The rapid start circuit eliminates the annoying flicker associated with starting of preheat systems. Rapid start circuits also simplify maintenance since no starter is used.

The rapid start lamp operates on the principle of utilizing a starting voltage which is insufficient to start the lamps while the cathodes are cold but is sufficient to start the lamps when the cathodes have been heated to emission temperature. This voltage range between starting cold and starting hot is a very narrow band of voltage which must be closely controlled in order to prevent either failure of the lamps to start or instant starting of the lamps with cold cathodes which is detrimental to the lamps.

In order to stay within this range of voltage, it is necessary to excite the gas within the lamps by means of an external voltage which is applied to the gas within the lamps to create ionization. This external excitation is created by means of the capacity that is present between the lamp and the reflector or channel. In order to act effectively, the fixture must be connected to a ground and the white lead of ballast connected to ground lead of power supply. Thus, it is stated on the label of rapid start ballasts "MOUNT LAMPS WITHIN 1/2 INCH (OR 1 INCH) OF GROUNDED METAL REFLECTOR."

The majority of new fluorescent installations today use ballasts of the rapid start design. The high output (800 mA), and very high output (1500 mA) lamps are of the rapid start design. Blue-blue, yellow-yellow, and red-red leads are the built-in filament windings which supply a voltage of 3.4 to 4 volts to the lamp cathodes. If the cathodes are not properly heated, premature lamp end blackening will result. The lack of heating could be due to:

1. Improper seating of the lamp within the socket.
2. Broken sockets.
3. Broken lamp pins.
4. Too great of socket spacing.
5. Damaged lamp cathode(s).
6. Ballast lead wire not properly connected to socket.
7. Low supply voltage.
8. Inadequate ballast filament voltage.
9. Improper wiring.

To determine if there is adequate voltage at the lamp cathodes, measure the voltage at the socket terminals with a volt-ohmmeter or an inexpensive tester which is available. The voltage at the sockets should read between 3.4 to 4 volts. If there is adequate voltage, then the lamp end blackening can be due to conditions 1, 2, 3, 4, or 5. If the voltage is not adequate, it can be due to one or more of conditions 6, 7, 8, or 9.

If conditions of random starting of rapid start lamps are experienced, be sure the fixture is properly grounded. As previously stated, for completely reliable starting in rapid start circuits, it is necessary to have a starting aid. The starting aid should be an electrically grounded metal strip at least 1 inch wide and extending the full length of the lamp. The lamp should be within 1/2 inch of the grounded strip for 40 watt lamps and smaller, and 1 inch for higher output lamps.

If under high humidity conditions, rapid start lamps start slowly, or do not start at all although the cathodes are properly heated, this may be due to dirt on the lamps, which is offsetting the silicon coating on the lamps or it may be due entirely to a poor silicon coating. If it is a new installation (in operation only a few months) which experiences random starting under high humidity conditions, it in most cases will be due to low supply voltage or poor silicon coating on the lamps. A poor coating can be determined by running water over the lamp bulb and seeing if small droplets of water form on the tubes as would result on a newly waxed car. If the water does not form droplets, the tube is improperly siliconized.

When random starting is experienced under high humidity conditions in an installation in operation for a longer period of time, this is usually due to dirt on the lamps. The lamps should be washed in water to remove the dirt.

Sometimes with a two-lamp rapid start series ballast, only one lamp will light to full brilliance and the other will not light. If the lamp between the red leads and yellow leads is lit and the other lamp is out, look for a pinched yellow lead. If the lamp between the red and yellow leads does not light and the other does, in all probability it is due to a short within the ballast.

3.10.5   Slimline And Instant Start

An electrostatic or high resistance type voltmeter with a scale of zero to 1000 volts is required for testing.

To determine starting voltage, the lamp must be removed and voltmeter connected between the respective primary and secondary leads of each lamp as designated on the ballast label. For series-sequence ballasts, the red lead must be in position while measuring the starting voltage of the remaining lamp. The following chart lists minimum starting voltage.

LAMP TYPE	*STARTING VOLTAGE (MINIMUM)
F24T12	270
F36T12	315
F40T12	385
F40T17	385
F42T6	405
F48T12	385
F64T6	540
F72T8	540
F72T12	475
F96T8	675
F96T12	565

*For single lamp measure voltage between red and white.

For two lamp (series sequence), measure voltage between red and white, insert lamp in red and white position, then read voltage between blue and black.

For two lamp (lead/lag), measure voltage between red and white and blue and white leads.

3.11.0 ELECTRONIC BALLASTS – A DRAMATIC IMPROVEMENT IN DESIGN AND PERFORMANCE

Present day electromagnetic ballasts operate at a voltage frequency of 60 Hertz (Hz)—60 cycles per second, which is the standard alternating current frequency provided in the United States. Electronic ballasts, on the other hand, convert this 60 Hz input to operate at frequencies of between 20 and 60 kilo-Hertz (kHz), 20,000 to 60,000 cycles per second, depending on the specific model.

The operating frequency of a ballast is often illustrated as a sine curve, as shown in *Figure 13*. The differences in the operating frequencies of both electromagnetic and electronic ballasts are immediately apparent in a comparative graph.

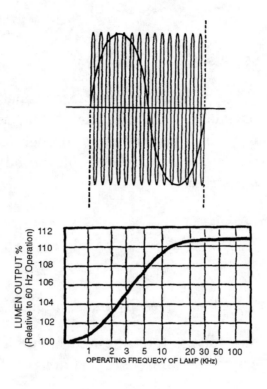

Figure 13.   Operating Frequency

3.11.1   Greater Energy Efficiency

Because electronic ballasts function at high frequency, the fluorescent lighting systems they operate can convert power to light more efficiently than systems run by standard electromagnetic ballasts. For example, electronic ballasts can produce about 10% more light from standard fluorescent lamps using the same power as electromagnetic ballasts.

However, the lighting industry considers the amount of light already produced by today's electromagnetic ballast-driven lighting systems to be consistent with generally accepted lighting levels. So, some electronic ballasts are designed to produce the same amount of light from standard fluorescent lamps as conventional electromagnetic ballasts—but using significantly less power, thereby cutting energy costs. For example, an electronic ballast operating two, four-foot energy-saving rapid start lamps requires input power of 60 watts to deliver the equivalent light output of a standard electromagnetic ballast that requires input power of 82 watts. This represents a 27% energy savings.

4.0.0　HIGH-INTENSITY DISCHARGE BALLAST

4.1.0　BALLASTS

4.1.1　Standby Lighting Systems

Quadri-volt ballasts are ideal for applications where an incandescent lamp is incorporated in the fixture. This provides immediate standby lighting in the event power is momentarily lost, extinguishing the HID lamp, which then must go through its restrike phase. The 120-volt tap can be connected, via a relay, to power a 120-volt incandescent lamp until the HID lamp has cooled sufficiently to restrike.

4.2.0　BALLAST CIRCUITRY

4.2.1　General

The ballast in an HID lighting system generally has two purposes:

1. To provide the proper starting voltage to strike and maintain the arc.
2. To provide the proper current to the lamp once the arc is established.

In addition to being designed to operate a particular type of HID lamp, a ballast design incorporates a basic circuitry to provide specific lamp/ballast operating characteristics. As an example, the effects of line voltage variations on resultant changes of lamp wattage are a function of the ballast circuit design. Requirements for a circuit that will provide a finer degree of lamp regulation generally result in a higher ballast cost.

For some types of lighting applications a particular ballast circuit has already been proven most cost-effective and is, therefore, the only circuit offered. For others, a ballast with optimum circuitry for the particular application must be selected from the two or three alternatives that are available.

4.2.2　Lamp Regulation Characteristics

One of the most important characteristics of each particular ballast circuit is the degree to which it controls the lamp wattage (light output) when the input line voltage changes. *Figure 14* compares the relationship of the three basic circuits as the input volts are changed.

As an example, the CWA line indicates that at 90% of line voltage, the ballast will operate the lamp at 95% of its nominal wattage. Similarly, at 110% of line voltage, the ballast will operate the lamp at 105% of nominal wattage.

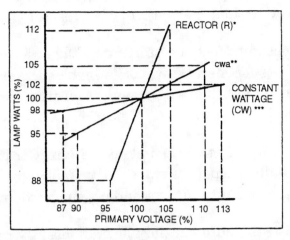

Figure 14. Watts Versus Voltage

4.3.0 CIRCUIT TYPES FOR MERCURY, METAL-HALIDE AND LOW-PRESSURE SODIUM LAMPS

When the input voltage to a fixture meets the starting voltage requirements of an HID lamp, a reactor ballast circuit may be employed to operate the lamp. The necessary lamp starting voltage comes from the input voltage to the ballast. Because most Mercury lamps are designed to start at 240 or 277 volts, the reactor ballast is the most economical way to ballast a Mercury lamp in systems operated at either of these two input voltages.

Both reactor and high reactance ballasts (described in the following section) provide the same degree of lamp wattage regulation. They are inherently normal power factor devices (50%).

With these ballasts, the input line voltage should be controlled to within ±5% because the resultant lamp wattage will vary ±12%. However, this fair degree of lamp regulation is acceptable in many applications. In addition, in the event of a momentary power drop where the line voltage dips below 75% (e.g., to 180 volts on a 240 volt system), the HID lamp may extinguish.

Where necessary to reduce the current draw, a capacitor may be utilized across the input terminals to provide high power factor (90%) operation. However, the addition of this capacitor will not change the ballast's lamp regulation characteristics. Additionally, because a reactor ballast draws substantially higher current during warm-up and/or open-circuit operation, the power distribution system must provide ample line capacity for this condition. As a result, there are fewer fixtures per circuit with reactor ballasts.

When the input voltage does not meet the starting voltage requirements of the HID lamp, such as 120, 208, or 480 volts for mercury vapor, a high reactance autotransformer ballast can be used to ballast the lamp. This will provide operating characteristics equal to the reactor.

The ballast, in addition to limiting the current to the lamp, transforms the input voltage to the required level. The ballast employs two coils, primary and secondary. It is called an autotransformer because the primary and secondary share common windings.

Also, like the reactor ballast, the autotransformer is inherently normal power factor (50%), but it may be corrected to high power factor (90%) with the addition of a capacitor across the input.

Its current draw and ability to withstand voltage dips are similar to that of the reactor.

This is the most commonly used circuit because it offers the best compromise between cost and performance. It is a high power factor device, utilizing a capacitor in series with the lamp. A ±10% line voltage variation will result in a ±5% change in lamp wattage for mercury, or ±10% change in wattage for metal halide. These regulation characteristics are greatly improved over the reactor and the high reactance circuits.

Additionally, the ballast input current during lamp warm-up does not exceed the current when the lamp is stabilized. The incidence of accidental lamp outages due to voltage dips is also greatly reduced because a CWA ballast can tolerate drops in line voltage of 30 to 40% before the lamp extinguishes (lamp dropout).

Sometimes referred to as premium constant wattage, this type of ballast will provide the highest lamp regulation available. Because there is no connection between the primary and secondary coils, this isolated circuit provides a safety factor against the danger of shock hazard.

Constant wattage ballasts are used with mercury vapor lamps and will accommodate a ±13 change in line voltage while yielding only a ±2 change in lamp watts. Incorporating a capacitor in series with the lamp, they are inherently high power factor, and their low input current at lamp start-up does not exceed their operating current. These units can tolerate up to a 50% dip in line voltage before lamp dropout.

Additionally, this same circuit is also used on ballasts for two-lamp series circuits. With the isolated feature, the screw shells of the two-lamp sockets can be connected together and grounded to provide an important safety feature in the fixture.

4.4.0 REGULATION OF HIGH-PRESSURE SODIUM LAMPS

4.4.1 Volt-Watt Traces

The voltage of a typical mercury or metal halide lamp remains fairly constant throughout its operational life. For this reason, regulation of these ballasts can be defined as a simple ±%. With a high pressure sodium lamp, however, the arc tube voltage increases significantly during the operational life of the lamp.

The high pressure sodium lamp ballast must therefore compensate for this changing lamp voltage in order to maintain a somewhat constant wattage even at nominal input.

Consequently, a simple ±% regulation is not an adequate definition for HPS lamp regulation. Instead, a boundary picture called a trapezoid is defined for this dynamic system which restricts the performance of the lamp and the ballast to certain acceptable limits that are established by the American National Standards Institute (ANSI). The ballast is designed to operate a high pressure sodium lamp throughout its life within this trapezoid for any input voltage within the rated input voltage range of the ballast.

For an input of 120 volts, the simplest and most economical way to ballast a 35 through 150 watt (55-volt) lamp is by utilizing reactor circuitry because these lamps require a 120 volt open circuit starting potential. Here, the reactor performs only the basic function of controlling current through the lamp. This is shown in *Figure 15*.

Circuit Types for High Pressure Sodium Lamps

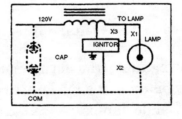

E307F15.EPS

Figure 15. Typical Reactor Circuit With Optional Capacitor

Although inherently normal power factor (50%), a capacitor may be used with a reactor ballast to provide high power factor (90%) operation. However, the addition of a capacitor will not improve the regulation of the ballast.

4.5.0 HIGH REACTANCE AUTOTRANSFORMER (HX)

For 35-watt through 150-watt (55-volt) lamps, where the input voltage does not meet the starting voltage requirements of the lamp, such as 208, 240, 277, or 480 volts, the high reactance autotransformer circuit is used and its operating characteristics are similar to the HPS reactor. In addition to limiting the current to the lamp, the autotransformer reduces the input to the 120 volts required to start and operate the lamp.

The ballast employs two coils, primary and secondary. It is called an autotransformer because the primary and secondary share common windings. This is shown in *Figure 16.*

High Reactance Autotransformer (HX)

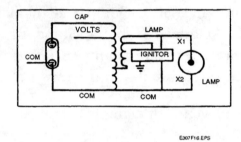

E307F16.EPS

Figure 16. Typical High Reactance Autotransformer Circuit

Also, like the reactor ballast, the autotransformer is inherently normal power factor (50%), but it is generally corrected to high power factor (90%) with the addition of a capacitor. However, this correction will not affect its lamp regulation characteristics.

Its current draw and ability to withstand voltage dips are similar to that of the reactor.

This lead circuit, the constant wattage autotransformer circuit, *Figure 17*, is the most popular of all because it offers excellent regulation at a moderate cost. It is similar to the CWA circuit used with mercury and metal halide lamps. A capacitor is utilized in series with the secondary coil of the ballast and the lamp. The power factor of CWA ballasts exceeds 90%.

Constant Wattage Autotransformer (CWA)

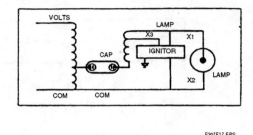

E307F17.EPS

Figure 17.   Typical Constant Wattage Autotransformer Circuit

Lamp regulation is much finer than with the reactor and high reactance circuits. At a ±10% variation in line voltage, the ballast will operate the lamp within its defined trapezoidal boundary.

Additionally, the ballast input current during lamp warm-up does not exceed the current when the lamp is stabilized. The incidence of accidental lamp outage is also reduced because a CWA ballast can tolerate an approximate 25% drop in line voltage before lamp dropout.

This lag circuit, *Figure 18*, provides much better regulation than the reactor, high reactance autotransformer, or constant wattage autotransformer circuits, but at an increase in ballast size, losses, and price. A ballast incorporating this circuit consists of three coils instead of the usual two or one, with the third coil and its capacitor stabilizing the lamp.

Regulated Lag (REG. LAG)

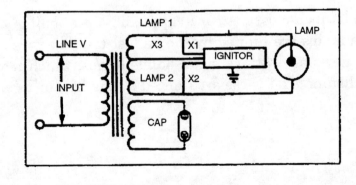

E307F18.EPS

Figure 18.   Typical Regulated Lag Circuit

In this circuit, the secondary windings are isolated from the primary, providing a degree of added safety during lamp change-out on live circuits.

4.6.0   BALLAST CIRCUITRY CHARACTERISTICS COMPARISON

This chart permits a quick comparison of the various attributes of the four basic ballast circuits.

	BALLAST CIRCUIT			
	Reactor (R)	Hi-Reactance Autotransformer (HX)	Constant Wattage Autotransformer (CWA)	Constant Wattage (CW)
Lamp Watts Regulation	Poor	Poor	Good	Excellent
Ballast Losses	Low	Low	Medium	High
Power Factor	50/90	50/90	90 Avg.	90–100
Starting and Open Circuit Currents	Greater	Greater	Less	Less
Lamp Life	Excellent	Excellent	Good	Good
Input Voltage Dip Withstand	Poor	Poor	Good	Excellent
Isolation	No	No	No	Yes

5.0.0   BALLAST-TO-LAMP REMOTE MOUNTING DISTANCES

5.1.0   MERCURY VAPOR AND METAL-HALIDE

5.1.1   Ballasts

The distances at which most mercury vapor and metal halide ballasts can be located from their respective lamps are limited by the ballast-to-lamp wire size. The exceptions are the ballasts for the new, low-wattage metal-halide lamps, which require an ignitor for starting. The mounting distances for these are limited by the ignitor as shown on the following page.

Use this chart to determine the minimum wire size required for the mercury and metal-halide lamps as shown:

LAMP			MAXIMUM ONE-WAY LENGTH OF WIRE BETWEEN LAMP AND BALLAST (FEET) (Voltage Drop Limited to 1% of Lamp Voltage) Minimum Wire Size				
Wattage	Mercury	Metal-Halide	#10	#12	#14	#16	#18
100	H38	-	750	470	295	185	115
175	H39	M57	425	265	165	105	65
250	H37	M58	300	190	120	75	45
1-400 or 2-400	H33	M59	200	125	75	50	30
700	H35	-	465	290	180	115	70
1000	H36	M47	325	205	125	80	50
1500	-	M48	225	140	85	55	35

5.2.0   CAPACITORS

The ballast limits the current through the lamp by providing a coil with a high reactance in series with the lamp. An inductive circuit of this type has an uncorrected power factor of from 40% to 60%; however, the power factor can be corrected to within 5% of unity by the addition of a capacitor. In any installation where there are to be a large number of ballasts, it is advisable to install ballasts with a high power factor.

5.2.1   General

All constant wattage autotransformers, high power factor reactors, and hi-reactance ballasts require a capacitor. With core and coil units, this capacitor is a separate component and must be properly connected electrically. In high power factor ADVANCE® outdoor weatherproof, indoor enclosed, F-can, and postline ballasts, the capacitor is already properly connected within the assembly. Two general types of capacitors are currently in widest use: the oil-filled and the dry type.

5.2.2   Oil-Filled Capacitors

Oil-filled capacitors furnished today contain a non-PCB oil and are equipped with UL component recognized internal interrupters to prevent can rupture and resultant oil leakage in the event of a failure. Additionally, capacitors utilized with mercury and metal halide CW and CWA ballast circuits have a UL-required discharge resistor connected across the terminals to discharge the capacitor after the power is extinguished for the safety of service personnel. In high pressure sodium lighting applications, a discharge resistor contained within the ignitor meets this safety requirement.

Some precautions must be taken when an oil-filled capacitor is installed. Underwriters' Laboratories, Inc. requires the fixture manufacturer to provide a clearance of at least 1/2 inch above the terminals to allow for expansion of the capacitor in the event of failure.

Whether furnished singly or as pairs prewired in parallel, capacitors must be properly wired in all installations. Proper wiring methods, as well as some common miswiring methods that must be avoided, are shown in *Figures 19* and *20*. If the capacitor is miswired, ballast or lamp failure could result.

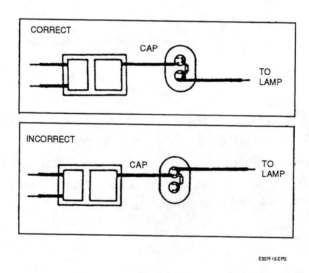

E307F19.EPS

Figure 19.   Single Capacitors

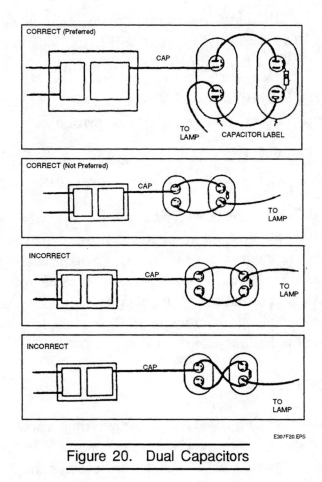

Figure 20. Dual Capacitors

5.3.0 DRY, METALIZED-FILM CAPACITORS

Dry, metalized-film capacitors are relatively new to the lighting industry and are not yet available in all ratings for all applications. However, they are rapidly gaining popularity because of their compact size and extreme ease of installation.

Unlike the oil-filled type of capacitor, the dry type is virtually foolproof to install, requiring only the wiring of the two leads. Because these units contain no oil, they are inherently "non-PCB." Additionally, dry type capacitors utilize a thermoplastic case, which does not require grounding.

ADVANCE® dry-type capacitors typically require only half the space used by oil-filled units. Clearance problems within a fixture are eliminated because dry type units have no exposed live parts nor oil-filled cans that may expand. The compact, lightweight, cylindrical shape can, therefore, fit more easily into the fixture.

5.4.0 CAPACITOR FAILURE

The older, PCB oil-filled capacitors, which have been discontinued but may still be found in older fixtures, generally fail shorted when they fail. A shorted capacitor affects ballast types and circuits as follows:

Mercury and Metal Halide Ballasts – A seemingly normal lamp operation, but a large increase in line current. This increased current will usually cause the ballast primary coil to burn up in anywhere from a few minutes to a few days.

High Pressure Sodium (HX & R) Ballasts – A direct short is created across the primary which may either open the circuit breaker or burn-up the primary.

High Pressure Sodium (CWA & CW) Ballasts – Lamp operation continues at a very low wattage and resultant low light output with no damage to the ballast.

The newer, non-PCB oil-filled capacitors, which are presently being furnished, contain integral interrupters so the majority of capacitor failures will result in an apparent open circuit. In CWA circuits (the most widely used), this will prevent the lamp from lighting. In HX-HPF circuits, capacitors failing open will result in NPF operation, which means the current drawn by the fixture approximately doubles. In turn, this could cause a fuse to blow, a circuit breaker to open, a ballast primary to fail, or it could have no detrimental effect (depending on the particular ballast used).

5.5.0 TROUBLESHOOTING

At times when an HID Lighting System becomes inoperative, a complex, thorough, troubleshooting procedure may prove overly time consuming. In these instances, a simple check of the power switches, when a bank of fixtures becomes inoperative, or a visual check of the lamp, when a singular fixture becomes inoperative, may provide the quickest response to the problem. At other times, where individual isolated fixtures are involved, it may be necessary to systematically isolate the problem and perform complete electrical tests in order to properly restore the lighting.

The four basic troubleshooting methods outlined in this booklet offer procedures which can be applied to cover virtually all situations:

1. Visual Inspection Check List – Quick visual checks for normal end-of-lamp life and application irregularities not requiring electrical testing.
2. Quick Fix For Restoring Lighting – Where lighting must be immediately restored.
3. Troubleshooting Flow Charts – Simplified diagrams to quickly locate the problem in any given lighting fixture based on the lamp characteristics.
   a. Lamp will not start.
   b. Lamp cycles.
   c. Lamp too bright or dim.
4. Electrical Tests – In-depth check of system by performing electrical tests.

5.6.0 VISUAL INSPECTION CHECKLIST/NORMAL END OF LAMP LIFE

5.6.1 Metal And Metal-Halide Lamps

These lamps at end-of-life are characterized by low light output and/or intermittent starting. Visual signs include blackening at the ends of the arc tube and electrode tip deterioration.

5.6.2 High-Pressure Sodium Lamps

Aged HPS lamps will tend to cycle at end-of-life. After start-up, they will cycle OFF and ON as the aged lamp requires more voltage to stabilize and operate the arc than the ballast is capable of providing.

Visual signs include a general blackening at the ends of the arc tube. The lamp may also exhibit a brownish tinge (sodium deposit) on the outer glass envelope.

5.6.3 Low-Pressure Sodium Lamps

At end-of-life these lamps retain their light output but starting becomes intermittent and then impossible. Visual signs include some blackening of the ends of the arc tube.

5.7.0 VISUAL INSPECTION CHECKLIST/ADDITIONAL CHECKS

5.7.1 Lamps

- Broken arc tube or outer lamp jacket.
- Lamp broken where glass meets the base.
- Broken or loose components in lamp envelope.
- Arc tube end blackening.
- Deposits inside outer glass envelope.
- Lamp type (H, M, S, or L number) and wattage must correspond to that required by ballast label.
- Lamp orientation designation (BU or BD) incorrect for application (base up, base down, etc.).

5.7.2 Lighting System Components

- Charred ballast coils.
- Damaged insulation or coils on ballast.
- Evidence of moisture or excessive heat.
- Loose, disconnected, pinched or frayed leads.
- Incorrect wiring.
- Swollen or ruptured capacitor.
- Damaged ignitor.

5.8.0   QUICK FIX FOR RESTORING LIGHTING

5.8.1   Visual Inspection

Visually inspect lamp, ballast, capacitor and ignitor (where used) for physical signs of failure, replacing any apparently defective components.

If either core and coiled ballast or capacitor appear abnormal, replace both.

5.8.2   Component Replacement Where No Visual Defects Appear

* Verify that the correct line voltage is being supplied to the fixture.
* Check power switches, circuit breakers, fuses, photo control, etc.
* Replace lamp.
* Replace ignitor (where used).
* Replace both ballast and capacitor.

5.9.0   ELECTRICAL TESTS

Note   Voltage and current measurements present the possibility of exposure to hazardous voltages and should be performed only by qualified personnel.

The following equipment is recommended for testing HID fixtures:
* RMS Voltmeter
* Ranges: 0-150-300-750 Volts AC
* Ammeter (Clamp-on type acceptable)
* Ranges: 0-1-5-10 Amperes AC
* Multimeter (with voltage and current ratings as shown above).

5.9.1   Line Voltage

Measure the line voltage at input to fixture to determine if the power supply conforms to the requirements of the lighting system. For constant wattage ballasts, the measured line voltage should be within 10% of the nameplate rating. For high reactance or reactor ballasts, the line voltage should be within 5% of the nameplate rating.

If the measured line voltage does not conform to the requirements of the lighting system, as specified in the ballast or fixture nameplate, electrical problems exist outside of the fixture which can result in non-starting or improper lamp operation.

Check fuses, breakers and switches when line voltage readings cannot be obtained. High, low or variable voltage readings may be due to load fluctuations on the same circuits.

5.9.2 Open Circuit Voltage

To determine if the ballast is supplying proper starting voltage to the lamp, an open circuit voltage test is required, as shown in *Figure 21*. The proper test procedure is:

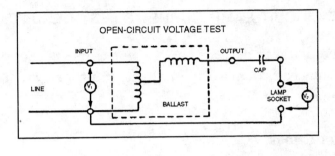

Figure 21. Open Circuit Voltage Test

Step 1 Measure input voltage (V_1) to verify rated input voltage is being applied.

Step 2 With the lamp out of the socket and the proper voltage applied to the ballast, read the voltage (V_2) between the socket pin and shell. Reading must be within test limits shown in *Table 1*.

	LAMP		RMS VOLTAGE
	Wattage	ANSI Number	
MERCURY BALLASTS	50	H46	225-255
	75	H43	225-255
	100	H38	225-255
	175	H39	225-255
	250	H37	225-255
	400	H33	225-255
	2-400 (ILO)	2-H33	225-255
	2-400 (Series)	2-H33	475-525
	700	H35	405-455
	1000	H36	405-455
METAL HALIDE BALLASTS	70	M85	210-250
	100	M90	250-300
	150	M81	220-260
	175	M57	285-320
	250	M80	230-270
	250	M58	285-320
	400	M59	285-320
	2-400 (ILO)	2-M59	285-320
	2-400 (Series)	2-M59	600-665
	1000	M47	400-445
	1500	M48	400-445
HIGH PRESSURE SODIUM BALLASTS *	35	S76	110-130
	50	S68	110-130
	70	S62	110-130
	100	S54	110-130
	150	S55	110-130
	150	S56	200-250
	200	S66	200-230
	250	S50	175-225
	310	S67	155-190
	400	S51	175-225
	1000	S52	420-480
LOW PRESSURE SODIUM BALLASTS	18	L69	300-325
	35	L70	455-505
	55	L71	455-505
	90	L72	455-525
	135	L73	645-715
	180	L74	645-715

*CAUTION: Always disconnect the ignitor before measuring the output voltage of HPS ballasts. High voltage starting pulses can damage commonly used multi-meters.

E307T01.EPS

Table 1. Lamp Voltage Levels

As an alternative, this test may also be performed simply by screwing an adapter into the lamp socket for easy access. Then hook up the voltmeter to this adapter. Reading must be within test limits.

5.10.0   WHEN SHORT CIRCUIT LAMP CURRENT TEST RESULTS IN NO READING

Further checks should be made to determine whether cause is attributable to lamp socket short, shorted or open capacitor, inoperative ballast, improper wiring, or open connection. Simple checks may be made as follows:

5.10.1   Shorted Socket Check

Step 1   Turn off power and remove lamp from socket.

Step 2   Check for internal short in lamp socket with continuity meter across two lamp leads.

Step 3   Should read NO continuity.

5.10.2   Capacitor Check

Step 1   Disconnect capacitor from circuit

Step 2   Discharge capacitor by shorting between terminals.

Step 3   Check capacitor with ohmmeter set at highest resistance scale.

If meter indicates a very low resistance which then gradually increases, the capacitor does not require replacement.

If meter indicates a very high resistance which does not diminish, it is open and should be replaced.

If meter indicates a very low resistance which does not increase, the capacitor is shorted and should be replaced.

5.10.3   Continuity Of Primary Coil

Step 1   Disconnect ballast from power supply and discharge the capacitor.

Step 2   Check for continuity of ballast primary coil between input leads.

5.10.4   Continuity Of Secondary Coil

Step 1   Disconnect ballast from power supply and discharge the capacitor.

Step 2   Check for continuity of ballast secondary coil between lamp and common leads.

5.10.5   Short Circuit Lamp Current

To assure the ballast is delivering the proper current under lamp starting conditions, a measurement may be taken by connecting an ammeter between the lamp socket center pin and the socket shell with rated input voltage applied to the ballast. If available, a socket adapter may be used. The circuit is shown in *Figure 22*.

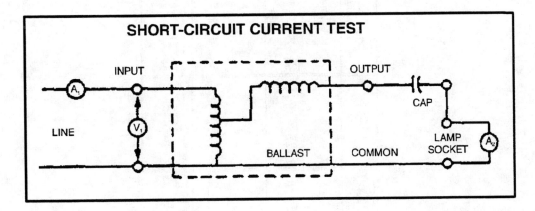

Figure 22.   Short Circuit Current Test

Step 1   Energize ballast with proper rated input voltage.

Step 2   Measure current with ammeter at A_1 and A_2.

Step 3   Readings must be within test limits shown in *Table 2*.

	LAMP		SECONDARY SHORT CIRCUIT CURRENT AMPS
	Wattage	ANSI Number	
MERCURY BALLASTS	50	H46	.85-1.15
	75	H43	.95-1.70
	100	H38	1.10-2.00
	175	H39	2.0-3.6
	250	H37	3.0-3.8
	400	H33	4.4-7.9
	2-400 (ILO)	2-H33	4.4-7.9
	2-400 (Series)	2-H33	4.2-5.40
	700	H35	3.9-5.85
	1000	H36	5.7-9.0
METAL HALIDE BALLASTS	70	M85	.85-1.30
	100	M90	1.15-1.76
	150	M81	1.75-2.60
	175	M57	1.5-1.90
	250	M80	2.9-4.3
	250	M58	2.2-2.85
	400	M59	3.5-4.5
	2-400 (ILO)	2-M59	3.5-4.5
	2-400 (Series)	2-M59	3.3-4.3
	1000	M47	4.8-6.15
	1500	M48	7.4-9.6
HIGH PRESSURE SODIUM BALLASTS *	35	S76	0.85-1.45
	50	S68	1.5-2.3
	70	S62	1.6-2.9
	100	S54	2.45-3.8
	150	S55	3.5-5.4
	150	S56	2.0-3.0
	200	S66	2.50-3.7
	250	S50	3.0-5.3
	310	S67	3.8-5.7
	400	S51	5.0-7.6
	1000	S52	5.5-8.1
LOW PRESSURE SODIUM BALLASTS	18	L69	0.30-.40
	35	L70	0.50-.78
	55	L71	0.52-.78
	90	L72	0.8-1.2
	135	L73	0.8-1.2
	180	L74	0.8-1.2

E307F02.EPS

Table 2. Short Circuit Lamp Current Test Limits

5.11.0 WHEN SHORT CIRCUIT LAMP CURRENT TEST RESULTS IN HIGH, LOW, OR NO READING

Further checks should be made to determine whether cause is attributable to improper supply voltage, shorted or open capacitor or inoperative ballast. Checks may be made as follows:

5.11.1 Supply Voltage Check

Step 1   Measure line voltage.

Step 2   If ballast is multi-voltage unit such as ADVANCE® quadri-volt, make certain input voltage connection is made to proper input voltage terminal or lead.

5.11.2 Capacitor Check

Step 1   Verify capacitor is as required and shown on ballast label.

Step 2   Perform capacitor check.

5.11.3 Ballast Check

Perform open circuit voltage test.

5.11.4 HPS Ignitors

Ignitors are used as a starting aid with all high pressure sodium and certain low-wattage metal halide lamps.

Measurement of the starting pulse voltage of an ignitor is beyond the capability of most instruments available in the field. In laboratory tests, an oscilloscope is used to measure pulse height and width. In the field, some simple tests may be performed to determine if the ignitor is operable:

Step 1 Replace the ignitor with another ignitor which is known to be operable. If the lamp then starts, the previous ignitor was either miswired or inoperative.

Step 2 Remove the high pressure sodium lamp and replace it with a known operable HPS lamp of proper wattage. If lamp lights, ignitor is operating properly.

Step 3 If lamp does not light, disconnect ignitor and proceed as follows:

   a. 35W to 150W (55V) HPS – Insert 120V incandescent lamp in socket. If lamp lights, ignitor requires replacement.
   b. 150W (100V) to 400W HPS – Install mercury lamp of comparable wattage. If mercury lamp starts, ignitor requires replacement.
   c. 1000 W – Replace ignitor.

Note   Ignitors are not interchangeable. Refer to ballast label for designation of proper ignitor to be used with ballast.

5.11.5 Probable Causes Of Inoperative Ballast

1. Normal end-of-life failure.
2. Operating incorrect lamps. Use of higher or lower wattage lamps than rated for ballast will cause premature ballast end-of-life.
3. Overheated due to heat from fixture or ambient temperature.
4. Voltage surge.
5. Miswiring or pinched wires.
6. Shorted or open capacitor.
7. Incorrect capacitor rating for ballast.
8. Capacitor miswired or wiring shorting against frame.

5.11.6 Probable Causes Of Shorted Or Open Capacitor

1. Normal end-of-life failure.
2. Overheated due to heat from fixture or ambient temperature.
3. Capacitor heat barrier inadvertently removed.
4. Incorrect voltage rating of capacitor.
5. Mechanical damage such as overtightened bracket.

5.11.7 Dimming And Flashing

Rapid start fluorescent lamps can be dimmed from full brightness to approximately zero output by a number of special electrical and electronic circuits. This has made it possible to greatly increase the flexibility of fluorescent lighting systems. However, a special rapid start ballast in conjunction with dimming control devices is required. See *Figure 23*.

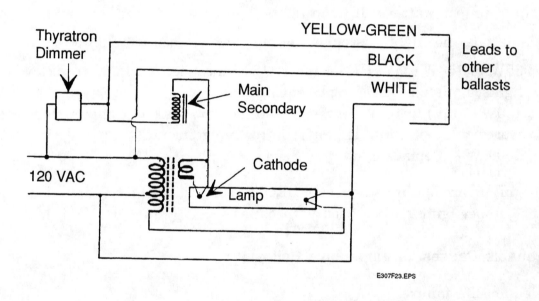

E307F23.EPS

Figure 23. A Dimmer Circuit Using A Special Ballast Transformer

Rapid start and cold-cathode fluorescent lamps can be flashed ON and OFF (as for sign use) without any appreciable loss in lamp life, since a special ballast can be used to provide continuous electrode heating while the current to the arc is interrupted. Additional information on sign lighting and other specialty lighting is presented in later modules. In the meantime, manufacturer's data are very helpful learning sources, and it is recommended that each trainee accumulate as much of this information as possible for current and future reference.

5.12.0 ADVANTAGES AND DISADVANTAGES OF FLUORESCENT LAMPS

5.12.1 Advantages

- High efficiency.
- Long life.
- A linear source of light.
- Variety of colors available.
- Relatively low surface brightness.
- Economy in operation.

5.12.2 Disadvantages

- Very sensitive to temperature and humidity.
- Radio interference.
- Difficult to control.
- Higher initial installation cost.

6.0.0 INTERIOR LIGHTING DESIGN

The basic requirement for any lighting design is to determine the amount of light that should be provided and the best means of providing it. However, since individual tastes and opinions vary greatly, there can be many suitable solutions to the same lighting problem. Some of these solutions will be dull and commonplace, while others will show imagination and resourcefulness. The lighting designer should always strive to select lighting equipment that will provide the highest visual comfort and performance that is consistent with the type of area to be lighted and the budget provided.

Lighting design for commercial, institutional, and industrial applications is a highly specialized field. It is best left to consulting engineers who are trained in the science of lighting and lighting design. For such installations, the electrician must only know how to interpret the working drawings and lighting-fixture schedules, plus have a knowledge of wiring practices in general. However, electricians are frequently called upon to design lighting layouts for residential applications, so every electrician should have a knowledge of basic residential lighting design. Knowing these basic principles of illumination will also help the electrician to better understand why and how more sophisticated lighting applications are designed.

7.0.0 RESIDENTIAL LIGHTING

Properly designed lighting is one of the greatest comforts and conveniences that any home owner can enjoy. In building new homes or remodeling old ones, the lighting should be considered equally as important as the heating/air-conditioning system, the furniture placement, and as one of the most important features of both interior and exterior decorations.

As a rule, residential lighting does not require a large quantity of elaborate calculations, as does a school or office building. However, electricians must apply their talent and ingenuity in selecting the best types of lighting fixtures for various locations in order to obtain a desirable effect, as well as the proper amount of illumination at the desired quality. Light has certain characteristics that can be used to change the apparent shape of a room, to create a feeling of separate areas within one room, or to alter architectural lines, form, color, pattern, or texture. Light also affects the mood and atmosphere within the area where it is used.

While calculations of any quantity are unnecessary, the electrician must use some guide until he or she has gained the necessary experience to improvise. The methods described in this section are suitable for selecting the proper amount of light as well as the proper types of lighting fixtures.

Let's review two definitions before continuing.

- **Lumen**: A lumen is the quantity of light that will strike a surface of one square foot, all points of which are a distance of one foot from a light source of one candlepower.
- **Footcandle**: A footcandle is a unit of measure which represents the intensity of illumination that will be produced on a surface that is one foot away from a source of one candlepower, and at right angles to the light rays from the source.

From the preceding statements, we can say that one lumen per square foot equals one footcandle. Thus, the following method will be called the lumens-per-square-foot method of residential lighting design.

In using this method, it is important to remember that lighter room colors reflect light and darker colors absorb light. This method is based on rooms with light colors; therefore, if the room surfaces are dark—like one that has its walls covered with dark wood-grain paneling—the total lumens should be multiplied by a factor of at least 1.25.

Table 3 gives the required lumens per square foot for various areas in the home and also the required illumination in footcandles for those who desire to use a different method of calculating required illumination. The recommended footcandle level is "fixed" and will apply regardless of the type of lighting fixtures used. However, the recommended lumens given in this table are based on the assumption that portable table lamps, surface-mounted fixtures, or efficient, structural lighting techniques will be used. If the majority of the lighting fixtures in an area will be recessed, the lumen figures in the table should be multiplied by 1.8. Note that this method produces only approximate results; yet, it is quite adequate for most residential lighting applications.

Area	Lumens Required per ft²	Average Footcandles Required
Living room	80	70
Dining room	45	30
Kitchen	80	70
Bathroom	65	50
Bedroom	70	30
Hallway	45	30
Laundry	70	50
Work bench	70	70

Table 3.   Recommended Lumens And Footcandles For Various Areas In The Home

7.1.0   LIVING ROOM

As the living room is the social heart of most homes, lighting should emphasize special architectural features such as fireplace, bookcases, paintings, etc. The same is true of draperies, walls, planters, or any other special room accents.

Dramatizing fireplaces with accent lights brings out texture of bricks, adds to overall room light level, and eliminates bright spots that cause subconscious irritation over a period of time. Use 75- to 150-watt lamps in wallwash-type fixtures—either recessed or surface-mounted—for this application.

While recessed downlights, cornice, or valance lighting all add life to draperies, they also supplement the general living room lighting level. Position downlights 2.5 to 3 feet apart and 8 to 10 inches from the wall. Valances are always used at windows, usually with draperies. They provide up-light, which reflects off the ceiling for general room lighting, and down-light for drapery accent. Cornices direct all their light downwind to dramatic interest to wall coverings, draperies, etc., and are good for low-ceiling rooms.

Pulldown fixtures or table lamps are used for reading areas. The pulldown fixtures are more dramatic, but the electrician must know the furniture arrangement prior to fixture placement.

As a final touch, add dimmers to vary the lighting levels exactly to the living room activities— low for a relaxed mood, bright for a merry, party mood.

The floor plan in *Figure 24* shows a living room for a small residence. Let's see how one electrician went about designing a simple, yet highly attractive and functional, lighting layout for this area.

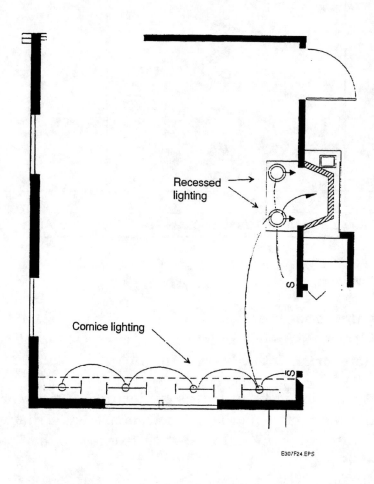

Recessed lighting

Cornice lighting

E307F24.EPS

Figure 24. Floor Plan Of Living Room Lighting Layout

Remember that more than one solution to a lighting design is usually available.

The first step is to scale the drawings to find the dimensions of the area in question. In doing so, we find that the area is 13.75 feet wide and 19 feet long. Thus, 13.75 x 19 = 261.25, which is rounded off to 261 square feet. *Table 3* recommends 80 lumens per square foot for the living area. Therefore, 80 × 261 = 20,880 lumens required in this area.

The next step is to refer to a lamp catalog in order to select lamps that will give the required lumens. A sample sheet of lamp data is shown in *Table 4*. At the same time, the designer should have a good idea of the type of lighting fixtures that will be used as well as their location.

General Service Lamps

Watts	Bulb/Base	Lumens (initial)	Life (Hours)
60W	A-19/Med.	870	1000
75W	A-19/Med.	1190	750
100W	A-19/Med.	1750	750
100W	A-21/Med.	1710	750
150W	A-21/Med.	2880	750
150W	A-23/Med.	2780	750
150W	PS-25/Med.	2680	750
200W	A-23/Med.	4010	750
200W	PS-30/Med.	3710	750
300W	PS-25/Med.	6360	750
300W	PS-30/Med.	6110	750

Silver Bowl Lamps

Watts	Bulb/Base	Lumens (Initial)	life (Hours)
100W	A-21/Med.	1450	1000
150W	PS-25/Med.	2370	1000
200W	PS-30/Med.	3320	1000
300W	PS-35/Med.	5410	1000
500W	PS-40/Med.	9530	1000

Tungsten-Halogen Lamps (Tubular)

Watts	Bulb/Base	Lumens (Initial)	life (Hours)
75W	T-3/Min.	1600	2000
200W	t-3/R.S.C.	3460	1500
250W	T-4/Min. Can.	4700	2000
300W	T-3/R.S.C.	5900	2000
400W	T-3/R.S.C.	7750	2000
500W	T-4/Min. Can.	9500	2000

NOTE: The description of fluorescent lamps are as follows:

CW = Cool White
CWX = Cool White Deluxe
WW = Warm White

Fluorescent-Tubular

Watts	Bulb/Base	Description	Lumens (Initial)	Life (Hours)
20W	T-12	CW	1300	9000
20W	T-12	CWX	850	9000
20W	T-12	WW	1300	9000
20W	T-12	WWX	820	9000
30W	T-12	CW	2300	15000
30W	T-12	CWX	1530	12000
30W	T-12	WW	2360	15000
30W	T-12	WWX	1480	12000
40W	T-12	CW	2150	18000
40W	T-12	CWX	2200	18000
40W	T-12	WW	3200	18000
40W	T-12	WWX	2150	18000
40W	T-12	Chroma 55	2020	18000
40W	T-12	Chroma 75	1990	18000
40W	T-12	Inc./Fluor.	1700	17000
40W U	T-12	CW(35/8)	2800	12000
40W U	T-12	WW(35/8)	2800	12000
40W U	T-12	CW(6")	2950	12000
40W U	T-12	WW(6")	3025	12000

Fluorescent-Circline

Watts	Bulb/Base	Description	Lumens (Initial)	Life (Hours)
22W	T-9	CW	950	7500
22W	T-9	CWX	755	7500
22W	T-9	WW	980	7500
22W	T-9	WWX	745	7500
32W	T-10	CW	1750	7500
32W	T-10	CWX	1250	7500
32W	T-10	WW	1800	7500
32W	T-10	WWX	1250	7500
40W	T-10	CW	2800	7500
40W	T-10	CWX	1780	7500
40W	T-10	WW	2350	7500
40W	T-10	WWX	1760	7500

E307T04.EPS

Table 4. Lumen Output For The Most Popular Lamps Used For Residential Lighting

Referring again to the floor plan of the living room, note that two recessed spotlights are mounted in ceiling above the fireplace. Each fixture contains two 75-watt R-30 lamps, rated at 860 lumens each, for a total of 1720 lumens.

However, since these are recessed into the ceiling (and not surface mounted), the total lumens will have to be reduced. This is accomplished by multiplying the total lumens by a factor of 0.555 to obtain the efficient lumens.

$$1720 \times 0.555 = 955 \text{ effective lumens}$$

This means that we now need 19,925 more lumens in this area to meet the recommended level of illumination.

The next section of this area will be the window area on the front side of the house. It was decided to use a wall-to-wall drapery cornice which would contain four 40-watt warm-white fluorescent lamps, rated at 2080 lumens each, for a total of 8320 lumens. Combining this figure with the 955 effective lumens from the recessed lamps gives a total of 9275 lumens, or 11,605 more lumens to account for.

Two 3-way (100-, 200-, 300-watt) bulbs in table lamps will be used on end tables, one on each side of a sofa for a total of 9460 lumens; this means that only 2145 lumens are unaccounted for. However, one 3-way (50-, 100-, 150-watt) bulb will be used in a lamp on a chairside table. Since this lamp is rated at 2190 lumens, we now have the total lumens required for the living room area.

It can be seen that this method of residential lighting calculation makes it possible to quickly determine the light source needed to achieve the recommended illumination level in each area of the home. Once you have calculated the required lamps for several areas in the homes, the process will become "second nature" in less time than you might imagine.

7.1.1   Kitchen Lighting

The lighting layout for the kitchen must always receive careful attention since the kitchen is the area where the whole family will spend a great amount of time.

Good kitchen lighting begins with general illumination—usually one or more ceiling mounted fixtures of a type that is close to the ceiling. The type of fixture for this general illumination should be a glare-free source that will direct light to every corner of the kitchen.

If a fluorescent lamp is selected for the source of illumination, it is recommended that deluxe cool-white lamps be used.

This type of fluorescent lamps contains more red than the standard cool-white lamp. It therefore emphasizes pink skin tones and is more flattering to people. This type of lamp also gives a good appearance to lean meat and emphasizes the fresh, crisp appearance of green vegetables. Standard cool-white fluorescent lamps are seldom recommended.

The ideal general lighting system for a residential kitchen would be a luminous ceiling, such as that shown in *Figure 25*.

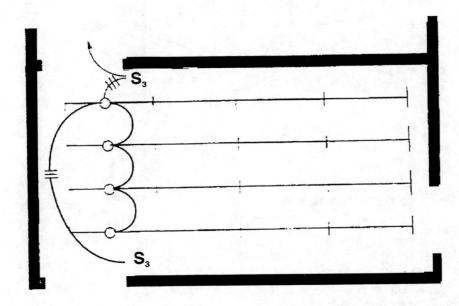

16 4-foot fluorescent
fixtures, spaced 2 feet
on center

Figure 25. Floor Plan Of A Luminous Ceiling

This kitchen floor plan shows bare fluorescent strips, with dimming control, above ceiling panels with attractive diffuser patterns. This arrangement, while the most expensive, provides a "skylight" effect which makes seeing easier.

Another kitchen floor-plan layout appears in *Figure 26*. First, single-tube fluorescent light fixtures were located under the wall cabinets and behind a shielding board. Then, warm-white fluorescent lamps were selected as the best color for lighting countertops. This shadow-free light not only accents the colorful countertops but also makes working at the counter much more pleasant.

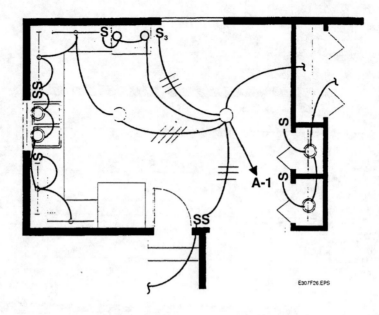

Figure 26. Kitchen/Dining Room Lighting For Small Residence

Two 75-watt R-30 floodlights installed in two recessed housings over the kitchen sink and space about 15 inches on center offer excellent light for work at the sink. However, a two-tube fluorescent fixture using warm-white fluorescent lamps and concealed by a shielding board, as shown in *Figure 27*, will work equally well.

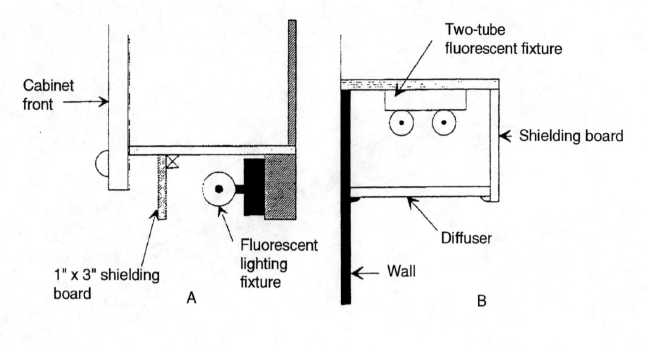

Figure 27. Sectional View Of Two Kitchen-Counter Lighting Arrangements

The light for the electric range is taken care of by a ventilating hood with self-contained lights. The hood contains two 60-watt incandescent lamps for proper illumination. If no hood is used, a 30-watt fluorescent light fixture mounted on the wall over the cooking area, as shown in *Figure 28* would be a good choice.

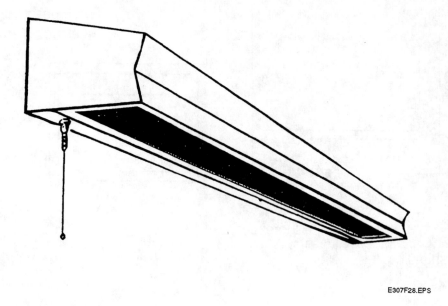

E307F28.EPS

Figure 28. Fluorescent Fixture Mounted On Wall Above Range

One surface-mounted ceiling fixture with an opal glass diffuser is used for general illumination. This fixture contains two 60-watt incandescent lamps. This light source accents and enriches the wood tones of the wall cabinets. In small kitchens, concealed fluorescent strip lighting mounted in a continuous cover around the perimeter will give the effect of a larger kitchen as well as provide an excellent source of general lighting.

In this particular example, the focal point of the kitchen is the dining area. Here, a versatile pull-down light completes the lighting layout. Using 150 watts of incandescent lamps, this fixture provides ample light on the table and also directs some of the light upward for a pleasing effect. This pull-down lamp is controlled by a wall-mounted dimmer for added versatility.

In homes with separate dining rooms, a chandelier mounted directly above the dining table and controlled by a dimmer switch becomes the centerpiece of the room while providing general illumination. The dimmer, of course, adds versatility since it can set the mood of the activity—low brilliance (candlelight effect) for formal dining or bright for an evening of cards. When chandeliers with exposed lamps are used, the dimmer is essential to avoid a garish and uncomfortable atmosphere. The size of the chandelier is also very important; it should be sized in proportion to the size of the dining area.

Good planning calls for supplementary lighting at the buffet and sideboard areas. For a contemporary design, use recessed accent lights in these areas. For a traditional setting, use wall brackets to match the chandelier. Additional supplementary lighting may be achieved with concealed fluorescent lighting in valances or cornices as discussed previously.

7.2.0   BATHROOMS

Lighting performs a wide variety of tasks in the bathroom of the modern residence. Proper light is needed for good grooming and hygiene practices.

The bathroom needs as much general lighting as any other room. If the bathroom is small, usually the mirror and tub/shower lights will suffice for general illumination. However, in the larger bathrooms, a bright central source is needed to transform it from the dim bath of the past to a smart, bright part of the house today. For good grooming, the lavatory-vanity should be lighted to remove all shadows from faces and from under chins for shaving. Two wall-mounted fixtures on each side of the mirror or lighted soffits or downlights above the mirror will give the best results. Over a vanity table, pendants or downlights for concentrated light with a decorative touch may be used with equal results. For safety and health, a moistureproof, ceiling mounted recessed fixture over the bath tub or shower should be included. Linen-closet lighting should also be considered. For best results, bathroom lighting should be from three sources, like the three lighting fixtures illuminating the mirror in *Figure 29a*. A theatrical effect may be obtained by using exposed-lamp fixtures across the top and sides of the mirror, as shown in *Figure 29b*.

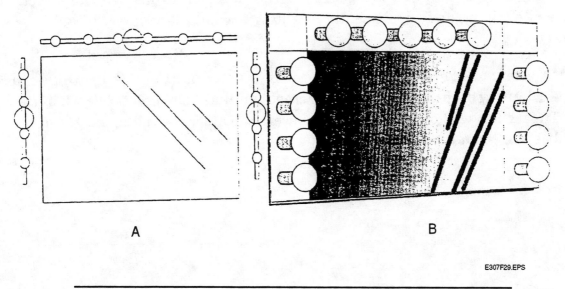

E307F29.EPS

Figure 29 (a&b).   Two Types Of Bathroom Mirror Lighting Fixtures

A small 7- to 15-watt night light is also recommended for the bath to permit occupants to see their way at night without turning on overhead lights that might disturb others. A ceiling fixture with a sunlamp is another convenience that adds warmth to the room. Such a lamp requires approximately two minutes to reach full ultraviolet output after it has been lighted and approximately three minutes to cool before it is relighted.

The sunlamp can be conveniently screwed into an ordinary household socket without the necessity of any other equipment. A good location for it in a residential bathroom is about two feet from the face, either over the shaving mirror or in a position where one would normally dry off after bathing.

7.3.0   BEDROOM LIGHTING

The majority of people spend at least one-third of their lives in their bedroom. Still, the bedroom is often overlooked in terms of decoration and lighting as most home owners would rather concentrate their efforts and money on areas that will be seen more by visitors. However, proper lighting is equally important in the bedroom for such activities as dressing, grooming, studying, reading, and for a relaxing environment in general.

Basically, bedroom lighting should be both decorative and functional with flexibility of control in order to create the desired lighting environment. For example, both reading and sewing (two common activities occurring in the bedroom) require good general illumination combined with supplemental light directed onto the page or fabric. Other activities, however, like casual conversation or watching television, require only general nonglaring room illumination, preferably controlled by a dimmer switch/control.

Proper lighting in and around the closet area can do much to help in the selection and appearance of clothing, and supplementary lighting around the vanity will aid in personal grooming.

One master bedroom lighting layout is shown in *Figure 30*. Cornice lighting is used to highlight a colorfully draped wall and also to create an illusion of greater depth in this small bedroom. The wall-to-wall cornice board also lowers the apparent ceiling height in the room, which makes the room seem wider.

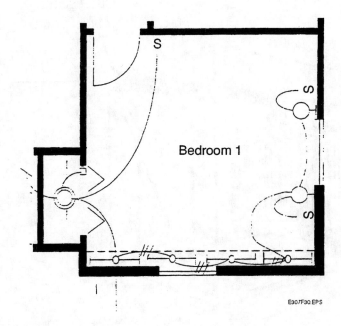

Bedroom 1

E307F30.EPS

Figure 30. Lighting Layout For Master Bedroom

Two wall-mounted "swing-away" lamps on each side of the bed furnish reading light, while a surface-mounted ceiling fixture in the center of the room furnishes general illumination. The owners indicated that they preferred matched vanity lamps (table lamps) for grooming. This was handled with duplex receptacles located near the vanity.

A single recessed lighting fixture in the closet provides adequate illumination for selecting clothes and identifying articles on the shelves. The light is controlled by a door switch which turns the light on when the door is opened and turns it out when the door is shut. This recessed lighting fixture, when combined with the general illumination of the bedroom, also illuminates a full-length mirror on the inside of the closet door.

If a closet is unusually long, two equally spaced recessed lighting fixtures may be required to provide adequate light distribution on the closet shelf; or, a fluorescent fixture mounted as shown in *Figure 31* is an excellent choice.

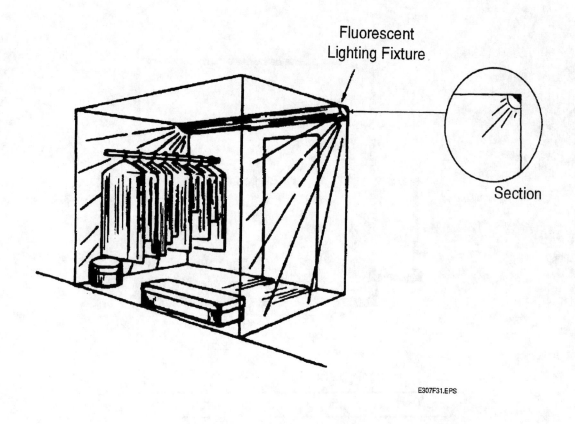

Fluorescent
Lighting Fixture

Section

E307F31.EPS

Figure 31.   Fluorescent Fixtures Used For Closet Lighting

To prevent closet lamps from coming in contact with clothing hung in the closet, which would be a potential fire hazard, certain requirements have been specified by the National Electrical Code.

A (lighting) fixture in a clothes closet shall be installed:

- On the wall above the closet door, provided the clearance between the fixture and a storage area where combustible material may be stored within the closet is not less than 18 inches.
- On the ceiling over an area which is unobstructed to the floor, maintaining an 18-inch clearance horizontally between the fixture and a storage area where combustible material may be stored within the closet.
- Pendants shall not be installed in clothes closets.

The recessed fixtures used in the closet of the residence in question have a solid Fresnel lens and are therefore considered to be located outside of the closet area. For this reason, they can be mounted anywhere on the ceiling or upper walls of the closet area.

7.4.0 BASEMENTS

If the basement area is to be unfinished and used for utility purposes only, inexpensive lighting outlets should be located to illuminate designated work areas or equipment locations. All mechanical equipment, such as the furnace, pumps, etc. should be properly illuminated for maintenance. Laundry or work areas should have general illumination as well as areas where specific tasks are performed. At least one light near the stairs should be controlled by two 3-way switches. One at the top of the stairs and one at the bottom. Other outlets may be pull-chain porcelain lampholders.

Today, many families have a portion of the basement converted into a family or recreation room. Lighting in this area should be designed for a relaxed, comfortable living atmosphere with the family's interests and activities as a starting point in design.

A typical well-designed lighting layout for a family room should include graceful blending of general lighting with supplemental lighting. For example, diffused incandescent lighting fixtures recessed in the ceiling furnish even, glare-free light throughout the room. The number of fixtures should be increased around game tables for added visual comfort.

Lamps concealed behind cornices near the ceiling enrich the natural beauty of paneled walls. This technique is particularly effective where the light shines down over books with colorful bindings. Fluorescent lamps installed end-to-end in a cove lighting system will not only furnish excellent general illumination for a family room, but will also give the impression of a higher ceiling; this is a very desirable effect in low-challenged family rooms.

Light for reading can be accomplished by either table or floor lamps. Post lamps with two or three bullet fixtures are also helpful. One of the bullet housings (containing a reflector may be aimed at the proper angle for reading while the other(s) may be aimed at the ceiling for indirect general lighting.

Directional light fixtures mounted on the ceiling can be used to display mantel decorations or to illuminate a painting.

Fluorescent fixtures such as cabinet fixtures mounted on the underside of a bookcase or shelf create an excellent lighting source for displaying family portraits, collectibles, hobby items, and the like.

Any family-room lighting scheme must be very flexible because most family rooms are in the scene of a variety of daily activities, and these activities require different atmospheres, which can be created by light. For instance, TV viewing requires softly lighted surroundings, while reading calls for a somewhat brighter setting with light directed on the printed pages. Game participants feel more comfortable in a uniformly lighted room with additional glare-free light directed onto the playing area. Casual conversation flourishes amid subdued, complexion-flattering light such as incandescent or warm-white fluorescent.

Low-level lights over the bar area should be just bright enough for mixing a drink or fixing a late-night snack.

A typical family room may be illuminated as follows. The general illumination is accomplished with recessed incandescent fixtures with Fresnel lenses. The electric circuit controlling these recessed fixtures is provided with a rheostat dimmer control to change the lighting level of the room as well as the atmosphere. A fluorescent lamp is installed in the center of built-in bookshelves to provide light on the counter below for writing, studying, etc. Wallwash fixtures may also be used near the bookcase to highlight the colorful bindings of the various books.

Small recessed incandescent lamps with star-shaped lenses may be installed above a bar, while Slimline fluorescent fixtures may be mounted on the inside and under the bar top to prove additional light on the work counter. If glass shelves are present behind the bar, these may be highlighted with fluorescent fixtures. With all of these lighting fixtures controlled by dimmers, many exciting effects can be achieved with this one lighting scheme.

7.5.0 LIGHTING CATALOGS

Manufacturers of lighting fixtures have colorful brochures and catalogs that can be extremely helpful to the electrician involved in residential lighting applications. Besides showing the types of fixtures available, and listing them by catalog number, examples of their use are also given. When you begin a new project, a glance through a dozen or so of these brochures and catalogs will often give you some excellent ideas on how to proceed.

Some of the larger lighting catalogs also provide design data for their fixtures that will prove helpful on more sophisticated commercial projects.

Brochures are normally available from lighting-fixture manufacturers at little or no charge. Your local lighting-fixture dealers will probably have free literature in their places of business. Visiting these dealers will also give you a chance to see the various types of fixtures on display.

8.0.0 RESIDENTIAL OUTDOOR LIGHTING

Outdoor lighting is a partner in modern residential living. When well planned, it creates a total home environment, combining maximum aesthetic appeal with efficiency. It welcomes guests and lights their way to the house entrance; it creates a hospitable look and turns the area surrounding the home into an extra living area in warm months; reveals the beauty of gardens, trees, and foliage; expands the hospitality and comfort of patios and porches; stretches the hours for outdoor recreation or work; provides sure seeing to safeguard persons against accidents at night; and helps protect the home from prowlers.

This section is designed to show the reader how to create appealing outdoor lighting and will cover the following:

- Capturing the mystery and subtle qualities of outdoor lighting.
- Lighting ground contours and focal points.
- Creating silhouetted forms and shadow patterns.
- Using colored light.
- Using outdoor lighting to make the home more attractive, safe and fun to live in.

Unlike some types of indoor lighting, designing outdoor lighting is mainly the process of using techniques gained from experience.

8.1.0   ENTRANCE LIGHTING

Entrance lighting makes a visitor's first impression a good one. At the same time, good entrance lighting flatters the home. Even with the modest residence covered in this module, the two wall-bracket lighting fixtures mounted on each side of the entrance door create a festive and somewhat luxurious look.

The two low-level lighting fixtures along the walk from the driveway and the front door clear a path through the darkness and helps see visitors safely to the front door. When used in combination with the wall-bracket lights, these two lights create a grand setting for the front lawn, making it seem larger and more beautiful. *Figure 32* shows how these fixtures appear on the plan.

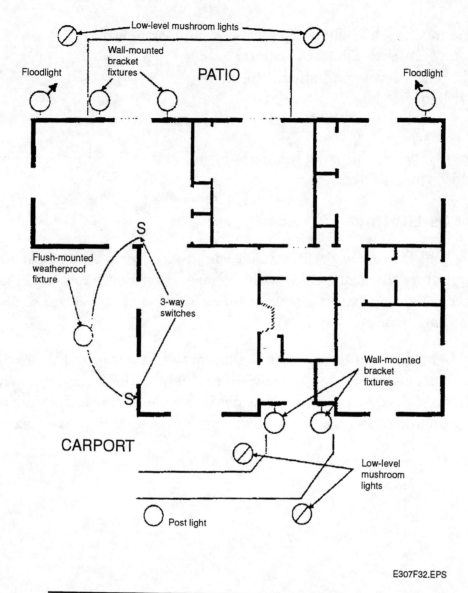

Figure 32. Outdoor Lighting Layout For A Typical Residence

The post light at the driveway entrance lights a small sign giving the owner's name and house number—a convenience first-time visitors to the home will appreciate.

Since our residence under consideration is a modest one, the outdoor lighting is not too elaborate. The four lighting fixtures previously described are all that were used for entrance lighting. However, in larger and more luxurious homes other outdoor entrance lighting may be appropriate:

- Lighting fixtures recessed in the soffit of the front of the house will act as wallwash fixtures, lighting a stone facade or bringing out the texture of beautiful brick.

- Ground-mounted uplights installed under trees on the lawn shining up through the tree branches will add elegance to any lawn and home.
- Spotlights recessed into the roof overhang can accent the house finish, light walks near the house, accent the entrance door, and dramatize painting or architectural detail— again extending a friendly welcome.

This same outdoor lighting that gives the home a proper introduction to friends also chases away shadows where danger could lurk. The lighting that enhances the architecture and landscape around the sides and rear of the home also forms a protective ring of light. Therefore, properly planned outdoor lighting tightens the security of any home at no additional cost. It is recommended that some or all of the outdoor lighting be controlled by either a photocell or a time switch, or a combination of both.

8.1.1 Porches, Patios, And Terraces

The carport of the residence under consideration has one ceiling mounted lighting fixture installed in the center of the area. A three-way switch with a weatherproof cover is located on the outside wall of the house, about where a person would get out of the car. This makes it convenient to turn the light on when the owners are coming home at night. The light may be turned off inside the house by means of another three-way switch.

The outdoor patio in back of our residence is lighted by a combination of wall-bracket fixtures and floodlights; low-level "mushroom" lighting fixtures are used along the outside edge of the patio for added flexibility.

This arrangement of outdoor lighting pushes back the barriers of darkness and opens up wide vistas of family fun and fulfillment. The owners of this house now enjoy the relaxing qualities of night in the comfort and convenience of a family recreation room.

A patio or terrace can be lighted in several ways, such as with a group of carefully positioned lighting fixtures installed under the roof overhang to evenly flood the terrace or patio with illumination, or with a pendant-type fixture hung from the terrace ceiling over a table on the terrace to provide glare-free light for card games, eating, or just engaging in conversation with friends.

When friends are being entertained on a patio or screened-in porch, a higher level of illumination on colorful shrubs and flowers than that used for the porch will make the porch seem more spacious besides showing off the floral growth in the yard.

E307F33.EPS

Figure 33. Example Of Low Horizontal Light Distribution

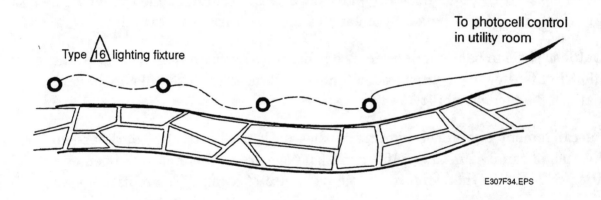

To photocell control
in utility room

Type /16\ lighting fixture

E307F34.EPS

Figure 34. Plan View Of Figure 33

In choosing outdoor lighting for any home, the designer should coordinate the fixtures, matching them in styling to the architectural character of the home. This will enhance the home's appearance and value. The following are basic:

- Wall-bracket fixtures should be used on both sides of the main entrance.
- The outlet boxes should be located approximately 66 inches above the top landing of the entrance steps.
- The lighting fixtures should be shielded with 8-inch or larger enclosures and contain a 50-watt or larger lamp in each fixture.

ELECTRICAL TRAINEE TASK MODULE 20308

Other entrances require only one fixture mounted at the same height and on the lock side of the door. Sometimes, however, it is desirable to have a fixture on each side of the door at other entrances, as in the case of our residence discussed earlier.

When it is not feasible to use two wall-bracket fixtures at the front door, one large fixture can be located directly above the door, or recessed fixtures can be mounted in the porch ceiling or roof overhang, as close to the door as possible. Each of these fixtures should have a minimum lamp size of 100 watts.

In addition to doorway lighting, a ceiling mounted lighting fixture should be located at the center of all breezeway and porch ceilings. Again, this fixture should be shielded with ceramic-enameled or opal glass and be a minimum of 10 inches in diameter. The fixture should contain lamps that are rated at a minimum of 150 watts total. If the ceiling is over a small portico, an 8-inch diameter fixture with a 60-watt bulb will suffice.

A post lantern should be located at the main entrance walk to mark the driveway or the sidewalk to the main entrance. The shielded fixture should be a minimum of 12 inches square or 12 inches in diameter. Avoid clear glass designs and exposed high-wattage lamps because the afterimage caused by viewing these will hamper seeing and can cause accidents.

A single-car garage should have one 8-inch diameter fixture or porcelain lampholder on each side of the outside of the garage door. In addition, there should be one 8-inch diameter fixture on each side of the driveway about six feet back from the garage to light passageways at the sides of the car.

For wide-coverage yard lighting, single or double weatherproof, adjustable floodlight units should be mounted under the eaves or roof overhang and aimed at the desired areas. By using 150-watt PAR-38 lamps, a relatively large area can be covered with light at a very reasonable cost.

At least one switch-controlled weatherproof receptacle should be located on the outside near the front entrance for holiday lighting. However, remember that NEC Section 210-8(a) states: "For residential occupancies, all 125-volt, single-phase, 15- and 20-ampere receptacle outlets installed outdoors shall have approved ground-fault circuit protection for personnel."

8.2.0   OUTDOOR LIGHTING EQUIPMENT

A wide variety of outdoor lighting equipment is readily available. Many factors related to the mechanical and electrical design are important, but the lighting effects desired are the primary considerations in selecting the equipment for outdoor lighting. Both daytime and nighttime appearance of the lighting fixtures should be considered very carefully. Also, make certain that fixture lamps and all electrical wiring are concealed from direct view.

The resistance to weather is another important consideration in selecting outdoor lighting fixtures, since the equipment will be subjected to sun, rain, wind, and snow. Aluminum, brass, copper, stainless steel, and even plastic are the materials most used in outdoor lighting fixtures.

Durability is a prime concern in the selection of outdoor lighting equipment, and the cost of the fixtures—like most everything else—is usually directly proportional to durability. The equipment must be able to stand up under a great amount of abuse throughout the year—not only against the weather but also the destructiveness of lawn mowers, snow throwers, and humans. Fixture design and color normally should blend with the landscape in the case of ground-mounted lawn or garden lighting fixtures. These should also be as inconspicuous as possible in the daytime as well as nighttime. Fixtures mounted on the house should follow the general architectural theme.

Personal taste is the final factor to be considered in the process of selecting outdoor lighting equipment. *Figure 35* shows typical examples of outdoor lighting fixtures.

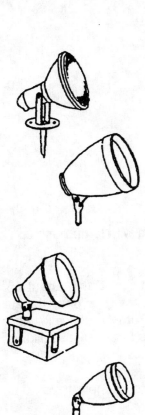

1. Adjustable holders - for PAR 38 projector lamps and others with spike for ground placement, cover plates for outlet boxes and attachment clamps for use on tree or pole.
USES: for all areas described in this chapter.

a. for PAR lamps. Color glass covers, louver and shield clip on bulb.

b. of metal, offers deep shielding, better appearance.

c. mercury floodlights - adjustable units mounted on enclosed ballast.

d. for R20 floodlights. Cover lens protects bulb.

e. Enclosed floodlights - often called "handy floodlights." Use up to 300 watts. Cover glass protects regular household and reflector bulbs.

f. Flush mounted fixtures for projector lamps. Specific housings available for 150 PAR 38 up to 500 PAR 64 and for mercury lamps. USE: in open areas without available natural shielding.

2. Mushroom unit - use with any height stem or post. Both side-suspended and center stem type available. Wide range of reflector width, depth and contour result in differences in complete bulb shielding. USES: general lighting on terrace with 4- to 5-foot stems; most visual tasks with 100 watt bulb, placed at rear corner of a chair, circulation areas; flowers and plants of all heights. Bulb wattage: as desired. (Many variations of the basic mushroom design are available).

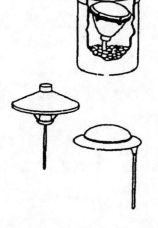

E307F35A.EPS

Figure 35. Typical Examples Of Outdoor Lighting Fixtures

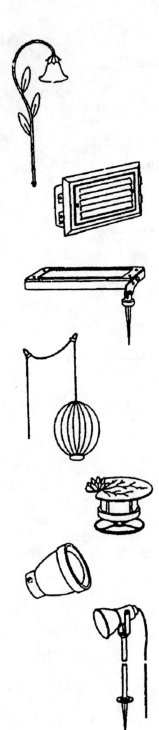

3. Bell-type reflector - suspended from fixed height stem with lamp base up. USES: flowers and plants in small area. Garden steps.

4. Recessed units with lens or louver control. Directs light down about 45° below horizontal. Bulb wattage: from 6 to 25 watts depending on unit. Locate from 4 inches to 24 inches above ground. USES: Paths and walks near buildings - 8 to 12 feet apart, steps - mounted in risers or adjacent building; terraces to light floor - 6 to 10 feet apart.

5. Weatherproof fluorescent units - wattage depends on length of unit. USES: where line of light is desired.

6. Diffusing plastic shade attached to suspended socket. Various sizes & shapes. USES: general lighting for terraces with roof or overhang - 40 to 75 watts in 10-inch diameter units - 100 to 150 watts larger sizes. Decorative: with 10 to 40 watt colored or white bulbs.

7. Underwater fixtures - lily pad shield attached to glass enclosed housing. Use 25 to 60 watt bulb.

8. Underwater fixtures - provide controlled light. Bulb size depends on unit. Available in both low voltage and 120 volts, depending on unit.

9. Telescopic poles for PAR 38 and enclosed floodlights. Fit into pipe sleeves driven into ground. USES: sports and area floodlighting.

E307F35B.EPS

Figure 35.   Examples Of Outdoor Lighting Fixtures (continued)

ELECTRICAL TRAINEE TASK MODULE 20308

Once again, manufacturers of outdoor lighting equipment have catalogs and brochures with many helpful design hints in them. If you anticipate encountering much outdoor lighting work, you should obtain as many of these publications as possible and study them. Then, keep for reference for future projects. More are available from electrical distributors or they may be ordered directly from the manufacturers.

9.0.0 HIGH-INTENSITY DISCHARGE LAMPS

9.1.0 OPERATING CHARACTERISTICS

High-intensity discharge, mercury, metal-halide, high-pressure sodium, and low-pressure sodium lamps all are electric discharge lamps. Light is produced by an arc discharge between two electrodes located at opposite ends of an arc tube within the lamp. The purpose of the ballast is to provide the proper starting and operating voltage and current to initiate and sustain this arc.

9.2.0 MERCURY LAMPS

This type of lamp has been used mainly for outdoor lighting and for industrial applications due to its poor color and high wattage. However, recent improvements in color, availability in lower wattages, and a higher output efficiency have made these lamps attractive for commercial indoor applications. See *Figure 36.*

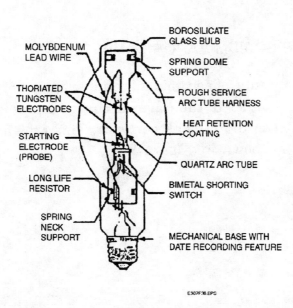

E307F36.EPS

Figure 36. A Mercury Vapor Lamp

The traditional mercury-vapor lamp produces light with a predominance of yellow and green rays and a small percentage of violet and blue.

Since this blue-green light distorts almost all colors, color correction has been added by coating the outer bulb with phosphors. The phosphors are activated by the ultraviolet light and radiate generally in the red bond, which is entirely absent in the basic lamp color. The light has now been corrected to make it acceptable for indoor use.

Mercury-vapor lamps operate by passing an arc through a high-pressure mercury vapor that is contained in an arc tube made of quartz or glass. This action produces visible light. As with all arc discharge lamps, ballasts are required to start the lamp, and thereafter to control the arc.

9.3.0 METAL-HALIDE LAMP

The metal-halide lamp is basically a mercury lamp that has been altered by adding iodine compounds to the mercury and argon gas in the arc tube. These iodine compounds are of metals such as indium, sodium, thallium, or dysprosium. The addition of these salts causes the emission of light which is of a better color than the basic mercury colors, although life and lumen maintenance may be decreased in the process.

Metal-halide lamps furnish approximately 75 to 90 lumens per watt of white light, white is much warmer than the mercury light and is suitable for many commercial indoor applications.

9.4.0 HIGH-PRESSURE SODIUM VAPOR LAMPS

This type of lamp uses an arc tube of ceramic material such as polycrystalline alumina. The lamp operates in a similar manner to the other discharge lamps, producing a warm yellow orange tone at the rate of over 100 lumens per watt. This type of lamp is mostly used for outdoor applications. This is shown in *Figure 37.*

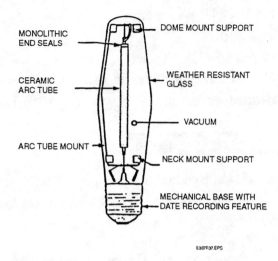

MONOLITHIC
END SEALS

DOME MOUNT SUPPORT

CERAMIC
ARC TUBE

WEATHER RESISTANT
GLASS

VACUUM

ARC TUBE MOUNT

NECK MOUNT SUPPORT

MECHANICAL BASE WITH
DATE RECORDING FEATURE

E307F37.EPS

Figure 37. High-Pressure Sodium Lamp

High-Pressure Sodium Lamps: They are the most efficient HID sources available. These lamps are used for general lighting applications where high efficiency and long life are desired and color rendering is not critical. Typical applications include street lighting, industrial high-bay, parking lot lighting, and building floodlighting.

Mogul and Medium Base: Are available in a broad range of mogul base general lighting lamps from 35 to 1000 watts for universal operation. Lamps from 35 to 150 watts are also available in medium base. All lamps are available in clear and coated bulbs.

Instant Restrike: Lamps are designed with two arc tubes to provide instant restrike capability in the event of a momentary power interruption without total loss of light.

Retrofit: These retrofit lamps are designed to operate on all mercury reactor ballasts when increased light and reduced energy is desired.

9.5.0 ADVANTAGES/DISADVANTAGES OF HIGH-INTENSITY DISCHARGE LAMPS

9.5.1 Advantages

- Long life.
- High lumen output.
- Have compactness of incandescent lamps.
- Not affected by ambient temperatures as are fluorescent lamps.
- Better degree of light control than with fluorescent lamps.

9.5.2 Disadvantages

- Color acceptability low with clear mercury.
- Sensitive to voltage variation.
- Long restarting time required.
- Light is extinguished if momentary current interruption occurs.
- Delay of four to seven minutes from starting time to full brightness.

10.0.0 ELECTROLUMINESCENT LAMPS

The fluorescent lamp produces light by exciting phosphors with ultraviolet light. With certain special phosphors, it is possible to produce light of the intermediate step. This process is known as electroluminescence.

Though still in early stages of development, this process has been successfully used in night lights for residential use, road sign illumination, control panel lighting, and other applications requiring low brightness and maintenance-free illumination.

10.1.0 ELECTROLUMINESCENCE

Electroluminescence is achieved by exciting phosphor under a pulsating electrical field. An electroluminescent lamp is a thin area source in which phosphor is sandwiched between two conductive layers, one (or both) of which is translucent. Additional transparent material is provided to protect it from external abuse *(Figure 38)*. Electricity (usually 120-VAC) is passed through the lamp to create an electrostatic field that excites phosphor producing light in green, blue, yellow, or white, depending on the type of phosphor; its intensity varies with applied voltage, frequency, and temperature. Green phosphor has the highest luminance. These lamps are available in ceramic and plastic form, flexible or stiff, with or without sticky-back, and are easily fabricated into shapes and sizes to meet needs that range from decorative to general, low-level illumination.

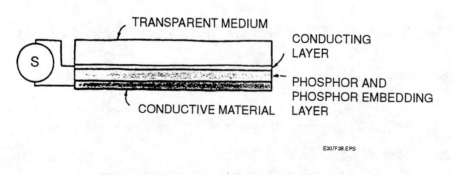

Figure 38.   Electroluminescent Lamp

Lighting fixtures, appliances, and fixed electric space-heating equipment must be wired according to provisions of the NEC. In recent years, the number and type of appliances, fixtures, and heating equipment available for new installations and remodeling work have increased dramatically.

Refrigerators, freezers, stoves, washers, dryers, food processors, blenders, toasters, microwave ovens, dishwashers, compactors, disposals, electric furnaces, wall heaters, and the complete range of fixed and portable electrical equipment in use today require careful study of applicable sections of the NEC for proper installation. Conductors, branch circuits, and overcurrent protection devices must be designed and installed properly for safe electrical operation of this equipment.

11.1.0 FIXTURES IN CORROSIVE LOCATIONS AND IN DUCTS OR HOODS

Fixtures installed in areas subject to corrosive conditions must be approved for such conditions (*Figure 39*). Fixtures installed in range hoods must meet several requirements as specified in Part C.

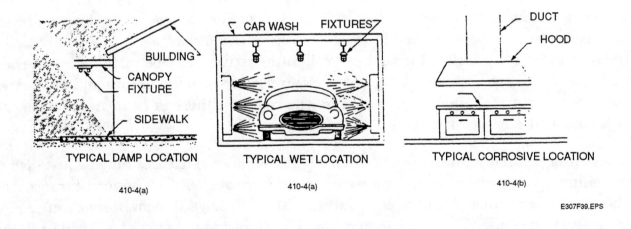

Figure 39. Fixtures Must Be Suitable For The Locations Where They Are Installed

11.2.0 HANGING FIXTURES

No parts of cord-connected fixtures, hanging fixtures, lighting track, pendants, or ceiling fans shall be located within a zone measured 3 feet horizontally and 8 feet vertically from the top of the bathtub rim. This prevents a person from being able to reach the fixture from the bathtub (*Figure 40*). Cord-connected fixtures are considered hanging fixtures and must comply with requirements of 410-4(d).

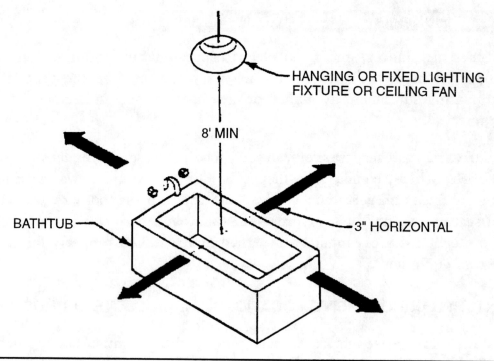

HANGING OR FIXED LIGHTING
FIXTURE OR CEILING FAN

8' MIN

BATHTUB

3" HORIZONTAL

Figure 40. Hanging Or Fixed Lighting Or Ceiling Fan Fixtures Must Be At Least 8 Feet Above The Top Of A Bathtub And Not Within 3 Feet Horizontally Of Any Edge Of The Tub

11.3.0   FIXTURES IN CLOTHES CLOSETS

Wall-mounted lighting fixtures in clothes closets shall be installed over the door to prevent clothes from being hung over the fixture and to keep the fixture away from combustible material. A completely enclosed incandescent lighting fixture may be installed on the wall above the door if there is at least 12 inches of clearance between the fixture and the nearest storage area. The fixture may be mounted on the ceiling if there is 12 inches of clearance from the nearest storage area.

Fluorescent or incandescent, flush or recessed (with solid lens) fixtures may be installed on closet ceilings. A recessed incandescent fixture shall have at least 6 inches of clearance from the nearest storage area. A fluorescent fixture also shall have 6 inches of clearance from the nearest storage area. Fixtures hanging from a cord (pendants) are not allowed in clothes closets *(Figure 41)*.

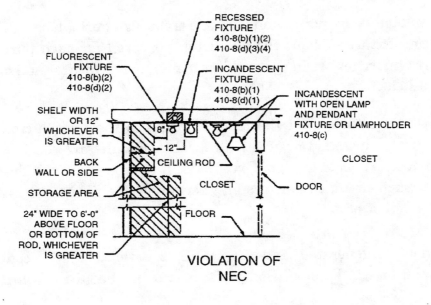

RECESSED
FIXTURE
410-8(b)(1)(2)
410-8(d)(3)(4)

FLUORESCENT
FIXTURE
410-8(b)(2)
410-8(d)(2)

INCANDESCENT
FIXTURE
410-8(b)(1)
410-8(d)(1)

SHELF WIDTH
OR 12"
WHICHEVER
IS GREATER

INCANDESCENT
WITH OPEN LAMP
AND PENDANT
FIXTURE OR LAMPHOLDER
410-8(c)

8"

CLOSET

12"

BACK
WALL OR SIDE

CEILING ROD

STORAGE AREA

CLOSET

DOOR

24" WIDE TO 6'-0"
ABOVE FLOOR
OR BOTTOM OF
ROD, WHICHEVER
IS GREATER

FLOOR

VIOLATION OF
NEC

E307F41.EPS

Figure 41.   Positioning Incandescent And Fluorescent Fixtures In Clothes Closets

11.4.0   SPACE FOR COVE LIGHTING

Space must be provided for ready access for maintenance and replacement of lamps or fixtures. In addition, lamps or fixtures must have space enough for ventilation to dissipate heat and thus protect against high-temperature deterioration in cove lighting *(Figure 42)*.

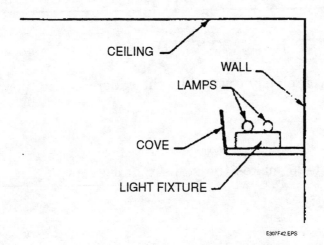

CEILING

WALL

LAMPS

COVE

LIGHT FIXTURE

E307F42.EPS

Figure 42.   Cove Lighting Fixtures Must Be Accessible For Servicing And Changing Lamps

11.5.0 TEMPERATURE LIMIT OF CONDUCTORS IN OUTLET BOXES

Branch-circuit conductors are not permitted to pass through outlet boxes that are an integral part of the lighting fixture without the box being approved for such purpose. The branch-circuit conductors and fixture wiring are subject to high temperatures and must have a rating to withstand the heat.

Fixture whips are used to connect the fixture wiring to the branch-circuit conductors. A remote junction box is installed between the branch circuit and the fixture with the tap conductors in a 4 foot to 6 foot length of Greenfield. If the fixture is suitable for 600 C supply conductors, the branch-circuit conductors may terminate the fixture wiring by being spliced in the fixture's junction box.

Note   If heat is generated into the feature's outlet box, the conductor's insulation can become fried and cause a fire hazard *(Figure 43)*.

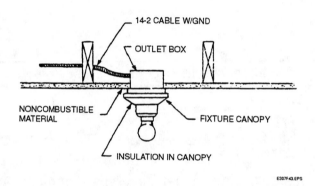

Figure 43. Fixtures Must Be Constructed So That Conductors Are Not Subjected To Temperatures Higher Than Their Rating

11.6.0 OUTLET BOXES TO BE COVERED

If the outlet box is not completely covered by the canopy of the fixture, it must be fitted with a cover.

Note   Section 370-25 makes the same requirement.

ELECTRICAL TRAINEE TASK MODULE 20308

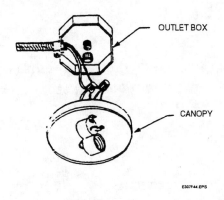

OUTLET BOX

CANOPY

E307F-44.EPS

Figure 44. An Additional Cover Is Not Required If The Fixture Canopy Covers The Outlet Box Opening

11.7.0 CONNECTION OF ELECTRIC-DISCHARGE LIGHTING FIXTURES

When fluorescent lighting fixtures are mounted independently of their outlet boxes, the connections must be made through metal raceways, metal-clad cable (BX), AC cable, MI cable, nonmetallic raceways, or nonmetallic sheathed cable (Romex). When the fixture is remote and suspended below the outlet box, the connection may be made with type AC cable (BX), nonmetallic sheathed cable (Romex), EMT, flexible metal conduit (Greenfield), flexible cord, or any approved means. Any fixture that is not supported by the outlet box that provides the branch-circuit conductors supplying the fixture wiring is governed by the rules of this section *(Figure 45)*.

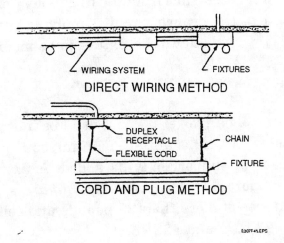

WIRING SYSTEM   FIXTURES

DIRECT WIRING METHOD

DUPLEX RECEPTACLE   CHAIN
FLEXIBLE CORD   FIXTURE

CORD AND PLUG METHOD

E307F-45.EPS

Figure 45. Two Methods Of Wiring Fluorescent Fixtures

When a flexible cord is used, it must be visible for its entire length.

11.8.0 SUPPORTS

A fixture must not be hung from the screw shell of a light socket if the fixture weighs over 6 pounds or exceeds 16 inches in any one dimension; other means of support must be devised *(Figure 46)*.

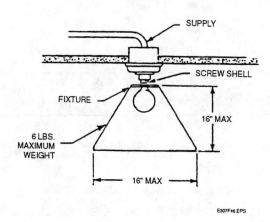

Figure 46. Fixtures Weighing Over 6 Pounds Or Exceeding 16 Inches In Any Dimension Must Not Be Supported By The Screw Shell Of A Lampholder

11.9.0 FIXTURE SUPPORTED BY METAL POLES

Metal poles supporting fixtures shall have not less than a 2 inches x 4 inches accessible handhole with a grounding terminal accessible from the handhole. Conductors may be routed through the pole and connected to the branch circuit supplying power to the fixture. See the exception (NEC 410-15[b][1]) which allows the fixture to be mounted on the metal pole.

11.10.0 MEANS OF SUPPORT

NEC 410-16(a) requires fixtures weighing over 50 pounds to be independently supported by ceiling joists or other ceiling members. Substantial mounting support must be provided for heavy fixtures. Additional means of support may be added. A fixture weighing under 50 pounds may be mounted and supported by the outlet box *(Figure 47)*. However, sometimes extra support is needed for fixtures of more than 25 pounds and ceiling fans cannot weigh more than 35 pounds per 370-23(b)(c) and 422-18.

NEC 410-16(b) requires suitable opening space in the back of the fixture to permit access to the branch conductors and outlet box. All parts of the fixture wiring must be accessible for inspection without it being necessary to disconnect any wiring.

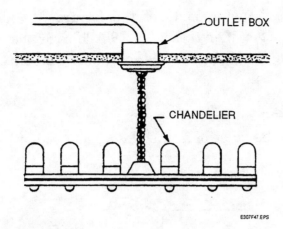

OUTLET BOX

CHANDELIER

E307F47.EPS

Figure 47.   Outlet Boxes May Support Fixtures Weighing 50 Pounds Or Less

NEC 410-16(c) requires fixtures supported by the framing member of suspended ceiling systems (gridwork-types) to have additional supports. Bolts, screws, rivets, or other mechanical means may be used.

Note   The ceiling itself must be properly installed *(Figure 48)*.

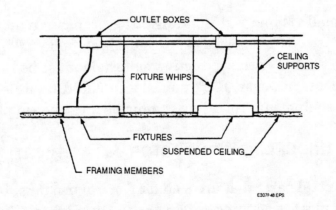

OUTLET BOXES

FIXTURE WHIPS

CEILING SUPPORTS

FIXTURES

SUSPENDED CEILING

FRAMING MEMBERS

E307F-48.EPS

Figure 48. Suspended Ceiling Framing Members
Must Be Securely Attached To The Building Structure

11.11.0 EXPOSED FIXTURE PARTS

Part A requires the exposed metal parts of the fixture (if any) to be connected to the equipment grounding conductor (if any) *(Figure 49)*. This connection of the fixture's exposed metal parts to the grounding conductor may be achieved by mounting hardware on the fixture or by a bonding jumper.

Part B states that if an equipment grounding conductor is not present in the circuit supplying the fixture, the fixture must be made of an insulated nonconductive material such as plastic.

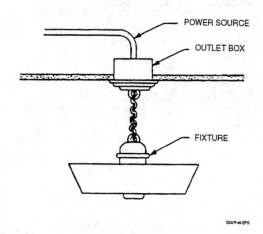

POWER SOURCE

OUTLET BOX

FIXTURE

E307-49.EPS

Figure 49. Exposed Metal Parts Of Fixtures Must Be Grounded Or Double-Insulated

In older houses and small commercial buildings, the branch circuit usually does not have a grounding conductor, so the rules of *Part B* must be applied. This does not mean that existing lighting fixtures made with conductive materials must be replaced. When they are replaced for other reasons, however, they must be replaced by a nonconductive type, or a branch circuit with an equipment grounding conductor must be supplied to the new fixture.

11.12.0 EQUIPMENT GROUNDING CONDUCTOR ATTACHMENT

Fixtures with exposed metal parts that are sold for general use must be provided with means of connecting the equipment grounding conductor to the fixture *(Figure 50)*. If the fixture is not equipped with a means of attachment, then bonding jumpers and clips, or other approved means where applicable, may be used.

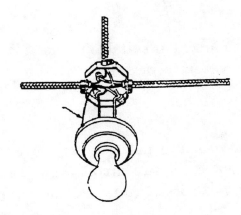

Figure 50. Newly Installed With Exposed Metal PartsMust Have An Equipment Grounding Conductor

11.13.0 METHODS OF GROUNDING

When fixtures are installed, they should be carefully checked to ensure they are effectively grounded. See 250-91(b) for methods of grounding.

11.14.0 CORD-CONNECTED SHOWCASES

Showcases not in fixed positions may be connected by flexible cord to permanently installed receptacles in groups of not more than six showcases. The six showcases also may be grouped together by flexible cord. The cord must not be exposed to physical damage, and the separation between cases must not exceed 2 inches. No more than 12 inches must exist between the first case and the supply receptacle.

11.15.0 CORD-CONNECTED LAMPHOLDERS AND FIXTURES

Chain-hung or independently mounted fluorescent fixtures may be connected by flexible cords and grounded-type attachment caps as long as the cord is visible for its entire length. Fluorescent fixtures mounted in suspended ceilings cannot be connected in this manner because such a connection would prevent the cord from being visible for its entire length.

11.16.0 FIXTURES AS RACEWAYS

Only those conductors associated with the lighting fixture may be run through the lighting fixture enclosure. Branch-circuit conductors passing within 3 inches of a ballast must have a temperature rating of at least 90°C or 75°C THW per Table 310-13.

BASIC LIGHTING

11.16.1 Exception No. 1

Lighting fixtures that are marked as suitable for use as a raceway may have the circuit run straight through as if they were a raceway.

11.16.2 Exception No. 2

Fixtures approved for end-to-end installation may have two circuits supplying the fixtures. One of the two circuits may be a 3-wire circuit while the other is a 2-wire circuit. However, three 2-wire circuits or two 3-wire circuits are not permitted.

11.16.3 Exception No. 3

Fixed fluorescent strip lighting assembled end-to-end may have an additional 2-wire circuit to supply one or more of the connected fixtures in the row.

11.17.0 TEMPERATURE

New recessed incandescent lighting fixtures must be equipped with thermal protection except where installed in concrete and identified for use without thermal protection. The thermal cutout (TC) type of fixture is designed to be installed with insulation at least 3 inches from the top and all sides of the fixture. If insulation covers the fixture, or if the fixture is overlamped (a 60-watt bulb removed and a 100-watt bulb installed), the thermal cutout will sense the additional heat and disconnect the fixture.

Fixtures rated for insulated ceilings (IC) are designed to be used in direct contact with insulation. They are down-rated in wattage and operate at an extremely cool temperature. Fixtures rated for suspended ceilings (SC) are designed to be used in commercial buildings. These fixtures are high-wattage, low-cost fixtures. Check with inspectors before installing this type of fixture.

11.18.0 CLEARANCE

Recessed lighting fixtures, other than at points of support, must have at least 1/2 inch space between the fixture enclosure and combustible material. Thermal insulation must be at least 3 inches from the fixture enclosure with free air circulation space on the sides and no insulation above the fixture. This circulation space allows heat to dissipate. Approved IC fixtures may be covered by insulation and mounted against combustible material at points of support. See NEC 410-65(c) and 410-66(a), Ex.

11.19.0 WIRING

Branch-circuit conductors that are not high temperature conductors (167°F) must not be run to the fixture and wired direct. Only conductors rated 167°F and above may be run to fixtures and wired direct. Conductors rated below 167°F may be run to a junction box installed at least 12 inches from the fixture.

11.20.0 RECESSED HIGH-INTENSITY DISCHARGE FIXTURES

The ballasts of fluorescent fixtures mounted indoors must have integral thermal protection. When the ballasts are replaced in existing fixtures, they must be replaced with integrally protected ballasts. Smaller fluorescent fixtures such as under-cabinet strips, desk lamps, and medicine chest lamps do not require thermal protection. These fixtures utilize small reactance-type ballasts instead of Class P ballasts.

Fixtures not approved for direct mounting to ceilings shall be set off 1.5 inches from combustible material.

11.21.0 FIXTURE MOUNTING

Fixtures must be approved and marked as suitable to be mounted directly on ceilings. Fixtures that are not marked suitable for direct mounting must be set off at least 1 foot from combustible material.

Note   A fixture is marked "suitable for surface-mounting on combustible low-density cellulose fiberboard" when it is suitable to be mounted directly on the surface of the ceiling.

12.0.0 RELAYS

Next to switches, relays play the most important part in the control of light. However, the design and application of relays is a study in itself. Consequently, Task Module 01211 covers contractors and relays in depth. Still, brief mention of relays is necessary to round out your study of lighting controls.

An electric relay is a device whereby an electric current causes the opening or closing of one or more pairs of contacts.

These contacts are usually capable of controlling much more power than is necessary to operate the relay itself. This is one of the main advantages of relays.

12.1.0 MAGNETIC RELAY

This type of relay, *Figure 51*, is the most common in use today for the control of lighting. It consists of a coil of wire wound around an iron core and an armature. The coil acts as an electromagnet, and when the proper amount of electric current is applied, it causes the armature to move. The armature, in turn, moves contact points together or apart, depending on how the relay is constructed. When the power is disconnected from the coil, a spring returns the armature to its original position. The following example demonstrates how a relay and a 15-amp single-pole switch can be used to control 60 amperes of lighting.

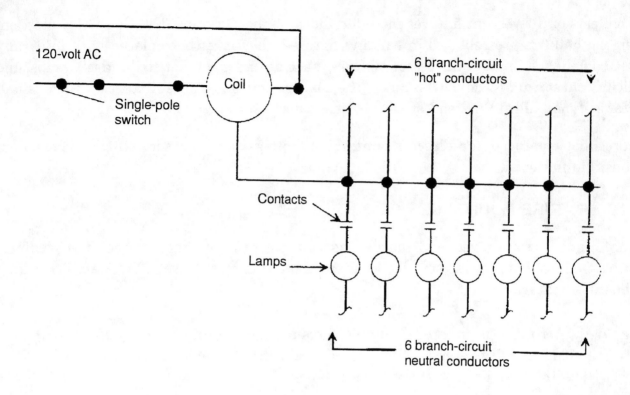

Figure 51. Wiring Diagram Of A Group Of Lights Controlled By A Magnetic Relay

We see a single-line diagram of a relay controlled by a single-pole switch. The relay operates six contacts, each of which is connected to a circuit with lamps totaling 1200 volt-amperes (watts)—totaling 10 amperes per circuit. When the single-pole switch is closed, the coil is energized and moves the armature which, in turn, closes all of the six contacts. When the contacts are closed, this completes the circuit to all of the lamps. Thus, all lamps will come on at the same time. Six single-pole switches could have been used in place of the relay, but simultaneous switching of the lamps would then not be possible.

12.2.0   REMOTE-CONTROL SWITCHES

One popular use of the relay in residential lighting systems is that of remote-control lighting. In this type of system, all relays are designed to operate on a 24-volt circuit and are used to control 120-volt lighting circuits. They are rated at 20 amperes, which is sufficient to control the full load of a normal lighting branch circuit, if desired.

ELECTRICAL TRAINEE TASK MODULE 20308

Remote-control switching makes it possible to install a switch wherever it is convenient and practical to do so or wherever there is an obvious need for having a switch—no matter how remote it is from the lamp or lamps it is to control. This method enables the lighting designs to achieve new advances in lighting control convenience at a reasonable cost.

One relay is required for each fixture or for each group of fixtures that are controlled together. Switch locations for remote-control follow the same rules as for conventional direct switching. However, since it is easy to add switches to control a given relay, no opportunities should be overlooked for adding a switch to improve the convenience of control.

Remote-controlled lighting also has the advantage of using selector switches at certain locations. For example, selector switches located in the master bedroom or in the kitchen of a home enable the owner to control every lighting fixture on the property from this location. The selector switch may turn on and off an outside or other light that customarily would be left on until bedtime and that might otherwise be forgotten.

12.3.0 DIMMERS

Dimming the lighting systems is usually desired for one of two reasons. First, dimming provides a gentle method of turning the lights ON and OFF. A common application of this is in an auditorium used for plays, movies, and similar functions. In order to avoid shock or surprise when the lights are suddenly turned ON or OFF, or to avoid discomfort to the dark-adapted eye when the lights are suddenly turned on, dimmers are used to make this transition gradual. Second, dimming also provides control of the quantity of illumination. It may be done to create various atmospheres and moods or to blend certain lights with others for various lighting effects.

12.3.1 Rheostat Dimmer

The oldest type of dimming equipment is the rheostat. This device was connected in series with an incandescent lamp circuit to vary the voltage that would reach the lamps. A dial on the rheostat varied the resistance within the circuit which, in turn, changed the voltage (usually zero to 120 volts). The less voltage reaching the lamps, the less light the lamps would give off.

The rheostat dimmer had its disadvantages and brought about the use of the variable autotransformers for the most popular method of dimming AC incandescent lamps. The variable autotransformer controls voltage from full line voltage to zero, or even above line voltage by means of an extension winding.

For large installations, several autotransformers are used and are usually installed in banks and have control handles for the manual operation of individual dimmers or of a group of dimmers. Such banks are suitable for small stages, church auditorium lighting, as well as for general lighting control. However, for very large installations with several circuits, the banks of variable autotransformers are motor driven.

Solid state electronic dimmers are rapidly replacing the older rheostat dimmers, but variable autotransformers are still holding their own.

12.4.0   DIMMING FLUORESCENT LAMPS

The advent of the rapid start principle of fluorescent lamp operation made possible the dimming of fluorescent lamps. A cathode-heating supply current is provided with a constant voltage, while at the same time the arc passing through the tube is varied to produce dimming. In order to maintain this constant cathode-heating current while varying the lamp current, a special ballast transformer is necessary. It should be connected to the dimmer circuit as shown in *Figure 52*.

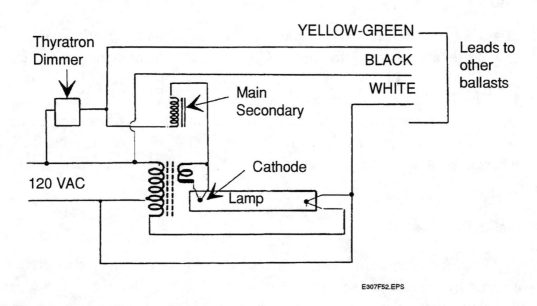

E307F52.EPS

Figure 52. A Dimmer Circuit Using A Special Ballast Transformer

In general, the diagram in *Figure 52* shows that 120-volt line voltage is fed into the primary of the ballast transformer which produces constant voltage on the secondary winding connected to the cathode of the fluorescent lamp. The current that passes the length of the lamp, however, must pass through the main secondary of the ballast transformer and also through the dimmer. Leads connect to other ballasts (controlling other lamps) in the system. Dimming ballasts of this type are commonly available for 40-watt rapid start fluorescent lamps.

12.5.0 FLASHING OF LAMPS

A flashing light is one of the best methods of getting attention, since it combines both brightness and motion. Its apparent motion or action has more attraction value for the human eye than any of the steadily glowing light sources. For these reasons, the application of flashing lights is especially desirable for advertising purposes. Economy is another reason for flashing. The time during which there is no light in a flashing action results in savings in electric power consumption and also increases the lamp life.

Incandescent, fluorescent, and neon lamps are the most-used types for flashing lights. The flashing may be accomplished by bimetallic thermal flashers but are most often flashed by means of motor-driven cam-actuated contact flashers.

According to the effects produced and according to the actions performed, the basic types of flashers are:

- OFF-ON
- Alternate-Speller
- Twinkler
- Border-Chaser

Used by themselves, these flashers can turn lights OFF and ON, spell out words, cause a lamp to twinkle, or cause the border-chasing effect around a sign. By combining two or more types of flashers, the sign designer can obtain an almost endless number of other attention-getting effects.

12.5.1   Off-On Flasher

The off-on flasher, as the name implies, turns the lights off and on at various intervals, producing both brightness and motion and conserving electrical energy—the amount depending on the time interval between flashing.

12.5.2   Alternate-Speller

The alternate-speller may be used to spell out the name of a product, manufacturer, agency, etc. By lighting sections of the lamps in a particular order, the effect of script writing may be accomplished. After the word is spelled out, it stays readily lit for a few seconds, goes off, then starts the spelling action again. This is probably the most common type of flasher. This type of flasher can also be modified to perform many different actions, such as OFF-ON, spelling, color changing, and animation of figures. Effects are obtained by changing the opening and closing time of the circuit contacts.

12.5.3 Flashing Fluorescent Lamps

Rapid start fluorescent lamps can be flashed without damage to the electrodes, provided these electrodes are kept heated to the right temperature at all times by using a special flasher ballast as shown in *Figure 53*.

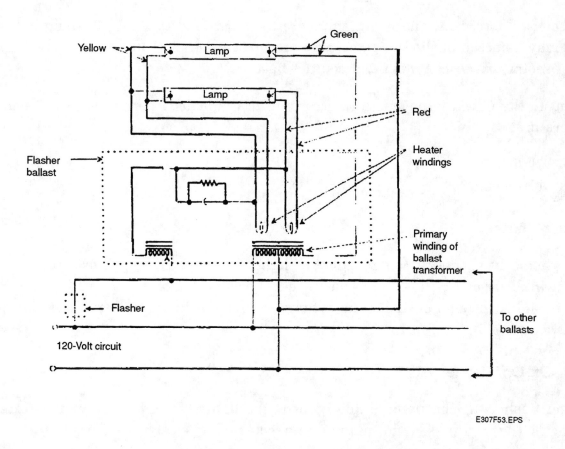

Figure 53. Wiring Diagram Of Fluorescent Flasher Control

The flasher control shown in *Figure 53* is connected to the 120-volt line and continues to other ballasts (which control other lamps). The two fluorescent lamps are connected to the flasher ballasts, and the flashing action is controlled from the flasher as indicated. Notice that this flasher diagram is quite similar to the dimming diagram in *Figure 52*.

It is also possible to obtain a dimming effect by inserting resistance in place of the flasher. A high dimming ratio cannot be achieved by this method, but interesting effects can be produced by flashing lamps from bright to dim.

12.6.0   STROBOSCOPIC EFFECT

The arc of a mercury lamp operating on a 60-cycle AC is completely extinguished 120 times per second. Thus, there is a tendency for the eye to see in flashes, with the result that a rapidly moving object may appear to move in a series of jerks (stroboscopic effect). This is often unnoticed, and in most installations it is not a serious disadvantage. Where necessary, stroboscopic effect may be greatly reduced by operating pairs of lamps on a lead-lag two-lamp ballast, or three lamps on separate phases of a three-phase supply. The use of incandescent lamps in combination with mercury lamps also lessens stroboscopic effect.

For industrial applications where high output is more important than color (such as the lighting of steel mills, aircraft plants, and foundries), high-output white or standard white semi-reflector lamps are often the most economical choice, particularly those types that may be operated directly from 460-volt circuits with inexpensive choke-type ballasts.

Wholly coated lamps have lower surface brightness than do semi-reflector lamps, and at relatively low mounting heights they may provide a more comfortable lighting installation, as well as better color rendition.

For commercial and nonindustrial interior applications, such as the lighting of gymnasiums, banks, high-challenged offices, transportation terminal buildings, and so forth, deluxe white lamps are recommended. The standard white color may be used where economy is more important than color quality. In installations where the unavoidable slight hum created by the ballasts might be objectionable, the ballasts are sometimes located outside of the room.

Summary

The proper control of lighting plays a very important role in the proper design of lighting systems. An otherwise good lighting design can become a poor design if improperly controlled. We urge the study of manufacturers' catalogs and wiring diagrams for more uses of the items described in this section. The catalogs are usually free of charge and contain information invaluable to electricians dealing with lighting installations.

References

For advanced study of topics covered in this Task Module, the following works are suggested.

American Electricians Handbook, Latest Edition, McGraw-Hill, New York, NY.
Lighting Handbook, 1992, Westinghouse.
Illustrated Dictionary for Electrical Workers, Delmar, 1991.
Illuminating Engineers Society, IES Handbook.
National Electrical Code, Latest Edition, NFPA, Quincy, MA.

SELF-CHECK REVIEW/PRACTICE QUESTIONS

1. Name the three basic types of lamps. *Incandescent, Gaseous discharge, Electroluminescent p 20*

2. Name the part of the eye that corresponds to the lens of a camera. *lens*

3. Name four factors that are involved in the process of seeing. *Size of object, Luminance of object, contrast b/w object & background, time*

4. Does it take more or less illumination to adequately light a room with dark-colored walls? *more*

5. What are the basic colors? *red, yellow, blue*

6. Which gives more light, a 100-watt incandescent lamp or a 40-watt fluorescent lamp? *40 watt Fluorescent*

7. What unit is used to measure the intensity of illumination that is produced on a surface? *footcandle*

8. Under what lamp classification does a quartz-iodine lamp fall? *incandescent*

9. Which of the electric discharge lamps is most widely used? *Fluorescent*

10. Which of the fluorescent lamps more closely matches the warm tones of incandescent lighting? *warm white*

11. What is the name of the device that is used in fluorescent fixtures that limits the current flow through the lamp to the value for which the lamp is designed? *ballast*

12. What is the recommended illumination level (in footcandles) for a residential kitchen? *70*

13. How many lumens per square foot are required to obtain the illumination level in Question 12? *80*

14. Which type of installation uses high-bay HID lighting the most? *industrial*

15. Name two ways in which the stroboscopic effect may be eliminated when using mercury lamps in industrial and shop applications. *by operating pairs of lamps on a lead-lag two lamp ballast, or three lamps on separate phases of a three phase supply.*

Motor Calculations

Module 20309

Electrical Trainee Task Module 20309

MOTOR CALCULATIONS

Objectives

Upon completion of this module, the trainee will be able to:

1. Size branch circuits and feeders for electric motors.
2. Size and select overcurrent protective devices for motors.
3. Size and select overload relays for electric motors.
4. Install overcurrent and overload protective devices for electric motors.
5. Calculate and install devices to improve the power factor at motor locations.
6. Size motor short-circuit protectors.
7. Size multimotor branch circuits.
8. Size motor disconnects.
9. Protect motor circuits with transformers.
10. Perform work on electric motors according to NEC requirements.

Prerequisites

Successful completion of the following Task Modules is required before beginning study of this Task Module: Core Curricula, Electrical Levels 1 and 2, Electrical Level 3, Modules 20301 through 20308.

Required Student Materials

1. Trainee Task Module
2. Copy of the latest edition of the National Electrical Code

COURSE MAP INFORMATION

This course map shows all of the *Wheels of Learning* Task Modules in the third level of the Electrical curricula. The suggested training order begins at the bottom and proceeds up. Skill levels increase as a trainee advances on the course map. The training order may be adjusted by the local Training Program Sponsor.

Course Map: Electrical, Level 3

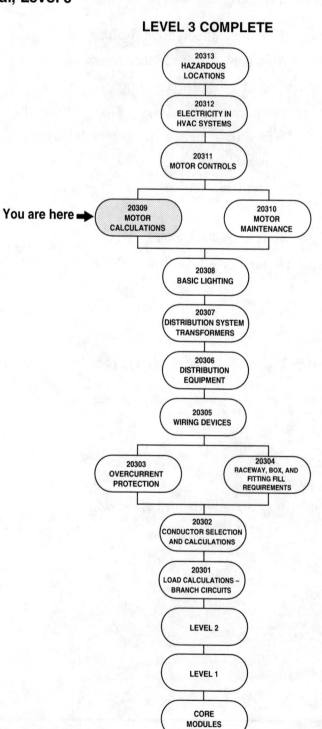

LEVEL 3 COMPLETE

20313
HAZARDOUS
LOCATIONS

20312
ELECTRICITY IN
HVAC SYSTEMS

20311
MOTOR CONTROLS

You are here ➤ 20309
MOTOR
CALCULATIONS

20310
MOTOR
MAINTENANCE

20308
BASIC LIGHTING

20307
DISTRIBUTION SYSTEM
TRANSFORMERS

20306
DISTRIBUTION
EQUIPMENT

20305
WIRING DEVICES

20303
OVERCURRENT
PROTECTION

20304
RACEWAY, BOX, AND
FITTING FILL
REQUIREMENTS

20302
CONDUCTOR SELECTION
AND CALCULATIONS

20301
LOAD CALCULATIONS –
BRANCH CIRCUITS

LEVEL 2

LEVEL 1

CORE
MODULES

CMAP309.EPS

TABLE OF CONTENTS

ambient temperature compensated: A device, such as an overload relay, which is not affected by the temperature surrounding it.

block diagram: A diagram showing the relationship of separate sub-units (blocks) in the control system.

circuit interrupter: A non-automatic manually-operated device designed to open, under abnormal conditions, a current-carrying circuit without injury to itself.

compelling circuit: A control circuit which requires correct sequencing of starting operations. Usually consists of one or more control relays with interlock circuits.

pick-up voltage or current: The voltage or current at which the device starts to operate.

polyphase: More than one phase, usually three-phase.

rating: A designated limit of operating characteristics based on definite conditions. Such operating characteristics as load, voltage, frequency, etc., may be given in the rating.

rating, continuous: The rating which defines the substantially constant load which can be carried for an indefinitely long time.

sealing, voltage or current: The voltage or current which is necessary to seat the armature of a magnetic circuit closing device to the position at which the contacts first touch each other.

almost like pick up voltage

** **service factor:** The number by which the horsepower rating is multiplied to determine the maximum safe load that a motor may be expected to carry continuously at its rated voltage and frequency.

starter, wye-delta: A starter that connects the motor leads in a wye configuration for reduced voltage starting and reconnects the leads in a delta configuration for run.

terminal: A point at which an electrical component may be connected to another electrical component.

torque: A force which produces or tends to produce rotation. Common units of measurement of torque are feet-pound, inch-pound, feet-ounce, and inch-ounce. A force of one pound applied to the handle of a crank, the center of which is displaced one foot from the center of the shaft, produces a torque of one pound-foot on the shaft, if the force is provided perpendicular and not along the crank.

torque, locked rotor: The minimum torque which a motor will develop at a standstill for all angular positions of the rotor, with rated voltage applied at rated frequency at a winding temperature of 25°C + 5°C. Also called torque or breakaway.

Electric motors have long been the workhorses of practically every kind of installation from residential appliances to heavy industrial machines. Many types of motors are available from small shaded-pole motors (used mostly in household fans) to huge synchronous motors for use in large industrial installations. There are several types in between to fill every conceivable niche. None, however, have the wide application possibilities of the three-phase motor. This is the type of motor that electricians will work with the most. Therefore, the majority of the material in this module will deal with three-phase motors.

There are three basic types of three-phase motors:

- The squirrel cage induction motor.
- The wound rotor induction motor.
- The synchronous motor.

The type of three-phase motor is determined by the rotor or rotating member (See Fig. 1). The stator winding for any of these motors is the same.

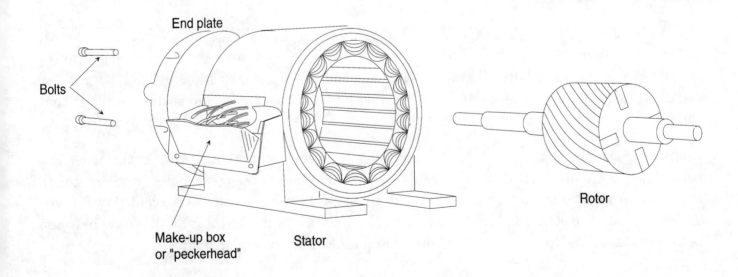

Figure 1. Basic parts of a three-phase motor.

The principle of operation for all three-phase motors is the rotating magnetic field. There are three factors that cause the magnetic field to rotate:

- The voltages of a three-phase electrical system are 120° out of phase with each other.
- The three voltages change polarity at regular intervals.
- The arrangement of the stator windings around the inside of the motor.

All three of these items will be covered in this module.

The NEC also plays an important role in the installation of electric motors. NEC Article 430 covers application and installation of motor circuits and motor control connections—including conductors, short-circuit and ground-fault protection, controllers, disconnects, and overload protection.

NEC Article 440 contains provisions for motor-driven air conditioning and refrigerating equipment—including the branch circuits and controllers for the equipment. It also takes into account the special considerations involved with sealed (hermetic-type) motor compressors, in which the motor operates under the cooling effect of the refrigeration. In referring to NEC Article 440, be aware that the rules in this NEC Article are *in addition to*, or are *amendments to*, the rules given in NEC Article 430.

Motors are also covered to some degree in NEC Articles 422 and 424.

2.0.0 MOTOR BASICS

The rotor of an AC squirrel-cage induction motor (Fig. 2) consists of a structure of steel laminations mounted on a shaft. Embedded in the rotor is the rotor winding, which is a series of copper or aluminum bars, short-circuited at each end by a metallic end ring. The stator consists of steel laminations mounted in a frame. Slots in the stator hold stator windings that can be either copper or aluminum wire coils or bars. These are connected to form a circuit.

Energizing the stator coils with an AC supply voltage causes current to flow in the coils. The current produces an electromagnetic field that, in turn, causes magnetic poles to be created in the stator iron. The strength and polarity of these poles vary as the AC current flows in one direction, then the other. This change causes the poles around the stator to alternate between being south and north poles, in effect producing a rotating magnetic field.

The rotating magnetic field cuts through the rotor, inducing a current in the rotor bars. This induced current only circulates in the rotor, which in turn causes a rotor magnetic field. As with two conventional bar magnets, the north pole of the rotor field attempts to line up with the south pole of the stator magnetic field, and the south pole to line up with the north pole.

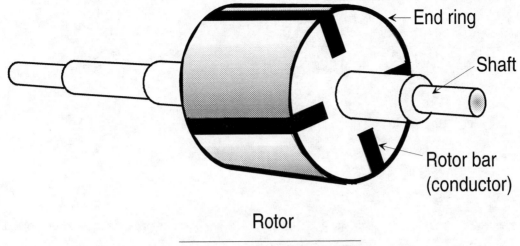

End ring

Shaft

Rotor bar
(conductor)

Rotor

Figure 2. Squirrel-cage rotor.

However, because the stator magnetic field is rotating, the rotor "chases" the stator field. The rotor field never quite catches up due to the need to furnish torque to the mechanical load.

2.1.0 SYNCHRONOUS SPEED

The speed at which the magnetic field rotates is known as the *synchronous* speed. The synchronous speed of a three-phase motor is determined by two factors:

- The number of stator poles.
- The frequency of the AC line.

Since 60 Hz is the standard frequency throughout the United States and Canada, the following gives the synchronous speeds for motors with different numbers of poles.

Hz in US = 60 Hz

$$RPM = \frac{f \cdot 120}{P}$$

# of poles

2 Poles	3600 RPM
4 Poles	1800 RPM
6 Poles	1200 RPM
8 Poles	900 RPM

From the above, the RPM of any three-phase, 60 Hz motor can be determined by counting the number of poles in the stator.

2.2.0 STATOR WINDINGS

The stator windings of three-phase motors are connected in either wye or delta. See Fig. 3. Some motor stators are designed to operate both ways; that is, some motors are started as a wye-connected motor to help reduce starting current, and then changed to a delta connection for running.

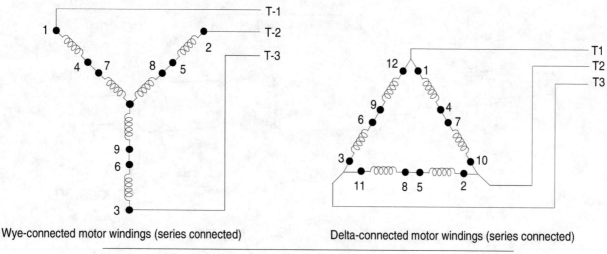

Wye-connected motor windings (series connected)   Delta-connected motor windings (series connected)

Figure 3. The two types of windings found in three-phase motors.

Many three-phase motors have dual-voltage stators. These stators are designed to be connected to, say, 240 volts or 480 volts. The leads of a dual-voltage stator use a standard numbering system. Figure 4 shows a dual-voltage wye-connected stator. Note that the 9 motor leads are numbered in a spiral. For use on the higher voltage, the leads are connected in series; for the lower voltage, the leads are connected in parallel. Therefore, for the higher voltage, leads 4 and 7, 5 and 8, and 6 and 9 are connected together. For the lower voltage, leads 4, 5, and 6 are connected together; further connections are 1 and 7, 2 and 8, and 3 and 9. These latter connections are then connected to the three-phase power source. Figure 5 shows the equivalent parallel circuit when the motor is connected for use on the lower voltage.

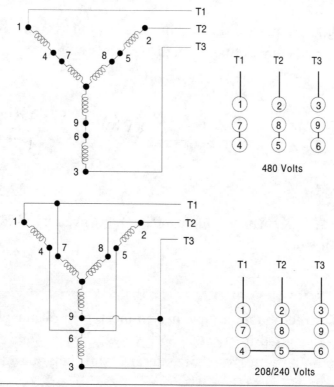

Figure 4. High- and low-voltage connections for wye-connected three-phase motors.

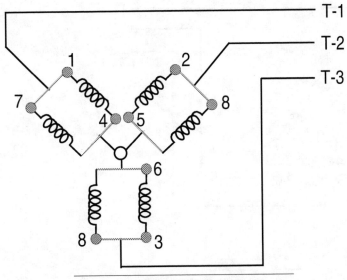

Figure 5. Equivalent parallel circuit.

The same standard numbering system is used for delta-connected motors, and many delta-wound motors also have 9 leads as shown in Fig. 6. However, there are only three circuits of three leads each. The high- and low-voltage connections for a three-phase, delta-wound, 9-lead, dual-voltage motor is shown in Fig. 7.

In some instances, a dual-voltage motor connected in delta will have 12 leads instead of 9. Figure 8 shows the high-voltage and low-voltage connections for dual-voltage, 12-lead, delta-wound motors.

2.2.1 Principles of Dual-Voltage Connections

When a motor is operated at 240 volts, the current draw of the motor is double the current draw of a 480-volt connection. For example, if a motor draws 10 amperes of current when

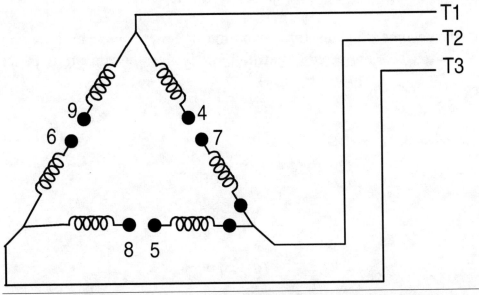

Figure 6. Arrangement of leads in a 9-lead, delta-wound, dual-voltage motor.

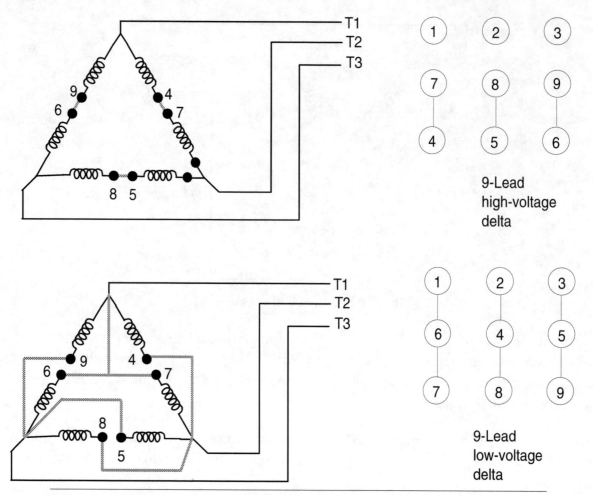

Figure 7. Lead connections for a three-phase, dual-voltage, delta-wound motor.

connected to 240 volts, it will draw only 5 amperes when connected to 480 volts. The reason for this is the difference of impedance in the windings between a 240-volt connection and a 480-volt connection. Remember that the low-voltage windings are always connected in parallel, while the high-voltage windings are connected in series.

For instance, let's assume that the stator windings of a motor have an impedance of 48 ohms. If the stator windings are connected in parallel, the total impedance may be found as follows:

$$Rt = \frac{R1 \times R2}{R1 + R2}$$

$$Rt = \frac{48 \times 48}{48 + 48}$$

$$Rt = \frac{2304}{96}$$

$$Rt = 24 \ ohms$$

ELECTRICAL TRAINEE TASK MODULE 20309

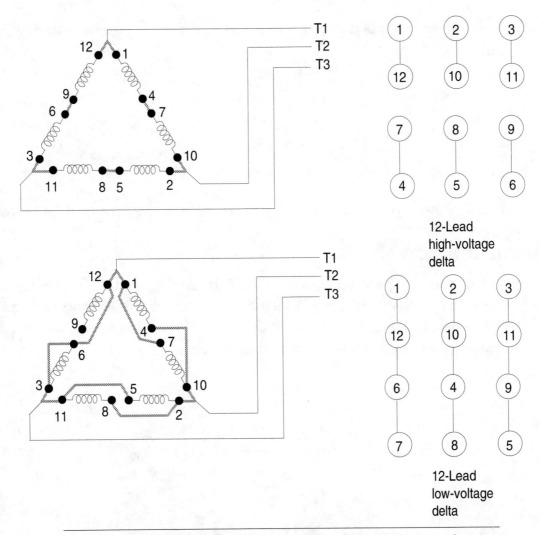

Figure 8. Connections for a 12-lead, dual-voltage, delta-wound motor.

Therefore, the total impedance of the motor winding connected in parallel is 24 ohms, and if 240 volts is applied to this connection, the following current will flow:

$$I = \frac{E}{R}$$

$$I = \frac{240}{24}$$

$$I = 10 \text{ amperes}$$

If the windings are connected in series for operation on 480 volts, the total impedance of the winding is:

$$Rt = R1 + R2$$

$$Rt = 48 + 48$$

$$Rt = 96 \text{ ohms}$$

Consequently, if 480 volts is applied to this winding, the following current will flow:

$$I = \frac{E}{R}$$

$$I = \frac{480}{96}$$

$$I = 5 \; amperes$$

From the above, it is obvious that twice the voltage means half the current flow, or vice-versa.

2.3.0   SPECIAL CONNECTIONS

Some three-phase motors designed for operation on voltages higher than 600 volts may have more than 9 or 12 leads. Motors with 15 or 18 leads are common in high-voltage installations. A 15-lead motor has 3 coils per phase as shown in Fig. 9. Notice that the leads are numbered in the same spiral sequence as a 9-lead, wye-wound motor.

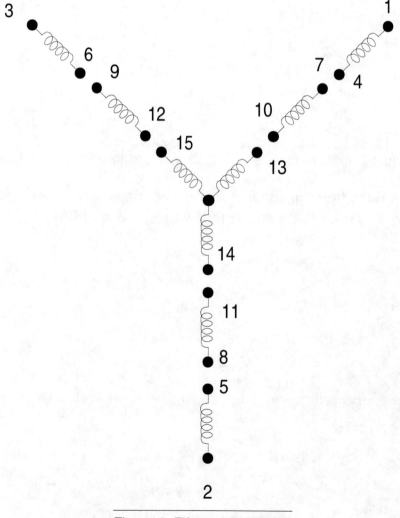

Figure 9. Fifteen-lead motor.

ELECTRICAL TRAINEE TASK MODULE 20309

3.0.0 CALCULATING MOTOR CIRCUIT CONDUCTORS

The basic elements that must be accounted for in any motor circuit are shown in Fig. 10. Although these elements are shown separately in this illustration, there are certain cases where the NEC permits a single device to serve more than one function. For example, in some cases, one switch can serve as both the disconnecting means and controller. In other cases, short-circuit protection and overload protection can be combined in a single circuit breaker or set of fuses.

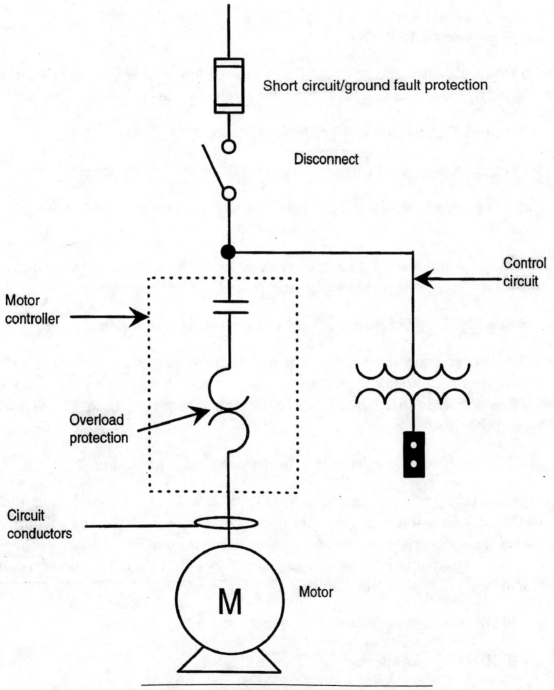

Figure 10. Basic elements of any motor circuit.

Note: The basic NEC rule for sizing conductors supplying a single-speed motor used for continuous duty specifies that conductors must have a current-carrying capacity of not less than 125% of the motor full-load current rating [NEC Section 430-22(a)].

Conductors on the line side of the controller supplying multi-speed motors must be based on the highest of the full-load current ratings shown on the motor nameplate.

Conductors between the controller and the motor must have a current-carrying rating based on the current rating for the speed of the motor each set of conductors is feeding.

A typical motor-control center and branch circuits feeding four different motors are shown in Fig. 11. Let's see how the feeder and branch-circuit conductors are sized for these motors.

Step 1.   Refer to NEC Table 430-150 for the full-load current of each motor.

Step 2.   Determine the full-load current of the largest motor in the group.

Step 3.   Calculate the sum of the full-load current ratings for the remaining motors in the group.

Step 4.   Multiply the full-load current of the largest motor by 1.25 (125%) and then add the sum of the remaining motors to your answer (NEC Section 430-24).

Step 5.   The combined total of Step 4 will give the *minimum* feeder size.

When sizing feeder conductors for motors, be aware that the procedure described in the five steps above will give the *minimum* conductor rating based on temperature rise only. Consequently, it is often necessary to increase the size of conductors to compensate for voltage drop and power loss in the circuit.

Now let's complete the conductor calculations for the motor circuits in Fig. 11.

Step 1.   A partial list of the motors listed in NEC Table 430-150 is shown in Fig. 12. Referring to this table, the motor horsepower is shown in the very left-hand column. Follow across the appropriate row until you come to the column titled "460V"—the voltage of the motor circuits in question. In doing so, we find that the ampere ratings for the motors in question are as follows:

> 50 HP   =   65 amperes
>
> 40 HP   =   52 amperes
>
> 10 HP   =   14 amperes

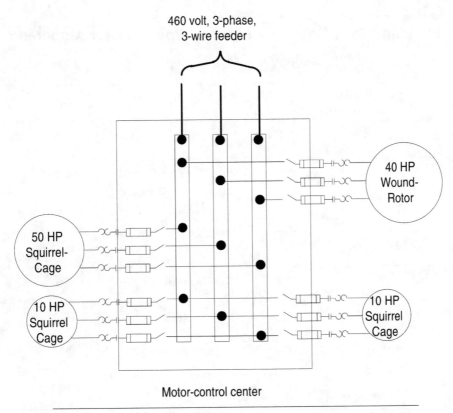

460 volt, 3-phase,
3-wire feeder

40 HP
Wound-
Rotor

50 HP
Squirrel-
Cage

10 HP
Squirrel
Cage

10 HP
Squirrel
Cage

Motor-control center

Figure 11. Motor branch circuits from motor-control center.

Step 2.   The largest motor in this group is the 50 HP squirrel-cage motor which has a full-load current of 65 amperes.

Step 3.   The sum of the remaining motors is as follows:

40 HP   =   52 amperes

10 HP   =   14 amperes

10 HP   =   14 amperes

80 amperes

(× 1.25)

Step 4.   Multiplying the full-load current of the largest motor and then adding the total amperage of the remaining motors results in the following:

(1.25)(65) + 80 = 161.25 amperes

Step 5.   Therefore, the minimum feeder size for the 460V, 3-phase, 3-wire motor control center will be 161.25 amperes. Referring to NEC Table 310-16, under the column headed "90°C.," the closest conductor size is 1/0 copper (rated at 170 amperes) or 3/0 aluminum (rated at 175 amperes).

Induction Type Squirrel-Cage and Wound-Rotor Amperes

HP	115V	200V	208V	230V	460V
½	4.4	2.5	2.4	2.2	1.1
¾	6.4	3.7	3.5	3.2	1.6
1	8.4	4.8	4.6	4.2	2.1
1½	12.0	6.9	6.6	6.0	3.0
2	13.6	7.8	7.5	6.8	3.4
3	-	11.0	10.6	9.6	4.8
5	-	17.5	16.7	15.2	7.6
7½	-	25.3	24.2	22	11
10	-	32.2	30.8	28	14
15	-	48.3	46.2	42	21
20	-	62.1	59.4	54	27
25	-	78.2	74.8	68	34
30	-	92	88	80	40
40	-	120	114	104	52
50		150	143	130	65

Figure 12. NEC Table 430-150 (partial).

The branch-circuit conductors feeding the individual motors are calculated somewhat differently. NEC Section 430-22(a) requires that the ampacity of branch-circuit conductors supplying a single continuous-duty motor must not be less than 125% of the motor's full-load current rating. Therefore, the current-carrying capacity of the branch-circuit conductors feeding the four motors in question are calculated as follows:

- 50 HP motor = 65 amperes × 1.25 = 81.25 amperes
- 40 HP motor = 52 amperes × 1.25 = 65.00 amperes
- 10 HP motor = 14 amperes × 1.25 = 17.5 amperes

Referring to NEC Table 310-16, the closest size 90°C THHN copper conductors that will be permitted to be used on these various branch circuits are as follows:

- 50 HP motor = 81.25 amperes requires No. 4 AWG THHN conductors.
- 40 HP motor = 65 amperes requires No. 6 AWG THHN conductors.
- 10 HP motor = 17.5 amperes requires No. 12 AWG THHN conductors.

Refer to Fig. 13 for a summary of the conductors used to feed the motor-control center in question, along with the branch-circuits supplying the individual motors.

If voltage drop and/or power loss must be taken into consideration, please refer to Task Module 20302, Conductor Selection and Calculations.

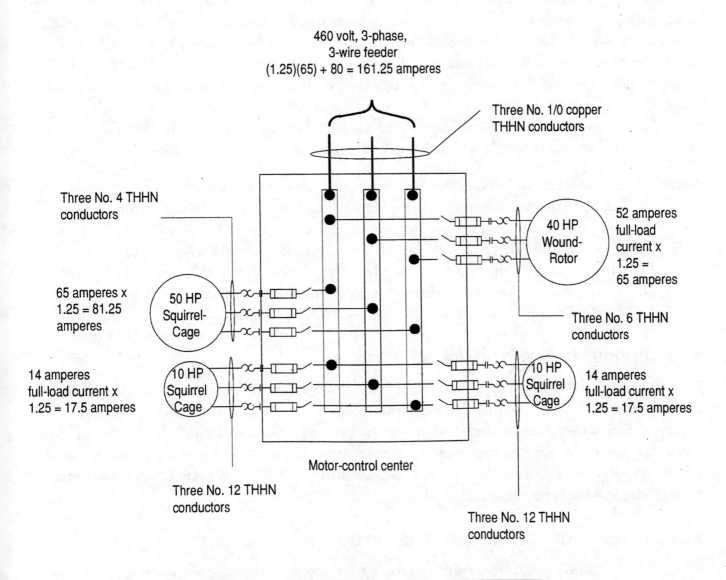

Figure 13. Sized branch-circuit and feeder conductors.

For motors with other voltages (up to 2300 volts) or for synchronous motors, refer to NEC Table 430-150.

In accordance with NEC Section 430-22(a) Exception 1, branch-circuit conductors serving motors used for short-time, intermittent, or other varying duty, must have an ampacity not less than the percentage of the motor nameplate current rating shown in NEC Table 430-22(a), Exception. However, to qualify as a short-time, intermittent motor, the nature of the apparatus that the motor drives must be arranged so that the motor cannot operate continuously with load under any condition of use. Otherwise, the motor must be considered continuous duty. Consequently, the majority of motors encountered in the electrical trade must be rated for continuous duty, and the branch-circuit conductors sized accordingly.

3.1.0 WOUND-ROTOR MOTORS

The primary full-load current of wound-rotor motors is listed in NEC Table 430-150 and is the same as squirrel-cage motors. Conductors connecting the secondary leads of wound-rotor induction motors to their controllers must have a current-carrying capacity at least equal to 125% of the motor's full-load secondary current if the motor is used for continuous duty. If the motor is used for less than continuous duty, the conductors must have a current-carrying capacity of not less than the percentage of the full-load secondary nameplate current given in NEC Table 430-22(a). Conductors from the controller of a wound-rotor induction motor to its starting resistors must have an ampacity in accordance with NEC Table 430-22(c).

Note: NEC Section 430-6 specifies that for general motor applications (excluding applications of torque motors and sealed hermetic-type refrigeration compressor motors), the values given in NEC Tables 430-147, 430-148, 430-149, and 430-150 should be used instead of the actual current rating marked on the motor nameplate when sizing conductors, switches, and overcurrent protection. Overload protection, however, is based on the marked motor nameplate.

3.2.0 CONDUCTORS FOR DC MOTORS

NEC Sections 430-22(a) Exception 2 and 430-29 cover the rules governing the sizing of conductors from a power source to a DC motor controller and from the controller to separate resistors for power accelerating and dynamic braking. Section 430-29, with its table of conductor ampacity percentages, assures proper application of DC constant-potential motor controls and power resistors. However, when selecting overload protection, the actual motor nameplate current rating must be used.

3.3.0 CONDUCTORS FOR MISCELLANEOUS MOTOR APPLICATIONS

NEC Section 430-6 should be referred to for torque motors, shaded-pole motors, permanent split-capacitor motors and AC adjustable-voltage motors.

NEC Section 430-6(b) specifically states that the motor's nameplate full-load current rating is used to size ground-fault protection for a torque motor. However, branch-circuit conductors and overcurrent protection is sized by the provisions listed in NEC Section 430-52(b) and the full-load current rating listed in NEC Tables 430-147 through 430-150 are used instead of the motor's nameplate rating.

For sealed (hermetic-type) refrigeration compressor motors, the actual nameplate full-load running current of the motor must be used in determining the current rating of the disconnecting means, the controller, branch-circuit conductor, overcurrent-protective devices, and motor overload protection.

4.0.0 MOTOR PROTECTIVE DEVICES

The NEC Section 430-51 through 430-58 requires that branch-circuit protection for motor controls must protect the circuit conductors, the control apparatus, and the motor itself against overcurrent due to short circuits or ground faults.

Motors and motor circuits have unique operating characteristics and circuit components. Therefore, these circuits must be dealt with differently from other types of loads. Generally, two levels of overcurrent protection are required for motor branch circuits:

- Overload protection—Motor running overload protection is intended to protect the system components and motor from damaging overload currents.
- Short-circuit protection (includes ground-fault protection)—Short-circuit protection is intended to protect the motor circuit components such as the conductors, switches, controllers, overload relays, motor, etc. against short-circuit currents or grounds. This level of protection is commonly referred to as motor branch-circuit protection applications. Dual-element fuses are designed to provide this protection provided they are sized correctly.

There are a variety of ways to protect a motor circuit—depending upon the user's objective. The ampere rating of a fuse selected for motor protection depends on whether the fuse is of the dual-element time-delay type or the non-time-delay type.

In general, NEC Table 430-152 specifies that short circuit/ground fault protection non-time-delay fuses can be sized at 300% of the motor full-load current for ordinary motors, while wound rotor or direct current motors may be sized 150% of the motor full-load current. Consequently, the sizes of non-time-delay fuses for the four motors previously mentioned are listed in Fig. 14. Because none of these sizes are standard, NEC Section 430-52(c) Exception 1 permits the size of the fuses to be increased to a standard size. However, where absolutely necessary to permit motor starting, the overcurrent device may be increased, but never more than 400% of the full-load current [NEC Section 430-52(c) Exception 2.a.]. In actual practice, most electricians would use a 200-ampere non-time-delay fuse for the 50 HP motor; 175-ampere fuse for the 40 HP motor, and 45-ampere fuses for the 10 HP motors. If any of these fuses do not allow

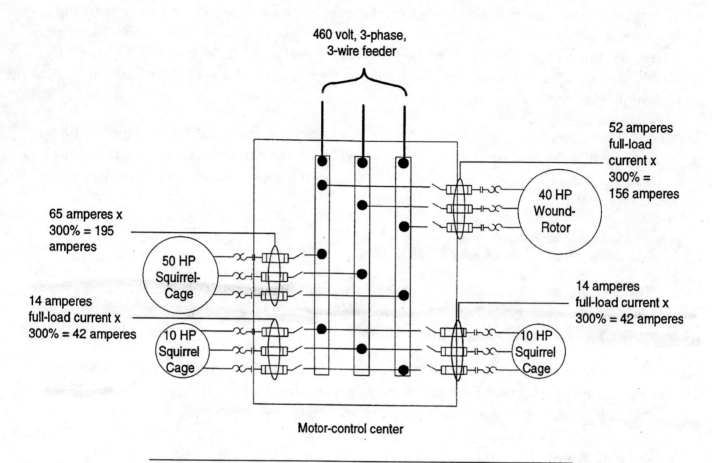

460 volt, 3-phase,
3-wire feeder

52 amperes
full-load
current x
300% =
156 amperes

40 HP
Wound-
Rotor

65 amperes x
300% = 195
amperes

50 HP
Squirrel-
Cage

14 amperes
full-load current x
300% = 42 amperes

10 HP
Squirrel
Cage

14 amperes
full-load current x
300% = 42 amperes

10 HP
Squirrel
Cage

Motor-control center

Figure 14: Ratings of non-time-delay fuses for typical motor circuits.

the motor to start without "blowing," the fuses for the 50 HP motor may be increased to a maximum of 260 amperes; the 40 HP motor to 208 amperes, and the 10 HP motors to 56 amperes. Standard sizes are 250, 200, and 50 amperes, respectively.

Per NEC Table 430-152, dual-element, time-delay fuses are able to withstand normal motor-starting current and can be sized closer to the actual motor rating than can non-time-delay fuses. If necessary for proper motor operation, dual (time-delay) fuses may be sized up to 175% of the motor's full-load current for all standard motors with the exception of wound rotor and direct current motors. These motors must not have fuses sized for more than 150% of the motor's full-load current rating. Where absolutely necessary for proper operation, the rating of dual-element (time-delay) fuses may be increased, but never more than 225% of the motor's full-load current rating [NEC Section 430-52(c) Exception 2.b.]. Sizing dual-element fuses at 175% for the four motors in Fig. 14 will be:

50 HP motor: 65 amperes × 175% = 113.75 amperes

40 HP motor: 52 amperes × 175% = 91 amperes

10 HP motors: 14 amperes × 175% = 24.5 amperes

The table in Fig. 15 gives generalized fuse application guidelines for motor branch circuits (NEC Article 430, Part C). In using this table, bear in mind that in many cases the maximum fuse size depends on the type of motor design letter, motor type, and starting method.

Type of Motor	Dual-Element, Time-Delay Fuses			Non-Time-Delay Fuses
	Desired Level of Protection			
	Motor Overload and Short-Circuit	Backup Overload and Short-Circuit	Short-Circuit Only (Based on NEC Tables 430-147 through 150 current ratings)	Short-Circuit Only (Based on NEC Tables 430-147 through 150 current ratings)
Service Factor 1.15 or Greater or 40°C Temp. Rise or Less	125% or less of motor nameplate current	125% or next standard size not to exceed 140% of motor nameplate current	150% to 175%	150% to 300%
Service Factor Less than 1.15 or Greater than 40°C Temp. Rise	115% or less of motor nameplate current	115% or next standard size not to exceed 130% of motor nameplate current	150% to 175%	150% to 300%

Fuses give overload and short-circuit protection

Overload relay gives overload protection and fuses provide backup overload protection

Overload relay provides overload protection and fuses provide only short-circuit protection

Overload relay provides overload protection and fuses provide only short-circuit protection

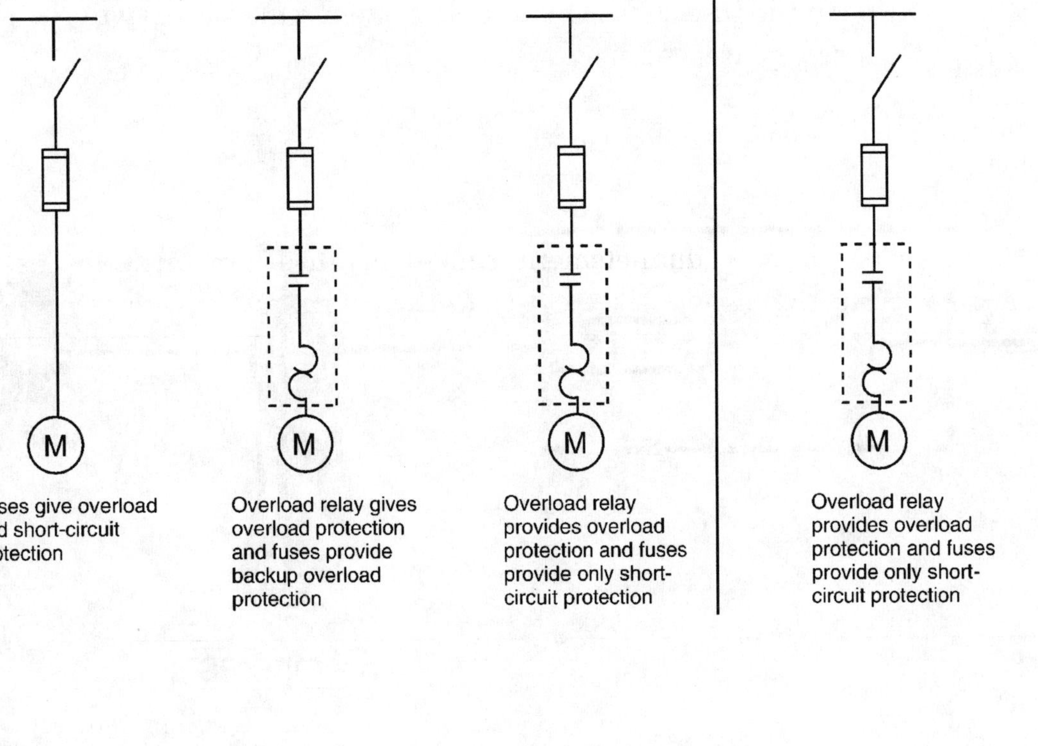

Figure 15. Sizing of fuses as a percentage of motor full-load current.

4.1.0 PRACTICAL APPLICATION

Often, for various reasons, motors are oversized for applications. For instance, a 5 HP motor is installed when the load demand is only 3 HP. In these cases a much higher degree of overload protection can be obtained by sizing the overload relay elements and/or Fusetron and Low-Peak Dual-Element Time-Delay fuses based on the actual full-load current draw. In existing installations, here's the procedure for providing the maximum overcurrent protection for oversized motors.

Step 1. With a clamp-on ammeter, determine running RMS current when the motor is at normal full-load as shown in Fig. 16. (Be sure this current does not exceed nameplate current rating). The advantage of this method is realized when a lightly loaded motor (especially those over 50 HP) experiences a single-phase condition. Even though the relays and fuses may be sized correctly based on the motor nameplate, circulating currents within the motor may cause damage.

If unable to meter the motor current, then take the current rating off the motor's nameplate.

Step 2. Size the overload relay elements and/or overcurrent protection based on this current. The table in Fig. 17 may be used to assist in sizing dual-element fuses.

Step 3. Use a labeling system to mark the type and ampere rating of the fuse that should be in the fuse clips. This simple system makes it easy to run spot checks for proper fuse replacements.

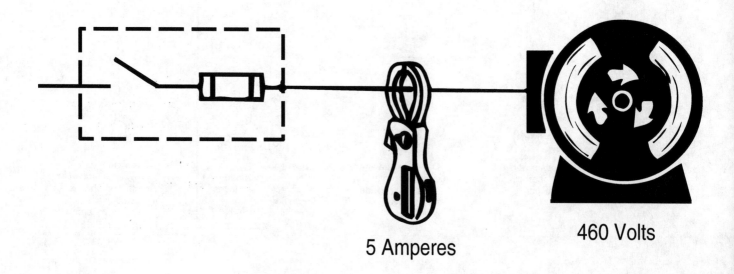

5 Amperes

460 Volts

Figure 16. Determining running RMS current with an ammeter.

Dual-Element Fuse Size	Motor Protection (Used without properly sized overload relays). Motor Full-Load Amps		Back-up Motor Protection (Used with properly sized overload relays). Motor Full-load Amps	
	Motor Service Factor of 1.15 or Greater or With Temp. Rise Not Over 40° C.	Motor Service Factor Less Than 1.15 or With Temp. Rise Not Over 40° C.	Motor Service Factor of 1.15 or Greater or With Temp. Rise Not Over 40° C.	Motor Service Factor of Less Than 1.15 or With Temp. Rise Not Over 40° C.
$\frac{1}{10}$	0.08 - 0.09	0.09 - 0.10	0 - 0.08	0 - 0.09
$\frac{1}{8}$	0.10 - 0.11	0.11 - 0.125	0.09 - 0.10	0.10 - 0.11
$\frac{5}{100}$	0.12 - 0.15	0.14 - 0.15	0.11 - 0.12	0.12 - 0.13
$\frac{2}{10}$	0.16 - 0.19	0.18 - 0.20	0.13 - 0.16	0.14 - 0.17
$\frac{1}{4}$	0.20 - 0.23	0.22 - 0.25	0.17 - 0.20	0.18 - 0.22
$\frac{3}{10}$	0.24 - 0.30	0.27 - 0.30	0.21 - 0.24	0.23 - 0.26
$\frac{4}{10}$	0.32 - 0.39	0.35 - 0.40	0.25 - 0.32	0.27 - 0.35
$\frac{1}{2}$	0.40 - 0.47	0.44 - 0.50	0.33 - 0.40	0.36 - 0.43
$\frac{6}{10}$	0.48 - 0.60	0.53 - 0.60	0.41 - 0.48	0.44 - 0.52
$\frac{8}{10}$	0.64 - 0.79	0.70 - 0.80	0.49 - 0.64	0.53 - 0.70
1	0.80 - 0.89	0.87 - 0.97	0.65 - 0.80	0.71 - 0.87
$1\frac{1}{8}$	0.90 - 0.99	0.98 - 1.08	0.81 - 0.90	0.88 - 0.98
$1\frac{1}{4}$	1.00 - 1.11	1.09 - 1.21	0.91 - 1.00	0.99 - 1.09
$1\frac{4}{10}$	1.12 - 1.19	1.22 - 1.30	1.01 - 1.12	1.10 - 1.22
$1\frac{1}{2}$	1.20 - 1.27	1.31 - 1.39	1.13 - 1.20	1.23 - 1.30
$1\frac{6}{10}$	1.28 - 1.43	1.40 - 1.56	1.21 - 1.28	1.31 - 1.39
$1\frac{8}{10}$	1.44 - 1.59	1.57 - 1.73	1.29 - 1.44	1.40 - 1.57
2	1.60 - 1.79	1.74 - 1.95	1.45 - 1.60	1.58 - 1.74
$2\frac{1}{4}$	1.80 - 1.99	1.96 - 2.17	1.61 - 1.80	1.75 - 1.96
$2\frac{1}{2}$	2.00 - 2.23	2.18 - 2.43	1.81 - 2.00	1.97 - 2.17

Figure 17. Selection of dual-element fuses for motor protection.

Dual-Element Fuse Size	Motor Protection (Used without properly sized overload relays). Motor Full-Load Amps		Back-up Motor Protection (Used with properly sized overload relays). Motor Full-load Amps	
	Motor Service Factor of 1.15 or Greater or With Temp. Rise Not Over 40° C.	Motor Service Factor Less Than 1.15 or With Temp. Rise Not Over 40° C.	Motor Service Factor of 1.15 or Greater or With Temp. Rise Not Over 40° C.	Motor Service Factor of Less Than 1.15 or With Temp. Rise Not Over 40° C.
$2^6/_{10}$	2.24 - 2.39	2.44 - 2.60	2.01 - 2.24	2.18 - 2.43
3	2.40 - 2.55	2.61 - 2.78	2.25 - 2.40	2.44 - 2.60
$3^2/_{10}$	2.56 - 2.79	2.79 - 3.04	2.41 - 2.56	2.61 - 2.78
$3^1/_2$	2.80 - 3.19	3.05 - 3.47	2.57 - 2.80	2.79 - 3.04
4	3.20 - 3.59	3.48—3.91	2.81 - 3.20	3.05 - 3.48
$4^1/_2$	3.60 - 3.99	3.92 - 4.34	3.21 - 3.60	3.49 - 3.91
5	4.00 - 4.47	4.35 - 4.86	3.61 - 4.00	3.92 - 4.35
$5^6/_{10}$	4.48 - 4.79	4.87 - 5.21	4.01 - 4.48	4.36 - 4.87
6	4.80 - 4.99	5.22 - 5.43	4.49 - 4.80	4.88 - 5.22
$6^1/_4$	5.00 - 5.59	5.44 - 6.08	4.81 - 5.00	5.23 - 5.43
7	5.60 - 5.99	6.09 - 6.52	5.01 - 5.60	5.44 - 6.09
$7^1/_2$	6.00 - 6.39	6.53 - 6.95	5.61 - 6.00	6.10 - 6.52
8	6.40 - 7.19	6.96 - 7.82	6.01 - 6.40	6.53 - 6.96
9	7.20 - 7.99	7.83 - 8.69	6.41 - 7.20	6.97 - 7.83
10	8.00 - 9.59	8.70 - 10.00	7.21 - 8.00	7.84 - 8 70
12	9.60 - 11.99	10.44 - 12.00	8.01 - 9.60	8.71 - 10.43
15	12.00 - 13.99	13.05 - 15.00	9.61 - 12.00	10.44 - 13.04
$17'/2$	14.00 - 15.99	15.22 - 17.39	12.01 - 14.00	13.05 - 15.21
20	16.00 - 19.99	17.40 - 20.00	14.01 - 16.00	15.22 - 17.39
25	20.00 - 23.99	21.74 - 25.00	16.01 - 20.00	17.40 - 21.74
30	24.00 - 27.99	26.09 - 30.00	20.01 - 24.00	21.75 - 26.09
35	28.00 - 31.99	30.44 - 34.78	24.01 - 28.00	26.10 - 30.43

Figure 17. Selection of dual-element fuses for motor protection. (*Cont.*)

Dual-Element Fuse Size	Motor Protection (Used without properly sized overload relays). Motor Full-Load Amps		Back-up Motor Protection (Used with properly sized overload relays). Motor Full-load Amps	
	Motor Service Factor of 1.15 or Greater or With Temp. Rise Not Over 40° C.	Motor Service Factor Less Than 1.15 or With Temp. Rise Not Over 40° C.	Motor Service Factor of 1.15 or Greater or With Temp. Rise Not Over 40° C.	Motor Service Factor of Less Than 1.15 or With Temp. Rise Not Over 40° C.
40	32.00 - 35.99	34.79 - 39.12	28.01 - 32.00	30.44 - 37.78
45	36.00 - 39.99	39.13 - 43.47	32.01 - 36.00	37.79 - 39.13
50	40.00 - 47.99	43.48 - 50.00	36.01 - 40.00	39.14 - 43.48
60	48.00 - 55.99	52.17 - 60.00	40.01 - 48.00	43.49 - 52.17
70	56.00 - 59.99	60.87 - 65.21	48.01 - 56.00	52.18 - 60.87
75	60.00 - 63.99	65.22 - 69.56	56.01 - 60.00	60.88 - 65.22
80	64.00 - 71 .99	69.57 - 78.25	60.01 - 64.00	65.23 - 69.57
90	72.00 - 79.99	78.26 - 86.95	64.01 - 72.00	69.58 - 78.26
100	80.00 - 87.99	86.96 - 95.64	72.01 - 80.00	78.27 - 86.96
110	88.00 - 99.99	95.65 - 108.69	80.01 - 88.00	86.97 - 95.65
125	100.00 - 119.99	108.70 - 125.00	88.01 - 100.00	95.66 - 108.70
1 50	120.00 - 139.99	131.30 - 150.00	100.01 - 1 20.00	108.71 - 30.43
175	140.00 - 159.99	152.17 - 173.90	120.01 - 140.00	130.44 - 152.17
200	160.00 - 179.99	173.91 - 195.64	140.01 - 160.00	152.18 - 173.91
225	180.00 - 199.99	195.65 - 217.38	160.01 - 180.00	173.92 - 195.62
250	200.00 - 239.99	217.39 - 250.00	180.01 - 200.00	195.63 - 217.39
300	240.00 - 279.99	260.87 - 300.00	200.01 - 240.00	217.40 - 260.87
350	280.00 - 319.99	304.35 - 347.82	240.01 - 280.00	260.88 - 304.35
400	320.00 - 359.99	347.83 - 391.29	280.01 - 320.00	304.36 - 347.83
450	360.00 - 399.99	391.30 - 434.77	320.01 - 360.00	347.84 - 391.30
500	400.00 - 479.99	434.78 - 500.00	360.01 - 400.00	391.31 - 434.78
600	480.00 - 600.00	521.74 - 600.00	400.01 - 480.00	434.79 - 521.74

Figure 17. Selection of dual-element fuses for motor protection. (*Cont.*)

HINT!   When installing the proper fuses in the switch to give the desired level of protection, it is often advisable to leave spare fuses on top of the disconnect, starter enclosure or in a cabinet adjacent to the motor-control center. In this way, should the fuses open, the problem can be corrected and the proper size of fuses readily reinstalled.

Individual motor disconnect switches must have an ampere rating of at least 115% of the motor full-load ampere rating [NEC Section 430-110(a)] or as specified in NEC Section 430-109 exceptions. The next larger size switches with fuse reducers may sometimes be required.

Abnormal installations may require dual-element fuses of a larger size than shown in the table in Fig. 17 providing only short-circuit protection. These applications include:

- Dual-element fuses in high ambient temperature environments.
- A motor started frequently or rapidly reversed.
- Motor is directly connected to a machine that cannot be brought up to full speed quickly. For example, centrifugal machines such as extractors and pulverizers, machines having large fly wheels such as large punch presses, etc.
- Motor has a design letter E with full-voltage start.

4.2.0   MOTOR OVERLOAD PROTECTION

A high quality electric motor, properly cooled and protected against overloads, can be expected to have a long life. The goal of proper motor protection is to prolong motor life and postpone the failure that ultimately takes place. Good electrical protection consists of providing both proper overload protection and current-limiting, short-circuit protection. AC motors and other types of high inrush loads require protective devices with special characteristics. Normal, full-load, running currents of motors are substantially less than the currents that result when motors start or are subjected to temporary mechanical overloads. This characteristic is illustrated by the typical motor-starting current curve shown in Fig. 18.

At the moment an AC motor circuit is energized, the starting current rapidly rises to many times normal current and the rotor begins to rotate. As the rotor accelerates and reaches running speed, the current declines to the normal running current. Thus, for a period of time, the overcurrent protective devices in the motor circuit must be able to tolerate the rather substantial temporary overload. Motor starting currents can vary substantially depending on the motor type, load type, starting methods, and other factors. For the initial first $\frac{1}{2}$ cycle, the momentary transient RMS current can be as high as 11 times or more. After this first half-cycle, the starting current subsides to 4 to 8 times (typically 6 times) the normal current for several seconds. This current is called the locked rotor current. When the motor reaches running speed, the current then subsides to its normal running level.

In summary, the special requirements for protection of motors requires that the motor overload protective device withstand the temporary overload caused by motor starting currents, and,

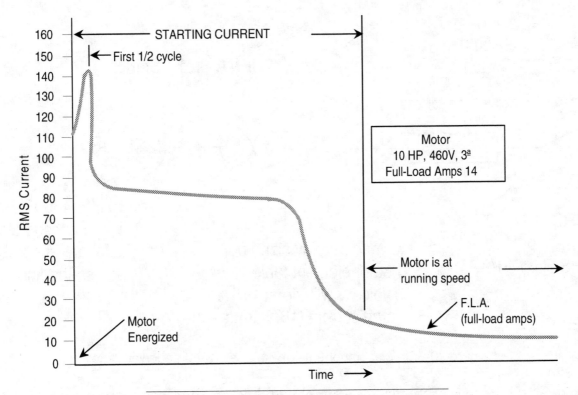

Figure 18. Motor starting-current characteristics.

at the same time, protect the motor from continuous or damaging overloads. The main types of devices used to effectively provide overload protection include:

- Overload relays
- Fuses
- Circuit breakers

There are numerous causes of overloads, but if the overload protective devices are properly responsive, such overloads can be removed before damage occurs. To insure this protection, the motor running protective devices should have time-current characteristics similar to motor damage curves but should be slightly faster. This is illustrated in Fig. 19.

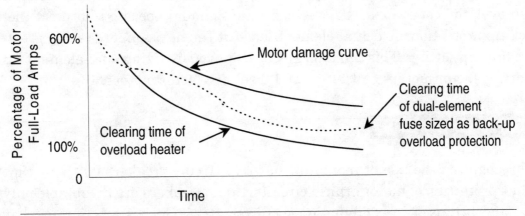

Figure 19. Clearing times of overload heaters, fuses, and motor damage curve.

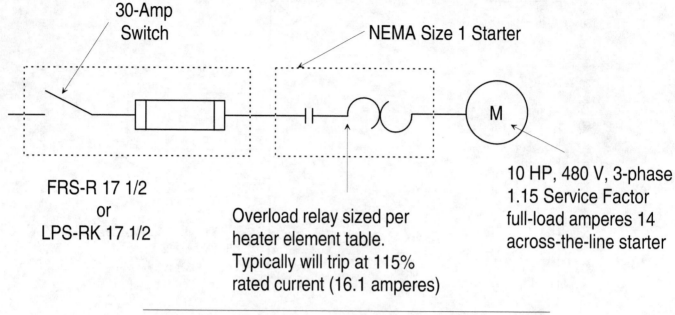

30-Amp
Switch

NEMA Size 1 Starter

FRS-R 17 1/2
or
LPS-RK 17 1/2

Overload relay sized per
heater element table.
Typically will trip at 115%
rated current (16.1 amperes)

M

10 HP, 480 V, 3-phase
1.15 Service Factor
full-load amperes 14
across-the-line starter

Figure 20. Circuit components of a typical 10-horsepower motor.

Let's take a 10 horsepower motor, for example, and determine the proper circuit components that should be employed. Refer to Fig. 20.

To begin, select the proper size overload relays. Typically, the overload relay is rated to trip at about 115% (average). The correct starter size (using NEMA standards) is a NEMA 1. The switch size that should be used is 30 amperes. Switch sizes are based on NEC requirements; dual-element, time-delay fuses allow the use of smaller switches.

For short-circuit protection on large motors with currents in excess of 600 amperes, LOW-PEAK time-delay fuses are recommended. Most motors of this size will have reduced voltage starters and the inrush currents are not as rigorous. LOW-PEAK, KRP-C fuses should be sized at approximately 150% to 175% of the motor full-load current.

Motor controllers with overload relays commonly used on motor circuits provide motor running overload protection. The overload relay setting or selection must comply with NEC Section 430-32. On overload conditions, the overload relays should operate to protect the motor. For motor back-up protection, size dual-element fuses at the next ampere rating greater than the overload relay trip setting. This can typically be achieved by sizing dual-element fuses at 125% for 1.15 service factor motors and 115% for 1.0 service factor motors.

5.0.0 CIRCUIT BREAKERS

The NEC recognizes the use of instantaneous-trip circuit breakers (without time delay) for short-circuit protection of motor branch circuits. Such breakers are acceptable only if they are adjustable and are used in combination motor starters. Such starters must have overload protection for each conductor and must be approved for the purpose. This permits the use of smaller circuit breakers than would be allowed if a standard thermal-magnetic circuit breaker was used. Smaller circuit breakers, in this case, offers faster operation for greater protection against grounds and short circuits. Fig. 21 shows a schematic diagram of magnetic-only circuit breakers used in a combination motor starter.

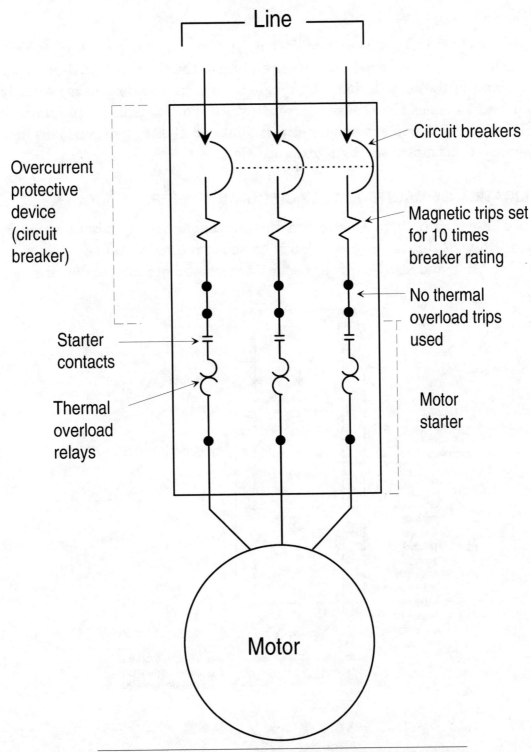

Figure 21. Listed combination starter arranged as per NEC.

The use of magnetic-only circuit breakers in motor branch circuits requires careful consideration due to the absence of overload protection up to the short-circuit trip rating that is normally available in thermal elements in circuit breakers. However, heaters in the motor starter protect the entire circuit and all equipment against overloads up to, and including, locked-rotor current. Heaters (thermal overload relays) are commonly set at 115% to 125% of the motor's full-load current.

In dealing with such circuits, an adjustable circuit breaker can be set to take over the interrupting task at currents above locked rotor and up to the short-circuit duty of the supply system at the point of the installation. The magnetic trip in such breakers typically can be adjusted from 3 to 13 times the breaker current rating. For example, a 100 ampere circuit breaker can be adjusted to trip anywhere between 300 and 1300 amperes. Consequently, the circuit breaker serves as motor short-circuit protection.

5.1.0 APPLICATION OF MAGNETIC-ONLY CIRCUIT BREAKERS

Let's compare the use of both thermal-magnetic and magnetic-only circuit breakers to the motor circuit in Fig. 22. In doing so, our job is to select a circuit breaker that will provide short-circuit protection and also qualify as the motor circuit disconnecting means.

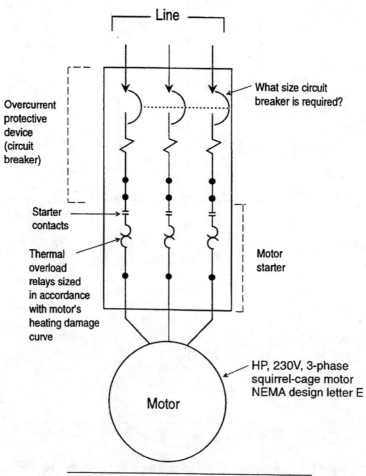

Figure 22. Typical 30 HP motor circuit.

Step 1.   Determine the motor's full-load current from NEC Table 430-150. This is found to be 80 amperes.

Step 2.   A circuit breaker suitable for use as a motor disconnecting means must have a current rating at least 115% of motor's full-load current. Thus,

$$1.15 \times 80 = 92 \text{ amperes}$$

Note: NEC Table 430-152 permits the use of an inverse-time circuit breaker rated not more than 250% of the motor full-load current. However, a circuit breaker could be rated as high as 400% of the motor's full-load current if necessary to "hold" the motor-starting current without opening.

Step 3.   From the above note, assuming that a circuit breaker rated at 250% of the motor full-load current will be used, perform the following calculation:

$$2.5 \times 80 \text{ amperes} = 200 \text{ amperes}$$

Step 4.   Select a regular thermal-magnetic circuit breaker with a 225 ampere frame and set to trip at 200 amperes.

Step 5.   Determine the initial starting current of the motor.

Step 6.   Refer to Fig. 22 and note that the NEMA design letter is E. Refer to NEC Table 430-152 and note that an instantaneous breaker for design letter E should be 1100% of the motor full-load current.

Determine circuit breaker rating by multiplying the full-load current by 1100%.

$$80 \times 1100\% \ (11.00) = 880 \text{ amperes}$$

The thermal-magnetic circuit breaker selected in step 4 will provide protection for grounds and short-circuits without interfering with motor-overload protection. Note, however, that the instantaneous trip setting of a 200 ampere circuit breaker will be about 10 times the current rating, or:

$$200 \times 10 = 2000 \text{ amperes}$$

Now consider the use of a 100-ampere circuit breaker with thermal and adjustable magnetic trips. The instantaneous trip setting at 10 times the normal current rating would be:

$$100 \times 10 = 1000 \text{ amperes}$$

Although this "1000-amperes instantaneous trip setting" is above the 880-ampere locked-rotor current of the 30 HP motor in question, starting current would probably trip the thermal element and open the circuit breaker.

This problem can be solved by removing the circuit breaker's thermal element and leaving only the magnetic element in the circuit breaker. Then the conditions of overload can be cleared by the overload devices (heaters) in the motor starter. If the setting of the instantaneous-trip circuit breaker will not hold under the starting load in NEC Section 430-52(c)(3), then Exception 1 will, under engineering evaluation, permit increasing the trip setting up to, but not exceeding, 1300% (1700% for NEMA design letter E).

Therefore, since it has been determined that the 30 HP motor in question has a full-load ampere rating of 80 amperes, the maximum trip must not be set higher than:

$$80 \times 17 = 1360 \text{ or } 1300 \text{ amperes}$$

This circuit breaker would qualify as the circuit disconnect because it has a rating higher than 115% of the motor's full-load current (92 amperes). However, the use of a magnetic-only circuit breaker does not protect against low-level grounds and short-circuits in the branch-circuit conductors on the line side of the motor starter overload relays, such an application must be made only where the circuit breaker and motor starter are installed as a combination motor starter in a single enclosure.

5.2.0 MOTOR SHORT-CIRCUIT PROTECTORS

Motor short-circuit protectors (MSCP) are fuse-like devices designed for use only in its own type of fusible-switch combination motor starter. The combination offers short-circuit protection, overload protection, disconnecting means, and motor control – all with assured coordination between the short-circuit interrupter and the overload devices.

The NEC recognizes MSCPs in NEC Section 430-40 and 430-52, provided the combination is identified for the purpose. This means that a combination motor starter equipped with an MSCP and listed by UL or another nationally recognized third-party testing lab as a package called an MSCP starter.

6.0.0 MULTIMOTOR BRANCH CIRCUITS

NEC Section 430-53(a) and 430-53(b) permits the use of more than one motor on a branch circuit provided the following conditions are met:

Two or more motors, each rated not more than 1 HP, and each drawing a full-load current not exceeding 6 amperes, may be used on a branch circuit protected at not more than 20 amperes at 125 volts or less, or 15 amperes at 600 volts or less. The rating of the branch circuit protective device marked on any of the controllers must not be exceeded. Individual overload protection is necessary in such circuits unless the motor is not permanently installed, or is manually started and is within sight from the controller location, or has sufficient winding impedance to prevent overheating due to locked-rotor current, or is part of an approved assembly which does not subject the motor to overloads and which incorporates protection for the motor against locked-rotor, or the motor cannot operate continuously under load.

Two or more motors of any rating, each having individual overload protection, may be connected to a single branch circuit that is protected by a short-circuit protective device (MSCP). The protective device must be selected in accordance with the maximum rating or setting that could protect an individual circuit to the motor of the smallest rating. This may be done only where it can be determined that the branch-circuit device so selected will not open under the most severe normal conditions of service that might be encountered. The permission of this NEC section offers wide application of more than one motor on a single circuit, particularly in the use of small integral-horsepower motors installed on 208-volt, 240-volt, and 480-volt, 3-phase industrial and commercial systems. Only such 3-phase motors have full-load operating currents low enough to permit more than one motor on circuits fed from 15-ampere protective devices.

Using these NEC rules, let's take a typical branch circuit (Fig. 23) with more than one motor connected and see how the calculations are made.

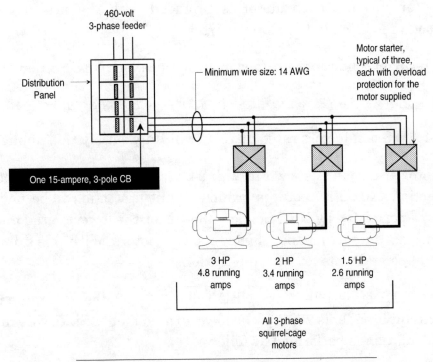

Figure 23. Several motors on one branch circuit.

Step 1. The full-load current of each motor is taken from NEC Table 430-150 as required by NEC Section 430-6(a).

Step 2. A circuit breaker must be chosen that does not exceed the maximum value of short-circuit protection (250%) required by NEC Section 430-52 and NEC Table 430-152 for the smallest motor in the group. In this case: 1.5 horsepower. Since the listed full-load current for the smallest motor (1.5 HP) is 2.6 amperes, the calculation is made as follows:

$$2.6 \text{ amperes} \times 2.5 \text{ } (250\%) = 6.5A$$

Note: NEC Section 430-52, Exception No. 1, allows the next higher size, rating or setting for a standard circuit breaker. Since a 15-ampere circuit breaker is the smallest standard rating recognized by NEC Section 240-6, a 15-ampere, 3-pole circuit breaker may be used.

Step 3. The total load of the motor currents must be calculated as follows:

$$4.8 + 3.4 + 2.6 = 10.8 \text{ amperes}$$

The total full-load current for the three motors (10.8 amperes) is well within the 15-ampere circuit breaker rating, which has sufficient time delay in its operation to permit starting of any one of these motors with the other two already operating. Torque characteristics of the loads on starting are not high. Therefore, the circuit breaker will not open under the most severe normal service.

Step 4. Make certain that each motor is provided with the properly-rated individual overload protection in the motor starter.

Step 5. Branch-circuit conductors are sized in accordance with NEC Section 430-24. In this case:

$$4.8 + 3.4 + 2.6 + (25\% \text{ the largest motor} - 4.8 \text{ amperes}) = 12 \text{ amperes.}$$

No. 14 AWG conductors rated at 75°C will fully satisfy this application.

Another multimotor situation is shown in Fig. 24. In this case, smaller motors are used. In general, NEC Section 430-53(b) requires branch-circuit protection to be no greater than the maximum amperes permitted by NEC Section 430-52 for the lowest rated motor of the group, which, in our case, is 1 ampere for the 0.5 horsepower motors. With this information in mind, let's size the circuit components for this application.

Step 1. From NEC Section 430-52 and NEC Table 430-152, the maximum protection rating for a circuit breaker is 250% of the lowest rated motor. Since this rating is 1 ampere, the calculation is performed as follows:

$$2.5 \times 1 = 2.5 \text{ amperes}$$

Note: Since 2.5 amperes is not a standard rating for a circuit breaker, according to NEC Section 240-6, NEC Section 430-52(c)(1) (Exception 1) permits the use of the next higher rating. Because 15 amperes is the lowest standard rating of circuit breakers, it is the next higher device rating above 2.5 amperes and satisfies NEC rules governing the rating of the branch-circuit protection.

These two previous applications permit the use of several motors up to the circuit capacity, based on NEC Sections 430-24 and 430-53(b) and on starting torque characteristics, operating duty cycles of the motors and their loads, and the time-delay of the circuit breaker. Such applications greatly reduce the number of circuit breakers, number of panels and the amount of wire used in the total system. One limitation, however, is placed on this practice in NEC Section 430-52(c)(2):

- Where maximum branch-circuit short-circuit and ground-fault protective device ratings are shown in the manufacturer's overload relay table for use with a motor controller or are otherwise marked on the equipment, they shall not be exceeded even if higher values are allowed as shown in the preceding examples.

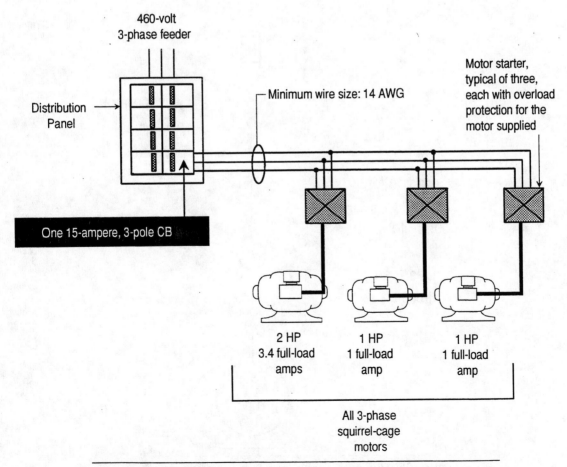

Figure 24. Several smaller motors supplied by one branch circuit.

Generally, the most effective method of power-factor correction is the installation of capacitors at the cause of the poor power factor—the induction motor. This not only increases power factor, but also releases system capacity, improves voltage stability and reduces power losses. See Task Module 20307 for a thorough explanation of power factor.

When power factor correction capacitors are used, the total corrective KVAR on the load side of the motor controller should not exceed the value required to raise the no-load power factor to unity. Corrective KVAR in excess of this value may cause over excitation that results in high transient voltages, currents and torques that can increase safety hazards to personnel and possibly damage the motor or driven equipment.

Do not connect power factor correction capacitors at motor terminals on elevator motors, multispeed motors, plugging or jogging applications or open transition, wye-delta, autotransformer starting and some part-winding start motors.

If possible, capacitors should be located at position No 2 in Fig. 25. This does not change the current flowing through motor overload protectors.

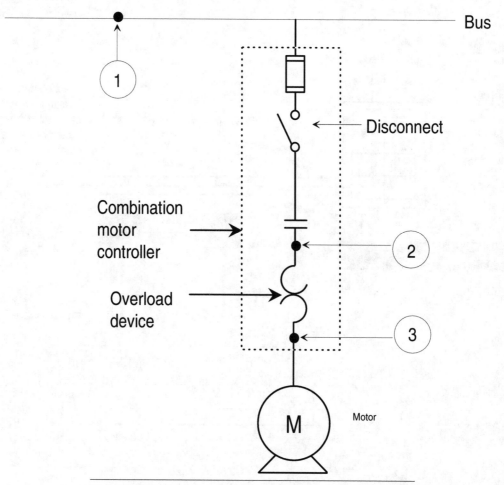

Figure 25. Placement of capacitors in motor circuits.

Connection of capacitors at position No. 3 requires a change of overload protectors. Capacitors should be located at position No. 1 for the following:

- Elevator motors
- Multispeed Motors
- Plugging or jogging applications
- Open transition, wye-delta, autotransformer starting
- Some part-winding motors

Note: Make sure that bus power factor is not increased above 95 percent under all loading conditions to avoid over excitation.

The table in Fig. 26 allows the determination of corrective kvar required where capacitors are individually connected at motor leads. These values should be considered the maximum capacitor rating when the motor and capacitor are switched as a unit. The figures given are for 3-phase, 60 Hz, NEMA Class B motors to raise full-load power factor to 95%.

Nominal Motor Speed in RPM

Induction Motor Horsepower Rating	3600		1800		1200		900		720		600	
	Capacitor Rating KVAR	Line Current Reduction %	Capacitor Rating KVAR	Line Current Reduction %	Capacitor Rating KVAR	Line Current Reduction %	Capacitor Rating KVAR	Line Current Reduction %	Capacitor Rating KVAR	Line Current Reduction %	Capacitor Rating KVAR	Line Current Reduction %
3	1.5	14	1.5	15	1.5	20	2	27	2.5	35	3.5	41
5	2	12	2	13	2	17	3	25	4	32	4.5	37
7½	2.5	11	2.5	12	3	15	4	22	5.5	30	6	34
10	3	10	3	11	3.5	14	5	21	6.5	27	7.5	31
15	4	9	4	10	5	13	6.5	18	8	23	9.5	27
20	5	9	5	10	6.5	12	7.5	16	9	21	12	25
25	6	9	6	10	7.5	11	9	15	11	20	14	23
30	7	8	7	9	9	11	10	14	12	18	16	22
40	9	8	9	9	11	10	12	13	15	16	20	20
50	12	8	11	9	13	10	15	12	19	15	24	19
60	14	8	14	8	15	10	18	11	22	15	27	19
75	17	8	16	8	18	10	21	10	26	14	32.5	18
100	22	8	21	8	25	9	27	10	32.5	13	40	17
125	27	8	26	8	30	9	32.5	10	40	13	47.5	16
150	32.5	8	30	8	35	9	37.5	10	47.5	12	52.5	15
200	40	8	37.5	8	42.5	9	47.5	10	60	12	65	14
250	50	8	45	7	52.5	8	57.5	9	70	11	77.5	13

Figure 26. Motor power-factor correction table.

Summary

The NEC plays an important role in the selection and application of motors—including branch-circuit conductors, disconnects, controller, overcurrent protection, and overload protection. For example, NEC Article 430 covers application and installation of motor circuits and motor-control connections—including conductors, short-circuit and ground-fault protection, controllers, disconnects, and overload protection.

NEC Article 440 contains provisions for motor-driven air conditioning and refrigerating equipment—including the branch circuits and controllers for the equipment. It also takes into account the special considerations involved with sealed (hermetic-type) motor compressors, in which the motor operates under the cooling effect of the refrigeration. In referring to NEC Article 440, be aware that the rules in this NEC Article are *in addition to*, or are *amendments to*, the rules given in NEC Article 430.

Anyone working with electric motors or electric-motor circuits and controllers must have a thorough knowledge of these NEC Articles.

References

For more advanced study of topics covered in this Task Module, the following works are suggested:

National Electrical Code Handbook, Latest Edition, NFPA, Quincy, MA
American Electricians Handbook, Latest Edition, Croft, McGraw-Hill, New York, NY

SELF-CHECK REVIEW/PRACTICE QUESTIONS

1. Which of the following best describes the construction of a squirrel-cage motor rotor?

   a. A structure of copper or aluminum wire coils
   b. Insulating fibers mounted on a shaft
   c. Copper bars mounted on a spindle
   d. Steel laminations mounted on a shaft

2. Which of the following best describes the construction of a motor stator?

   a. A structure of copper or aluminum wire coils
   b. Insulating fibers mounted on a shaft
   c. Copper bars mounted on a spindle
   d. Steel laminations mounted on a shaft

3. What is the speed of a 60 Hz, three-phase induction motor with two poles?

   a. 3600 RPM
   b. 1800 RPM
   c. 1200 RPM
   d. 900 RPM

4. What are the two most common three-phase motor connections?

   a. Zig and zag
   b. Scott-T and Scott-L
   c. Wye and delta
   d. Star and zig-zag

5. What is the most common number of motor leads found on a three-phase, wye-wound motor?

   a. 3
   b. 6
   c. 9
   d. 12

6. What is the total impedance of the stator windings in a three-phase motor if the windings are connected in parallel and the impedance of each winding is 96 ohms?

    a. 48 ohms
    b. 96 ohms
    c. 192 ohms
    d. 220 ohms

7. Which of the following motors is the most likely to have 15 to 18 motor leads?

    a. Single-phase capacitor-start motor
    b. 120-volt shaded-pole motor
    c. 480-volt three-phase squirrel-cage motor
    d. 2100-volt three-phase synchronous motor

8. If a squirrel-cage induction motor draws 2 amperes of current at 240 volts, what will the amperage be if connected for use on 120 volts?

    a. 1
    b. 2
    c. 3
    d. 4

9. If a three-phase wound-rotor motor, with its coils connected in parallel for use on 240 volts draws 5.2 amperes, what current will be drawn if the coils are connected in series for use on 480 volts?

    a. 1.5 amperes
    b. 2.6 amperes
    c. 3.4 amperes
    d. 11 amperes

10. What is the main purpose of motor overload protection?

    a. To protect the motor against short-circuits
    b. To protect the motor against ground-faults
    c. To protect the motor from damaging overload currents
    d. To protect the motor against locked-rotor currents

PERFORMANCE/LABORATORY EXERCISE

1. Remove the end bells from a three-phase, 60 Hz induction motor and determine its RPMs.

2. Calculate the conductor sizes and short-circuit protection required for the motor circuit in Fig. 27.

3. Calculate the conductor sizes and short-circuit protection required for the motor circuit in Fig. 28.

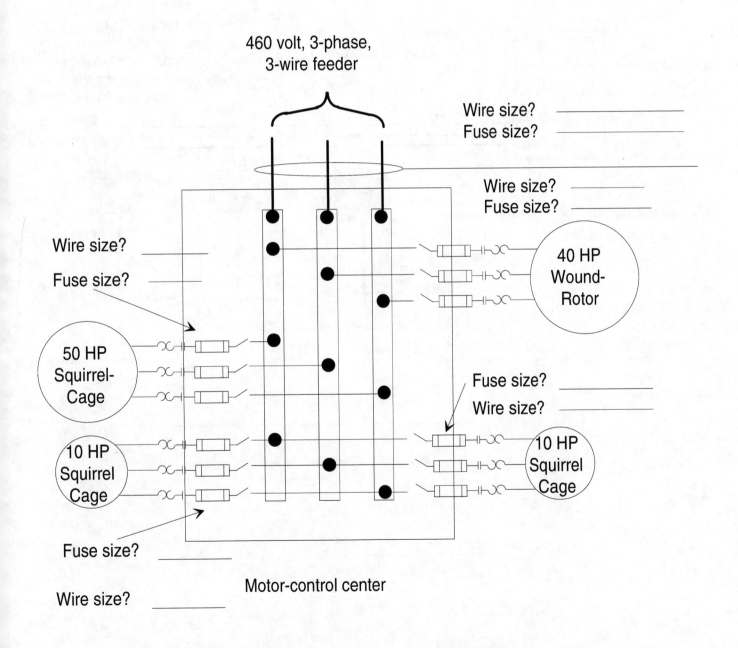

Figure 27. Motor circuit with individual branch circuits.

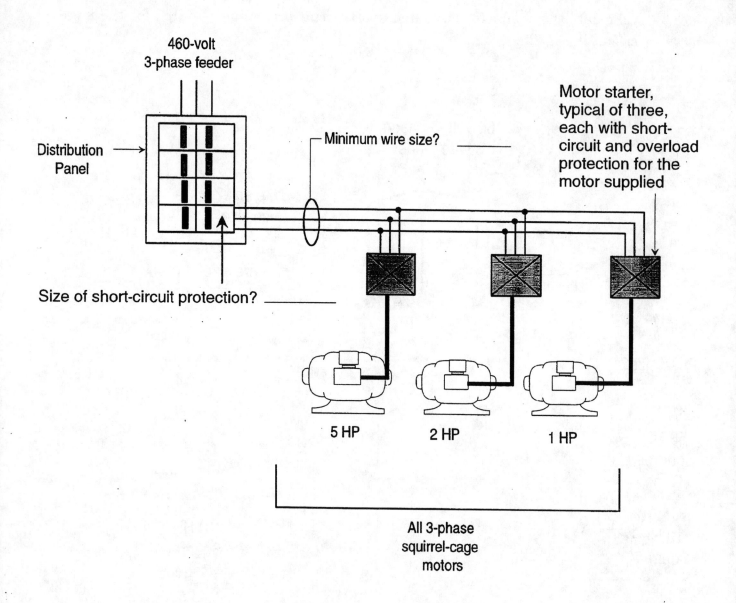

Figure 28. Several motors connected to one branch circuit.

Answers to Self-Check Questions

1. d

2. a

3. a

4. c

5. c

6. a

7. d

8. d

9. b

10. c

Motor Maintenance

Module 20310

MOTOR MAINTENANCE

Objectives

Upon completion of this module, the trainee will be able to:

1. Properly store motors and generators.
2. Test motors and generators.
3. Make connections for specific types of motors and generators.
4. Clean and dry open frame motors.
5. Lubricate motors that require this type of maintenance.
6. Collect and record motor data.
7. Select tools for motor maintenance.
8. Select instruments for motor testing.
9. Make conductor terminations and splices.

Prerequisites

Successful completion of the following Task Modules is required before beginning study of this Task Module: Core Curricula, Electrical Levels 1 and 2, Electrical Level 3, Modules 20301 through 20308.

Required Student Materials

1. Trainee Task Module
2. Copy of the latest edition of the National Electrical Code

COURSE MAP INFORMATION

This course map shows all of the *Wheels of Learning* Task Modules in the third level of the Electrical curricula. The suggested training order begins at the bottom and proceeds up. Skill levels increase as a trainee advances on the course map. The training order may be adjusted by the local Training Program Sponsor.

Course Map: Electrical, Level 3

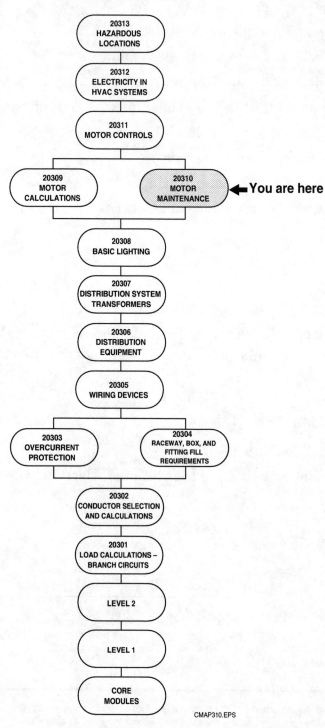

LEVEL 3 COMPLETE

20313
HAZARDOUS LOCATIONS

20312
ELECTRICITY IN HVAC SYSTEMS

20311
MOTOR CONTROLS

20309
MOTOR CALCULATIONS

20310
MOTOR MAINTENANCE ◄── You are here

20308
BASIC LIGHTING

20307
DISTRIBUTION SYSTEM TRANSFORMERS

20306
DISTRIBUTION EQUIPMENT

20305
WIRING DEVICES

20303
OVERCURRENT PROTECTION

20304
RACEWAY, BOX, AND FITTING FILL REQUIREMENTS

20302
CONDUCTOR SELECTION AND CALCULATIONS

20301
LOAD CALCULATIONS – BRANCH CIRCUITS

LEVEL 2

LEVEL 1

CORE MODULES

CMAP310.EPS

TABLE OF CONTENTS

Trade Terms Introduced in This Module

armature: 1) Rotating machine: the member in which alternating voltage is generated. 2) Electromagnetic: the member that is moved by magnetic force.

brush: A conductor between the stationary and rotating parts of a machine, usually of carbon.

brush holders: Adjustable arms for holding the commutator brushes of a generator against the commutator, feeding them forward to maintain proper contact as they wear, and permitting them to be lifted from contact when necessary.

brush loss: The loss in watts due to the resistance of the brush contact against the surface of the commutator.

commutating pole: An electromagnetic bar inserted between the pole pieces of a generator to offset the cross magnetization of the armature currents.

commutator: Device used on electric motors or generators to maintain a unidirectional current.

compound wound: A generator or motor having a part of a series-field winding wound on top of a part of a shunt-field winding on each of the main pole pieces.

direct drive: A compact arrangement of driving a generator by direct connection with the prime mover or the load, avoiding the use of shafts and belts.

drum armature: A generator or motor armature having its coils wound longitudinally or parallel to its axis.

generator: 1) A rotating machine that is used to convert mechanical to electrical energy. 2) Automotive-mechanical to direct current. 3) General-apparatus, equipment, etc. to convert or change energy from one form to another.

interpole: A small field pole placed between the main field poles and electrically connected in series with the armature of an electric rotating machine.

shunt-wound motor: Used when the motor speed must be constant, irrespective of variation in load.

slip: The difference between the speed of a rotating magnetic field and the speed of its rotor.

slip rings: The means by which the current is conducted to a revolving electrical circuit.

starting winding: Winding in an electric motor used only during the brief period when the motor is starting.

AC motor failure accounts for a high percentage of electrical repair work. The care given to an electric motor while it is being stored, and also while operating, affects the life and usefulness of the motor. A motor that received good maintenance practices will outlast a poorly treated motor many times over. Actually, if a motor is initially installed correctly and it has been properly selected for the job, very little maintenance is necessary—provided it does receive a little care at regular intervals. The basic care consists of:

- Cleanliness
- Lubrication

Cleanliness: The frequency of cleaning motors will depend on the type of environment in which it is used. In general, keep both the interior and exterior of the motor free from dirt, water, oil, and grease. Motors operating in dirty areas should be periodically disassembled and thoroughly cleaned.

If the motor is totally enclosed—fan-cooled or nonventilated—and is equipped with automatic drain plugs as shown in Fig. 1, they should be free of oil, grease, paint, grit, and dirt so they do not clog up.

Lubrication: Most motors are properly lubricated at the time of manufacture, and it is not necessary to lubricate them at the time of installation. However, if a motor has been in storage for a period of 6 months or longer, it should be relubricated before starting.

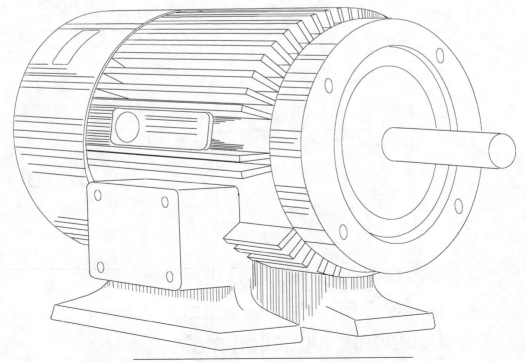

Figure 1. Totally enclosed, fan-cooled motor.

To lubricate conventional motors:

Step 1.   Stop the motor.

Step 2.   Wipe clean all grease fittings (filler and drain).

Step 3.   Remove the filler and drain plugs (A and B in Fig. 2).

Step 4.   Free the drain hole of any hard grease (use a piece of wire if necessary).

Step 5.   Add grease using a low-pressure grease gun.

Step 6.   Start the motor and let it run for approximately 30 minutes.

Step 7.   Stop the motor, wipe off any drained grease, and replace the filler and drain plugs.

Step 8.   The motor is now ready for operation.

Every four years (every year in the case of severe duty) motors with open bearings should be thoroughly cleaned, washed, and repacked with grease. Figure 3 shows the relubrication period. Standard conditions reflect in this chart (Fig. 3) mean 8-hr/day operation, normal or light loading, and at 100°F maximum ambient. Severe conditions are 24-hr/day operation, or shock loadings, vibration, or in dirt or dust, or run in areas at 100-150°F ambient, and extreme conditions are defined as heavy shock or vibration, dirt or dust, or run at high ambients.

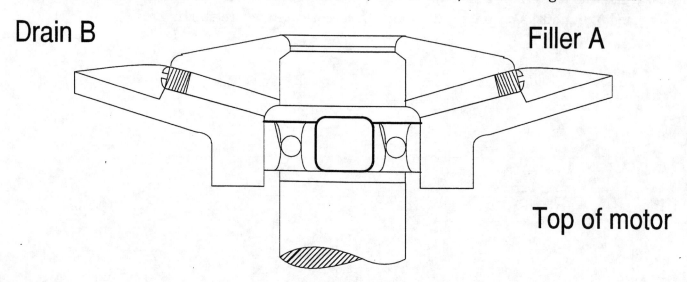

Drain B

Filler A

Top of motor

Location of filler and drain plugs in end-bell bearings of standard induction motor

Figure 2. Location of motor filler and drain plugs.

Frame Size @ 900, 1200 & Var. Speed	Relub. Period @ Std. Conditions	Severe Conditions	Extreme Conditions
140-180	4.5 Years	18 Months	9 Months
210-280	4 Years	16 Months	8 Months
320-400	3.5 Years	14 Months	7 Months
440-508	3.0 Years	12 Months	6 Months
510	2.5 Years	11½ Months	6 Months
Frame Size @ 1800 RPM	**Std. Conditions**	**Severe Conditions**	**Extreme Conditions**
140-180	3 Years	1 Year	6 Months
210-280	2.5 Years	10½ Months	5½ Months
320-400	2.0 Years	9 Months	4½ Months
440-508	1.5 Years	8 Months	4 Months
510	1 Year	6 Months	3½ Months
All Motors over 1800 RPM	6 Months	3 Months	3 Months

Figure 3. Relubrication periods for various sizes and types of motors.

The quantity of grease is important. The grease cavity should be filled one third to one half full. Always remember that too much grease is as detrimental as insufficient grease. Figure 4 shows the amount of grease required, and Fig. 5 gives the recommended greases. However, always check with the motor manufacturers for their recommendations and specifications for greasing motors.

WARNING! The amount of grease added to motor bearings is very important. Only enough grease should be added to replace the grease used by the bearing. Too much grease can be as harmful as insufficient grease.

Bearing Number	Amount in Cubic Inches	Approximate Equivalent Teaspoons
203	.15	.5
205	.27	.9
206	.34	1.1
207	.43	1.4
208	.52	1.7
209	.61	2.0
210	.72	2.4
212	.95	3.1
213	1.07	3.6
216	1.49	4.9
219	2.8	7.2
222	3.0	10.0
307	.53	1.8
308	.66	2.2
309	.81	2.7
310	.97	3.2
311	1.14	3.8
312	1.33	4.4
313	1.54	5.1
314	1.76	5.9
316	2.24	7.4
318	2.78	9.2

Figure 4. Typical amount of grease required when regreasing electric motors.

Insulation Class Shown on Nameplate	Grease Designation	Grease Supplier
B or F	Chevron SRI-2	Standard Oil of California or equivalent

Figure 5. Recommended greases for motor lubrication.

2.0.0 PRACTICAL MAINTENANCE TECHNIQUES

The key to long, trouble-free motor life—once the motor has been sized and installed properly—is proper maintenance. Maintaining a motor in good operating condition requires periodic inspection to determine if any faults exist and then promptly correcting these faults. The frequency and thoroughness of these inspections depend on such factors as the following:

- Number of hours and days the motor operates
- Importance of the motor in the production scheme
- Nature of service
- Environmental conditions

Each week every motor in operation should be inspected to see if the windings are exposed to any dripping water, acid, or alcoholic fumes as well as excessive dust, chips, or lint on or about the motor. Make certain that objects that will cause problems with the motor's ventilation system are not placed too near the motor and do not come into direct contact with the motor's moving parts.

In sleeve-bearing motors, check the oil level frequently (at least once a week) and fill the oil cups to the specified line with the recommended lubricant. If the journal diameter is less than 2 inches, always stop the motor before checking the oil level. For special lubricating systems, such as forced, flood and disc, and wool-packed lubrication, follow the manufacturer's recommendations. Oil should be added to the bearing housing only when the motor is stopped, and then a check should be made to ensure that no oil creeps along the shaft toward the windings where it may harm the insulation.

Always be alert to any unusual noise which may be caused by metal-to-metal contact (bad bearings, etc.), and also learn to detect any abnormal odor which might indicate scorching insulation varnish.

Feel the bearing housing each week for evidence of vibration and listen for any unusual noise. A standard screwdriver with the blade on the bearing housing and the handle clasped with the hand while the ear is positioned so as to rest on the cupped hand will magnify the noise. Also inspect the bearing housings for the possibility of creeping grease on the inside of the motor which might harm the insulation.

Commutators and brushes should be checked for sparking and should be observed through several cycles if the motor is on cycle duty. A stable copper oxide carbon film—as distinguished from a pure copper surface—on the commutator is an essential requirement for good commutation. Such a film, however, may vary in color from copper to straw or from chocolate brown to black. The commutator should be clean and smooth and have a high polish to prevent problems. All brushes should be checked for wear, and connections should be checked for looseness. The commutator surface may be cleaned by using a piece of dry canvas or other hard, nonlimiting material which is wound around and securely fastened to a wooden stick and held against the rotating commutator.

The air gap on sleeve-bearing motors should be checked frequently, especially if the motor has recently been rewound or otherwise repaired. After new bearings have been installed, for example, make sure that the average reading is within 10%, provided the reading should be less than .020 in. Check the air passages through punchings and make sure they are free of all foreign matter.

Compressed air may be used to blow motor windings clean, provided too much pressure is not used. Industrial-type vacuum cleaners have also been used with success. Before performing either of these cleaning operations, however, make certain that the motor is disconnected from the line. Then the windings may be wiped off with a dry cloth. In doing so, check for moisture, and see if any water has accumulated in the bottom of the motor frame. Check also to see if any oil or grease has worked its way up to the rotor or armature windings. If so, clean with AWA 1,1,1 or a similar cleaning solution.

When performing any of the preceding maintenance operations, check other motor parts and accessories such as the belt, gears, flexible couplings, chain, and sprockets for excessive wear or improper location. Also check the starter and that the motor comes up to proper speed each time it is started.

Once every month or so maintenance personnel should check the shunt, series, and commutating field windings for tightness. Do this by trying to move the field spools on the poles, as drying out may have caused some play. If this condition exists, the motor(s) should be serviced immediately. Also check the motor cable connections for rightness and tighten if necessary.

At the same time the preceding checks are made, also check the brushes in their holders for fit and free play. The brush spring pressure should also be checked. Tighten the brush studs in the holders to take up slack from the drying out of washers, making sure that studs are not displaced, particularly on dc motors. All worn or damaged brushes should be replaced at this time. Look for chipped toes or heels and for heat cracks during the inspection.

Each month examine the commutator surface for high bars and high mica or evidence of scratches or roughness. See that the risers are clean and have not been damaged in any way.

Where motors are subjected to hard use, all ball or roller bearing motors should be serviced by purging out the old grease through the drain hole and applying new grease once each month or more frequently if circumstances dictate the need. After each grease change, check to make sure grease or oil is not leaking out of the bearing housing. If so, correct this condition before starting the motor, or the insulation may be damaged.

Check the sleeve bearings for wear about six to eight times each year. Clean out oil wells if there is evidence of dirt or sludge. Flush with lighter oil before refilling.

For motors with enclosed gears, open the drain plug and check the oil flow for the presence of metal scale, sand, grit, or water. If the condition of the oil is bad, drain, flush, and refill as recommended by the manufacturer of the motor. Rock the rotor to see if slack or backlash is increasing.

Loads being driven by motors have a tendency to change from time to time due to wear on the machine or the product being processed through the machine. Therefore, all loads should be checked from time to time for a changed condition, bad adjustment, and poor handling or control.

During the monthly inspection, note if belt-tightening adjustments are all used up. If they are, the belts may be shortened. Also see if the belts run steadily and close to the inside edge of the pulley. On chain-driven machines, check the chain for evidence of wear and stretch and clean the chain thoroughly. Check the chain-lubricating system and note the incline of the slanting base to make sure it does not cause oil rings to rub on the housing.

Once or twice each year, all motors in operation should be given a thorough inspection consisting of the following:

Windings: Check the insulation resistance by using the instruments and techniques described later. The windings should also be given a visual inspection; look for dry cracks and other evidence of a need for coating insulating material. Clean all surfaces thoroughly, especially ventilating passages. Also examine the frame to see if any mold is present or if water is standing in the bottom. Either will suggest dampness and may require that the windings be dried out, varnished, and baked.

Air gap and bearings: Check the air gap to make sure that average readings are within 10% provided the readings should be less than .020 in. All bearings, ball, roller, and sleeve, should be thoroughly checked and defective ones replaced. Waste-packed and wick-oiled bearings should have waste or wicks renewed if they have become glazed or filled with metal, grit, or dirt, making sure that new waste bears well against the shaft.

Squirrel cage rotors: Check for broken parts or loose bars as well as evidence of local heating. If the fan blades are not cast in place, check for loose blades. Also look for marks on the rotor surface which indicate the presence of foreign matter in the air gap or else that the motor has a worn bearing or bearings.

Wound rotors: Wound rotors should be cleaned thoroughly, especially around collector rings, washers, and connections. Tighten all connections. If rings appear to be rough, spotted, or eccentric, they should be refinished by qualified personnel. Make certain that all top sticks or wedges are tight; tighten those which are not.

Armatures: Clean all armature air passages thoroughly. In doing so, look for oil or grease creeping along the shaft back to the bearing. Check the commutator for its surface condition, high bars, high mica, or eccentricity. If necessary, turn down the commutator to secure a smooth fresh surface. This operation is performed in a lathe of suitable size as shown in Fig. 6.

For armatures with drilled center holes, put a lathe dog on the shaft opposite the commutator and tighten it. If it is necessary to put the lathe dog on a bearing surface, put a piece of thin copper around the shaft so it will not be injured. Put a faceplate on the spindle end of the lathe along with centers in both the headstock and tailstock. Apply some white lead or oil on the tailstock center and then place the commutator between centers and tighten the tailstock, but not so tight as to spread the end of the shaft.

Use a sharp-pointed lathe cutting tool in the tool holder to turn the commutator down, running lathe at medium speed, say, around 700 rpm. Finish the job with a fine file and abrasive paper.

Armatures without drilled center holes will necessitate the use of chucks. The armature shaft opposite the commutator should be chucked in a three-jaw universal chuck in the headstock.

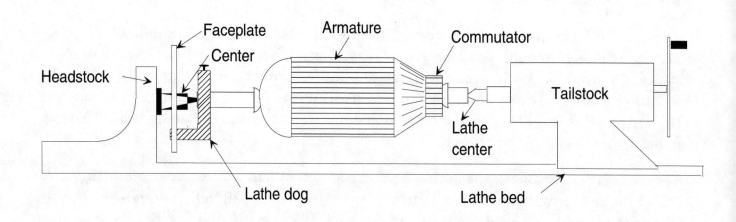

Figure 6. Motor armature secured in lathe between centers.

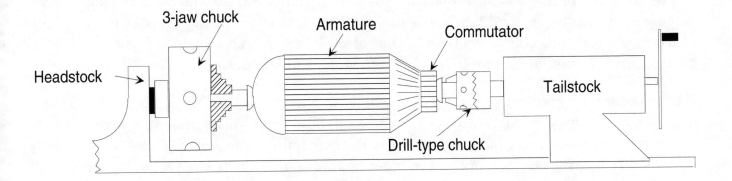

Figure 7. Motor armature secured in lathe with lathe chucks.

Chuck a bearing of proper size in a drill-type chuck and place it in the tailstock of the lathe. Oil this bearing and then place the commutator end shaft into it. See Fig. 7. The commutator is then turned in the same manner as described previously.

To summarize, when turning down an armature, always use a pointed tool with a sharp and smooth edge to obtain the cleanest cut possible. Take only a fine cut each time to prevent tearing the commutator. To finish the job, smooth down the surface with a soft file while the armature is revolving in the lathe between centers. While the armature is still turning, some workers like to polish its surface with various sizes of abrasive paper.

When the armature has been turned, clean between the bars if necessary and then test with a growler or other test instrument to determine if any shorts are present. The vibration of a hacksaw blade on any coil means that the coil is shorted at the leads or commutator. Clean between the commutator bars and test again. As soon as the armature tests okay, it is ready to put back into service.

Motor loads should be re-evaluated from time to time, as they will vary for several reasons. Use testing instruments and take an ampere reading on the motor, first with no load and then at a full load—or through an entire cycle. This should give a fair check as to the mechanical condition of the driven machine.

Without proper maintenance, no motor can be expected to perform its best for any length of time or to remain in service as long as it should. Although motor maintenance is costly, it is far less expensive than continually replacing motors or overhauling them frequently.

AC motors account for a high percentage of electrical repair work. And most of these failures can be traced to faulty bearings. Consequently, most industrial establishments place heavy emphasis on the proper handling, repair and maintenance of various types of sleeve and ball bearings. As a result, plants are finding that electric motors last longer and perform better when a carefully planned motor lubrication schedule is followed.

If ac motor failure occurs, the first step is to find out why the motor failed. There are various causes of motor breakdowns, such as excessive load, binding or misalignment of motor drives, wet or dirty surroundings and bearing failure. Bearing failure can occur in newer motors with high-quality bearings as frequently as in older motors equipped with less reliable bearings. A notable exception is motors equipped with sealed bearings are much less prone to failure.

Considering this information, particularly the fact that sealed bearings are shielded from contamination and do not require lubrication, it stands to reason that contamination of bearings is one of the major causes of bearing failure. Consequently, most plant maintenance departments are updating the lubrication methods and bearing maintenance techniques, emphasizing cleanliness in procedures for all types of motor bearings.

3.1.0 TYPES OF BEARINGS

There are many types of bearings, but ball bearings seem to be the most common. This type of bearing is found on various-size motors and their construction may be:

- Open
- Single shielded
- Double shielded
- Sealed
- Double row and other special types

Open bearings, as the name implies, are open construction and must be installed in a sealed housing. These bearings are less apt to cause churning of grease, and are therefore used mostly on large motors.

The single shield bearing has a shield on one side to preclude grease from the motor windings. Double-shielded bearings have a shield on both sides of the bearing. This type of bearing is less susceptible to contamination and, because of its design, reduces the possibility of over-greasing. Sealed bearings have, on each side of the bearing, double shields which form an excellent seal. This bearing requires no maintenance, affords protection from contamination at all times, and does not require regreasing. It is usually used on small or medium-size motors.

The largest motors usually are furnished with oil-ring sleeve bearings. And some of the fractional-horsepower motors are equipped with plain sleeve bearings.

Each bearing type has characteristics which make it the best choice for a certain application. Replacement should be made with the same type bearings. The following list of functions provide a basic understanding of bearing application, a guide to analysis of bearing troubles due to misapplication, and emphasize the importance of proper replacement.

Figure 8 shows several types of bearings used in electric motors. The following is a brief description of each:

Self-aligning ball bearings: The self-aligning ball bearing, with two rows of balls rolling on the spherical surface of the outer ring, compensates for angular misalignment resulting from errors in mounting, shaft deflection, and distortion of the foundation. It is impossible for this bearing to exert any bending influence on the shaft—a most important consideration in applications requiring extreme accuracy, at high speeds. Self-aligning ball bearings are used for radial loads and moderate thrust loads in either direction.

Single-row, deep-groove ball bearings: The single-row, deep-groove ball bearing will sustain, in addition to radial load, a substantial thrust load in either direction, even at very high speeds. This advantage results from the intimate contact existing between the balls and the deep, continuous groove in each ring. When using this type of bearing, careful alignment between the shaft and housing is essential. This bearing is also available with seals and shields, which serves to exclude dirt and retain lubricant.

Angular-contact ball bearings: The angular-contact ball bearing supports a heavy thrust load in one direction, sometimes combined with a moderate radial load. A steep contact angle, assuring the highest thrust capacity and axial rigidity, is obtained by a high thrust-supporting shoulder on the inner ring and a similar high shoulder on the opposite side of the outer ring. These bearings can be mounted singly or, when the sides are flush ground, in tandem for constant thrust in one direction; mounted in pairs, also when sides are flush ground, for a combined load, either face-to-face or back-to-back.

Double-row, deep-groove ball bearings: The double-row, deep-groove ball bearing embodies the same principle of design as the single-row bearing. However, this bearing has a lower axial displacement than occurs in the single-row design, substantial thrust capacity in either direction, and high radial capacity due to the two rows of balls.

Spherical-roller bearings: The spherical-roller bearing, due to the number, size and shape of the rollers, and the accuracy with which they are guided, has maximum capacity. Since the bearing is inherently self-aligning, angular misalignment between the shaft and housing has no detrimental effect, and the full capacity is always available for useful work. The design and proportion are such that, in addition to radial load, thrust loads may be carried in either direction.

 Self-aligning ball bearing

 Spherical-roller bearing

 Single-row, deep-groove ball bearing

 Cylindrical-roller bearing

 Angular-contact ball bearing

 Ball-thrust bearing

 Spherical-roller thrust bearing

 Double-row, deep-groove ball bearing

 Tapered-roller bearing

Figure 8. Various bearing types.

Cylindrical-roller bearings: This type of bearing has high radial capacity and provides accurate guiding of the rollers, resulting in a close approach to true rolling. Consequent low friction permits operation at high speed. Those types which have flanges on one ring only, allow a limited free axial movement of the shaft in relation to the housing. They are easy to dismount even when both rings are mounted with a tight fit. The double-row type is particularly suitable for machine-tool spindles.

Ball-thrust bearings: The ball-thrust bearing is designed for thrust load in one direction only. The load line through the balls in parallel to the axis of the shaft, resulting in high thrust capacity and minimum axial deflection. Flat seats are preferred for heavy loads or for close axial positioning of the shaft.

Spherical-roller thrust bearings: The spherical-roller thrust bearing is designed to carry heavy thrust loads, or combined loads which are predominantly thrust. This bearing has a single row of rollers which roll on a spherical outer race with full self-alignment. The cage, centered by an inner ring sleeve, is constructed so that lubricant is pumped directly against the inner ring's unusually high guide flange. This bearing operates best with relatively heavy-oil lubrication.

Tapered-roller bearings: Since the axes of its rollers and raceways form an angle with the shaft axis, the tapered-roller bearing is especially suitable for carrying radial and axial loads acting simultaneously. A bearing of this type usually must be adjusted toward another bearing capable of carrying thrust loads in the opposite direction. Tapered roller bearings are separable—their cones (inner rings) with rollers and their cups (outer rings) are mounted separately.

The do's and don'ts for ball-bearing assembly, maintenance, inspection and lubrication is shown in Fig. 9. Refer to this list often when working with electric motors.

3.2.0 FREQUENCY OF LUBRICATION

Frequency of motor lubrication depends not only on the type of bearing but also on the motor application.

Small and medium-size motors equipped with ball bearings (except sealed bearings) are greased every three to six years if the motor duty is normal. On severe applications (high temperature, wet or dirty locations, or corrosive atmospheres), lubrication may be required more often. In severe applications, past experience and condition of the grease are the best guides as to frequency of lubrication.

Lubrication in sleeve bearings should be changed at least once a year. When the motor duty is severe or the oil appears dirty, it should be changed sooner.

3.2.1 Lubrication Procedure

For effective motor lubrication, cleanliness and use of the proper lubricant are of paramount importance.

DO'S

DO work with clean tools, in clean surroundings.

DO remove all outside dirt from housing before exposing bearing.

DO treat a used bearing as carefully as a new one.

DO use clean solvents and flushing oils.

DO lay bearings out on clean paper or cloth.

DO protect disassembled bearings from dirt and moisture.

DO use clean, lint-free rags to wipe bearings.

DO keep bearings wrapped in oil-proof paper when not in use.

DO clean outside of housing before replacing bearings.

DO keep bearing lubricants clean when applying and cover containers when not in use.

DO be sure shaft size is within specified tolerances recommended for the bearing.

DO store bearings in original unopened cartons in a dry place.

DO use a clean, short-bristle brush with firmly embedded bristles to remove dirt, scale or chips.

DO be certain that, when installed, the bearing is square with and held firmly against the shaft shoulder.

DO follow lubricating instructions supplied with the machinery. Use only grease where grease is specified; use only oil where oil is specified. Be sure to use the exact kind of lubricant called for.

DO handle grease with clean paddles or grease runs. Store grease in clean containers. Keep grease containers covered.

DON'TS

DON'T work under the handicap of poor tools, rough bench, or dirty surroundings.

DON'T use dirty, brittle or chipped tools.

DON'T handle bearings with dirty, moist hands.

DON'T spin uncleaned bearings.

DON'T spin any bearings with compressed air.

DON'T use same container for cleaning and final rinse of bearings.

DON'T scratch or nick bearing surfaces.

DON'T remove grease or oil from new bearings.

DON'T use incorrect kind or amount of lubricant.

DON'T use a bearing as a gauge to check either the housing bore or the shaft fit.

DON'T install a bearing on a shaft that shows excessive wear.

DON'T open carton until bearing is ready for installation.

DON'T judge the condition of a bearing until after it has been cleaned.

DON'T pound directly on a bearing or ring, when installing, as this may cause damage to shaft and bearing.

DON'T overfill when lubricating. Excess grease and oil will ooze out of the overfilled housings past seals and closures, collect dirt and cause trouble. Too much lubricant will also cause overheating, particularly where bearings operate at high speeds.

DON'T permit any machine to stand inoperative for months without turning it over periodically. This prevents moisture which may condense in a standing bearing from causing corrosion.

Figure 9. Do's and don'ts for ball-bearing assembly, maintenance, and lubrication.

When greasing a ball-bearing motor, the bearing housing, grease gun and fittings are wiped clean. Great care must be taken to keep dirt out of the bearing when greasing. Next, the relief plug is removed from the bottom of the bearing housing. This is done to prevent excessive pressure from building up inside the bearing housing during greasing. Grease is then added, with the motor running if possible, until it begins to flow from the relief hole. Allow the motor to run from 5 to 10 minutes to expel excess grease. Then the relief plug is replaced and the bearing housing is cleaned.

It is important to avoid over-greasing. When too much grease is forced into a bearing, a churning of the lubricant occurs, resulting in high temperature and eventual bearing failure.

On motors that do not have a relief hole, grease should be applied sparingly. If possible, disassemble the motor and repack the bearing housing with the proper amount of grease. During this procedure, always maintain strict cleanliness.

The importance of adherence to these procedures cannot be overemphasized. Contamination and overgreasing of bearings are the major causes of bearing failure.

For sleeve bearings, use only the recommended oil for particular service conditions. Observing careful cleanliness, old oil is removed and new oil is added until the oil level reaches the "full" line on the oil sight gauge. This is done only when the motor is not running.

3.3.0 TESTING BEARINGS

Two of the most effective tests are what might be called the "feel" test and the "sound" test. When performing the "feel" test, if, while the motor is running, the bearing housing feels overly hot to the touch, it is probably malfunctioning.

Note: Some bearings may operate safely up to about 85° C.

During the "sound" test, listen for foreign noises coming from the motor. Also, one end of a steel rod (about 3 feet long and $\frac{1}{2}$ inch in diameter) may be placed on the bearing housing while the other end is held against the ear. The rod then acts as an amplifier, transmitting unusual sounds such as thumping or grinding, which would indicate a failing bearing. Special listening devices, such as the transistorized stethoscope, can also be used for the purpose.

Additional checks are usually in the form of checking the air gap on sleeve-bearing motors periodically. These tests, performed with a feeler gauge, indicate when a bearing begins to wear. Four measurements should be taken about 90 degrees apart around the rotor periphery. These measurements are recorded and compared with earlier readings, providing a check on the condition of bearings.

Motors should also be checked for end play. Ball-bearing motors should have about $\frac{1}{32}$ inch to $\frac{1}{16}$ inch end play. Sleeve-bearing motors may have up to $\frac{1}{2}$ inch end play.

On large sleeve bearings the oil level should be checked periodically, and the oil is visually inspected for contamination. If it is possible, the oil rings should be checked when the motor is operating.

Other inspections include checking for misaligned or bent shafts and for excessive belt pressure.

4.0.0 TROUBLESHOOTING MOTORS

To detect defects in electric motors, the windings are normally tested for ground faults, opens, shorts, and reverses. The exact method of performing these tests will depend on the type of motor being serviced. However, regardless of the motor type, a knowledge of some important terms is necessary before maintenance personnel can approach their work satisfactorily.

Ground: A winding becomes grounded when it makes an electrical contact with the iron of the motor. The usual causes of grounds include the following: Bolts securing the end plates come into contact with the winding; the wires press against the laminations at the corners of the slots, which is likely to occur if the slot insulation tears or cracks during winding; and the centrifugal switch may be grounded to the end plate.

Open circuits: Loose or dirty connections as well as a broken wire can cause an open circuit in an electric motor.

Shorts: Two or more turns of the coil that contact each other electrically will cause a short circuit. This condition may develop in a new winding if the winding is tight and much pounding is necessary to place the wires in position. In other cases, excessive heat developed from overloads will make the insulation defective and will cause shorts. A short circuit is usually detected by observing smoke from the windings as the motor operates or when the motor draws excessive current at no load.

4.1.0 TOOLS FOR TROUBLESHOOTING

Before actually getting into troubleshooting techniques, let's take a look at some of the customary tools used for troubleshooting motors of all types.

In addition to small portable devices such as voltmeters, ammeters, brush-spring tension testers and a transistorized stethoscope for checking motor bearings, maintenance equipment should include a 500-volt insulation-resistance tester, a spark-gap oil dielectric tester, and a portable oil-filtering unit.

The use of some of these tools has been covered in previous modules, so only the ones that have not been previously covered will be discussed here.

Transistorized stethoscope: This type of testing instrument—equipped with a transistor-amplifier—is used to ascertain the condition of motor bearings. A little practice in interpreting what is heard through it may be required, but in general, it is relatively simple to use. If, when the stethoscope is applied to a motor bearing, a purring sound is heard, the bearing is usually normal. On the other hand, a thumping sound or a rough grinding sound indicates a failing bearing.

Insulation-resistance tests: High-voltage cables, such as those rated at 2300 volts should be tested periodically. Circuits carrying 480 volts should be tested annually; this includes transformers, motors, motor starter, generators, and switches. Also, certain high-power process equipment such as electric furnaces and die-casting machines should receive insulation-resistance tests annually.

When performing an insulation-resistance test, first make a careful safety check and ensure that all circuits and equipment are rated at the voltage of the megger. Furthermore, all equipment scheduled for testing must be disconnected from all power sources. All safety switches should be opened and locked out to make certain that motor starters or other control equipment cannot accidentally energize the apparatus.

Before continuing, also check the megger and other testing instruments for functioning. This is accomplished by first testing for an infinity reading by operating the megger with the test leads not connected. A second test is made with the megger leads shorted and the megger handle turned very slowly; this time, the meter should read "zero."

Following these checks, insulation-resistance readings should be obtained by testing between a conductor and ground or between two conductors, or both. Insulation readings from conductor to ground are obtained by connecting the "line" test lead to the conductor and the "earth" test lead to ground. For the test between conductors, the test leads are connected to the two conductors to be tested.

After the proper connections have been made, the megger handle is turned at an even speed for about one minute. At the end of this time, the insulation-resistance value is recorded.

Because temperature and humidity have profound effects on insulation-resistance readings, temperature and humidity at the apparatus should be recorded immediately after the test. In addition, pertinent considerations should be noted such as condition of the immediate area—wet location or excessive dust, whether the apparatus has been in operation prior to the test or at rest for a prolonged period of time.

After the readings have been recorded, they are corrected for temperature using a temperature-correction chart supplied with most meggers. As a rule-of-thumb, most maintenance departments feel that 600-volt winding insulation is acceptable if the corrected resistance value is one megohm or more. High resistance readings which show a continuing downward trend over a period of time indicate failing insulation.

The chart in Fig. 10 gives a list of practical tools and equipment for effective electrical maintenance, not only for motors, but for other electrical apparatus as well. Note that the tool is listed in the left column, while the application is shown in the right column.

Tools or Equipment	Application
Multi-meters, voltmeters, ohmmeters, clamp-on ammeters, wattmeters, clamp-on P.F. meter	Measure circuit voltage, resistance, current and power. Useful for circuit tracing and troubleshooting.
Potential and current transformers, meter shunts	Increase range of test instruments to permit reading of high-voltage and high-current circuits.
Tachometer	Checks rotating machinery speeds.
Recording meters, instruments	Provide permanent record of voltage, current, power, temperature, etc. on charts for analytic study.
Insulation resistance tester, thermometer, psychrometer	Test and monitor insulation resistance; use thermometer and psychrometer for temperature-humidity correction.
Portable oil dielectric tester; portable oil filter	Test OCB, transformer oil or other insulating oils. Recondition used oil.
Transistorized stethoscope	Detect faulty rotating machinery bearings; leaky valves.
Air gap feeler gages	Check motor or generator air gap between rotor and stator.
Cleaning solvent	Removes grease or dirt from motor windings or other electrical parts.
Hand stones, (rough, medium fine); grinding rig; canvas strip	Grinding, smoothing and finishing commutators or slip rings.
Spring tension scale	Checks brush pressure on dc motor commutators or on ac motor skip rings; tests electrical contact pressure on relays, starters or contactors.

Figure 10. Tools for effective electrical maintenance.

4.2.0 GROUNDED COILS

The usual effect of one grounded coil in a winding is the repeated blowing of a fuse, or tripping of the circuit breaker, when the line switch is closed, that is, providing the machine frame and the line are both grounded. Two or more grounds will give the same result and will also short out part of the winding in that phase in which the grounds occur. A quick and simple test to determine whether or not a ground exists in the winding can be made with a conventional continuity tester. In testing with such an instrument, first make certain that the line switch is open and locked out, causing the motor leads to be *de-energized*. Place one test lead on the frame of the motor and the other in turn on each of the line wires leading from the motor. If there is a grounded coil at any point in the winding, the lamp of the continuity tester will light, or in the case of a meter, the dial will swing toward *infinity*.

To locate the phase that is grounded, test each phase separately. In a three-phase winding it will be necessary to disconnect the star or delta connections, if accessible. After the grounded phase is located the pole-group connections in that phase can be disconnected and each group tested separately. When the leads are placed one on the frame and the other on the grounded coil group, the lamp will indicate the ground in this group by lighting. The stub connections between the coils and this group may then be disconnected and each coil tested separately until the exact coil that is grounded is located.

Sometimes moisture in the insulation around the coils on old and defective insulation will cause a high-resistance ground that is difficult to detect with a test lamp. A megger can be used to detect such faults, but in many cases a megger may not be available. If not, use a test outfit consisting of a headphone set (telephone receiver) and several dry cell batteries connected in series as shown in Fig. 11. Such a test set will detect a ground of very high resistance, and this set will often be found very effective when the ordinary test lamp fails to locate the trouble.

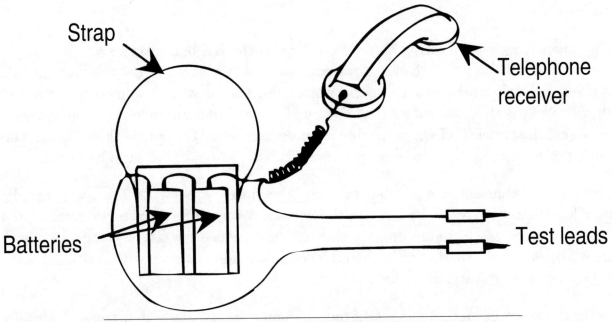

Figure 11. When used for testing, a clicking sound indicates a fault.

Armature windings and the commutator of a motor may be tested for grounds in a similar manner. On some motors, the brush holders are grounded to the end plate. Consequently, before the armature is tested for grounds, the brushes must be lifted away from the commutator.

When a grounded coil is located, it should be either removed and reinsulated or cut out of the circuit. At times, however, it may be inconvenient to stop a motor long enough for a complete rewinding or permanent repairs. In such cases, when trouble develops, it is often necessary to make a temporary repair until a later time when the motor may be taken out of service long enough for rewinding or permanent repairs.

To temporarily repair a defective coil, a jumper wire of the same size as that used in the coils is connected to the bottom lead of the coil immediately adjacent to the defective coil and run across to the top lead of the coil on the other side of the defective coil, leaving the defective coil entirely out of the circuit. The defective coil should then be cut at the back of the winding and the leads taped so as not to function when the motor is started again. If the defective coil is grounded, it should also be disconnected from the other coils.

4.3.0 SHORTED COILS

Shorted turns within coils are usually the result of failure of the insulation on the wires. This is frequently caused by the wires being crossed and having excessive pressure applied on the crossed conductors when the coils are being inserted in the slot. Quite often it is caused by using too much force in driving the coils down in the slots. In the case of windings that have been in service for several years, failure of the insulation may be caused by oil, moisture, etc. If a shorted coil is left in a winding, it will usually burn out in a short time and if it is not located and repaired promptly will probably cause a ground and the burning out of a number of other coils.

One inexpensive way of locating a shorted coil is by the use of a growler and a thin piece of steel. Figure 12 shows a sketch of a growler in use in a stator. Note that the poles are shaped to fit the curvature of the teeth inside the stator core. The growler should be placed in the core as shown, and the thin piece of steel should be placed the distance of one coil span away from the center of the growler. Then, by moving the growler around the bore of the stator and always keeping the steel strip the same distance away from it, all of the coils can be tested.

If any of the coils has one or more shorted turns, the piece of steel will vibrate very rapidly and cause a loud humming noise. By locating the two slots over which the steel vibrates, both sides of the shorted coil can be found. If more than two slots cause the steel to vibrate, they should all be marked, and all shorted coils should be removed and replaced with new ones or cut out of the circuit as previously described.

Sometimes one coil or a complete coil group becomes short-circuited at the end connections. The test for this fault is the same as that for a shorted coil. If all the coils in one group are

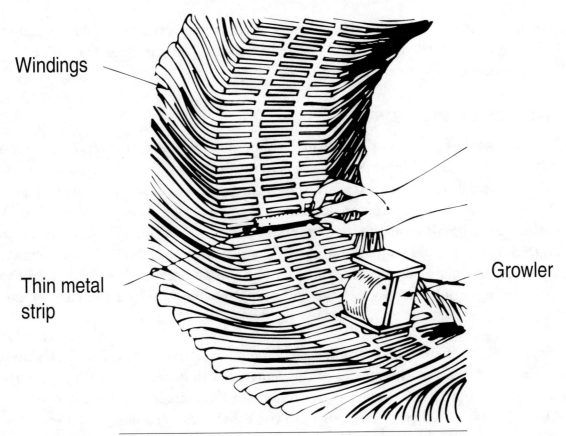

Windings

Thin metal strip

Growler

Figure 12. Growler used to test a stator of an AC motor.

shorted, it will generally be indicated by the vibration of the steel strip over several consecutive slots, corresponding to the number of coils in the group.

The end connections should be carefully examined, and those that appear to have poor insulation should be moved during the time that the test is being made. It will often be found that when the shorted end connections are moved during the test the vibration of the steel will stop. If these ends are reinsulated, the trouble should be eliminated.

4.4.0   OPEN COILS

When one or more coils become open-circuited by a break in the turns or a poor connection at the end, they can be tested with a continuity tester as previously explained. If this test is made at the ends of each winding, an open can be detected by the lamp failing to light. The insulation should be removed from the pole-group connections, and each group should be tested separately.

An open circuit in the starting winding may be difficult to locate, since the problem may be in the centrifugal switch as well as the winding itself. In fact, the centrifugal switch is probably more apt to cause trouble than the winding since parts become worn, defective and, more likely, dirty. Insufficient pressure of the rotating part of centrifugal switches against the stationary part will prevent the contacts from closing and thereby produce an open circuit.

If the trouble is a loose connection at the coil ends, it can be repaired by resoldering the splices, but if it is within the coil, the coil should either be replaced or a jumper should be connected around it until a better repair can be made.

4.5.0   REVERSED CONNECTIONS

Reversed coils cause the current to flow through them in the wrong direction. This fault usually manifests itself—as do most irregularities in winding connections—by a disturbance of the magnetic circuit, which results in excessive noise and vibration. The fault can be located by the use of a magnetic compass and some source of low-voltage direct current. This voltage should be adjusted so it will send about one-fourth to one-sixth of the full-load current through the winding, and the dc leads should be placed on the start and finish of one phase. If the winding is three-phase, star-connected, this would be at the start of one phase and the star point. If the winding is delta-connected, the delta must be disconnected and each phase tested separately.

Place a compass on the inside of the stator and test each of the coil groups in that phase. If the phase is connected correctly, the needle of the compass will reverse definitely as it is moved from one coil group to another. However, if any one of the coils is reversed, the reversed coil will build up a field in the direction opposite to the others, thus causing a neutralizing effect which will be indicated by the compass needle refusing to point definitely to that group. If there are only two coils per group, there will be no indication if one of them is reversed, as that group will be completely neutralized.

When an entire coil group is reversed, it causes the current to flow in the wrong direction in the whole group. The test for this fault is the same as that for reversed coils. The winding should be magnetized with direct current, and when the compass needle is passed around the coil groups, they should indicate alternately N.S., N.S., etc. If one of the groups is reversed, three consecutive groups will be of the same polarity. The remedy for either reversed coil groups or reversed coils is to make a visual check of the connections at that part of the winding, locate the wrong connection, and reconnect it properly.

When the wrong number of coils are connected in two or more groups, the trouble can be located by counting the number of ends on each group. If any mistakes are found, they should be remedied by reconnecting properly.

4.6.0   REVERSED PHASE

Sometimes in a three-phase winding a complete phase is reversed by either having taken the starts from the wrong coils or by connecting one of the windings in the wrong relation to the others when making the star or delta connections. If the winding is connected delta, disconnect any one of the points where the phases are connected together and pass current through the three windings in series. Place a compass on the inside of the stator and test each coil group by slowly moving the compass one complete revolution around the stator.

TROUBLESHOOTING CHART

AC Motors

Malfunction	Probable Cause	Corrective Action
Slow speed	Open primary circuit.	Locate fault with testing device and repair.
Slow to accelerate	Excess loading.	Reduce load.
	Poor circuit.	Check for high resistance.
	Defective squirrel-cage rotor.	Replace.
	Applied voltage too low.	Get power company to increase voltage tap.
Wrong rotation	Wrong sequence of phases.	Reverse connections at motor or at switchboard.
Motor overheats	Check for overload.	Reduce load.
	Wrong blowers or air shields.	May be clogged with dirt and prevent proper ventilation of motor.
	Motor may have one phase open.	Check to make sure that all leads are well connected.
	Grounded coil.	Locate and repair.
	Unbalanced terminal voltage.	Check to make sure that all leads are well connected.
	Grounded coil.	Locate and repair.
	Unbalanced terminal voltage.	Check for faulty leads.
	Shorted stator coil.	Repair and then check wattmeter reading.
	Faulty connection.	Indicate by high resistance.
	High voltage.	Check terminals of motor with voltmeter.
	Low voltage.	Same as above.
Motor stalls	Wrong application.	Change type or size. Consult manufacturer.
	Overloaded motor.	Reduce load.
	Low motor voltage.	See that nameplate voltage is maintained.
	Open Circuit.	Fuses blown.
	Incorrect control resistance of wound rotor.	Check control sequence. Replace broken resistors. Repair open circuits.
Motor does not start	One phase open.	See that no phase is open. Reduce load.
	Defective rotor.	Look for broken bars or rings.
	Poor stator coil connection.	Remove end bells.

Figure 13. General troubleshooting chart for motors.

Malfunction	Probable Cause	Corrective Action
Motor runs, then quits	Power failure.	Check for loose connections to line, to fuses and to control.
Slow speed	Not applied properly.	Consult supplier for proper type.
	Voltage too low at motor terminals because of line drop.	Use higher voltage on transformer terminals or reduce load.
	If wound rotor, improper control operation of secondary.	Correct secondary control.
	Starting load too high.	Check load that the motor is supposed to carry upon starting.
	Low pull-in torque of synchronous motor.	Change rotor starting resistance or change rotor design.
	Check that all brushes are riding on rings.	Check secondary connections. Leave no leads poorly connected.
	Broken rotor bars.	Look for cracks near the rings. A new rotor may be required.
Motor vibrates	Motor misaligned.	Realign.
	Weak foundations.	Strengthen base.
	Coupling out of balance.	Balance coupling.
	Driven equipment unbalanced.	Rebalance driven equipment.
	Defective ball bearing.	Replace bearing.
	Bearing not in line.	Line up properly.
	Balancing weights shifted.	Rebalance rotor.
	Wound rotor coils replaced.	Rebalance rotor.
	Polyphase motor running single phase.	Check for open circuit.
	Excessive end play.	Adjust bearing or add washer.
Unbalanced line current	Unequal terminal volts.	Check leads and connections.
	Single phase operation.	Check for open circuit.
	Poor rotor contacts in control wound rotor resistance.	Check control devices.
	Brushes not in proper position in wound rotor.	See that brushes are properly seated and shunts in good condition.
	Fan rubbing air shield.	Remove interference.
	Fan striking insulation.	Clear fan.
	Loose on bedplate.	Tighten holding bolts.

Figure 13. General troubleshooting chart for motors. (Cont.)

Malfunction	Probable Cause	Corrective Action
Magnetic noise	Air gap not uniform.	Check and correct bracket fits or bearing.

Figure 13. General troubleshooting chart for motors. (*Cont.*)

The reversals of the needle in moving the compass one revolution around the stator should be three times the number of poles in the winding.

In testing a star- or wye-connected winding, connect the three starts together and place them on one dc lead. Then connect the other dc lead and star point, thus passing the current through all three windings in parallel. Test with a compass as explained for the delta winding. The result should then be the same, or the reversals of the needle in making one revolution around the stator should again be three times the number of poles in the winding.

These tests for reversed phases apply to full-pitch windings only. If the winding is fractional pitch, a careful visual check should be made to determine whether there is a reversed phase or mistake in connecting the star or delta connections.

The troubleshooting chart in Fig. 13 may be used by qualified personnel who have the proper tools and equipment. These instructions do not cover all details or variations in equipment, nor do they provide for every possible condition to be met in actual practice.

5.0.0 TROUBLESHOOTING SPLIT-PHASE MOTORS

If a split-phase motor fails to start, the trouble may be due to one or more of the following faults:

- Tight or "frozen" bearings
- Worn bearings, allowing the rotor to drag on the stator
- Bent rotor shaft
- One or both bearings out of alignment
- Open circuit in either starting or running windings
- Defective centrifugal switch
- Improper connections in either winding
- Grounds in either winding or both
- Shorts between the two windings

Tight or worn bearings: Tight or worn bearings may be due to the lubricating system failing, or when new bearings are installed, they may run hot if the shaft is not kept well oiled.

If the bearings are worn to such an extent that they allow the rotor to drag on the stator, this will usually prevent the rotor from starting. The inside of the stator laminations will be worn bright where they are rubbed by the rotor. When this condition exists, it can generally be easily detected by close observation of the stator field and rotor surface when the rotor is removed.

Bent shaft and bearings out of line: A bent rotor shaft will usually cause the rotor to bind when in a certain position and then run freely until it comes back to the same position again. An accurate test for a bent shaft can be made by placing the rotor between centers on a lathe and turning the rotor slowly while a tool or marker is held in the tool post close to the surface of the rotor. If the rotor wobbles, it is an indication of a bent shaft.

Bearings out of alignment are usually caused by uneven tightening of the end-shield plates. When placing end shields or brackets on a motor, the bolts should be tightened alternately, first drawing up two bolts which are diametrically opposite. These two should be drawn up only a few turns and the other kept tightened an equal amount all the way around. When the end shields are drawn up as far as possible with the bolts, they should be tapped tightly against the frame with a mallet and the bolts tightened again.

Open circuits and defective centrifugal switches: Open circuits in either the starting or running winding will cause the motor to fail to start. This fault can be detected by testing in series with the start and finish of each winding with a test lamp or ohmmeter.

A defective centrifugal switch will often cause considerable trouble that is difficult to locate unless one has good knowledge of the operating characteristics of these switches. If the switch fails to close when the rotor stops, the motor will not start when the line switch is closed. Failure of the switch to close is generally caused by dirt, grit, or some other foreign matter getting into the switch. The switch should be thoroughly cleaned with a degreasing solution such as AWA1,1,1 and then inspected for weak or broken springs.

If the winding is on the rotor, the brushes sometimes stick in the holders and fail to make good contact with the slip rings. This causes sparking at the brushes. There will probably also be a certain place where the rotor will not start until it is moved far enough for the brush to make contact on the ring. The brush holders should be cleaned and the brushes carefully fitted so they move more freely with a minimum of friction between the brush and the holders. If a centrifugal switch fails to open when the motor is started, the motor will probably growl and continue to run slowly, causing the starting winding to burn out if not promptly disconnected from the line. In most cases, however, the "heaters" in the motor control will take care of this before any serious damage occurs. This fault is likely to be caused by dirt or hardened grease in the switch.

Reversed connections and grounds: Reversed connections are caused by improperly connecting a coil or group of coils. The wrong connections can be found and corrected by making a careful check on the connections and reconnecting those that are found at fault. The test with a dc power source and a compass can also be used for locating reversed coils. Test the starting and running windings separately, exciting only one winding at a time, with direct current. The compass should show alternate poles around the winding.

The operation of a motor that has a ground in the winding will depend on where the ground is and whether or not the frame is grounded. If the frame is grounded, then when the ground occurs in the winding, it will usually blow a fuse or trip the overcurrent device.

A test for grounds can be made with a test lamp or continuity tester. One test lead should be placed on the frame and the other on a lead to the winding. If there is no ground, the lamp will not light, nor will any deflection be present when a meter is used. If the light does light, it indicates a ground due to a defect somewhere in the insulation.

Short circuits: Short circuits between any two windings can be detected by the use of a test lamp or continuity tester. Place one of the test leads on one wire of the starting winding and the other test lead on the wire of the running winding. If these windings are properly insulated from each other, the lamp should not light. If it does, it is a certain indication that a short exists between the windings. Such a short will usually cause part of the starting winding to burn out. The starting winding is always wound on top of the running winding, so if it becomes burned out due to a defective centrifugal switch or a short circuit, the starting winding can be conveniently removed and replaced without disturbing the running winding.

6.0.0 STORING MOTORS

There are many reasons for storing motors, but the two major ones are:

- The project on which they are to be used is not complete.
- Spare motors are often kept as back-ups on most industrial installations.

The first consideration when storing motors for any length of time is the location. A dry location should be selected if at all possible—one that does not undergo severe changes in temperature over a 24-hour period. When the ambient temperature changes frequently during a 24-hour period, condensation is certain to form on the motor, and moisture is one of the worst enemies of motor insulation. Therefore, guarding against moisture is one of the chief concerns when storing motors of any type.

A means for transporting the motors from the place of storage to the place where it will be used, or else shifted around in the storage area, is also of importance. Motors should *not* be lifted by their rotating shafts. Doing so can damage the alignment of the rotor in relationship to the stator. Even picking up the smaller fractional horsepower motors by the shaft is not recommended. Many workers have received bad cuts from the sharp keyways on motor shafts when picked up with the bare hands.

CAUTION: Never lift a motor by its shaft; especially ones that can be lifted manually. The sharp edges of keyways are not unlike sharp kitchen knives which can cause deep cuts in the body.

When an electric motor is received at the job site, always refer to the manufacturer's instructions and follow them to the letter. Failure to do so could result in serious injury to the workers and motor alike.

Once the motor has been uncrated, check to see if any damage has occurred during handling. Be sure that the motor shaft and armature turn freely. This is also a good time to check to determine if the motor has been exposed to dirt, grease, grit, or excessive moisture in either shipment of storage. Motors in storage should have shafts turned over once each month to redistribute grease in the bearings.

Note: Motors in storage should have shafts turned over once each month to redistribute grease in the bearings.

WARNING!   Never start a motor which has been wet without having it thoroughly dried.

The measure of insulation resistance is a good dampness test. Clean the motor of any dirt or grit before putting it back in service.

Eyebolts of lifting lugs on motors are intended only for lifting the motor and factory motor-mounted standard accessories. These lifting provisions should never be used when lifting or handling the motor when the motor is attached to other equipment as a single unit.

The eyebolt lifting-capacity rating is based on a lifting alignment coincident with the eyebolt centerline. The eyebolt capacity reduces as deviation from this alignment increases.

The following is a list of items that must be considered when storing motors for any length of time:

- Make sure motors are kept clean.
- Make sure motors are kept dry.
- Supply supplemental heating in the storage area if necessary.
- Motors should be stored in an orderly fashion; that is, grouped by horsepower, etc.
- Motor armatures should be rotated periodically.
- Lubrication should be checked periodically.
- Protect shafts and keyways during storage and also while transporting motors from one location to another.
- Test motor-winding resistance upon receiving; test again after setting in storage.

Electrical workers will sometimes come across a motor with no identification (no nameplate or lead tags) which must be put back into service or else repaired. The experienced electrician should know how to positively identify the motor's characteristics, even with no written data.

The NEMA Standard method of motor identification is easy to remember by drawing the coils to form a wye. Identify one outside coil end with the Number one (1), and then draw a decreasing spiral and number each coil end in sequence as shown in Fig. 14.

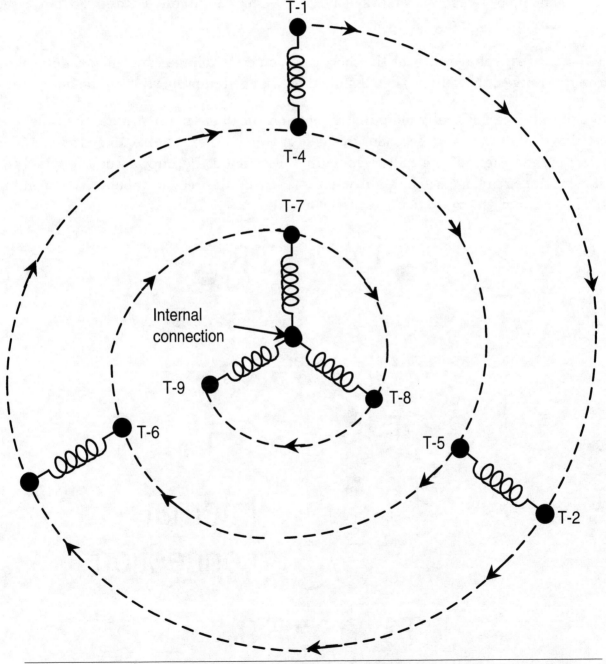

Figure 14.   Identify one outside coil and then draw a decreasing spiral and number each coil.

By using an ohmmeter or other continuity tester, the individual circuits can be located as follows.

Step 1. Connect one probe of the tester to any lead, and check for continuity to each of the other 8 leads. A reading from only one other lead indicates one of the two wire circuits. A reading to two other leads indicates the three wire circuit that makes up the internal wye connection.

Step 2. Continue checking and isolating leads until all four circuits have been located. Tag the wires of the three lead circuits T-7, T-8, and T-9 in any order. The other leads should be temporarily marked T-1 and T-4 for one circuit, T-2 and T-5 for the second circuit, T-3 and T-6 for the third and final circuit.

The following test voltages are for the most common dual voltage range of 230/460 volts. For other motor ranges, the voltages listed should be changed in proportion to the motor rating.

As all the coils are physically mounted in slots on the same motor frame, the coils will act almost like the primary and secondary coils of a transformer. Figure 15 shows a simplified electrical arrangement of the coils. Depending on which coil group power is applied to, the resulting voltage readings will be additive, subtractive, balanced, or unbalanced depending on physical location with regard to the coils themselves.

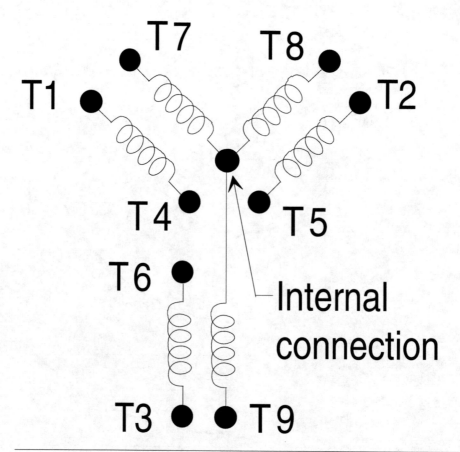

Figure 15. Simplified electrical arrangement of wye-wound motor coils.

Step 3.   The motor may be started on 230 volts by connecting leads T-7, T-8, and T-9 to the three phase source. If the motor is too large to be connected directly to the line, the voltage should be reduced by using a reduced voltage starter or other suitable means.

Step 4.   Start the motor with no load connected and bring up to normal speed.

Step 5.   With the motor running, a voltage will be induced in each of the open two-wire circuits that were tagged T-1 and T-4, T-2 and T-5, T-3 and T-6. With a voltmeter, check the voltage reading of each circuit. The voltage should be approximately 125-130 volts and should be the same on each circuit.

Note: The voltages referred to during the testing are only for reference and will vary greatly from motor to motor, depending on size, design and manufacturer. If the test calls for equal voltages of 125-130 and the reading is only 80-90, that is okay as long as the voltage readings are nearly equal.

Step 6.   With the motor still running, carefully connect the lead that was temporarily marked T-4 with the T-7 and line lead. Read the voltage between T-1 and T-8 and also between T-1 to T-9. If both readings are of the same value and are approximately 330-340 volts, leads T-1 and T-4 may be disconnected and permanently marked T-1 and T-4.

Step 7.   If the two voltage readings are of the same value and are approximately 125-130, disconnect and interchange leads T-1 and T-4 and mark permanently (original T-1 changed to T-4 and original T-4 changed to T-1).

Step 8.   If readings between T-1 and T-8, and between T-1 and T-9 are of unequal values, disconnect T-4 from T-7 and reconnect T-4 to the junction of T-8 and line.

Step 9.   Measure the voltage now between T-1 and T-7 and also between T-1 and T-9. If the voltages are equal and approximately 330-340 volts, tag T-1 is permanently marked T-2 and T-4 is marked T-5 and disconnected. If the readings taken are equal but are approximately 125-130 volts, leads T-1 and T-4 are disconnected, interchanged, and marked T-2 and T-5 (T-1 changed to T-5, and T-4 changed to T-2). If both voltage readings are different, T-4 lead is disconnected from T-8 and moved to T-9. Voltage readings are taken again (between T-1 and T-7, T-1 and T-8) and the leads permanently marked T-3 and T-6 when equal readings of approximately 330-340 volts are obtained.

Step 10.   The same procedure is followed for the other two circuits that were temporarily marked T-2 and T-5, and T-3 and T-6 until a position is found where both voltage readings are equal and approximately 330-340 volts and the tags change to correspond to the standard lead markings as shown in Fig. 16.

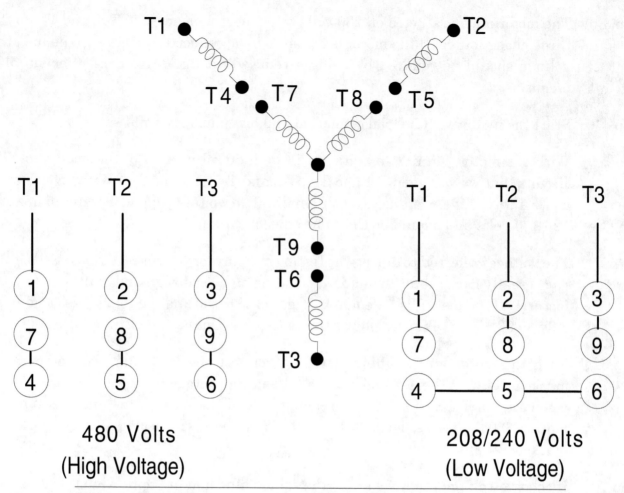

Figure 16. NEMA Standard lead markings for dual-voltage wye-wound motors.

Step 11.   Once all leads have been properly and permanently tagged, leads T-4, and T-5 and
T-6 are connected together and voltage readings are taken between T-1, T-2, and
T-3. The voltages should be equal and approximately 230 volts.

Step 12.   As an additional check, the motor is shut down and leads T-7, T-8, and T-9 are
disconnected, and leads T-1, T-2, and T-3 are connected to the line. Connect T-1 to
the line lead T-7 was connected to, T-2 to the same line as T-8 was previously
connected, and T-3 to the same lead that T-9 was connected to. With T-4, T-5, and
T-6 still connected together to form a wye connection, the motor can again be started
without a load. If all lead markings are correct, the motor rotation with leads T-1,
T-2, and T-3 connected will be the same as when T-7, T-8, and T-9 were connected.

The motor is now ready for service and is connected in series for high voltage or parallel for
low as indicated by the NEMA Standard connections shown in Fig. 16.

Note: This procedure may not work on some wye wound motors with concentric
coils.

7.1.0 THREE-PHASE DELTA-WOUND MOTORS

Most dual voltage, delta wound motors also have 9 leads, as indicated in Fig. 17, but there are only three circuits of three leads each.

Continuity tests are used to find the three coil groups as was done for the wye wound motor. Once the coil groups are located and isolated, further resistance checks must be made to locate the common wire in each coil group. As the resistance of some delta wound motors is VERY low, a digital ohmmeter, wheatstone bridge, or other sensitive device may be needed.

Each coil group consists of two coils tied together with three leads brought out to the motor junction or terminal box. Reading the resistances carefully between each of the three leads shows that the readings from one of the leads to each of the other two leads will be the same (equal), but the resistance reading between those two leads will be double the previous readings, Fig. 18 may help clarify the technique.

The common lead found in the first coil group is permanently marked T-1, and the other two leads temporarily marked T-4 and T-9. The common lead of the next coil group is found and permanently marked T-2 and the other leads temporarily marked T-5 and T-7. The common lead of the last coil group is located and marked T-3 with the other leads being temporarily marked T-6 and T-8.

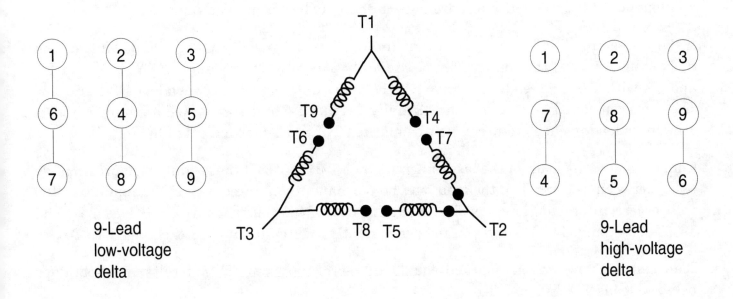

Figure 17. NEMA Standard lead markings for dual-voltage delta-wound motors.

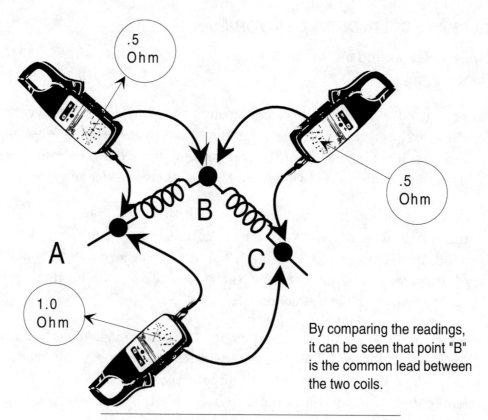

.5
Ohm

.5
Ohm

B

A

C

1.0
Ohm

By comparing the readings,
it can be seen that point "B"
is the common lead between
the two coils.

Figure 18. Using ohmmeter to test motor leads.

After the leads have been marked, the motor may be connected to a 230V three phase line using leads T-1, T-4, and T-9. Lead T-7 is connected to line and T-4, and the motor is started with no load connected. Voltage readings are taken between T-1 and T-2. If the voltage is approximately 460, the markings are correct and may be permanently marked.

If the voltage reading is 400 volts or less, interchange T-5 and T-7 OR T-4 and T-9 and read the voltage again. If the voltage is approximately 230 volts, interchange BOTH T-5 with T-7 and T-4 with T-9. The readings should now be approximately 460 between leads T-1 and T-2. The leads connected together now are actually T-4 and T-7 and are marked permanently. The remaining lead in each group can now be marked T-9 and T-5 as indicated by Fig. 17.

Connect one of the leads of the last coil group (not T-3) to T-9. If the reading is approximately 460V between T-1 and T-3, the lead may be permanently marked T-6. If the reading is 400 volts or less, interchange T-6 and T-8. A reading now of 460 volts should exist between T-1 and T-3. T-6 is changed to T-8 and marked permanently and temporary T-8 is changed To T-6.

If all leads are now correctly marked, equal readings of approximately 460 volts can be obtained between leads T-1, T-2, and T-3.

To double check the markings, the motor is shut off and reconnected using T-2, T-5, and T-7. T-2 is connected to the same line lead as T-1, lead T-5 is connected where T-4 was, and T-7 is hooked where T-9 was previously connected. When started, the motor should rotate the same direction as before.

ELECTRICAL TASK MODULE 20310

Stop the motor and connect leads T-3, T-6, and T-8 to the line leads previously connected to T-2, T-5, and T-7 respectively, and when the motor is started it should still rotate in the same direction.

The motor is now ready for service and is connected in series for high or parallel for low voltage as indicated by the NEMA Standard connections shown in Fig. 17.

Summary

The first step towards a reliable maintenance program is to prepare records. And, while this chore often is distasteful, it also is essential to tabulate the mass of pertinent data which cannot possibly be retained by memory alone. Obviously, records can take many forms and must be tailored to specific needs and resources. But, as a minimum, records on each motor should include:

- A complete description, including age and nameplate data.
- Location and application, keeping such notations up-to-date if motors are transferred to different areas or used for different purposes.
- Notations of scheduled preventive maintenance and previous repair work performed.
- Location of duplicate or interchangeable motors.
- An estimate of the motor's importance in the productive process to which it relates.

In determining which motors are likely to fail first, it is well to remember that motor failures generally are caused by either loading, age, vibration, contamination or commutation problems.

Advanced motor maintenance techniques will be presented in your Level 4 training.

References

For more advanced study of topics covered in this Task Module, the following works are suggested:

National Electrical Code Handbook, Latest Edition, NFPA, Quincy, MA
American Electricians Handbook, McGraw-Hill, Latest Edition, New York, NY
Electric Motor Repair, Robert Rosenberg, Construction Book Store, Gainesville, FL

SELF-CHECK REVIEW/PRACTICE QUESTIONS

1. Which of the following statements is true concerning motors in storage?

   a. Rotors should not be moved or turned until motor is put in use
   b. Rotors should be turned once a month to distribute bearing grease
   c. Rotors should be turned once a year to distribute bearing grease
   d. Rotors should be turned once every 3 years to keep them from rusting

2. Which of the following is the best attachment point for lifting heavy motors?

   a. Motor's eyebolt
   b. Motor's shaft
   c. Motor's base
   d. Motor's end bells

3. Which of the following is the worst enemy of motor windings?

   a. Undersized fuses or circuit breakers
   b. Undersized heaters
   c. Low voltage
   d. Moisture

4. Which of following is the name of the motor conductors that are used only during the brief period when the motor is starting?

   a. Motor leads
   b. Drum armature
   c. Starting winding
   d. Shunt-field winding

5. Which of the following provides minimum resistance and aligns the motor rotor while turning?

   a. Compensator
   b. Brushes
   c. End bells
   d. Bearings

6. Which of the following is true concerning lubricating motor bearings?

   a. Always add a little extra grease; more than needed
   b. Too much grease can be as harmful as insufficient grease
   c. No grease is better than too much grease
   d. Too much grease is better than insufficient grease

7. Which of the following is not a concern of good motor maintenance?

    a. Number of hours and days the motor operates
    b. Manufacturer of the motor
    c. Environmental conditions
    d. Importance of the motor in the production scheme

8. Which of the following is usually an indication of a bad motor bearing?

    a. Hot bearing housing
    b. Cold bearing housing
    c. No unusual vibration
    d. No unusual noise while motor is running

9. When using compressed air to clean motors, which of the following is one precaution that should be taken?

    a. Make sure the air is warmer than the ambient motor temperature
    b. Make sure too much pressure is not used
    c. Make sure the air is colder than the ambient motor temperature
    d. Use only a high-velocity nozzle

10. When cleaning wound rotor motors, which of the following parts should receive the most attention?

    a. Lifting eyebolt
    b. End bells
    c. Collector rings
    d. Peckerhead

PERFORMANCE/LABORATORY EXERCISE

1. Using an untagged three-phase motor with no nameplate supplied by your instructor, determine as many of the motor's characteristics as possible:

2. Connect the motor leads (in the above motor) for operation on its lower voltage.

3. Connect the motor leads (same motor) for operation on its higher voltage.

Answers to Self-Check Questions

1. b

2. a

3. d

4. c

5. d

6. b

7. b

8. a

9. b

10. c

Motor Controls

Module 20311

MOTOR CONTROLS

Objectives

Upon completion of this module, the trainee will be able to:

1. Describe the operating principles of motor controls and control circuits.
2. Select motor controls for specific applications.
3. Connect motor controllers for specific applications.
4. Design and install motor-control circuits for specific applications.
5. Explain NEC regulations governing the installation of motor controls.
6. Follow NEC requirements when installing motor-control circuits.
7. Interpret motor-control diagrams.
8. Size and select heaters and other protective devices for motor controls.
9. Connect control transformers in conjunction with motor-control circuits.

Prerequisites

Successful completion of the following Task Modules is required before beginning study of this Task Module: Core Curricula, Electrical Levels 1 and 2, Electrical Level 3, Modules 20301 through 20310.

Required Student Materials

1. Trainee Task Module
2. Copy of the latest edition of the National Electrical Code

COURSE MAP INFORMATION

This course map shows all of the *Wheels of Learning* Task Modules in the third level of the Electrical curricula. The suggested training order begins at the bottom and proceeds up. Skill levels increase as a trainee advances on the course map. The training order may be adjusted by the local Training Program Sponsor.

Course Map: Electrical, Level 3

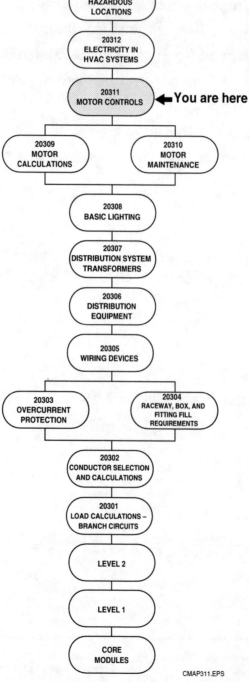

LEVEL 3 COMPLETE

20313 HAZARDOUS LOCATIONS

20312 ELECTRICITY IN HVAC SYSTEMS

20311 MOTOR CONTROLS ◄ You are here

20309 MOTOR CALCULATIONS

20310 MOTOR MAINTENANCE

20308 BASIC LIGHTING

20307 DISTRIBUTION SYSTEM TRANSFORMERS

20306 DISTRIBUTION EQUIPMENT

20305 WIRING DEVICES

20303 OVERCURRENT PROTECTION

20304 RACEWAY, BOX, AND FITTING FILL REQUIREMENTS

20302 CONDUCTOR SELECTION AND CALCULATIONS

20301 LOAD CALCULATIONS – BRANCH CIRCUITS

LEVEL 2

LEVEL 1

CORE MODULES

CMAP311.EPS

TABLE OF CONTENTS

Trade Terms Introduced in This Module

ambient temperature: The temperature of the air where a piece of equipment is situated.

bimetal strip: Temperature regulating or indicating device that works on the principle that two dissimilar metals with unequal expansion rates, welded together, will bend as temperature changes.

control: Automatic or manual device used to stop, start, and/or regulate flow of gas, liquid, and/or electricity.

controller: A device or group of devices that serves to govern in some predetermined manner the electric power delivered to the apparatus to which it is connected.

float switch: A switch that is opened and closed by a float that rises and falls with the level of the liquid in a tank.

inching: Momentary activation of machinery used for inspection or maintenance. Controls for such activation usually consists of *start, stop,* and *inch*. The device used to operate such machinery is usually the AC induction motor—either single- or three-phase.

induction motor: An AC motor that does not run exactly in step with the alternations. Currents supplied are led through the stator coils only; the rotor is rotated by currents induced by the varying field set up by the stator coils.

interlock: A safety device to ensure that a piece of apparatus will not operate until certain conditions have been satisfied.

jogging: The repeated starting and stopping of a motor at frequent intervals for short periods of time. See *inching*.

limit control: Control used to open or close electrical circuits as temperature or pressure limits are reached.

limiter: A device in which some characteristic of the output is automatically prevented from exceeding in predetermined value.

magnetic coil: The winding of an electromagnet. A coil of wire wound in one direction, producing a dense magnetic field capable of attracting iron or steel when carrying a current of electricity.

make and break: The term may be applied to several electrical devices. Primarily, there is a pair of contact points, one stationary, and the other operated by a cam that makes the break in a circuit between these points.

motor control: Device to start and/or stop a motor at certain temperature or pressure conditions.

motor control center: A grouping of motor controls such as starters.

multispeed motor: A motor capable of being driven at any one of two or more different speeds independent of the load.

NC: Normally closed.

NO: Normally open.

nonautomatic: Used to describe an action requiring personal intervention for its control.

phase protection: A means of preventing damage in an electric motor through overheating in the event a fuse blows or a wire breaks when the motor is running. In order to provide the protection, phase-failure and -reversal relays are used.

plugging: Braking an induction motor by reversing the phase sequence of the power to the motor. This reversal causes the motor to develop a counter torque, which results in the exertion of a retarding force. Plugging is used to secure both rapid stop and quick reversal.

relay: A device designed to abruptly change a circuit because of a specified control input.

relay, overcurrent: A relay designed to open a circuit when current in excess of a particular setting flows through the sensor.

starter: 1) An electric controller for accelerating a motor from rest to normal speed and for stopping the motor. 2) A device used to start an electric discharge lamp.

temperature rise: The difference between the winding temperature of the motor when running and the ambient temperature.

thermally protected (as applied to motors): When the words thermally protected appear on the nameplate of a motor or motor-compressor, it means that the motor is provided with a thermal protector designed to protect the motor from overloads.

thermal protector: A protective device that is assembled as an integral part of a motor or motor-compressor and that, when properly applied, protects the motor against dangerous overheating due to overload and failure to start.

time (duty) rating: Ratings based on a fixed operating time (5, 15, 30, 60 minutes) after which the motor must be allowed to cool.

1.0.0 INTRODUCTION TO MOTOR CONTROLS

Electric motors provide one of the principal sources for driving all types of equipment and machinery. Every motor in use, however, must be controlled, if only to start and stop it, before it becomes of any value.

Motor controllers cover a wide range of types and sizes, from a simple toggle switch to a complex system with such components as relays, timers, and switches. The common function, however, is the same in any case, that is, to control some operation of an electric motor. A motor controller will include some or all of the following functions:

- Starting and stopping
- Overload protection
- Overcurrent protection
- Reversing
- Changing speed
- Jogging
- Plugging
- Sequence control
- Pilot light indication

The controller can also provide the control for auxiliary equipment such as brakes, clutches, solenoids, heaters, and signals, and may be used to control a single motor or a group of motors.

The term *motor starter* is often used and means practically the same thing as a *controller*. Strictly, a motor starter is the simplest form of controller and is capable of starting and stopping the motor and providing it with overload protection.

See Fig. 1 for a review of the electrical symbols used in the wiring diagrams in this module.

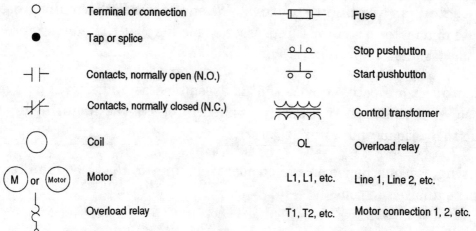

Figure 1. Symbols used for drawing wiring diagrams.

2.0.0 TYPES OF MOTOR CONTROLLERS

A large variety of motor controllers are available that will handle almost every conceivable application. However, all of them can be grouped in the following categories.

2.1.0 PLUG-AND-RECEPTACLE

NEC Section 430-81 defines a controller as any switch or device normally used to start and stop a motor by making and breaking the motor circuit current. The simplest form of controller allowed by the NEC is an attachment plug and receptacle. See Fig. 2. However, such an arrangement is limited to portable motors rated at $\frac{1}{3}$ horsepower or less.

Referring again to Fig. 2, note that drawing (A) is a pictorial view of a portable motor with a cord-and-plug assembly attached. If this motor is portable and less than $\frac{1}{3}$ horsepower, then the plug and receptacle may act as the motor's controller as permitted in NEC Section 430-81(c).

Drawing (B) in Fig. 2 is the same circuit, but this time depicted in the form of a wiring diagram. Note that symbols have been used to represent the various circuit items rather than actually drawing the items in life form; yet they are arranged on the basis of their physical relationship to each other. This simplifies the drawing — both from a drafters point of view and also those who must interpret the drawing.

Another form of drawing for this same circuit is shown in (C) of Fig. 2. This type of drawing has become known as the *ladder diagram*; it's a schematic representation of the electrical circuit in question — the same as the drawing in (B). However, ladder diagrams are drawn in an H format, with the energized power conductors represented by vertical lines and the individual circuits represented by horizontal lines. Rather than physically representing the circuit items as in drawing (B), a ladder diagram arranges the conductors and electrical components according to their electrical function in the circuit; that is, schematically. Therefore, ladder diagrams merely represent the current paths (shown as the rungs of a ladder) to each of the controlled or energized output devices.

Where stationary motors rated at $\frac{1}{8}$ HP or less that are normally left running (clock motors, fly fans, and the like), and are so constructed that they cannot be damaged by overload or failure to start, the branch-circuit protective device may serve as the controller. Consequently, the branch-circuit breaker or fusible disconnect serves as both branch-circuit overcurrent protection and motor controller. Such a circuit appears in Fig. 3 on page 9 of this module.

2.2.0 MANUAL STARTERS

A manual starter is a motor controller whose contact mechanism is operated by a mechanical linkage from a toggle handle or push button, which is in turn operated by hand. A thermal unit and direct-acting overload mechanism provide motor running overload protection. Basically, a manual starter is an ON-OFF switch with overload relays.

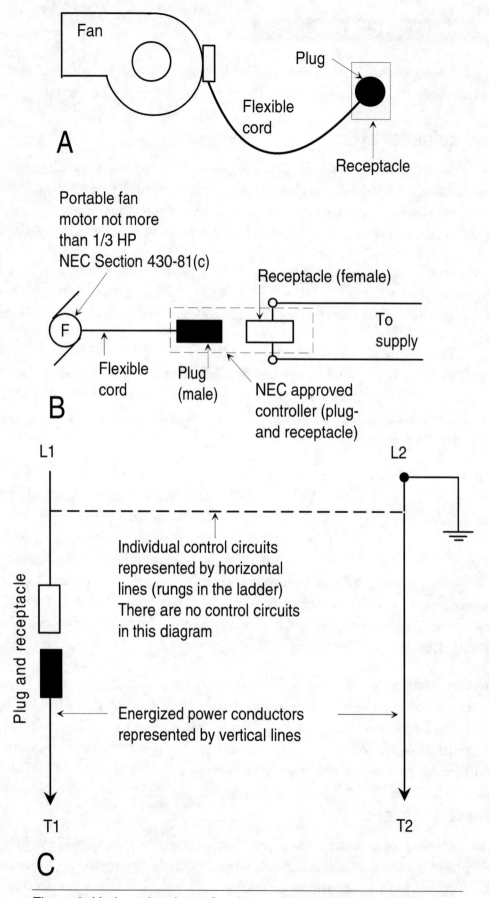

Fan

Plug

Flexible
cord

Receptacle

A

Portable fan
motor not more
than 1/3 HP
NEC Section 430-81(c)

Receptacle (female)

To
supply

B

Flexible
cord

Plug
(male)

NEC approved
controller (plug-
and receptacle)

L1

L2

Individual control circuits
represented by horizontal
lines (rungs in the ladder)
There are no control circuits
in this diagram

Plug and receptacle

Energized power conductors
represented by vertical lines

T1

T2

C

Figure 2. Various drawings of a plug-and-receptacle motor controller.

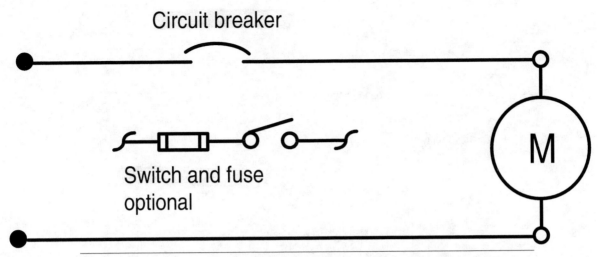

Figure 3. Branch-circuit protective device serving as the motor controller.

Manual starters are used mostly on small machine tools, fans and blowers, pumps, compressors, and conveyors. They have the lowest cost of all motor starters, have a simple mechanism, and provide quiet operation with no AC magnet hum. The contacts, however, remain closed and the lever stays in the ON position in the event of a power failure, causing the motor to automatically restart when the power returns. Therefore, low-voltage protection and low-voltage release are not possible with these manually operated starters. However, this action is an advantage when the starter is applied to motors that run continuously.

2.2.1   Fractional Horsepower Manual Starters

Fractional-horsepower manual starters are designed to control and provide overload protection for motors of 1 hp or less on 120- or 240-volt single-phase circuits. They are available in single- and two-pole versions and are operated by a toggle handle on the front. When a serious overload occurs, the thermal unit trips to open the starter contacts, disconnecting the motor from the line. The contacts cannot be reclosed until the overload relay has been reset by moving the handle to the full OFF position, after allowing about 2 minutes for the thermal unit to cool. The open-type starter will fit into a standard outlet box and can be used with a standard flush plate. The compact construction of this type of device makes it possible to mount it directly on the driven machinery and in various other places where the available space is small. Figure 4 shows fractional horsepower (FHP) manual motor-starter wiring diagrams for both 120- and 240-volt single-phase motors, along with a pictorial representation.

Note that the single-pole FHP starter has only one contact to trip and disconnect the motor from the line; the grounded or neutral conductor is not opened when the handle is in the OFF position. This single-pole starter also has one overload relay connected in series with the ungrounded conductor.

The two-pole FHP starter has two contacts to open both phases when connected to a 240-volt circuit. When the toggle handle is in the off position, no current flows to the motor. However,

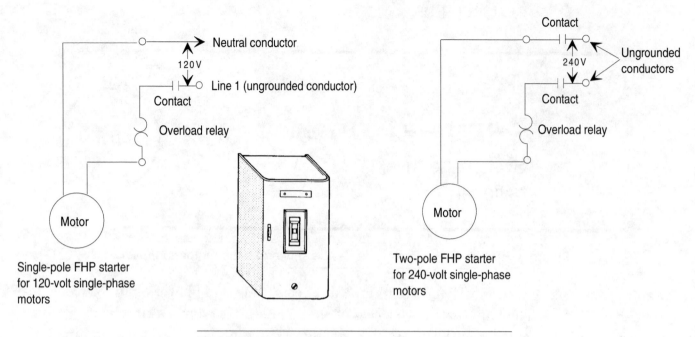

Figure 4. Wiring diagrams of FHP manual starters.

only one overload relay is needed, since one will shut down the motor if the relay detects an overload and opens.

2.2.2   Manual Motor-Starting Switches

Manual motor starting switches provide ON-OFF control of single- or three-phase AC motors where overload protection is not required or is separately provided. Two- or three-pole switches are available with ratings up to 10 hp, 600 volts, three phase. The continuous current rating is 30 amperes at 250 volts maximum and 20 amperes at 600 volts maximum. The toggle operation of the manual switch is similar to the fractional-horsepower starter, and typical applications of the switch include pumps, fans, conveyors, and other electrical machinery that have separate motor protection. They are particularly suited to switch nonmotor loads, such as resistance heaters.

2.2.3   Integral Horsepower Manual Starters

The integral horsepower manual starter is available in two- and three-pole versions to control single-phase motors up to 5 hp and polyphase motors up to 10 hp, respectively.

Two-pole starters have one overload relay and three-pole starters usually have three overload relays. When an overload relay trips, the starter mechanism unlatches, opening the contacts to stop the motor. The contacts cannot be reclosed until the starter mechanism has been reset by pressing the STOP button or moving the handle to the RESET position, after allowing time for the thermal unit to cool.

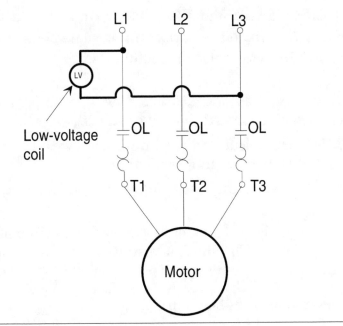

Figure 5. Integral HP manual starter with low-voltage protection.

Integral horsepower manual starters with low-voltage protection prevent automatic start-up of motors after a power loss. This is accomplished with a continuous-duty solenoid, which is energized whenever the line-side voltage is present. If the line voltage is lost or disconnected, the solenoid de-energizes, opening the starter contacts. The contacts will not automatically close when the voltage is restored to the line. To close the contacts, the device must be manually reset. This manual starter will not function unless the line terminals are energized. This is a safety feature that can protect personnel or equipment from damage and is used on such equipment as conveyors, grinders, metal-working machines, mixers, woodworking, etc. Figure 5 shows a wiring diagram of an integral HP manual starter with low-voltage protection.

2.3.0 MAGNETIC CONTROLLERS

Magnetic motor controllers use electromagnetic energy for closing switches. The electromagnet consists of a coil of wire placed on an iron core. When current flows through the coil, the iron of the magnet becomes magnetized and attracts the iron bar, called the *armature*. An interruption of the current flow through the coil of wire causes the armature to drop out due to the presence of an air gap in the magnetic circuit.

Line-voltage magnetic motor starters are electromechanical devices that provide a safe, convenient, and economic means for starting and stopping motors, and they have the disadvantage of being controlled remotely. The great bulk of motor controllers are of this type. Therefore, the operating principles and applications of magnet motor controllers should be fully understood.

In the construction of a magnetic controller, the armature is mechanically connected to a set of contacts so that, when the armature moves to its closed position, the contacts also close.

When the coil has been energized and the armature has moved to the closed position, the controller is said to be *picked up* and the armature is seated or sealed-in. Some of the magnet and armature assemblies in current use are as follows:

- *Clapper type:* In this type, the armature is hinged. As it pivots to seal in, the movable contacts close against the stationary contacts.
- *Vertical action:* The action is a straight line motion with the armature and contacts being guided so that they move in a vertical plane.
- *Horizontal action:* Both armature and contacts move in a straight line through a horizontal plane.
- *Bell crank:* A bell crank lever transforms the vertical action of the armature into a horizontal contact motion. The shock of armature pickup is not transmitted to the contacts, resulting in minimum contact bounce and longer contact life.

These four types of assemblies are shown in Fig. 6.

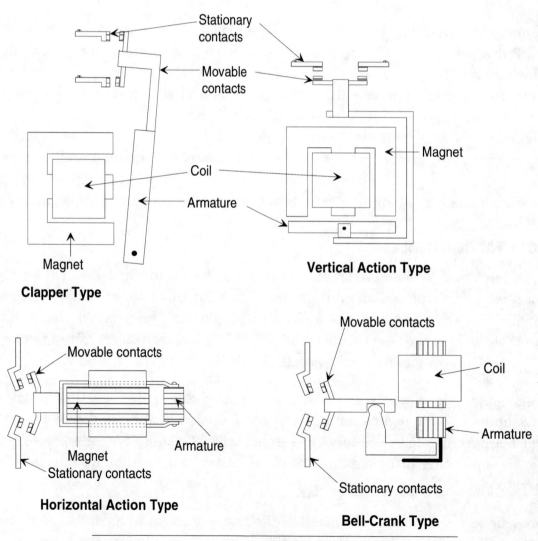

Figure 6. Several types of magnetic-armature assemblies.

The magnetic circuit of a controller consists of the magnet assembly, the coil, and the armature. It is so named from a comparison with an electrical circuit. The coil and the current flowing in it causes magnetic flux to be set up through the iron in a similar manner to a voltage causing current to flow through a system of conductors. The changing magnetic flux produced by alternating currents results in a temperature rise in the magnetic circuit. The heating effect is reduced by laminating the magnet assembly and armature by placing a coil of many turns of wire around a soft iron core, the magnetic flux set up by the energized coil tends to be concentrated; therefore, the magnetic field effect is strengthened. Since the iron core is the path of least resistance to the flow of the magnetic lines of force, magnetic attraction will concentrate according to the shape of the magnet.

The magnetic assembly is the stationary part of the magnetic circuit. The coil is supported by and surrounds part of the magnet assembly in order to induce magnetic flux into the magnetic circuit.

The armature is the moving part of the magnetic circuit. When it has been attracted into its sealed-in position, it completes the magnetic circuit. To provide maximum pull and to help ensure quietness, the faces of the armature and the magnetic assembly are ground to a very close tolerance.

When a controller's armature has sealed-in, it is held closely against the magnet assembly. However, a small gap is always deliberately left in the iron circuit. When the coil becomes de-energized, some magnetic flux (residual magnetism) always remains, and if it were not for the gap in the iron circuit, the residual magnetism might be sufficient to hold the armature in the sealed-in position. See Fig. 7.

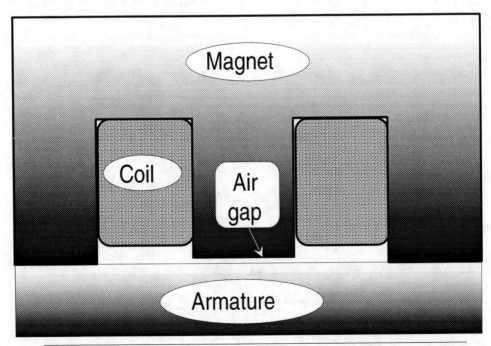

Figure 7. A small air gap is always deliberately left in the iron circuit.

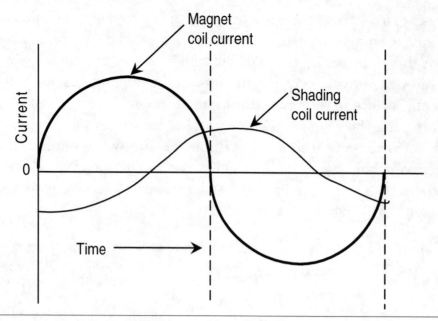

Figure 8. Auxiliary flux produces a magnetic pull out of phase from the main flux.

The shaded-pole principle is used to provide a time delay in the decay of flux in DC coils, but it is used more frequently to prevent a chatter and wear in the moving parts of AC magnets. A shading coil is a single turn of conducting material mounted in the face of the magnet assembly or armature. The alternating main magnetic flux induces currents in the shading coil, and these currents set up auxiliary magnetic flux that is out of phase from the pull due to the main flux, and this keeps the armature sealed-in when the main flux falls to zero (which occurs 120 times per second with 60-cycle AC). Without the shading coil, the armature would tend to open each time the main flux goes through zero. Excessive noise, wear on magnet faces, and heat would result. See Fig. 8. A magnet assembly and armature showing shading coils is shown in Fig. 9.

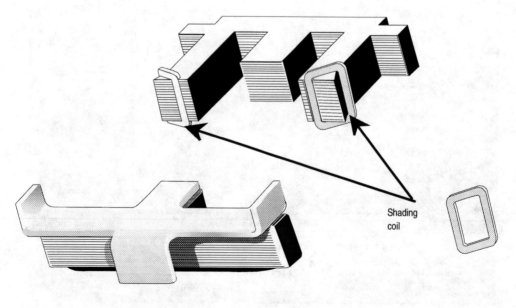

Figure 9. Magnet assembly and armature along with shading coils.

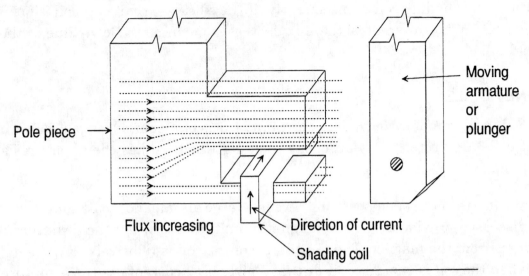

Flux increasing

Pole piece

Direction of current

Shading coil

Moving armature or plunger

Figure 10. Section of pole face with current in clockwise direction.

For a further explanation of shaded-coil principles, look at Fig. 10 where an exaggerated view of a pole face with a copper band or short-circuited coil of low resistance connected around a portion of the pole tip is shown. When the flux is increasing in the pole from left to right, the induced current in the coil is in a clockwise direction.

The magnetomotive force produced by the coil opposes the direction of the flux of the main field. Therefore, the flux density in the shaded portion of the iron will be considerably less, and the flux density in the unshaded portion of the iron will be more than would be the case without the shading coil.

Figure 11 shows the pole with the flux still moving from left to right but decreasing in value. Now the current in the coil is in a counterclockwise direction. The magnetomotive force produced by the coil is in the same direction as the main unshaded portion but less than it

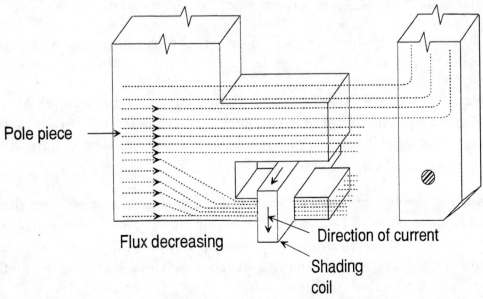

Pole piece

Flux decreasing

Direction of current

Shading coil

Figure 11. Section of pole face with current in counterclockwise direction.

would be without the shading coil. Consequently, if the electric circuit of a coil is opened, the current decreases rapidly to zero, but the flux decreases much more slowly due to the action of the shading coil.

3.0.0   MAGNET COILS

The magnet coil used in motor controllers has many turns of insulated copper wire wound on a spool. Most coils are protected by an epoxy molding which makes them very resistant to mechanical damage.

When the controller is in the open position there is a large air gap (not to be confused with the built-in gap discussed previously) in the magnet circuit; this is when the armature is at its furthest distance from the magnet. The impedance of the coil is relatively low, due to the air gap, so that when the coil is energized, it draws a fairly high current. As the armature moves closer to the magnet assembly, the air gap is progressively reduced, and with it, the coil current, until the armature has sealed in. The final current is referred to as the sealed current. The inrush current is approximately 6 to 10 times the sealed current. The ratio varies with individual designs. After the controller has been energized for some time, the coil will become hot. This will cause the coil current to fall to approximately 80% of its value when cold.

AC magnetic coils should never be connected in series. If one device were to seal-in ahead of the other, the increased circuit impedance will reduce the coil current so that the "slow" device will not pick up or, having picked up, will not seal. Consequently, AC coils are always connected in parallel.

Magnet coil data is usually given in volt-amperes (VA). For example, given a magnetic starter whose coils are rated at 600 VA inrush and 60 VA sealed, the inrush current of a 120-volt coil is 600/120 or 5 amperes. The same starter with a 480-volt coil will only draw 600/480 or 1.25 amperes inrush and 60/480 or .125 amperes sealed.

Pick-up voltage: The minimum voltage which will cause the armature to start to move is called the pick-up voltage.

Sealed-in voltage: The seal-in voltage is the minimum control voltage required to cause the armature to seat against the pole faces of the magnet. On devices using a vertical action magnet and armature, the seal-in voltage is higher than the pick-up voltage to provide additional magnetic pull to insure good contact pressure.

Control devices using the bell-crank armature and magnet arrangement are unique in that they have different force characteristics. Devices using this operating principle are designed to have a lower seal-in voltage than pick-up voltage. Contact life is extended, and contact damage under abnormal voltage conditions is reduced, for if the voltage is sufficient to pick-up, it is also high enough to seat the armature.

If the control voltage is reduced sufficiently, the controller will open. The voltage at which this happens is called the *drop-out* voltage. It is somewhat lower than the seal-in voltage.

3.1.0 VOLTAGE VARIATION

NEMA standards require that the magnetic device operate properly at varying control voltages from a high of 110% to a low of 85% of rated coil voltage. This range, established by coil design, insures that the coil will withstand given temperature rises at voltages up to 10% over rated voltage, and that the armature will pick up and seal in, even though the voltage may drop to 15% under the nominal rating.

3.1.1 Effects of Voltage Variation

If the voltage applied to the coil is too high, the coil will draw more than its designed current. Excessive heat will be produced and will cause early failure of the coil insulation. The magnetic pull will be too high, which will cause the armature to slam home with excessive force. The magnet faces will wear rapidly, leading to a shortened life for the controller. In addition, contact bounce may be excessive, resulting in reduced contact life.

Low control voltage produces low coil currents and reduced magnetic pull. On devices with vertical action assemblies, if the voltage is greater than pick-up voltage, but less than seal-in voltage, the controller may pick up but will not seal. With this condition, the coil current will not fall to the sealed value. As the coil is not designed to carry continuously a current greater than its sealed current, it will quickly get very hot and burn out. The armature will also chatter. In addition to the noise, wear on the magnet faces result.

In both vertical action and bell-crank construction, if the armature does not seal, the contacts will not close with adequate pressure. Excessive heat, with arcing and possible welding of the contacts, will occur as the controller attempts to carry current with insufficient contact pressure.

3.2.0 AC HUM

All AC devices which incorporate a magnetic effect produce a characteristic hum. This hum or noise is due mainly to the changing magnetic pull (as the flux changes) inducing mechanical vibrations. Contactors, starters and relays could become excessively noisy as a result of some of the following operating conditions:

- Broken shading coil.
- Operating voltage too low.
- Misalignment between the armature and magnet assembly—the armature is then unable to seat properly.
- Wrong coil.

- Dirt, rust, filings, etc. on the magnet faces—the armature is unable to seal in completely.

- Jamming or binding of moving parts so that full travel of the armature is prevented.

- Incorrect mounting of the controller, as on a thin piece of plywood fastened to a wall; such mounting may cause a "sounding board" effect.

4.0.0   POWER CIRCUITS IN MOTOR STARTERS

The power circuit of a starter includes the stationary and movable contacts, and the thermal unit or heater portion of the overload relay assembly. The number of contacts (or "poles") is determined by the electrical service. In a 3-phase, 3-wire system, for example, a 3-pole starter is required. See Fig. 12.

To be suitable for a given motor application, the magnetic starter selected should equal or exceed the motor horsepower and full-load current ratings. For example, let's assume that we want to select a motor starter for a 50-hp motor to be supplied by a 240-volt, 3-phase service, and the full-load current of the motor is 125 amperes. Referring to the table in Fig. 13, it can be seen that a NEMA Size 4 starter would be required for normal motor duty. If the motor is to be used for jogging or plugging duty, a NEMA Size 5 starter should be chosen.

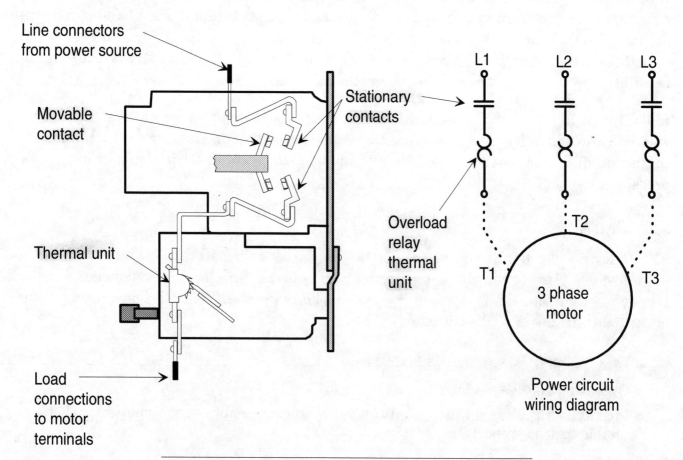

Figure 12. Power circuit in a typical 3-pole magnetic starter.

CAUTION! For three-phase motors having locked-rotor kVA per horsepower in excess of that for the motor code letters in the right table (Fig. 13), do not apply the controller at its maximum rating without consulting the manufacturer. In most cases, the next higher horsepower rated controller should be used. Per NEC Section 430-83(a) Exception 1, controllers for NEMA design E motors of more than 2 HP shall be marked for use with a design E motor or shall have an HP rating of no less than 1.4 times the rating of a motor rated 3 through 100 HP, or no less than 1.3 times the rating of a motor over 100 HP.

Power circuit contacts handle the motor load. The ability of the contacts to carry the full-load current without exceeding a rated temperature rise, and their isolation from adjacent contacts, corresponds to NEMA Standards established to categorize the NEMA Size of the starter. The starter must also be capable of interrupting the motor circuit under locked rotor current conditions.

NEMA Size	Volts	Maximum Horsepower Rating—Non plugging and Nonjogging Duty		Maximum Horsepower Rating—Plugging and Jogging Duty		Continuous Current Rating, amperes—600 Volt Max.	Service-Limit Current Rating, Amperes *	Tungsten and Infrared Lamp Load, Amperes—250 Volts Max. *	Resistance heating Loads, kW—other than Infrared Lamp Loads		KVA Rating for Switching Transformer Primaries at 50 or 60 Cycles		3-Phase Rating for Switching Capacitors
		Single Phase	Poly-Phase	Single Phase	Poly-Phase				Single Phase	Poly-phase	Single Phase	Poly-Phase	Kvar
00	115	⅓	...	...	...	9	11	5	...	...	...	...	...
	200	...	1½	...	...	9	11	5	...	...	...	...	...
	230	1	1½	...	...	9	11	5	...	...	...	...	...
	380	...	1½	...	...	9	11	...	...	...	...	...	...
	460	...	2	...	...	9	11	...	...	...	...	...	...
	575	...	2	...	...	9	11	...	...	...	...	...	...
0	115	1	...	½	...	18	21	10	...	...	0.9	1.2	...
	200	...	3	...	1½	18	21	10	...	...	...	1.4	...
	230	2	3	1	1½	18	21	10	...	...	1.4	1.7	...
	380	...	5	...	1½	18	21	...	...	...	...	2.0	...
	460	...	5	...	2	18	21	...	...	...	1.9	2.5	...

Figure 13. Electrical ratings for AC magnetic contactors and starters.

NEMA Size	Volts	Maximum Horsepower Rating—Non plugging and Nonjogging Duty		Maximum Horsepower Rating—Plugging and Jogging Duty		Continuous Current Rating, amperes—600 Volt Max.	Service-Limit Current Rating, Amperes *	Tungsten and Infrared Lamp Load, Amperes—250 Volts Max. *	Resistance heating Loads, kW—other than Infrared Lamp Loads		KVA Rating for Switching Transformer Primaries at 50 or 60 Cycles		3-Phase Rating for Switching Capacitors
		Single Phase	Poly-Phase	Single Phase	Poly-Phase				Single Phase	Poly-phase	Single Phase	Poly-Phase	Kvar
1	115	2	...	1	...	27	32	15	3	5	1.4	1.7	...
	200	...	7½	...	3	27	32	15	...	9.1	...	3.5	...
	230	3	7½	2	3	27	32	15	6	10	1.9	4.1	...
	380	...	10	...	5	27	32	...	...	16.5	...	4.3	...
	460	...	10	...	5	27	32	...	12	20	3	5.3	...
	575	...	10	...	5	27	32	...	15	25	3	5.3	...
1P	115	3	...	1½	...	36	42	24	...	...	...	...	...
	230	5	...	3	...	36	42	24	...	...	...	...	...
2	115	3	...	2	...	45	52	30	5	8.5	1.0	4.1	...
	200	...	10	...	7½	45	52	30	...	15.4	...	6.6	11.3
	230	7½	15	5	10	45	52	30	10	17	4.6	7.6	13
	380	...	25	...	15	45	52	...	...	28	...	9.9	21
	460	...	25	...	15	45	52	...	20	34	5.7	12	26
	575	...	25	...	15	45	52	...	25	43	5.7	12	33
3	115	7½	...	...	...	90	104	60	10	17	4.6	7.6	...
	200	...	25	...	15	90	104	60	...	31	...	13	23.4
	230	15	30	...	20	90	104	60	20	34	8.6	15	27
	380	...	50	...	30	90	104	...	...	56	...	19	43.7
	460	...	50	...	30	90	104	...	40	68	14	23	53
	575	...	50	...	30	90	104	...	50	86	14	23	67

Figure 13. Electrical ratings for AC magnetic contactors and starters. (*Cont.*)

NEMA Size	Volts	Maximum Horsepower Rating—Non plugging and Nonjogging Duty		Maximum Horsepower Rating—Plugging and Jogging Duty		Continuous Current Rating, amperes—600 Volt Max.	Service-Limit Current Rating, Amperes *	Tungsten and Infrared Lamp Load, Amperes—250 Volts Max. *	Resistance heating Loads, kW—other than Infrared Lamp Loads		KVA Rating for Switching Transformer Primaries at 50 or 60 Cycles		3-Phase Rating for Switching Capacitors
		Single Phase	Poly-Phase	Single Phase	Poly-Phase				Single Phase	Poly-phase	Single Phase	Poly-Phase	Kvar
4	200	...	40	...	25	135	156	120	...	45	...	20	34
	230	...	50	...	30	135	156	120	30	52	11	23	40
	380	...	75	...	50	135	156	...	...	86.7	...	38	66
	460	...	100	...	60	135	156	...	60	105	22	46	80
	575	...	100	...	60	135	156	...	75	130	22	46	100
5	200	...	75	...	60	270	311	240	...	91	...	40	69
	230	...	100	...	75	270	311	240	60	105	28	46	80
	380	...	150	...	125	270	311	...	...	173	...	75	132
	460	...	200	...	150	270	311	...	120	210	40	91	160
	575	...	200	...	150	270	311	...	150	260	40	91	200
6	200	...	150	...	125	540	621	480	...	182	...	79	139
	230	...	200	...	150	540	621	480	120	210	57	91	160
	380	...	300	...	250	540	621	...	...	342	...	148	264
	460	...	400	...	300	540	621	...	240	415	86	180	320
	575	...	400	...	300	540	621	...	300	515	86	180	400
7	230	...	300	...	...	810	932	720	180	315	...	...	240
	460	...	600	...	...	810	932	...	360	625	...	...	480
	575	...	600	...	...	810	932	...	450	775	...	...	600
8	230	...	450	...	...	1215	1400	1080	...	...	...	...	360
	460	...	900	...	...	1215	1400	...	...	...	...	...	720

* Per NEMA Standards paragraph IC 1-21A, 20, the service-limit current represents the maximum rms current, in amperes, which the controller may be expected to carry for protracted periods in normal service.

Figure 13. Electrical ratings for AC magnetic contactors and starters. (*Cont.*)

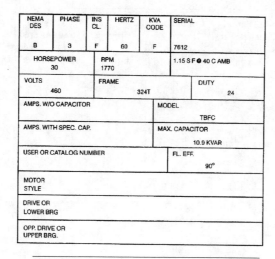

NEMA DES	PHASE	INS CL.	HERTZ	KVA CODE	SERIAL
B	3	F	60	F	7612

HORSEPOWER 30	RPM 1770		1.15 S F @ 40 C AMB

VOLTS 460	FRAME	324T	DUTY	24

AMPS. W/O CAPACITOR		MODEL	TBFC

AMPS. WITH SPEC. CAP.		MAX. CAPACITOR	10.9 KVAR

USER OR CATALOG NUMBER		FL. EFF.	90°

MOTOR STYLE	
DRIVE OR LOWER BRG	
OPP. DRIVE OR UPPER BRG.	

Figure 14. Typical motor nameplate.

4.1.0 MOTOR NAMEPLATES

Much information about sizing motor starters can be found from the motor's nameplate. Consequently, a review of motor nameplates is in order.

A typical motor nameplate is shown in Fig. 14. A nameplate is one of the most important parts of a motor since it gives the motor's electrical and mechanical characteristics; that is, the horsepower, voltage, rpms, etc. Always refer to the motor's nameplate before connecting it to an electric system or selecting the motor starter and related components. The same is true when performing preventative maintenance or troubleshooting motors.

Referring again to the motor nameplate in Fig. 14, note that the manufacturer's name and logo is at the top of the plate; these items, of course, will change with each manufacturer. The line directly below the manufacturer's name identifies the motor for use on AC systems as opposed to DC or AC-DC systems. The model number identifies that particular motor from any other. The type or class specifies the insulation used to ensure the motor will perform at the rated horsepower and service-factor load. The phase indicates whether the motor has been designed for single- or three-phase use.

When selecting motor starters for a given motor, the motor code letter and the NEMA design letter on the nameplate plays an important role. The chart in Fig. 15 may be used as a guide when selecting motor controllers. For design letter E motors, refer to NEC Section 430-83(a) Exception 1.

Controller HP Rating	Maximum Allowable Motor Code Letter
1½	L
3 - 5	K
7½ & above	H

Figure 15. Using motor code letters to size motor controllers.

5.0.0 OVERLOAD PROTECTION

Overload protection for an electric motor is necessary to prevent burnout and to ensure maximum operating life. Electric motors will, if permitted, operate at an output of more than rated capacity. Conditions of motor overload may be caused by an overload on driven machinery, by a low line voltage, or by an open line in a polyphase system, which results in single-phase operation. Under any condition of overload, a motor draws excessive current that causes overheating. Since motor winding insulation deteriorates when subjected to overheating, there are established limits on motor operating temperatures. To protect a motor from overheating, overload relays are employed on a motor control to limit the amount of current drawn. This is overload protection, or running protection.

The ideal overload protection for a motor is an element with current-sensing properties very similar to the heating curve of the motor (see Fig. 16), which would act to open the motor circuit when full-load current is exceeded. The operation of the protective device should be such that the motor is allowed to carry harmless overloads, but is quickly removed from the line when an overload has persisted too long.

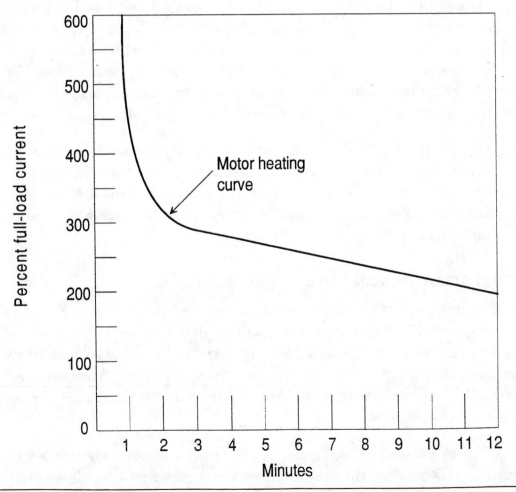

Figure 16. Sensing properties of overload protection should follow heating curve of motor.

Single element (non-time delay) fuses are not designed to provide overload protection. Their basic function is to protect against short circuits (overcurrent protection). Motors draw a high inrush current when starting and conventional single-element fuses have no way of distinguishing between this temporary and harmless inrush current and a damaging overload. Such fuses, chosen on the basis of motor full-load current, would blow every time the motor is started. On the other hand, if a fuse is chosen large enough to pass the starting or inrush current, it would not protect the motor against small, harmful overloads that might occur later.

Dual-element or time-delay fuses can provide motor overload protection, but suffer the disadvantages of being nonrenewable and must be replaced.

The overload relay is the heart of motor protection. It has inverse trip-time characteristics, permitting it to hold in during the accelerating period (when inrush current is drawn), yet providing protection on small overloads above the full-load current when the motor is running. Unlike dual-element fuses, overload relays are renewable and can withstand repeated trip and reset cycles without need of replacement. They cannot, however, take the place of overcurrent protective equipment.

The overload relay consists of a current-sensing unit connected in the line to the motor, plus a mechanism, actuated by the sensing unit, that serves to directly or indirectly break the circuit. In a manual starter, an overload trips a mechanical latch and causes the starter contacts to open and disconnect the motor from the line. In magnetic starters, an overload opens a set of contacts within the overload relay itself. These contacts are wired in series with the starter coil in the control circuit of the magnetic starter. Breaking the coil circuit causes the starter contacts to open, disconnecting the motor from the line.

Overload relays can be classified as being either thermal or magnetic. Magnetic overload relays react only to current excesses and are not affected by temperature. As the name implies, thermal overload relays rely on the rising temperatures caused by the overload current to trip the overload mechanism. Thermal overload relays can be further subdivided into two types, melting alloy and bimetallic.

5.1.0   MELTING ALLOY THERMAL OVERLOAD RELAYS

The melting alloy assembly of the heater element overload relay and solder pot is shown in Fig. 17. Excessive overload motor current passes through the heater element, thereby melting an eutectic alloy solder pot. The ratchet wheel will then be allowed to turn in the molten pool, and a tripping action of the starter control circuit results, stopping the motor. A cooling off period is required to allow the solder pot to "freeze" before the overload relay assembly may be reset and motor service restored.

Melting alloy thermal units are interchangeable and of a one-piece construction, which ensures a constant relationship between the heater element and solder pot and allows factory calibration, making them virtually tamper-proof in the field. These important features are not possible

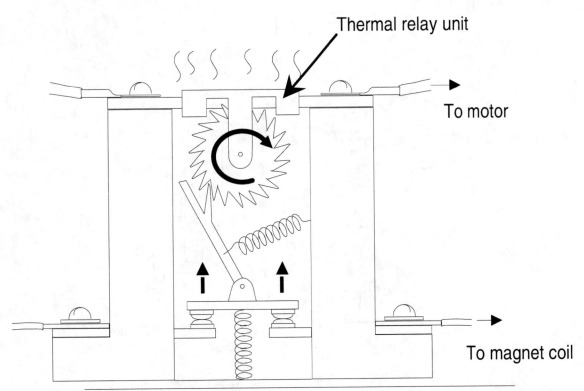

Thermal relay unit

To motor

To magnet coil

Figure 17. Operating characteristics of a melting alloy overload relay.

with any other type of overload relay construction. A wide selection of these interchangeable thermal units is available to give exact overload protection of any full-load current to a motor. An internal view of a melting-alloy relay is shown in Fig. 17. Again, as heat melts the alloy, the ratchet wheel is free to turn; the spring, in turn, pushes the contacts open and shuts down the motor.

5.2.0   BIMETALLIC THERMAL OVERLOAD RELAYS

Bimetallic overload relays are designed specifically for two general types of application: the automatic reset feature is of decided advantage when devices are mounted in locations not easily accessible for manual operation and, second, these relays can easily be adjusted to trip within a range of 85 to 115 percent of the nominal trip rating of the heater unit. This feature is useful when the recommended heater size might result in unnecessary tripping, while the next larger size would not give adequate protection. Ambient temperatures affect overload relays operating on the principle of heat. Figure 18 shows a bimetallic overload relay with side cover removed.

5.3.0   AMBIENT COMPENSATION

Ambient-compensated bimetallic overload relays were designed for one particular situation, that is, when the motor is at a constant temperature and the controller is located separately in a varying temperature. In this case, if a standard thermal overload relay was used, it would not trip consistently at the same level of motor current if the controller temperature changed. This thermal overload relay is always affected by the surrounding temperature. To compensate

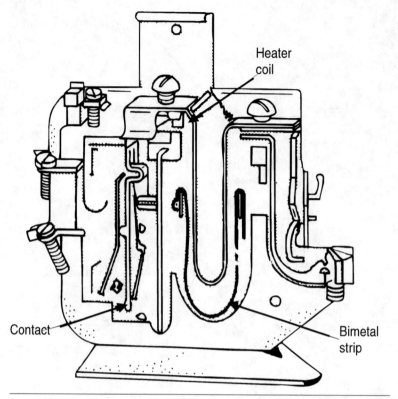

Figure 18. Bimetallic overload relay with side cover removed.

for the temperature variations the controller may see, an ambient-compensated overload relay is applied. Its trip point is not affected by temperature and it performs consistently at the same value of current.

Melting alloy and bimetallic overload relays are designed to approximate the heat actually generated in the motor. As the motor temperature increases, so does the temperature of the thermal unit. The motor and relay heating curves (see Fig.19) show this relationship. From this graph, we can see that, no matter how high the current drawn, the overload relay will provide protection, yet the relay will not trip out unnecessarily.

5.4.0   SELECTING OVERLOAD RELAYS

When selecting thermal overload relays, the following must be considered:

- Motor full-load current
- Type of motor
- Difference in ambient temperature between motor and controller

Motors of the same horsepower and speed do not all have the same full-load current, and the motor nameplate must always be checked to obtain the full-load amperes for a particular motor. Do not use a published table. Thermal unit selection tables are published on the basis of continuous-duty motors, with 1.15 service factor, operating under normal conditions. The

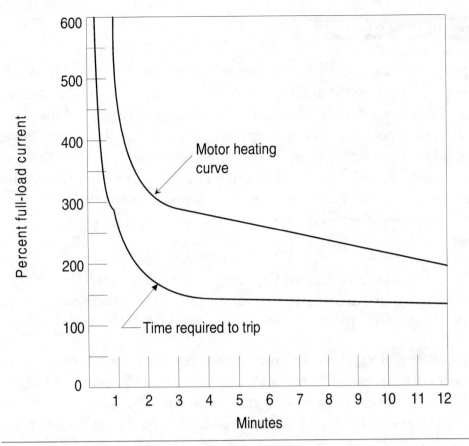

Figure 19. Comparison of motor heating curve and overload relay trip curve.

tables are shown in the catalog of manufacturers and also appear on the inside of the door or cover of the motor controller. These selections will properly protect the motor and allow the motor to develop its full horsepower, allowing for the service factor, if the ambient temperature is the same at the motor as at the controller. If the temperatures are not the same, or if the motor service factor is less than 1.15, a special procedure is required to select the proper thermal unit. Standard overload relay contacts are closed under normal conditions and open when the relay trips. An alarm signal is sometimes required to indicate when a motor has stopped due to an overload trip. Also, with some machines, particularly those associated with continuous processing, it may be required to signal an overload condition, rather than have the motor and process stop automatically. This is done by fitting the overload relay with a set of contacts that close when the relay trips, thus completing the alarm circuit. These contacts are appropriately called alarm contacts.

A magnetic overload relay has a movable magnetic core inside a coil that carries the motor current. The flux set up inside the coil pulls the core upward. When the core rises far enough, it trips a set of contacts on the top of the relay. The movement of the core is slowed by a piston working in an oil-filled dashpot mounted below the coil. This produces an inverse-time characteristic. The effective tripping current is adjusted by moving the core on a threaded rod. The tripping time is varied by uncovering oil bypass holes in the piston. Because of the time and current adjustments, the magnetic overload relay is sometimes used to protect motors having long accelerating times or unusual duty cycles.

The correct selection and installation of an enclosure for a particular application can contribute considerably to the length of life and trouble-free operation. To shield electrically live parts from accidental contact, some form of enclosure is always necessary. This function is usually filled by a general-purpose, sheet-steel cabinet. Frequently, however, dust, moisture, or explosive gases make it necessary to employ a special enclosure to protect the motor controller from corrosion or the surrounding equipment from explosion. In selecting and installing control apparatus, it is always necessary to consider carefully the conditions under which the apparatus must operate; there are many applications where a general-purpose enclosure does not afford protection.

Underwriters' Laboratories has defined the requirements for protective enclosures according to the hazardous conditions, and the National Electrical Manufacturers Association (NEMA) has standardized enclosures from these requirements:

NEMA 1: general purpose. The general-purpose enclosure is intended primarily to prevent accidental contact with the enclosed apparatus. It is suitable for general-purpose applications indoors where it is not exposed to unusual service conditions. A NEMA 1 enclosure serves as protection against dust and light and indirect splashing, but is not dust-tight.

NEMA 3: dust-tight, raintight. This enclosure is intended to provide suitable protection against specified weather hazards. A NEMA 3 enclosure is suitable for application outdoors, such as construction work. It is also sleet-resistant.

NEMA 3R: rainproof, sleet resistant. This enclosure protects against interference in operation of the contained equipment due to rain, and resists damage from exposure to sleet. It is designed with conduit hubs and external mounting as well as drainage provisions.

NEMA 4: watertight. A watertight enclosure is designed to meet a hose test which consists of a stream of water from a hose with a 1-inch nozzle, delivering at least 65 gallons per minute. The water is directed on the enclosure from a distance of not less than 10 feet and for a period of 5 minutes. During this period, it may be directed in one or more directions, as desired. There should be no leakage of water into the enclosure under these conditions.

NEMA 4X: watertight, corrosion-resistant. These enclosures are generally constructed along the lines of NEMA 4 enclosures except that they are made of a material that is highly resistant to corrosion. For this reason, they are ideal in applications such as meatpacking and chemical plants, where contaminants would ordinarily destroy a steel enclosure over a period of time.

NEMA 7: hazardous locations, class I. These enclosures are designed to meet the application requirements of the *National Electrical Code* for Class I hazardous locations: "Class I locations are those in which flammable gases or vapors are or may be present in the air in quantities sufficient to produce explosive or ignitable mixtures." In this type of equipment, the circuit interruption occurs in air.

NEMA 9: hazardous locations, class II. These enclosures are designed to meet the application requirements of the NE Code for Class II hazardous locations. "Class II locations are those which are hazardous because of the presence of combustible dust." The letter or letters following the type number indicates the particular group or groups of hazardous locations (as defined in the NEC) for which the enclosure is designed. The designation is incomplete without a suffix letter or letters.

NEMA 12: industrial use. This type of enclosure is designed for use in those industries where it is desired to exclude such materials as dust, lint, fibers and flyings, oil seepage, or coolant seepage. There are no conduit openings or knockouts in the enclosure, and mounting is by means of flanges or mounting feet.

NEMA 13: oil-tight, dust-tight. NEMA 13 enclosures are generally made of cast iron, gasketed, or permit use in the same environments as NEMA 12 devices. The essential difference is that due to its cast housing, a conduit entry is provided as an integral part of the NEMA 13 enclosure, and mounting is by means of blind holes rather than mounting brackets.

7.0.0 MOTOR-CONTROL CIRCUITS

A complete wiring diagram is best used when making the initial connections of a circuit or when tracing a fault in a circuit. It shows the devices in symbol form and indicates the actual connections of all wires between the devices. Ladder diagrams use the same symbols to represent the individual devices, but indicate by only one line the fact that these devices are in the same circuit. Such schematic diagrams are simple and can be quickly prepared when principles of motor control are being studied.

A motor-control circuit is represented by its complete wiring diagram in Fig. 20 and by a ladder diagram in Fig. 21. In the wiring diagram (Fig. 20), the three supply conductors are indicated by L_1 (Line1), L_2, and L_3, and the motor terminals by T_1, T_2, and T_3. Each line has a terminal overload protective device (O.L.) connected in series with the normally open line contactors M_1, M_2, and M_3, which are controlled by the magnetic coil (C). Each contactor has a pair of contacts that close or open during operation. The control station, consisting of start-stop pushbuttons is connected across line L_1 and L_2. An auxiliary contactor M is connected in series with the stop pushbutton and in parallel with the start pushbutton. The control circuit also has a normally closed overload contactor (O.C.) connected in series with the starter coil (C).

The same connections are represented in Fig. 21 by a ladder diagram. As mentioned previously, it is customary to draw the two main lines L_1 and L_2 vertically, and to connect them with a horizontal line representing the control circuit with a control station containing two pushbuttons, an auxiliary interlocking contactor, the normally closed overload contactor OC, and the

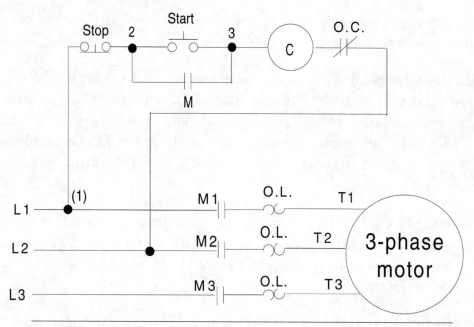

Figure 20. Complete wiring diagram of simple motor-control circuit.

starter coil C. The diagram depicts the situation when the control circuit is not energized and the normally open contactors are open. There is no complete path for the current unless the start button is pushed. Always read the ladder diagram from left to right; that is, from L_1 to L_2.

When the start pushbutton is momentarily pressed down, the path is complete from L_1 through the closed stop button, through the start button, the normally closed overload contactor, and the coil C to line L_2. Current will flow through this circuit and energize the coil C. Coil C closes the auxiliary, or sealing contactor. The spring-actuated start button may be released but the auxiliary contacts of contactor M interlock (or seal) the circuit and keep it closed as long as the coil C is energized.

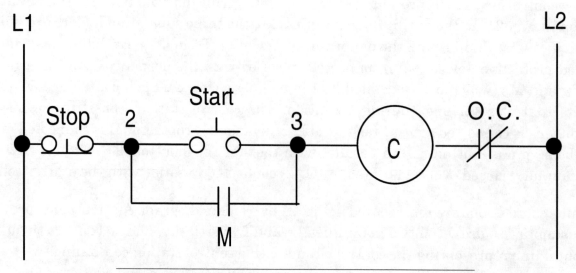

Figure 21. Ladder diagram of the control circuit in Fig. 20.

When the contacts of the control device close, they complete the coil circuit of the motor starter, causing it to pick up and connect the motor to the lines. When the control device contacts open, the starter is de-energized, stopping the motor.

The line contactors M1, M2, and M3 in Fig. 20 close when the coil C is energized causing the circuit to the motor terminals to be completed. The wiring diagrams in Figs. 20 and 21 do not show the motor starter, the speed controller, or similar control devices — only the pushbutton arrangement for starting and stopping the motor. They may be referred to as basic motor-control circuits.

When the stop pushbutton is pressed momentarily, the circuit is opened, the coil C is deenergized, and the auxiliary contact and the line contacts open. There is no path for current to the motor, and the motor stops.

A similar wiring diagram is shown in Fig. 22; this time with a slight modification of the control circuit. The only change in the control circuit is that the auxiliary contactor M interlocks the circuits to the motor side of the line, or to T2 instead of L2. The function of the control circuit and pushbuttons is the same as for the circuit in Figs. 20 and 21. A ladder diagram of the control circuit in Fig. 22 is shown in Fig. 23.

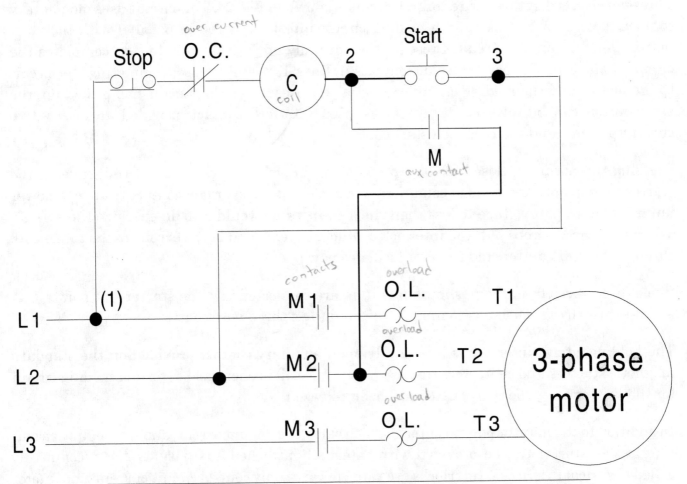

Figure 22. Control circuit interlocking the circuits to the motor side of the line.

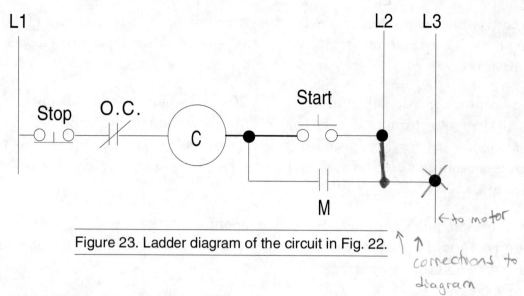

Figure 23. Ladder diagram of the circuit in Fig. 22.

to motor

corrections to diagram

Two-wire control provides low-voltage release but not low-voltage protection. When wired as illustrated, the starter will function automatically in response to the direction of the control device, without the attention of an operator. In this type of connection, a holding circuit interlock is not necessary.

Three-wire control: A three-wire control circuit is shown in Fig. 24. This circuit uses momentary contact, start-stop buttons, and a holding circuit interlock, wired in parallel with the start button to maintain the circuit. Pressing the normally open (NO) start button completes the circuit to the coil. The power circuit contacts in lines 1, 2, and 3 close, completing the circuit to the motor, and the holding circuit contact also closes. Once the starter has picked up, the start button can be released, as the now-closed interlock contact provided an alternative current path around the reopened start contact.

Pressing the normally closed (NC) stop button will open the circuit to the coil, causing the starter to drop out. An overload condition, which caused the overload contact to open, a power failure, or a drop in voltage to less than the seal-in value would also de-energize the starter. When the starter drops out, the interlock contact reopens, and both current paths to the coil, through the start button and the interlock, are now open.

Since three wires from the pushbutton station are connected into the starter—at points 1, 2, and 3—this wiring scheme is commonly referred to as three-wire control.

The holding circuit interlock is a normally open auxiliary contact provided on the standard magnetic starters and contactors. It closes when the coil is energized to form a holding circuit for the starter after the start button has been released.

In addition to the main or power contacts which carry the motor current, and the holding circuit interlock, a starter can be provided with externally attached auxiliary contacts, commonly called electrical interlocks. Interlocks are rated to carry only control circuit currents, not motor currents. Both NO and NC versions are available. Among a wide variety of applications,

ELECTRICAL TRAINEE TASK MODULE 20311

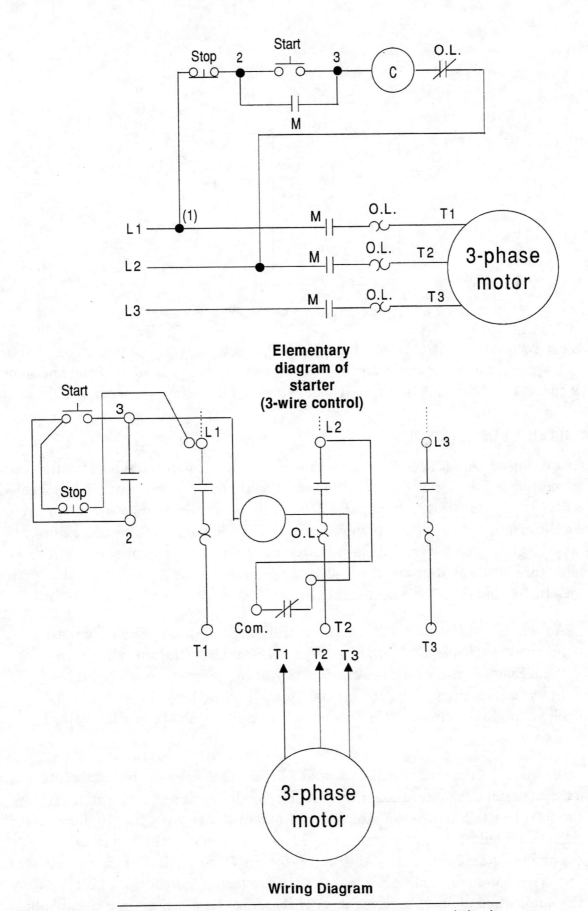

Elementary diagram of starter (3-wire control)

Wiring Diagram

Figure 24. Wiring diagrams of three-wire motor-control circuit.

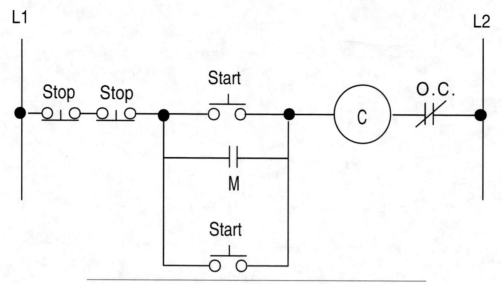

Figure 25. Control circuit with two pushbutton stations.

interlocks can be used to control other magnetic devices where sequence operation is desired; to electrically prevent another controller from being energized at the same time; and to make and break circuits to indicating or alarm devices such as pilot lights, bells, or other signals.

7.1.0   MULTIPLE PUSHBUTTONS

If motors are required to be started from more than one location, additional pushbutton stations may be connected to the circuit. In doing so, additional start buttons must be connected in parallel with the original start buttons, and the additional stop buttons must be connected in series with the original stop button as shown in Fig. 25. The auxiliary contactor must also be in parallel with the second start button if buttons should be only momentarily pressed for starting. For three control stations, there should be three start buttons in parallel with the auxiliary contactor and three stop buttons in series.

NOTE   Any control device connected in the control circuit to *start* the motor must be connected in *parallel with* the start button and be of the normally open type. Every device which has the function of *stopping* the motor must be in *series* with the stop button and be normally closed. By the addition of elements to the circuit in this manner, a complex control circuit may be obtained.

Another modification of the control circuit is used for inching the motor, that is, for intermittent operation through unequal short time intervals. The modified circuit is shown in Fig. 26. The stop button is connected across the start button but in series with the auxiliary interlock contactor. The stop button is latched open and keeps the circuit of the interlocking contactor open. Releasing the start button stops the motor, because the stop button keeps the interlock branch open. The motor runs only as long as the start button is held down. This type of control enables the application of short movement, or inching, of the machine in order to adjust the machine.

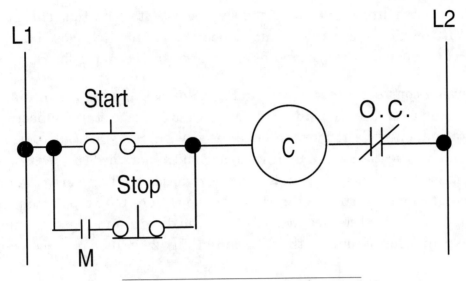

Figure 26. Control for inching a motor.

7.2.0 REVERSING MOTOR ROTATION

When the control circuit is being designed for reversing the motor direction as well as for starting the motor, like the one shown in Fig. 27, a reverse-start button with its interlock contactor, normally closed overload contactor, and reverse-contact coil must be added to the basic forward-control circuit. Since the added reverse-start button is a starting device, it must be connected in parallel with the original forward start button. The reverse-start button is connected behind the stop button so that the same stop button can stop both the forward and the reverse rotation of the motor. When the forward-start button is momentarily pressed, the forward coil FC is energized and closes the normally open contactor FC_1 and opens the normally closed contactor FC_2 in the reverse control circuit. The now open contactor FC, in the reverse control circuit makes sure that even if the reverse-start button is pressed, the reverse circuit will remain open and not active. After the motor has been stopped by the stop button, it may

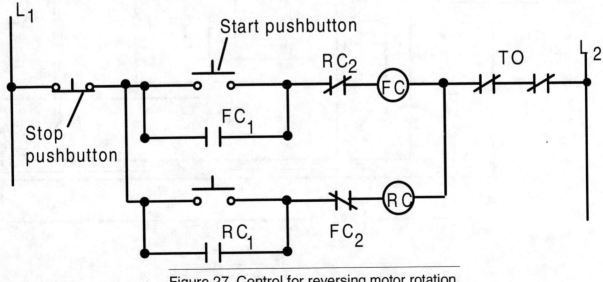

Figure 27. Control for reversing motor rotation.

be started in the reverse direction by pressing the reverse-start button. This action energizes the reverse coil RC, which keeps the normally open contactor RC$_1$ closed and the normally closed contactor RC$_2$ open, so that the motor cannot be started in the forward direction.

A modified reversing-control circuit is shown in Fig. 28 drawn as a full wiring diagram. A ladder diagram of the same circuit is shown in Fig. 29. Instead of two starting buttons for forward and reverse rotation, one selector switch (1) is used with one start button (2) and one stop button (3). When the selector switch is first turned to the forward (F) position and the start pushbutton is pressed, the motor runs in the forward direction. To reverse the direction of the motor, the selector switch is turned to the reverse (R) position. The stop button stops the motor in either direction. For actual connections of wires, the diagram in Fig. 28 is more useful, but for studying the principles of control, the diagram in Fig. 29 will be more convenient.

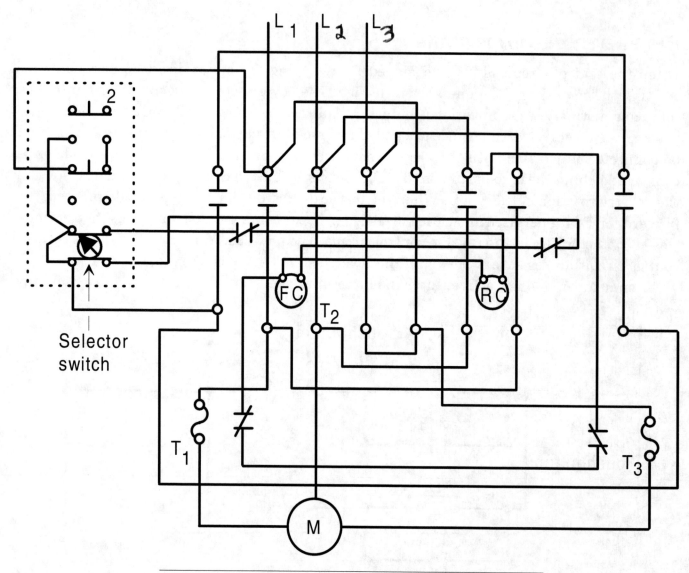

Figure 28. Wiring diagram of a reversing motor-control circuit.

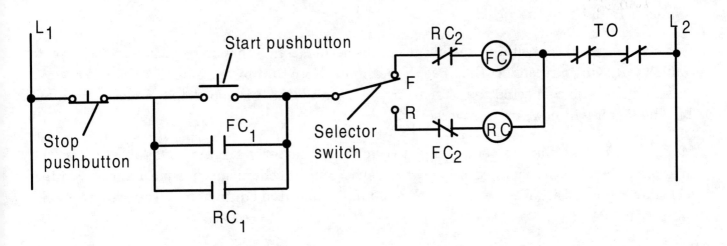

Figure 29. Ladder diagram of the control circuit in Fig. 28.

The circuit in Fig. 30 shows a three-pole reversing starter used to control a three-phase motor. Three-phase squirrel-cage motors can be reversed by reconnecting any two of the three line connections to the motor. By interwiring two contactors, an electromagnetic method of making the reconnection can be obtained.

As seen in the power circuit (Fig. 30), the contacts (F) of the forward contactor—when closed—connect lines 1, 2, and 3 to the motor terminals T1, T2, and T3, respectively. As long

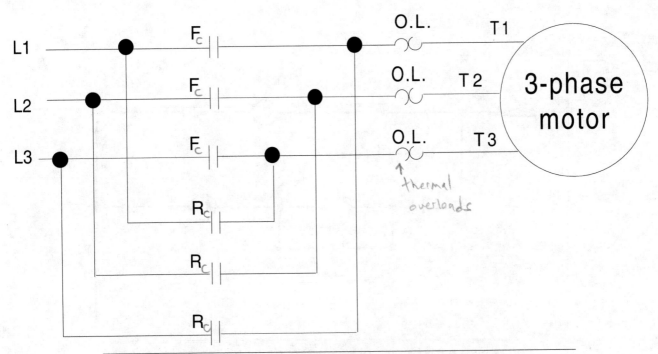

Figure 30. Three-pole reversing starter used to control a three-phase motor.

as the forward contacts are closed, mechanical and electrical interlocks prevent the reverse contactor from being energized.

When the forward contactor is de-energized, the second contactor can be picked up, closing its contacts (R), which reconnect the lines to the motor. Note that by running through the reverse contacts, line 1 is connected to motor terminal T_3, and line 3 is connected to motor terminal T_1. The motor will now run in reverse.

Manual reversing starters (employing two manual starters) are also available. As in the magnetic version, the forward and reverse switching mechanisms are mechanically interlocked, but since coils are not used in the manually operated equipment, electrical interlocks are not furnished.

7.3.0 CONTROL OF TWO MOTORS

The ladder diagram in Fig. 31 shows the basic components of a control circuit for two motors. One motor must be already running before the other can be started, and both should stop at the same time. Basic control circuits for both motors are connected between the lines L_1 and L_2. When the first start button closes the control circuit of the first motor, its control coil C_1 closes not only the interlock contacts M_1 in the first control circuit but also a normally open contactor M, in the control circuit of the second motor. If the start button for the second motor is now pushed, it energizes the coil C, and closes the normally open contactor M_2, in the second control circuit. Now, both motors will run until the stop button is pressed. The second motor is prevented from starting when its own start button is pressed unless the first motor is already running. Both motors can be stopped at the same time by using the stop button.

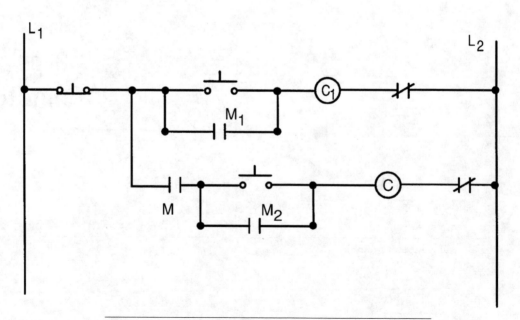

Figure 31. Control of two motors running alternately.

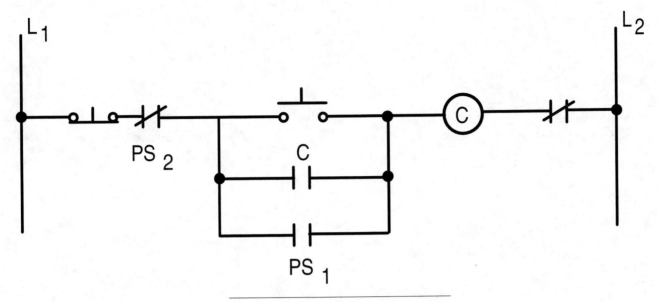

Figure 32. Automatic motor control.

7.4.0 AUTOMATIC CONTROL CIRCUIT

When completely automatic control of a motor is required, two pilot switches PS1, and PS2, are connected into the basic control circuit, as shown in Fig. 32. The starting pilot switch PS1 is connected in parallel with the start button. The pilot switch may be a floating switch, a pressure switch, or a time switch, and will close automatically when a preset condition occurs. For the floating switch, a certain level of the liquid closes the switch; for the pressure switch, a gas pressure of a certain value closes the switch; and for the time switch, a timer closes the switch after a certain preset time interval.

When the automatic starting switch closes the control circuit, the coil C is energized and closes the interlock contactor C. The automatic stopping switch PS2, is normally closed, and like the automatic starting switch, opens automatically at a preset condition and stops the control circuit.

8.0.0 MISCELLANEOUS STARTERS

There are several variations of the motor-control circuits previously described. There are also different types of motor starters. Some of these starters are becoming obsolete and are seldom used on new construction. However, electricians will often work on existing installations where some of the older motor starters are still in use, and an identifying knowledge of them is warranted.

8.1.0 ACROSS-THE-LINE MOTOR STARTERS

Induction-type and synchronous AC motors have a high starting current; therefore, they are mostly started with a reduced voltage, although there are many quite large AC motors which

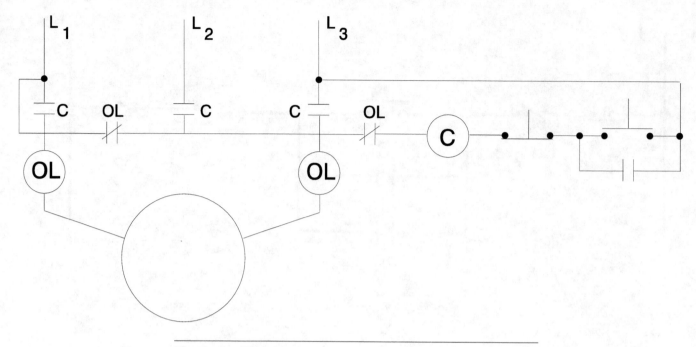

Figure 33. Full-voltage starter for a three-phase AC motor.

are started across the line, that is, by applying the full rated voltage. A connection diagram for a full-voltage starter of a three-phase induction motor is shown in Fig. 33. Overload protection relays *OL* are in two of the three lines and the motor is started by the basic control circuit, which uses a control station with start and stop buttons.

8.2.0   REDUCED-VOLTAGE STARTING OF AC MOTORS

The starter shown in Fig. 34 uses autotransformers to provide reduced-voltage starting of a squirrel-cage induction motor. The motor (*1*) is connected to the three lines L_1, L_2, and L_3 by means of the movable lever (*2*). When the lever is moved from the middle, or *OFF* position, to the *START* position, the autotransformers (*3*) are connected in the circuit by means of the oil-immersed contacts (*4*). The low-voltage taps from the autotransformers are connected to the motor and the motor starts. When the motor is up to speed, the lever is thrown from the *START* position to the *RUN* position, disconnecting the autotransformers and connecting the supply directly to the motor by the contacts (*5*).

The lever is held closed by the holding coil (6), which is connected across one phase of the supply through the relay contacts (7) of the relays and the stop button. The relays are built with dashpots, which delay the operation of the relays during the starting period and for momentary overloads. Operation of the relays, or of the stop pushbutton, opens the holding coil circuit and allows the lever to come to the *OFF* position, breaking the circuit made by the relay contacts (7).

Autotransformers with several taps may be used for multispeed control of a squirrel-cage motor. All squirrel-cage motors use controllers in the primary or stator winding.

ELECTRICAL TRAINEE TASK MODULE 20311

8.3.0 RESISTANCE MAGNETIC STARTERS

A resistance-type starter is shown in Fig. 35 as used with a three-phase squirrel-cage induction motor. The motor is started by pressing the start button to energize the coil C, which closes the three line contactors C_1, C_2, and C_3 and the interlock C_4 to maintain power on the coil after the start button is released.

When the contactor is closed, the timing relay TR becomes energized, and this relay immediately starts to measure the time for which it is adjusted. At the end of the timing period, the contacts TR close to energize the coil A of the accelerating contactor A. When the three-pole contactor A closes, the starting resistors will be short-circuited and the motor will be connected directly to the AC power lines. The heater elements of either of the thermal-type overload relays OL_1 and OL_2 open the normally closed contactor OL in case of overload.

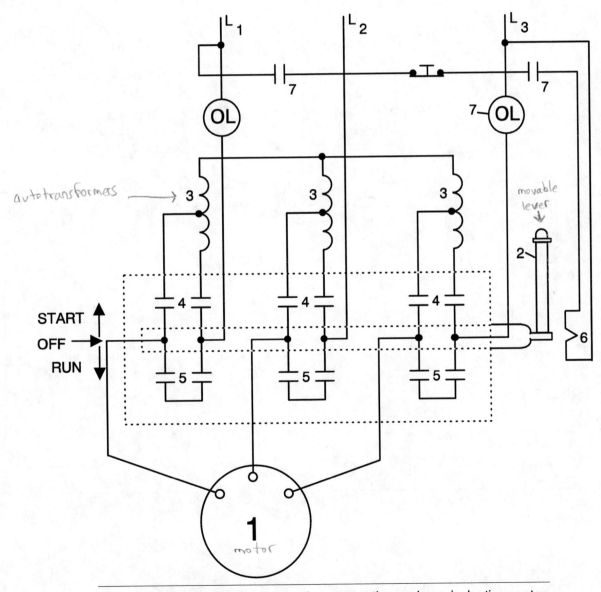

Figure 34. Autotransformers used to start a three-phase induction motor.

8.4.0 STARTER FOR WOUND INDUCTION MOTORS

The wound three-phase induction motor differs from the squirrel-cage induction motor in that it uses wound coils in the rotor instead of bars, and three sets of brushes to collect the current from the three collector rings, which are not used in the squirrel-cage motor. To start a wound motor, it is necessary to connect the brushes to external resistors, as shown in Fig. 36. The three-phase star-connected stator 1 is connected to a three-phase source 2, and the three-phase star-connected rotor 3, or secondary, is connected through the slip rings and brushes to the three resistors 4, which are also star connected. They increase the resistance of the rotor which in turn increases the torque of the rotor and decreases its speed. When the resistance is gradually cut out, the speed gradually increases and the torque decreases. With no secondary resistance, the wound motor operates at maxi multispeed. The starter and controller are built

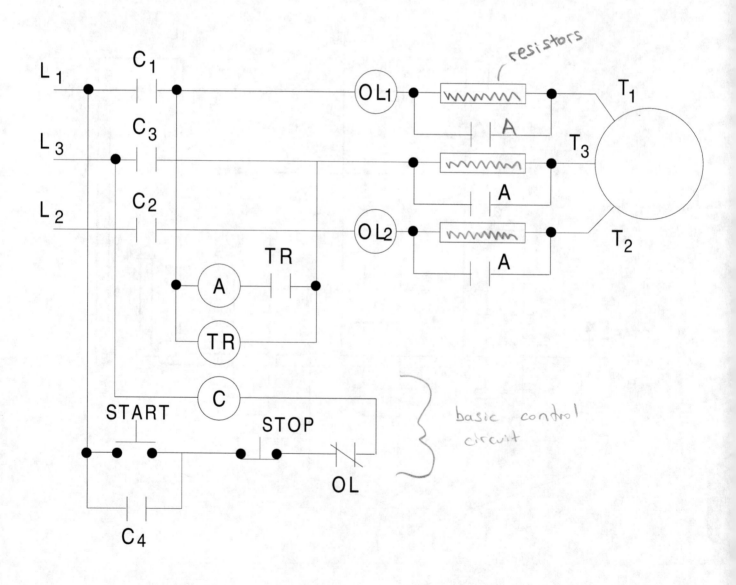

Figure 35. Resistance starter used for squirrel-cage induction motor.

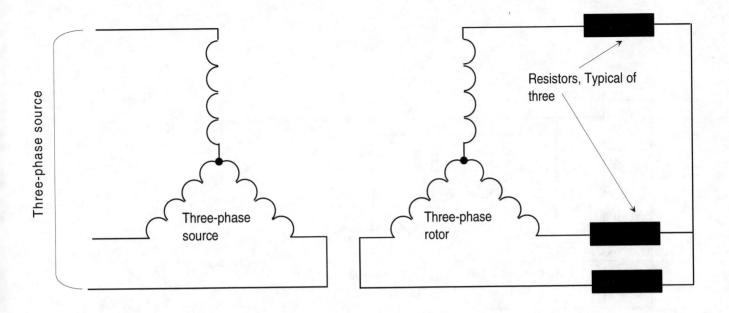

Figure 36. Wound motor and secondary resistors.

with combinations of basic control circuits. Manuals provided by manufacturers show control circuits for each specific type.

8.5.0 STARTING AND BRAKING SYNCHRONOUS MOTORS

The main electrical connections for a large high-voltage synchronous motor are shown in Fig. 37. The high-voltage three-phase lines L_1, L_2, and L_3 supply 13,800 volts to the synchronous motor (1) through a main (running) circuit breaker (2). This breaker is electrically interlocked with the starting breakers (3) and (4), and when the starting breakers are closed, the running breaker must be open. The starting breakers connect to the line the starting autotransformer (5), which reduces the voltage to about 50 percent, or 6900 volts, and energizes the motor for the start. When the motor comes up to speed, the starting breaker (4) is open and the running breaker connects the motor directly to the high-voltage line. The primary of a current transformer (6) is connected to one supply line and its secondary is connected to an ammeter (7) to measure the motor current at any time. The field winding is energized by the exciter (9) through the field rheostat (10). The exciter may be a shunt-wound DC generator with its own field rheostat (not shown). The field-discharge resistor (11) and the field switch (12) serve a double purpose in that they provide a low-resistance path for the induced currents in the motor field when the motor is started and discharge the induced currents when the field breaker is opened.

Dynamic braking is obtained by braking resistors (13) through the braking contactors (14). The field is left energized and the motor is disconnected from the line and connected to the

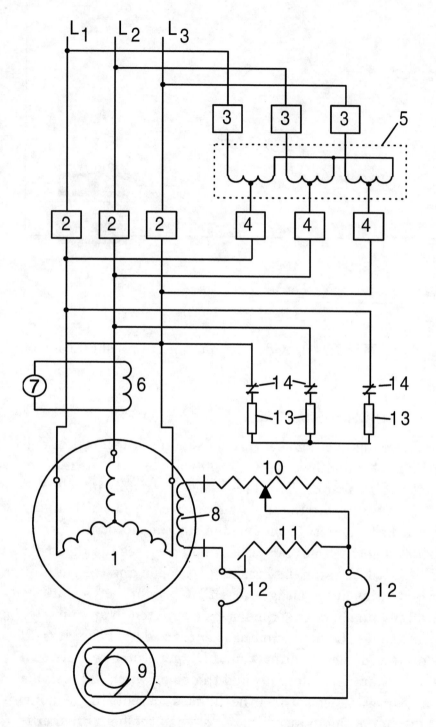

1. synchronous motor
2. running circuit breaker
3, 4. starting circuit breaker
5. starting autotransformer
6. current transformer
7. ammeter
8. field winding
9. exciter
10. field rheostat
11. field discharge resistor
12. field switch
13. braking resistors
14. braking contactors

Figure 37. Connections for three-phase synchronous motor.

braking resistors. By using quick-acting control sequences, the motor can be stopped within approximately one second, without shock or high stresses.

9.0.0 CONTROL RELAYS

Control relays: A relay is an electromagnetic device whose contacts are used in control circuits of magnetic starters, contactors, solenoids, timers, and other relays. They are generally used to amplify the contact capability or to multiply the switching functions of a pilot device.

The wiring diagrams in Fig. 38 demonstrate how a relay amplifies contact capacity. Fig. 39A represents a current amplification. Relay and starter coil voltages are the same, but the ampere rating of the temperature switch is too low to handle the current drawn by the starter coil (M). A relay is interposed between the temperature switch and the starter coil. The current drawn by the relay coil (CR) is within the rating of the temperature switch, and the relay contact (CR) has a rating adequate for the current drawn by the starter coil.

Figure 39B represents a voltage amplification. A condition may exist in which the voltage rating of the temperature switch is too low to permit its direct use in a starter control circuit operating at a higher voltage. In this application, the coil of the interposing relay and the pilot device are wired to a low-voltage source of power compatible with the rating of the pilot device. The relay contact, with its higher voltage rating, is then used to control the operation of the starter.

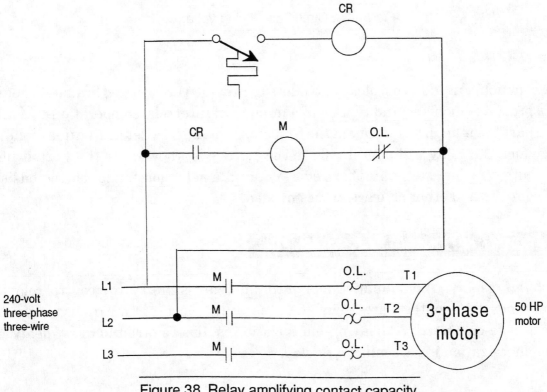

Figure 38. Relay amplifying contact capacity.

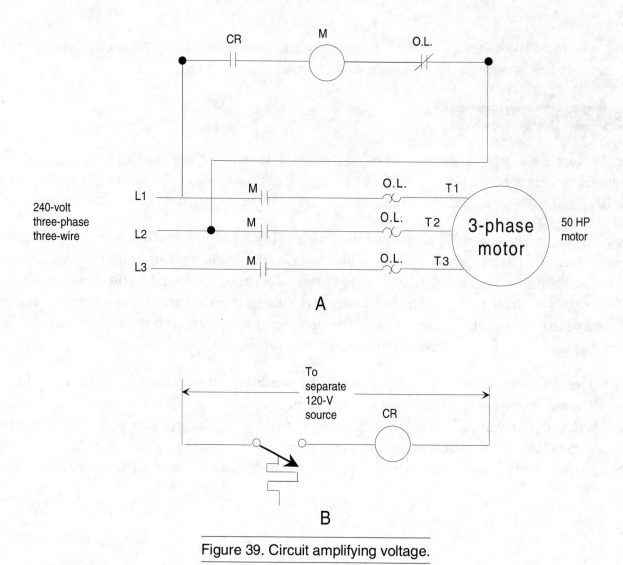

Figure 39. Circuit amplifying voltage.

Relays are commonly used in complex controllers to provide the logic or "brains" to set up and initiate the proper sequencing and control of a number of interrelated operations. In selecting a relay for a particular application, one of the first steps should be a determination of the control voltage at which the relay will operate. Once the voltage is known, the relays that have the necessary contact rating can be further reviewed, and a selection made, on the basis of the number of contacts and other characteristics needed.

10.0.0   ADDITIONAL CONTROLLING EQUIPMENT

Timers and timing relays: A pneumatic timer or timing relay is similar to a control relay, except that certain kinds of its contacts are designed to operate at a preset time interval after the coil is energized or de-energized. A delay on energization is also referred to as *on delay*. A time delay on de-energization is also called *off delay*.

A timed function is useful in such applications as the lubrication system of a large machine, in which a small oil pump must deliver lubricant to the bearings of the main motor for a set period of time before the main motor starts.

In pneumatic timers, the timing is accomplished by the transfer of air through a restricted orifice. The amount of restriction is controlled by an adjustable needle valve, permitting changes to be made in the timing period.

Drum switch: A drum switch is a manually operated three-position three pole switch which carries a horsepower rating and is used for manual reversing of single- or three-phase motors. Drum switches are available in several sizes and can be spring-return-to-off (momentary contact) or maintained contact. Separate overload protection, by manual or magnetic starters, must usually be provided, as drum switches do not include this feature.

Pushbutton station: A control station may contain pushbuttons, selector switches, and pilot lights. Pushbuttons may be momentary- or maintained-contact. Selector switches are usually maintained-contact, or can be spring-return to give momentary-contact operation. Standard-duty stations will handle the coil currents of contactors up to size 4. Heavy-duty stations have higher contact ratings and provide greater flexibility through a wider variety of operators and interchangeability of units.

Foot switch: A foot switch is a control device operated by a foot pedal used where the process or machine requires that the operator have both hands free. Foot switches usually have momentary contacts but are available with latches which enable them to be used as maintained-contact devices.

Limit switch: A limit switch is a control device that converts mechanical motion into an electrical control signal. Its main function is to limit movement, usually by opening a control circuit when the limit of travel is reached. Limit switches may be momentary-contact (spring-return) or maintained-contact types. Among other applications, limit switches can be used to start, stop, reverse, slow down, speed up, or recycle machine operation.

Snap Switch: Snap switches for motor control purposes are enclosed, precision switches which require low operating forces and have a high repeat accuracy. They are used as interlocks and as the switch mechanism for control devices such as precision limit switches and pressure switches. They are also available with integral operators for use as compact limit switches, door operated interlocks, and so on. Single-pole, double-throw and two-pole, double-throw versions are available.

Pressure Switch: The control of pumps, air compressors, and machine tools requires control devices that respond to the pressure of a medium such as wire, air, or oil. The control device that does this is a pressure switch. It has a set of contacts which are operated by the movement of a piston, bellows, or diaphragm against a set of springs. The spring pressure determines the pressures at which the switch closes and opens its contacts.

The table in Fig. 40 lists troubles encountered with motor controls, their causes, and remedies. This table is of a general nature and covers only the main causes of trouble.

Misapplication of a device can be a cause of serious trouble; however, rather than list this cause repeatedly, it should be noted here that misapplication is a major cause of motor control trouble and should always be questioned when a device is not functioning properly.

Actual physical damage or broken parts can usually be found quickly and replaced. Damage due to water or flood conditions requires special treatment.

Contacts: There are two types of contact wear: electrical and mechanical. The majority of wear to contact tips is due to electrical wear. The mechanical wear is insignificant and requires no further mention.

Arcing causes electrical wear by eroding the contacts. During arcing, a small part of each contact melts, then vaporizes and is blown away.

When a device is new, the contacts are smooth and have a uniform silver color. As the device is used, the contacts become pitted and the color may change to blue, brown, and black. These colors result from the normal formation of metal oxide on the contact's surface and are not detrimental to contact life and performance. Therefore, contacts should not be filed to restore the original color. This practice only shortens contact life and may cause welding.

The contacts should be replaced under the following conditions:

Insufficient contact material: This is when the amount of contact material remaining is inadequate. When less than $\frac{1}{64}$ inch remains, replace the contacts.

Irregular surface wear: This type of wear is normal. However, if a corner of the contact material is worn away and a contact may mate with the opposing contact support member, the contacts should be replaced. This condition can result in contact welding.

Pitting: Under normal wear, contact pitting should be uniform. This condition occurs during arcing, as described above. The contacts should be replaced if the pitting becomes excessive and little contact material remains.

Curling of contact surface: This condition results from severe service, producing high contact temperatures, and causes separation of the contact material from the contact support member.

The measurement procedure for checking the contact tip material requires a continuity checker and a $\frac{1}{32}$-inch feeler gauge. The procedure is as follows:

Step 1.   Place the feeler gauge between the armature and the magnet frame, with the armature held tightly against the magnet frame.

Step 2.   Check the continuity of each phase.

If there is continuity in all phases, the contacts are in good condition. If not, all contacts should be replaced. Even though the contacts pass condition 1, any of the other conditions would necessitate replacement of the contacts. Contacts should be replaced only when necessary; too frequent replacement is a waste of money and natural resources.

Trouble	Possible Cause	Remedy
Contacts	Broken pole shader.	Replace.
Contact chatter	Poor contact in control circuit.	Improve contact or use holding circuit interlock (3-wire control).
	Low voltage	Correct voltage condition. Check momentary voltage dip during starting.
Welding or freezing	Abnormal inrush of current.	Use larger contactor or check for grounds.
	Rapid jogging.	Install larger device rated for jogging service.
	Insufficient tip pressure.	Replace contact springs; check contact carrier for damage.
	Low voltage preventing magnet from sealing.	Correct voltage condition. Check momentary voltage dip during starting.
	Foreign matter preventing contacts from closing.	Clean contacts with approved solvent.
	Short circuit.	Remove short fault and check to be sure fuse or breaker size is correct.
Short contact life or overheating of tips	Filing or dressing.	Do not file silver-faced contacts. Rough spots or discoloration will not harm contacts.

Figure 40. Motor-control troubleshooting chart.

Trouble	Possible Cause	Remedy
Short contact life or overheating of tips	Interrupting excessively high.	Install larger device or check currents for grounds, shorts, or excessive motor currents. Use silver-faced contacts.
	Excessive jogging.	Install larger device rated for jogging.
	Weak contact pressure.	Adjust or replace contact springs.
	Dirt or foreign matter on contact surface.	Clean contacts with approved solvent.
	Short circuits.	Remove short fault and check for proper fuse or breaker size.
	Loose connection.	Clean and tighten.
	Sustained overload.	Install larger device or check for excessive load current.
Coil overheated	Overvoltage or high ambient temperature.	Check application and circuit.
	Incorrect coil.	Check rating and if incorrect replace with proper coil.
	Shorted turns caused by mechanical damage or corrosion.	Replace coil.
	Undervoltage, failure of magnet to seal in.	Correct system voltage.
	Dirt or rust on pole faces increasing air gap.	Clean pole faces.
Overload relays tripping	Sustained overload.	Check for grounds, shorts, or excessive currents.
	Loose connection on load wires.	Clean and tighten.
	Incorrect heater.	Relay should be replaced with correct size heater unit.
Failure to trip causing motor burn-out	Mechanical binding, dirt, corrosion, etc.	Clean or replace.

Figure 40. Motor-control troubleshooting chart. (*Cont.*)

Trouble	Possible Cause	Remedy
Failure to trip causing motor burn-out	Wrong heater or heaters omitted and jumper wires used.	Check ratings. Apply proper heater.
	Motor and relay in different temperatures.	Adjust relay rating accordingly.
	Wrong calibration or improper calibration adjustment.	Consult factory.
Magnetic & mechanical parts	Broken shading coil.	Replace shading coil.
Noisy Magnet humming	Magnet faces not mating.	Replace magnet assembly realign.
	Dirt or rust on magnet faces.	Clean and realign.
	Low voltage.	Check system voltage and voltage dips during starting.
Failure to pick-up and seal	Low voltage.	Check system voltage and voltage dips during starting.
	Coil open or shorted.	Replace.
	Wrong coil.	Check coil number.
	Mechanical obstruction.	With power off check for free movement of contact and armature assembly.
Failure to drop-out	Gummy substance on pole faces.	Clean with solvent.
	Voltage not removed.	Check coil circuit.
	Worn or rusted parts causing binding.	Replace parts.
	Residual magnetism due to lack of air gap in magnet path.	Replace worn magnet parts.

Figure 40. Motor-control troubleshooting chart. (*Cont.*)

Summary

The first of the motor-control arrangements is a plug and receptacle; next comes a fusible disconnect or circuit breaker, and then the manual and fractional-horsepower starters. The magnetic-contactor controller, however, is the type most used in electrical installations. This latter type of controller opens or closes circuits automatically when their control coil is energized. The contactors may be normally open or normally closed.

Protective devices such as overload relays, low-voltage protection devices and low-voltage release devices are an important part of a motor controller. An overload relay will open the contactors in motor circuits when current is too high; a low-voltage protective device will prevent the motor from starting as long as the full-rated voltage is not available, and manual restarting of the motor is necessary after the low-voltage protective device has operated; a low-voltage release device will disconnect the motor during a voltage dip, but the motor will start automatically when the normal voltage returns.

Controllers also contain braking arrangements, accelerators, and reversing switches which reverse the rotation of the motor.

The intent of this module is to familiarize electrical workers with terms and concepts which are fundamental to an understanding of motor control equipment and its applications. A knowledge of the definitions, symbols, diagrams and illustrations will give the trainee a sound background in the language and basic principles associated with motor controls and their application on electrical construction projects.

References

For more advanced study of topics covered in this Task Module, the following works are suggested:

National Electrical Code Handbook, Latest Edition, NFPA, Quincy, MA
American Electricians Handbook, Latest Edition, Croft, McGraw-Hill, New York, NY
Electricians Guide to AC Motor Controls, R.A. Cox, COXCO, Box 3822, Spokane, WA

SELF-CHECK REVIEW/PRACTICE QUESTIONS

1. What is the term used to describe the repeated starting and stopping of a motor at frequent intervals for short periods of time?

   a. Plugging
   b. Jogging or inching
   c. Air gapping
   d. Limiting

2. What is the term used to describe the action of momentarily reversing a motor that is already running in one direction to bring it to a rapid stop?

   a. Plugging
   b. Jogging
   c. Inching
   d. Gapping

3. What protective devices are used to provide overcurrent protection for motors?

   a. Circuit breakers or fuses
   b. Lightning arresters
   c. Heaters
   d. Overload relays

4. What protective devices are used to provide overload protection for motors?

   a. General-purpose current-limiting fuses
   b. Back-up current-limiting fuses
   c. Overload relays
   d. Timing relays

5. What are the two main functioning devices in a bimetallic thermal overload relay?

   a. Bimetallic strip and a current-carrying heater element
   b. A fuse capable of interrupting all currents from the maximum rated inter-rupting current down to the rated minimum interrupting current
   c. Timer and contactor
   d. A vented fuse with a time-delay element

6. What is the name given to the part of a melting alloy thermal overload relay that provides accurate response to overload current?

    a. Bimetal element
    b. Solder pot
    c. Air gap
    d. Terminals

7. Which of the following is *not* a consideration when selecting overload relays?

    a. Type of motor
    b. Full-load current rating of the motor
    c. Possible difference in ambient temperature between the motor and the controller
    d. Manufacturer of the motor

8. How many sets of contacts are used in a conventional fractional horsepower manual motor starter for a 240-V, single-phase motor?

    a. 1
    b. 2
    c. 3
    d. 4

9. How many thermal overload devices are used in the manual starter in Question 8?

    a. 1
    b. 2
    c. 3
    d. 4

10. What is the main purpose of manual starters with low-voltage protection?

    a. To prevent motor from stopping during a voltage dip
    b. To prevent the motor from slowing down during a voltage dip
    c. To prevent motors from overheating during a voltage dip
    d. To prevent automatic start-up of motors after a power loss

11. What is the main feature that distinguishes a magnetic starter from a manual starter?

    a. It is colored gray
    b. An electromagnet
    c. It has a threadless lug
    d. It is rated in motor horsepower

12. When a controller's armature has sealed-in and fits closely against the magnet assembly, what is always deliberately left in the iron-core circuit?

    a. More wire than needed
    b. Air gap
    c. Some minor defects
    d. A gap of 2 inches or more

13. Which of the following is not a cause of excessive magnetic hum in a magnetic motor starter?

    a. A full shading coil
    b. Incorrect mounting
    c. Misalignment between the armature and magnet assembly
    d. Bent parts

14. What is the main purpose of control relays when used in motor starters?

    a. To amplify the contact capability or multiply the switching functions of a pilot device
    b. Back-up current-limiting fuses
    c. Back-up overcurrent and overload devices
    d. To prevent the motor-starter from functioning automatically

15. What is the name of the device that converts mechanical motion into an electrical control signal?

    a. Mechanical interlock
    b. Pilot light
    c. Limit switch
    d. Contactor

PERFORMANCE/LABORATORY EXERCISE

1. Make all required connections for a FHP 120-volt, single-phase motor starter, including the motor connections. Energize the circuit and test for functioning.

2. Make all connections for a magnetic motor controller, controlled by two pushbutton stations, including the connections for the holding circuit interlock.

3. Make the final connections to a 240-volt motor controller in the circuit in Fig. 41. The controller will be operated by one pushbutton station (start and stop).

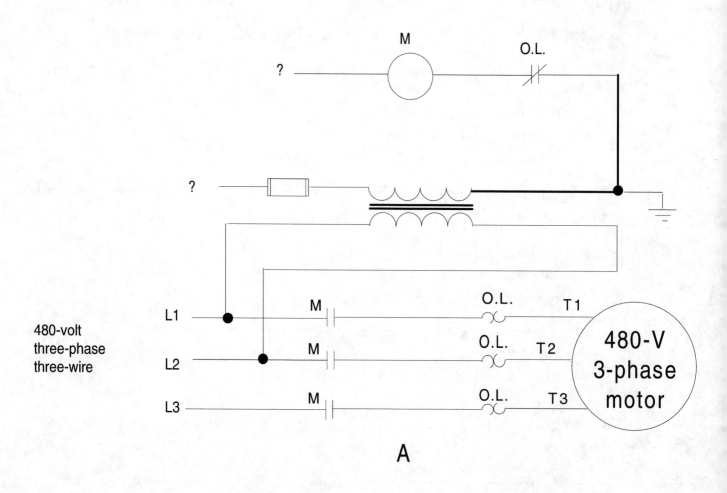

Figure 41. Wiring diagram for Performance/Laboratory Exercise.

Answers to Self-Check Questions

1. b

2. a

3. a

4. c

5. a

6. b

7. d

8. b

9. a

10. d

11. b

12. b

13. a

14. a

15. c

Electricity in HVAC Systems

Module 20312

Electrical Trainee Task Module 20312

ELECTRICITY IN HVAC SYSTEMS

Objectives

Upon completion of this module, the trainee will be able to:

1. Describe the basic operating principles of air conditioning systems.
2. Explain how refrigeration systems operate.
3. Interpret nameplate data on HVAC equipment.
4. Relate his or her knowledge of electric motors to the operation of HVAC equipment.
5. Explain the role of the National Electrical Code in HVAC power and control wiring.
6. Describe the operating principles of compressors as they relate to refrigeration.
7. Troubleshoot HVAC systems.
8. Explain NEC air conditioning requirements for computer rooms.
9. Install electrical circuits and related components to HVAC equipment according to NEC Articles 220, 424, and 440.

Prerequisites

Successful completion of the following Task Modules is required before beginning study of this Task Module: Core Curricula, Electrical Levels 1 and 2, Electrical Level 3, Modules 20301 through 20311.

The trainee should also read NEC Article 424.

Required Student Materials

1. Trainee Task Module
2. Copy of the latest edition of the National Electrical Code

Optional Material

NFPA 90A-(ANSI) *Installation of Air Conditioning and Ventilating Systems*, NFPA, Quincy, MA

Power Handbook, Latest Edition, McGraw-Hill, New York, NY

COURSE MAP INFORMATION

This course map shows all of the *Wheels of Learning* Task Modules in the third level of the Electrical curricula. The suggested training order begins at the bottom and proceeds up. Skill levels increase as a trainee advances on the course map. The training order may be adjusted by the local Training Program Sponsor.

Course Map: Electrical, Level 3

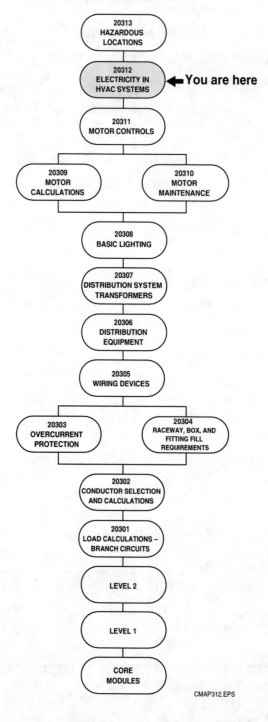

LEVEL 3 COMPLETE

20313
HAZARDOUS
LOCATIONS

20312
ELECTRICITY IN
HVAC SYSTEMS ◄— You are here

20311
MOTOR CONTROLS

20309
MOTOR
CALCULATIONS

20310
MOTOR
MAINTENANCE

20308
BASIC LIGHTING

20307
DISTRIBUTION SYSTEM
TRANSFORMERS

20306
DISTRIBUTION
EQUIPMENT

20305
WIRING DEVICES

20303
OVERCURRENT
PROTECTION

20304
RACEWAY, BOX, AND
FITTING FILL
REQUIREMENTS

20302
CONDUCTOR SELECTION
AND CALCULATIONS

20301
LOAD CALCULATIONS –
BRANCH CIRCUITS

LEVEL 2

LEVEL 1

CORE
MODULES

CMAP312.EPS

TABLE OF CONTENTS

Trade Terms Introduced in This Module

air cleaner: Device used for removal of airborne impurities.

air conditioning: A process that heats, cools, cleans, and circulates air and controls its moisture content. Ideally, it performs all of these functions simultaneously and on a year-round basis.

air diffuser: Air distribution outlet designed to direct airflow into desired patterns.

air flow: The distribution or movement of air.

air horsepower (AHP): Work done in moving a given volume or weight of air at a given speed.

ambient temperature: Temperature of fluid or air that surrounds an object on all sides.

area (A): The square feet of any plane surface or cross section of a duct, the air inlet or outlet from a room, or the circular plane of rotation of a propeller.

automatic: Operating by own mechanism when actuated by some impersonal influence; nonmanual; self-acting.

back pressure: Pressure in the low side of a refrigerating system; also called suction pressure or low-side pressure.

balance point: The lowest outside temperature at which the refrigeration cycle of a heat pump will supply the heating requirements without the aid of a supplementary heat source.

barometer: Instrument for measuring atmospheric pressure.

barometric pressure: Pressure expressed by the height in inches of a column of mercury, exerted by the weight of the earth's atmosphere on any surface.

bimetal strip: Temperature regulating or indicating device that works on the principle that two dissimilar metals with unequal expansion rates, welded together, will bend as temperature changes.

brake horsepower (BHP): Work done in driving any fan. This load, plus drive losses from belts and pulleys, is the work done by a fan's electric motor. It is always a higher number than air horsepower.

British thermal unit (Btu): Quantity of heat required to raise the temperature of 1 pound of water 1 degree Fahrenheit.

calibrate: Compare with a standard.

calorie: Heat required to raise temperature of 1 gram of water 1 degree centigrade.

Centigrade scale: Temperature scale used in metric system. Freezing point of water is 0 degrees; boiling point is 100 degrees.

check valve: A valve that allows the refrigerant to flow in one direction only.

coefficient of performance (COP): A ratio of heat output of a heat pump and the electrical input at a given outdoor temperature. Determined by the equation:

$$COP = \frac{Energy\ Output\ in\ Btuh}{Energy\ Input\ in\ Watts \times 3.413}$$

compressor: A component of a refrigerating system which pumps the refrigerant under different pressure through the indoor and outdoor units.

convection: The transfer of heat to a fluid by conduction as the fluid moves past the heat source.

cubic feet per minute (CFM): The physical volume (not weight) of air moved by a fan.

cycle: An interval of space or time in which one set of events or phenomena is completed.

damper: Valve for controlling air flow.

density: The actual weight of air in pounds per cubic foot.

evaporator: Part of a refrigerating mechanism in which the refrigerant vaporizes and absorbs heat.

flapper valve: The type of valve used in refrigeration compressors that allows gaseous refrigerants to flow in only one direction.

HVAC: Abbreviation for heating, ventilating, and air conditioning. Sometimes pronounced *H-vac*; at other times by merely pronouncing the letters *H-V-A-C*.

mechanical efficiency (ME): A decimal number or a percentage representing the ratio between air horsepower divided by brake horsepower of a fan, which is always less than 1.000 or 100 percent.

pressure: Any fan produces a total pressure. This is the sum of the velocity pressure, which is always positive, and the static pressure, which is usually positive on the outlet side and negative on the inlet side of any fan. Thus, total pressure equals velocity pressure plus static pressure. Velocity pressure results only when air is in motion. Static pressure, in its general effect, may be likened to friction and may be described as the pressure that tends to explode a duct (if positive) or collapse it (if negative). All these pressures are expressed in inches of a column of water that they will support.

reversing valve: A solenoid/valve assembly which by automatically changing the direction of the flow of the refrigerant changes the heat pump's cycle from heating to cooling or vice versa.

revolution per minute (rpm): The speed at which a fan or motor turns (revolves).

rotary blower: An encased rotating fan such as that used for forced draft in furnaces.

space heater: A heater for occupied spaces.

standard air: Most fan rating charts or curves are shown at standard air to provide a uniform basis for comparison. By definition, standard air has a density of 0.075 pounds per cubic foot, which is the weight of 1 cubic foot of dry air at a temperature of 70°F and a barometric pressure of 29.92 inches of mercury. Outside of laboratory-controlled conditions, standard air seldom exists in HVAC applications, but the existing environment is usually close enough to standard for practical purposes.

temperature: The dry-bulb temperature of either ambient air or exhaust-stream air. Most fans will work satisfactorily at temperatures up to about 104°F. If the temperature is higher, calculations must be made to determine what effect the temperature will have on satisfactory operation.

therm: Quantity of heat equivalent to 100,000 Btu.

ton of refrigeration: Quantity of refrigeration that can remove heat at the rate of 12,000 Btu's per hour.

thermistor: An electronic device that makes use of the change of resistivity of a semiconductor with change in temperature.

thermometer: Device for measuring temperature.

thermostat: Device responsive to ambient temperature conditions.

thermostatic expansion valve: A control valve operated by temperature and pressure within an evaporator coil, which controls the flow of refrigerant.

timer-thermostat: Thermostat control that includes a clock mechanism. Unit automatically controls room temperature and changes it according to the time of day.

valve, expansion: Type of refrigerant control that maintains a pressure difference between high side and low side pressure in a refrigerating mechanism. The valve operates by pressure in the low or suction side.

velocity: The speed in feet per minute at which air is moving at any location, such as through a duct, inlet damper, outlet shutter, or at the fan discharge point.

1.0.0 INTRODUCTION TO HVAC SYSTEMS

The fundamental concepts of air conditioning are not understood or even considered by the millions who enjoy the comfort heating and cooling produces. Nevertheless, it is readily accepted as part of the American scene.

Air conditioning makes it possible to change the condition of the air in an enclosed area. Because modern people spend most of their lives in enclosed areas, air conditioning is more important, and can produce a greater sustained beneficial influence on humans, than even outdoor weather. People work harder and more efficiently, play longer, and enjoy leisure more comfortably because of air conditioning. Scientific achievements and applications have been outstanding.

1. Military centers that track and intercept hostile missiles are able to operate continuously only because air is maintained at suitable temperatures. Without air conditioning, the mechanical brains in these centers would cease to operate in a matter of minutes because of the intense self-generated heat.

2. Atomic submarines can remain submerged almost indefinitely due, in part, to air conditioning.

3. Modern medicines such as the Salk vaccine are prepared in scientifically controlled atmospheres.

4. Human exploration of outer space has been greatly simplified by air conditioning.

Heating, ventilating, and air conditioning (HVAC) systems in the United States and other nations continue to grow at a phenomenal rate. Most new buildings in the United States now utilize some type of both heating and cooling systems; whereas, fifty years ago the luxury of air conditioning was limited mainly to theaters and similar establishments.

Even most of the older buildings, which had previously been without a complete comfort-conditioned atmosphere, have been renovated in the past ten years so that most commercial buildings in the United States now utilize complete, modern HVAC systems.

HVAC systems, however, would not be possible without the use of electricity. Electricity and the various components, controls, and materials that make up the system is the life-blood for the majority of HVAC and refrigeration systems.

This module is designed to provide up-to-date information on the electrical systems used to power and control HVAC and refrigeration equipment of all types. Furthermore, it provides information and techniques to help keep these systems operating. When a system does fail, data in this module gives troubleshooting techniques and how to make the repairs.

In summary, this module is a quick reference for those actively engaged in the field of electrical repairs for heating, ventilating, air conditioning and refrigeration, a learning method for trainees, and a refresher for those with wide experience in the field.

2.0.0 BODY COMFORT

The normal temperature of the human body is 98.6 degrees F. This temperature is sometimes called subsurface or deep tissue temperature as opposed to skin or surface temperature. An understanding of the manner by which the body maintains this temperature will help in understanding the manner by which the air conditioning process helps to keep the body comfortable.

All food taken into the body contains heat in the form of calories. The "large" or "great" calorie, which is used to express the heat value of food, is the amount of heat required to raise one kilogram of water 1 degree C. As calories are taken into the body, they are converted into energy that is stored for future use. The conversion process generates heat. All body movements not only use up the stored energy, they also add to the heat generated by the conversion process.

For body comfort, all the heat produced must be given off by the body. Because the body consistently produces more heat than it requires, heat must constantly be given off or "removed." The constant removal of body heat takes place through three natural processes that usually occur simultaneously. These processes are convection, radiation, and evaporation. See Fig. 1.

2.1.0 CONVECTION

The convection process of removing heat is based on two phenomena: (1) Heat flows from a hot to a cold surface. For example, heat flows from the body to surrounding air that is below body skin temperature. (2) Heat rises. This becomes evident when observing the smoke from a burning cigarette.

When these two phenomena are applied to the body process of removing heat, the following changes occur: (1) The body gives off heat to the cooler surrounding air. (2) The surrounding air becomes warm and moves upward. (3) As the warm air moves upward, more cool air takes its place, and the convection cycle is completed.

2.2.0 RADIATION

Radiation is the process by which heat moves from a heat source (sun, fire, etc.) to an object by means of heat rays. This principle is based on the previously noted phenomenon that heat moves from a hot to a cold surface. Radiation takes place independent of convection, however, and does not require air movement to complete the heat transfer. It is not affected by air temperature either, although it is affected by the temperature of surrounding surfaces.

Convective heat exchange:
While drifting downhill, the biker
is being cooled by giving up
heat to the air flowing
over the body.

This is an example of "forced"
convective cooling, since air is
being forced over the body
the same as if it were
placed in front
of a fan.

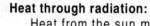

Heat through radiation:
Heat from the sun moves
by radiation to the cooler
body surface (and road
surface).

This is an example of "heat transfer"
by radiation effected by thermal waves
which heat the surrounding surfaces
by absorbtion but not the atmosphere.

Evaporative Cooling
An uphill climb causes the
body to sweat. Perspiration
evaporates from the biker's
body removing heat.

As in this example, an "evaporative"
cooling system does not recirculate
air because it contains too much water.

Figure 1. The processes of convection, radiation, and evaporation.

ELECTRICAL TRAINEE TASK MODULE 20312

The body quickly experiences the effects of sun radiation when it moves from a shady to a sunny area. It again experiences radiation effects when the body surface closest to a fire becomes warm while more distant surfaces remain cool. Just as the heat from the sun and fire moves by radiation to a colder surface, the heat from the body moves to a colder surface.

2.3.0 EVAPORATION

Evaporation is the process by which moisture becomes vapor. As moisture evaporates from a warm surface, it removes heat and thus cools the surface. This process takes place constantly on the body surface. Moisture is given off through the pores of the skin and, as the moisture evaporates, it removes heat from the body.

Perspiration, which appears as drops of moisture on the body, indicates that the body is producing more heat than can be removed by convection, radiation, and normal evaporation. Figure 1 illustrates convection, radiation, and evaporation.

2.4.0 CONDITIONS THAT AFFECT BODY HEAT

Temperature is one of the greatest factors that affects body comfort; therefore, most efforts in designing an air conditioning system are directed toward temperature. In general, temperature affects air conditioning systems in the following ways:

- Cool air increases the rate of convection; warm air slows it down.
- Cool air lowers the temperature of surrounding surfaces and, therefore, increases the rate of radiation; warm air raises the surrounding surface temperature and, therefore, decreases the radiation rate.

Moisture in the air is measured in terms of humidity. For example, 50 percent relative humidity means that the air contains one-half the amount of moisture that it is capable of holding. To simplify the measurement of humidity, a unit called a grain of water vapor is used. A grain is a small amount; in fact, there are approximately 2,800 grains in one cup of water and 7,000 grains in one pound of water.

The following puts this information into practical use: Assume that a room has a temperature of 70 degrees F and four grains of water vapor for each cubic foot of space. If the room temperature remains at 70 degrees F and water vapor is added, the air in the room eventually reaches the point at which it cannot absorb more water. At this point, the air is saturated, and one cubic foot of room space now holds eight grains of water vapor. At 70 degrees F, 8 grains per cubic foot represents 100 percent relative humidity. The original room condition of 4 grains at 70 degrees F represents 50 percent relative humidity:

$$\frac{4\ grains}{8\ grains} = 0.50,\ or\ 50\ percent$$

Relative humidity, then, is obtained by dividing the actual number of grains of moisture present in a cubic foot of room air at a given temperature by the maximum number of grains that the cubic foot of air can hold when it is saturated.

The relative humidity changes when the temperature changes. For example, at 80 degrees F, the relative humidity is:

$$\frac{4}{11} = 0.37, \text{ or 37 percent}$$

If, instead of increasing the temperature to 80 degrees F, the actual moisture content of the air is decreased from 4 grains to 3 grains per cubic foot at 70 degrees F, the relative humidity is:

$$\frac{3}{8}, \text{ or 37 percent again}$$

From the preceding examples, the means of changing relative humidity should become evident.

1.   To increase relative humidity, increase the actual moisture content of the air or decrease the air temperature.

2.   To decrease relative humidity, decrease the actual moisture content of the air or increase the air temperature.

A low relative humidity permits heat to be given off from the body by evaporation. This occurs because the air at low humidity is relatively "dry" and thus can readily absorb moisture. A high relative humidity has the opposite effect: It slows down the evaporation process and thus decreases the speed at which heat can be removed by evaporation. An acceptable comfort range for the human body is 72 degrees F to 80 degrees F at 45 percent to 50 percent relative humidity.

Another factor that affects the ability of the body to give off heat is the movement of air around the body. As air movement increases the following changes occur:

1.   The evaporation process of removing body heat speeds up because moisture in the air near the body is carried away at a faster rate.

2.   The convection process increases because the layer of warm air surrounding the body is carried away more rapidly.

3.   The radiation process tends to accelerate because the heat on the surrounding surfaces is removed at a faster rate, causing heat to radiate from the body at a faster rate.

As air movement decreases, the evaporation, convection, and radiation processes decrease.

3.0.0   TYPICAL AIR CYCLE

Indoor air can be too cold, too hot, too wet, too dry, too drafty, and too still. These conditions are changed by "treating" the air. Cold air is heated, hot air is cooled, moisture is added to dry air...removed from damp air, and fans are used to create adequate air movement. Each of these "treatments" are provided in the air conditioning air cycle. See Fig. 2.

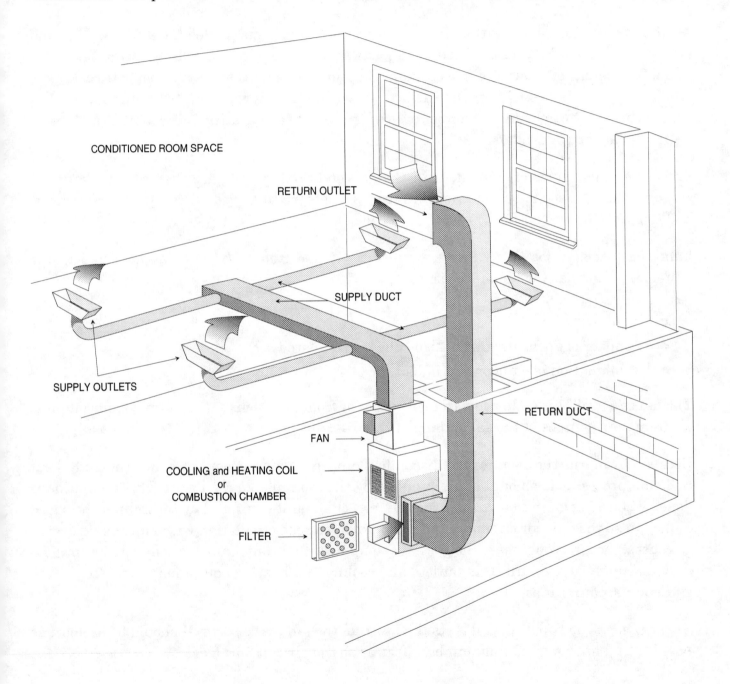

Figure 2. Typical air cycle.

3.1.0 CYCLE DESCRIPTION

The description begins with the fan because it is the one piece of equipment that starts the air through the cycle. A fan forces air into ductwork connected to openings in the room. These openings are commonly called outlets or terminals. The ductwork directs the air to the room through the outlets. The air enters the room and either heats or cools as required. Dust particles from the room enter the air stream and are carried along with it.

Air then flows from the room through a second outlet (sometimes called the return outlet) and enters the return ductwork, where dust particles are removed by a filter. After the air is cleaned, it is either heated or cooled depending on the condition in the room. If cool air is required, the air is passed over the surface of a cooling coil; if warm air is required, the air is passed through a combustion chamber or over the surface of a heating coil. Finally, the treated air flows back to the fan, and the cycle is completed.

Thus, the major parts of equipment in the air conditioning cycle are the fan, supply ducts, supply outlets, space to be conditioned, return outlets, return ducts, filter, heating chamber, and cooling coil.

Fan: The principal job of the fan is to move air to and from a room. In an air conditioning system, the air that the fan moves is made up of:

- All outdoor air
- All indoor or room air (this is also called recirculated air)
- A combination of outdoor and indoor air

The fan can "pull" air exclusively from outdoors or from the room, but in most systems, it pulls air from both sources at the same time.

Because drafts in the room cause discomfort, and poor air movement slows the body heat rejection process, the amount of air supplied by the fan must be regulated. This regulation is done by choosing a fan that can deliver the correct amount of air and by controlling the speed of the fan so that the air stream in the room provides good circulation without causing drafts. Of course, the fan is only one of the pieces of equipment that contributes to body comfort; others, such as supply and return room outlets and cooling and heating equipment, are described in subsequent paragraphs.

Supply Duct: The supply duct directs the air from the fan to the room. It should be as short as possible and have a minimum number of turns so the air can flow freely.

Supply Outlets: Supply outlets help to distribute the air evenly in a room. Some outlets "fan" the air, others direct it in a jet stream, still others can do a combination of both. Because supply outlets can either fan or jet the air stream, they are able to exert some control on the direction of the air delivered by the fan. This directional control combined with the location and the number of outlets in the room contributes a great deal to the comfort or discomfort effect of the air pattern.

Room Space: The room is one of the most important parts of the air cycle description. The dictionary states that a room is an "enclosed space" set apart by partitions. If this enclosed space did not exist, it would be impossible to complete the air cycle because conditioned air from the supply outlets would flow into the atmosphere. The enclosed space, therefore, is all important. In fact, the material and the quality of workmanship used to enclose the space are also important because they help to control the loss of heat or cold that is confined in it.

Return Outlets: Return outlets are openings in the room surface that are used to allow room air to enter the return duct. They are usually located at the opposite extreme of a wall or room from the supply outlet. For example, if the supply duct is on the ceiling or on the wall near the ceiling, the return duct may be located on the floor or on the wall near the floor. This is not true in all cases, however, because some systems have both supply and return outlets near the floor or near the ceiling.

Keep in mind that the main function of the return outlet is to allow air to pass from the room.

Filters: Filters are usually located at some point in the return air duct. They are made of many materials, from spun glass to composition plastic. Other types operate on the electrostatic principle and actually attract and capture dust and dirt particles through the use of electricity.

The end purpose of all filters is to clean the air by removing dust and dirt particles.

Cooling Coil and Heating Coil or Combustion Chamber: The cooling coil and the heating coil or combustion chamber can be located either ahead of or after the fan but should always be located *after* the filter. A filter ahead of the coil is necessary to prevent excessive dirt, dust, and dirt particles from covering the coil's surface.

4.0.0   REFRIGERATION AND COMPRESSORS

Refrigeration has been defined as the process of transferring heat from one location to another.

A diagram of a conventional refrigeration system is shown in Fig. 3. This diagram is typical of all refrigeration units regardless of whether they are for use in residential, automotive, or other applications. Please refer to this diagram often while reviewing the following paragraphs.

In general, heat is picked up by the boiling refrigerant at the evaporator and then rejected at the condenser; this heat must then be carried away by air, water, the evaporation of water, or some other means. The compressor provides the energy for the system's operation, while the function of the controls or expansion valve is to permit the compressor to maintain a pressure difference. A more detailed study of this system will be covered later.

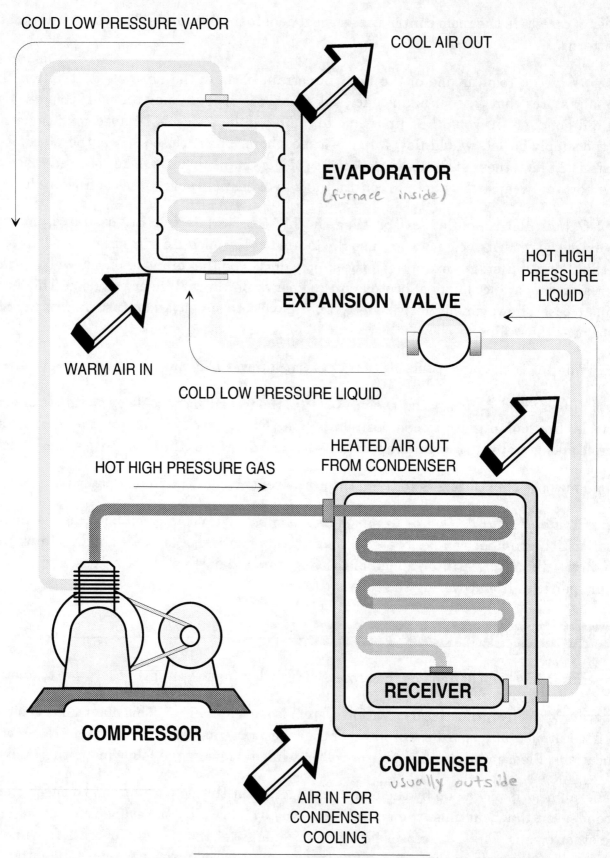

COLD LOW PRESSURE VAPOR

COOL AIR OUT

EVAPORATOR
(furnace inside)

HOT HIGH
PRESSURE
LIQUID

EXPANSION VALVE

WARM AIR IN

COLD LOW PRESSURE LIQUID

HEATED AIR OUT
FROM CONDENSER

HOT HIGH PRESSURE GAS

RECEIVER

COMPRESSOR

CONDENSER
usually outside

AIR IN FOR
CONDENSER
COOLING

Figure 3. Typical refrigeration system.

Although all types of refrigeration and air conditioning systems operate on the same basic principle, many operate in entirely different environments. For example, a residential air conditioner operates on a steady source of electrical power (120-240 volts, 60 cycles) and is seldom exposed to extremes in operating conditions.

4.1.0 THEORY OF REFRIGERATION

In order for an electrician to approach the job of electrical maintenance and repair of air conditioning systems more intelligently, a brief explanation of the theory of refrigeration is in order. With this foundation, the trainee will be in a much better position to troubleshoot any air conditioning or refrigeration system, and correct the problem in the shortest period of time.

Refrigeration is a process which involves the transfer of heat from any given article or space. In the mechanical air conditioning system this transfer of heat is accomplished by means of a refrigerant, which is conditioned so that it will absorb heat from the space to be cooled, and after this absorption, conditioned so it will give up this heat to an outside water, air or other supply.

To understand the operation of a mechanical refrigerating system, it is necessary to understand a few of the fundamental laws of heat, especially the conditions under which liquid refrigerant will vaporize and gaseous refrigerant will liquefy.

Heat is a form of energy which will cause certain changes in substances; that is, heat may cause the substance to become warmer, to melt, or to evaporate, depending upon the quantity of heat absorbed and the state of the substance. For example, at room temperature water is a liquid. But when water is made hot enough, it becomes a gas known as steam. If water is made cold enough, we all know that it will turn into a solid called ice.

Heat is not a substance and therefore cannot be measured by volume or weight. Rather, it must be measured by the effect that it produces. The unit most commonly used to measure heat is the Btu (British thermal unit), defined as the amount of heat necessary to raise the temperature of one pound of water by 1 degree F. The unit in the average home is probably rated somewhere around 24,000 Btuh—capable of producing 24,000 Btu's per hour.

Temperature is one of the many effects produced by heat—a measure not of the heat in a substance but of its intensity. Temperature indicates how warm or how cold a substance is.

4.2.0 TRANSFER OF HEAT

The basic law in the transfer of heat is that it can flow only from a body of a higher temperature to a body of a lower temperature—never in the reverse direction. Therefore, the result of such a transfer is that the colder body gains heat and the warmer body loses heat. When the colder body receives heat, two changes can take place: (1) its temperature may rise or (2) its state may change from solid to liquid or liquid to gas. When a warmer body loses heat, two changes can also take place: (1) its temperature may lower or (2) its state may change. This change will be from a gas to a liquid or from a liquid to a solid.

When the state of a body is changed, either from a solid to a liquid, a liquid to a gas, or vice versa, the temperature of the body remains constant during the time the change is taking place. For example, assume that a glass jar is filled with a solid piece of ice. The temperature of the ice will have to be less than 32 degrees F to remain as a solid. However, if heat is applied to the jar, the ice will warm up to 32 degrees F and then begin to melt. Although heat continues to be added, the temperature of the ice will not rise above 32 degrees F until all the ice has melted. After the ice has melted entirely, the temperature of the water will continue to rise (as long as the heat is applied) until the water reaches 212 degrees F At this temperature, the water will start to boil but the water temperature will remain at 212 degrees F even though heat is still applied to the jar. Continued heating of the boiling water will result in vaporizing the water into steam. Figure 4 shows the temperature changes resulting from heating ice.

Water, however, does not always boil at 212 degrees F; the boiling point varies according to the pressure to which it is subjected. To illustrate, water in a boiler under 100 psi (pounds per sq. in.) pressure boils at approximately 338 degrees F, while water under reduced pressure (less than atmospheric) can be made to boil at below its freezing point!

The boiling points of different liquids vary a great deal. Under normal atmospheric pressure, water will boil at 212 degrees F, alcohol at 173 degrees F, mercury at 674.5 degree F, methyl chloride at -11 degrees F, ammonia at -29 degrees F, Freon-12 at -22 degrees F, and Freon-22 at -41.4 degrees F. The latter three liquids have been used extensively in refrigeration units. Like water, if the pressure of these liquids is increased, the temperature at which they boil is raised, and if the pressure is decreased, the boiling point will be lowered.

To boil, a liquid must receive heat from a substance of a higher temperature. To boil water at normal atmospheric pressure, for example, the source of heat must have a temperature higher than 212 degrees F. However, Freon-12 will boil at atmospheric pressure if the source of heat is only slightly higher than -22 degrees F. In fact, melting ice at 32 degrees F is really warm when compared with the boiling point of Freon-12 at -22 degrees F. For this reason, Freon-12 at atmospheric pressure can be easily boiled and vaporized at the temperature of melting ice. Heat will flow to the Freon-12, causing it to vaporize into a gas.

4.3.0 THE REFRIGERATING PROCESS

By placing the refrigerant in a suitable container and then placing the container in contact with a warmer substance, the liquid refrigerant will be vaporized by the warmer substance. Although the liquid refrigerant is receiving heat from the warmer substance, the temperature of the refrigerant will not increase as long as the pressure remains constant. This is due to the fact that heating a boiling liquid does not increase its temperature but instead vaporizes the liquid more quickly. With the pressure remaining constant, the liquid refrigerant will continue to boil at a constant temperature as long as it receives heat from the warmer substance.

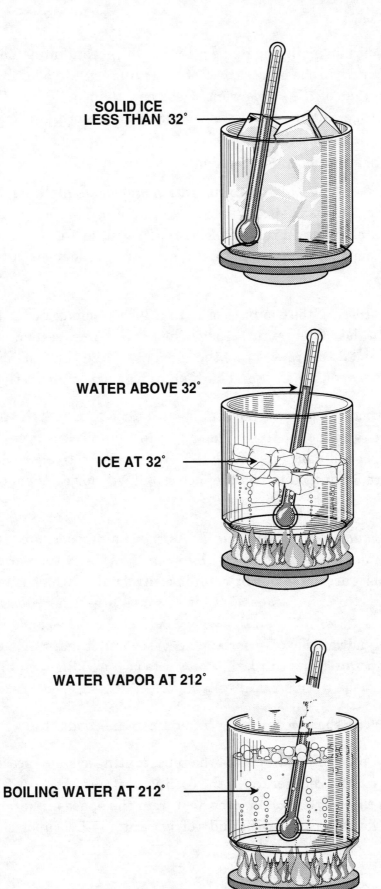

SOLID ICE
LESS THAN 32°

WATER ABOVE 32°

ICE AT 32°

WATER VAPOR AT 212°

BOILING WATER AT 212°

Figure 4. Temperature changes resulting from heating ice.

The warmer substance, in giving up its heat to the refrigerant, must naturally undergo some change. Unless it freezes in the process, loss of heat will result in a lowering of its temperature until it is equal to that of the refrigerant. Heat can then no longer flow from the warmer substance to the refrigerant and the boiling or vaporizing process will stop.

4.4.0   Operation of a Refrigeration System

Refer again to Fig. 3 on page 14. This illustration shows a simplified diagram of a complete refrigeration system. This is basically the same system used in all types of cooling units like your home refrigerator, home air conditioners, and, of course, the air conditioning system used in cars. The basic components consist of a compressor, a condenser, an expansion valve, and an evaporator.

Compressor: The purpose of the compressor in Fig. 3 is threefold. First, it pumps or withdraws heat-laded gas from the cooling unit through the suction line; second, it compresses this gas to a high pressure and discharges it into the condenser; and third, its discharge valve acts as the dividing point between the low- and high-pressure sides of the refrigerating system.

Condenser: The function of the condenser is to reject the heat in the refrigerant to the surrounding air or to a water supply, condensing it to a liquid. One type of air-cooled condenser is the fin-and-tube, or radiator, type. With this type, comparatively cool air is passed through a compact, finned radiator by a fan on the motor pulley; your car's engine cooling system is a good example of such a unit.

Expansion Valve (liquid refrigerant control): As the compressor pumps the refrigerant from the cooling unit it must be replenished with low-pressure, low-temperature liquid capable of absorbing heat. This is accomplished by a liquid control valve, which is known as an expansion valve. This valve has three functions. First, it acts as a pressure-reducing valve, reducing the pressure of the high-pressure liquid that was condensed by the condenser to a low-pressure, low-temperature liquid capable of absorbing heat; second, it maintains a constant refrigerant pressure in the evaporator; and third, the valve acts as a dividing point between the high- and low-pressure sides of the refrigerating system.

Receiver: The reservoir that accumulates liquid from the condenser.

Evaporator: The evaporator is that part of the refrigerating system directly connected with the refrigerating process. The refrigerant in the cooling unit absorbs heat from the space or air to be refrigerated. As the evaporator absorbs heat from the space or material to be refrigerated, the low-pressure, low-temperature liquid refrigerant in the vaporizing tubes (or coils) is vaporized.

4.5.0   HEAT PUMPS

A heat pump is an air conditioning unit consisting of a combination of components that supply heating in one cycle and cooling when the cycle is reversed. Heat-pump applications are becoming quite popular in both residential and commercial applications and electricians working on these types of projects will encounter this type of HVAC system frequently.

Heat-pump applications are most efficient in moderate or warm climates where the heating demand is low, but such equipment is also used in cooler northern climates where the heating demand might be considerably higher. However, when used in cooler climates, the heat generated from the heat pump is normally supplemented with auxiliary electric-heating elements. When the heat pump alone is unable to satisfy the demand of the thermostatic controls, the auxiliary heating "kicks in" to help—often in stages or steps as might be required.

A heat pump is able to use heat from almost any source. Earth, air and water are natural heat sources. For example, at an eight-foot depth, earth temperature remains relatively constant year-round—about 52 degrees F. Water, although frozen at the surface, remains liquid at a certain depth; even cool wintry air contains some heat that may be utilized by a heat pump. With these natural heat sources as a base, the heat pump pumps, moves and transfers this heat to a space to be conditioned.

There are several types of heat pumps. For example, a water-to-air heat pump uses water as the heat source and conditions the space with air. A water-to-water heat pump again uses water as the heat source and conditions the space with hot or chilled water. The most popular type of heat pump, however, is the air-to-air system. Since air is universally available, this method can be applied in any area.

4.5.1   Heating Cycle

During the pump's heating cycle, heat is taken from outside air, conditioned, and released to inside air. Figure 5 shows how this action is accomplished.

In general, the heat pump takes heat from outside air and in the process changes the liquid refrigerant to a gas. This gas is pumped to the compressor which compresses the gas and raises its temperature. This high-temperature, high-pressure gas continues to the indoor coil where the gas gives up heat to the fan-forced air that is used to condition the room or space. In doing so, the gas is changed to liquid. This liquid flows to the expansion valve which expands the liquid and reduces its temperature. The refrigerant then flows back into the outdoor coil and the process is repeated.

4.5.2   Cooling Cycle

The heat-pump cooling cycle is just the opposite of the heating cycle. During the cooling cycle, heat is taken from the inside air and released to the outside air. Heat-pump components are the same, but the refrigerant flow is reversed. This reverse-flow is readily possible through the

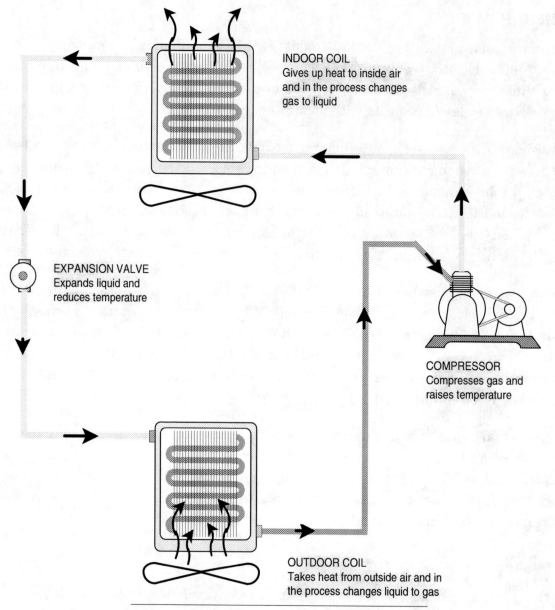

INDOOR COIL
Gives up heat to inside air
and in the process changes
gas to liquid

EXPANSION VALVE
Expands liquid and
reduces temperature

COMPRESSOR
Compresses gas and
raises temperature

OUTDOOR COIL
Takes heat from outside air and in
the process changes liquid to gas

Figure 5. Heat pump operating in heating cycle.

coils but not through the expansion valve and the compressor. Consequently, additional controls are necessary to enable a heat pump to operate in both modes. There are several ways to accomplish this dual function. For example, two expansion valves can be installed in the system so that the required flow is possible in both cycles. However, since both valves are in the same refrigerant line, one or the other will impede refrigerant flow depending upon the cycle in operation. This problem is solved by using a bypass check valve around each expansion valve. Such check valves are installed (and controlled) so that refrigerant can flow through the heating-cycle expansion valve only during the heating cycle and through the cooling-cycle expansion valve only during the cooling cycle.

In addition, a 4-way valve and suitable refrigerant piping assures a one-way flow through the compressor. This valve is positioned in one direction for heating and then in the opposite direction for cooling. The basic system is shown in Fig. 6.

ELECTRICAL TRAINEE TASK MODULE 20312

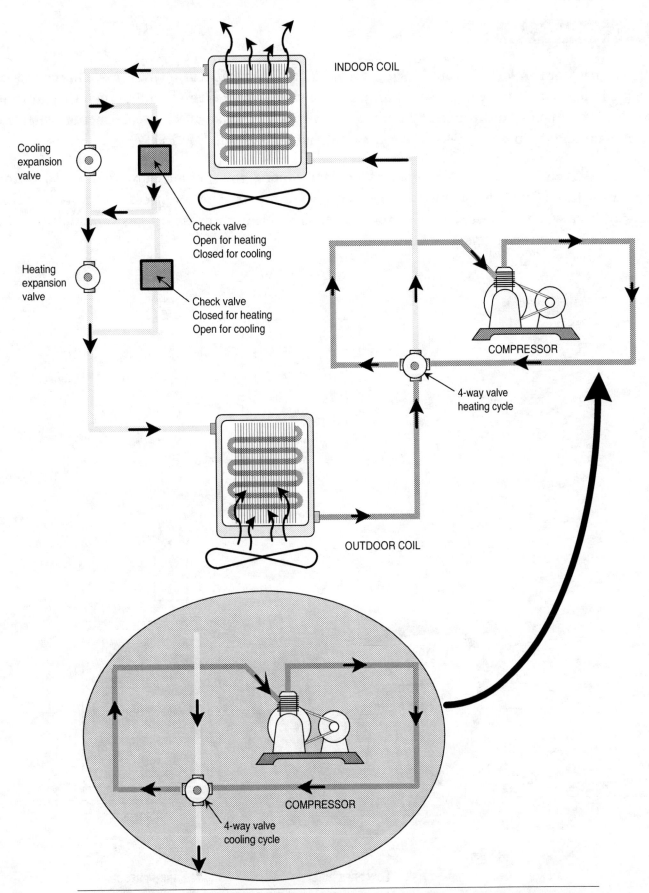

INDOOR COIL

Cooling expansion valve

Check valve
Open for heating
Closed for cooling

Heating expansion valve

Check valve
Closed for heating
Open for cooling

COMPRESSOR

4-way valve
heating cycle

OUTDOOR COIL

COMPRESSOR

4-way valve
cooling cycle

Figure 6. Heat-pump system with basic controls for both heating and cooling cycles.

5.0.0 COMPRESSORS

The compressor is referred to as the "heart" of mechanical refrigeration systems. This comparison is made because the compressor pumps refrigerant through the system much the same as the heart pumps blood through the body. The compressor also works in conjunction with the other components in a refrigeration or air conditioning system. See Fig. 7.

Starting at the low side of the evaporator, the vapor—now at both low temperature and pressure—flows through the suction line to the compressor. The compressor compresses this gas which raises its pressure and temperature. The hot, high pressure gas then flows to the condenser where it condenses to form a liquid.

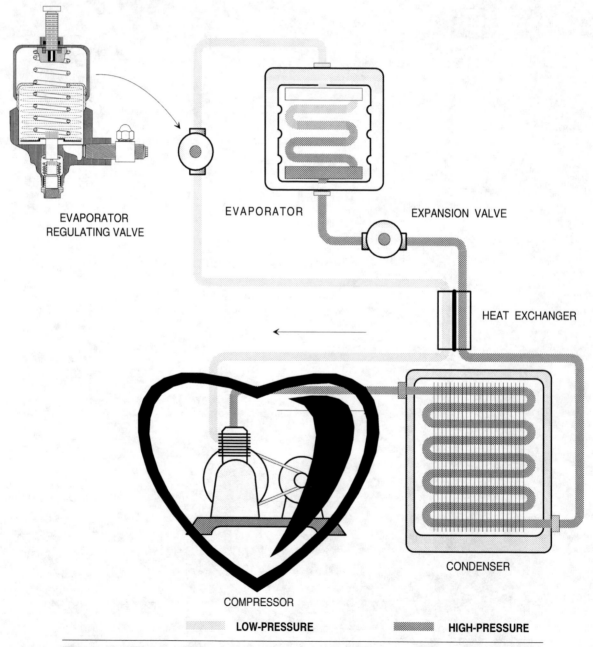

Figure 7. The compressor is known as the "heart" of the refrigeration system.

The compressor also lowers the pressure in the evaporator. This causes the refrigerant to boil at the reduced pressure and temperature. The heat from the space to be cooled flows into the evaporator because of the resulting low temperature. This heat vaporizes the liquid refrigerant. The refrigerant vapor which contains the absorbed heat from the evaporator is pumped back to the compressor. During this part of the cycle, the gas temperature is raised. This high temperature vapor is then discharged from the compressor. The heat from the hot gas flows into water or air which passes through or around the condenser. As a result, the refrigerant condenses to a liquid.

Stated briefly, the function of a compressor is to maintain a pressure difference between the high and low sides of the system. In this process certain conditions are created:

- The pressure and temperature of the refrigerant in the evaporator are lowered, allowing the refrigerant to boil and absorb heat from its surroundings.
- The pressure and temperature of the refrigerant in the condenser are raised allowing the refrigerant to give up heat at existing temperatures to whatever medium is used to absorb the heat.

In general, there are two basic operations performed by the compressor (Fig. 8): first, it draws the refrigerant from the cooling coil (suction) and secondly, forces it into the condenser (discharge). Therefore, the two basic operations of the compressor are:

- Suction
- Discharge

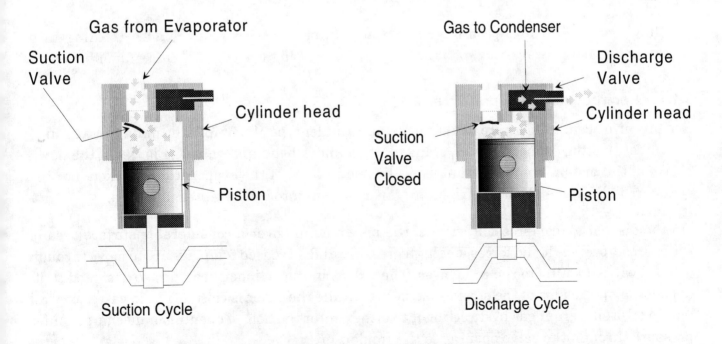

Figure 8. The two basic operations of a compressor.

During the suction cycle, the compressor reduces the pressure in the cooling coil and maintains it at a level low enough to permit the refrigerant to "boil" or vaporize and consequently absorb heat in the process.

NOTE   Refrigerant boils at a relatively low temperature when pressure is reduced.

By discharging or forcing refrigerant vapor into the condenser, the compressor increases the pressure of the refrigerant. In doing so, the compressor actually increases the refrigerant vapor temperature. This makes it easier for the condenser to do its job.

5.1.0   TYPES OF COMPRESSORS

Functions And Operating Parts: The variety of refrigerants and the size, location and application of the systems, are some of the factors which create the need for many types of compressors. Since refrigerant properties differ, one compressor may be required to handle large volumes of vapor at small pressure drops, and another, small volumes of vapor at large pressure drops. While the selection of such compressors normally fall under the design of mechanical engineers, the HVAC electrical technician should have a working knowledge of how compressors are selected to perform repair and maintenance work on HVAC systems.

There are three main classifications of compressors: (1) reciprocating, (2) rotary, and (3) centrifugal. The action of the mechanical parts of the compressor determines its classification.

1.   In a reciprocating compressor, a piston travels back and forth (reciprocates) in a cylinder.

2.   In a rotary compressor an eccentric rotates within a cylinder.

3.   In a centrifugal compressor a rotor (impeller) with many blades rotating in a housing, draws in vapor and discharges it at high velocity by centrifugal force.

5.1.1   Reciprocating Compressors

Reciprocating compressors are usually a piston-cylinder type of "pump." The main parts include a cylinder, piston, connecting rod, crankshaft, cylinder head and valves (Fig.). On the down stroke of the piston, the refrigerant from the suction line of the evaporator is sucked into the cylinder. This causes the hot refrigerant vapor to rush into this low pressure area.

On the discharge (compression) stroke, the piston acting over a considerable surface area of the gas, compresses it and forces it at high pressure and increased temperature to move through a small valve opening to the condenser. The valves in the cylinder head are so designed that, depending on the part of the stroke, one is open while the other is closed. These valves control part of the refrigerant gas by directing it to either enter the hollow opening or discharge under pressure through the valve opening to the condenser.

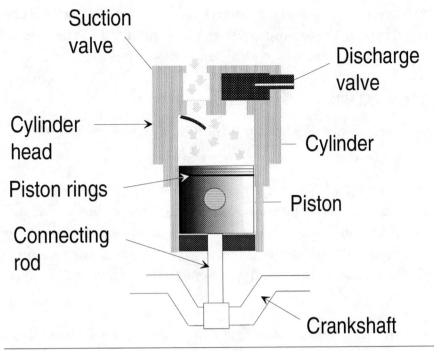

Figure 9. Cross-section and basic parts of a refrigeration compressor.

Returning from the top of the stroke, the piston again draws in the hot refrigerant vapor and the cycle continues. The connecting rod is attached to a rotating crankshaft and serves to change rotary motion to straight line (rectilinear) motion.

The valve that controls the flow of refrigerant from the suction line into the cylinder head is known as the "suction" valve; the valve leading to the discharge line as the "discharge" valve. The rings on the piston prevent the gas from escaping between the piston and cylinder walls and improve the operating efficiency.

The crankcase around the compressor contains part of the bearing surfaces for the crankshaft and stores oil that is used to lubricate the crankshaft and the connecting rod.

There are many types of reciprocating compressors. One of the most common ways of classifying them is by the number of cylinders. While most refrigerating compressors are one (single) cylinder, some models have two cylinders. The two cylinders run more smoothly and are more compact. Larger installations require compressors with three to ten cylinders or more.

The arrangement of the cylinders is still another method of classifying compressors. Some are "vertical"; others are "horizontal", "45 degree inclined", "V-type", "W-type", "radial", and the like. Any one of these may be single, double, or any other number of cylinders, depending on size and the nature of the installation.

On large installations the size of the lines requires the use of hard copper tubing. Vibration produced by the rotary, reciprocating motion requires the use of hard copper tubing. Vibration produced by the rotary, reciprocating and oscillating motion of the parts may cause excessive noises and may even break the soldered connections. Vibration absorbers of corrugated copper tubing with a braided bronze protecting cover may be installed in the liquid and suction lines to prevent condenser vibration from traveling into these lines.

5.2.0 COMPRESSOR VALVES

Two common types of valves are used in compressors: the nonflexing ring plate and the flexing disc.

The ring plate is a thin ring which is held closed over the circular discharge gas inlet in the top of the cylinder by springs. The suction valve is a ring plate that fits around the outside and just below the top of the cylinder. The valve is held closed by small springs. When the refrigerant vapor pressure inside the cylinder is greater than the spring tension, the valve opens on the up stroke of the piston to allow vapor to pass through the large discharge ports to the discharge outlet.

The suction valve opens on the down stroke of the piston because the cylinder pressure is less than the vapor pressure in the suction line. The pressure in the crankcase and in the lower portion of the compressor is the same as the suction pressure at the inlet side.

On the up stroke of the piston the suction valve closes and the pressure within the cylinder causes the discharge valve to open. The high pressure vapor then passes into the compressor cylinder head through the center holes in the valve cage as well as around the discharge valve cage. The metallic noise produced by the opening and closing of the discharge valve is cut down by a cushion of plastic material installed in the valve cage.

If a slug of liquid refrigerant were to enter a cylinder, the head might be blown off because of the small clearance space at the point where the piston reaches the top of its stroke. Instead of being fastened in position, the cylinder head is held firmly in place by a strong spring. When a slug of noncondensable liquid enters the cylinder, the entire head lifts and passes the liquid to the discharge outlet. When there is only compressible vapor in the cylinder, the pressure produced during ordinary operation is not great enough to lift the safety head.

The refrigerant vapor travels through the strainer to the cylinder head where the suction valve is located. A metal section inside the cylinder head separates the suction and discharge valves. The screen housing has a small opening in the bottom which allows oil to be carried by the suction vapor back into the crankcase.

5.2.1 Flexing Disc or Reed Type Valves

Small modern refrigeration compressors use high grade steel reed or disc valves and are especially adaptable to high speed compressors.

The operation, seating and tightness of the valves is important. When the valves leak, gas which has been compressed is lost and the temperature of the discharge gas to the condenser is increased. The hot gas leaking by the valves and piston raises the temperature of the suction gas. Then, as the warmer suction gas is compressed, the discharge gas becomes hotter. On the next stroke, the higher temperature has leaks back resulting in a still higher discharge temperature with a marked decrease in compressor capacity and efficiency.

5.3.0 COOLING COMPRESSOR HEADS

During compression, the temperature of the refrigerant vapor rises. The temperature is controlled by cooling the upper part of the cylinder walls and the cylinder head to minimize the work required for compression and to keep the cylinder head from overheating. This cooling is done on ammonia compressors by jacketing a cylinder wall and head jacket through which water is circulated.

When fluorinated hydrocarbon is discharge from compressors, its temperature is much lower than that of ammonia. Consequently, water jackets are rarely provided on fluorinated hydro-carbon compressors. Instead, the cylinder walls and cylinder head are designed with fins to facilitate the transfer of heat to the surrounding air.

5.4.0 SAFETY SPRINGS

There are times when liquid refrigerant or oil floods over into the suction line and compressor. This may be due to faulty adjustment of the expansion valve, a leaking float valve, or other similar difficulty. If the quantity of liquid is large and if it cannot get through the valve ports, serious damage may result.

The valve, the valve retainer, discharge valve coil springs and shoulder screws are mounted on the valve plate. The coil springs are strong enough to hold the valve retainer down during normal operation. Under these conditions, the discharge valve opens and allows gas to pass into the discharge chamber. The lift of the valve is limited by the valve retainer.

Whenever liquid or oil becomes trapped between the top of the piston and valve plate, a hydraulic pressure created in the cylinder forces the valve retainer to lift. This lift of the valve allows the liquid to discharge into the head where it either vaporizes or passes into the condenser. When the liquid has been cleared from the compressor, the valve retainer reseats.

5.5.0 SHAFT SEALS

Open compressors are made with the crankshaft extending through the crankcase for direct motor, V-belt, gear, or chain drive. A crankshaft seal is required to prevent refrigerant leakage. Leakage may take place under both static and moving conditions at the point at which the crankshaft passes through the housing. In practically all cases the crankcase is exposed to the circulating refrigerant vapor.

With horizontal double-acting compressors the piston rods slide back and through a stuffing box. This is usually sealed with asbestos and graphite or a metallic or semimetallic packing.

Most compressors that use a rotating shaft which projects out of the crankcase, utilize a bellows or siphon seal.

5.6.0 SERVICE VALVES

Service operations may be speeded up and simplified when conventional compressors are equipped with suction and discharge service valves. The discharge valve is fastened to the head and the suction valve to the body or head of the compressor.

The valve is known as a "back seating" type. This means that when the stem is turned all the way back, the gage port is closed. The valve stem is usually back seated so the line connection is open to the compressor. When the valve stem is in as far as it will go, the suction or discharge line is shut off while the gage port is open to the compressor. At midpoint the stem is so located that the compressor is open to the line connection, the gage port, or both. Because of this design the gage connection must be plugged except when a gage is being used.

6.0.0 HERMETIC COMPRESSORS

Extreme accuracy of finish and close dimensional tolerances are essential in the design and construction of hermetic compressors. There are two types that are in popular use:

- Reciprocating compressor
- Rotary compressor

Hermetic and open-type compressors are similar. The main difference is that the electric motor of the hermetic compressor is encased in the crankcase or within a sealed housing containing the compressor. With such a design, the compressor is driven directly by the motor, revolves at motor speed, and requires no shaft seal. Single cylinder models are available for small units while two-cylinder ones are used for larger units.

The compressor and motor are enclosed in a steel casing (dome or hat). The stationary field (stator) of the motor may be pressed into half the dome. The compressor is secured to this stator. This unit is usually mounted on springs or rubber mounts which dampen or absorb vibration. See Fig. 10.

One major problem with hermetic units that directly affects the HVAC electrical technician is the cooling of the electric motor. In one design the stator is pressed into the dome to help cool the motor. This provides easy heat transfer from the windings to the case.

A second design provides a way of passing the returning gas around the motor windings before the gas is compressed in the compressor. While the cool gas removes a great deal of heat, this design has the disadvantage of reducing compressor efficiency because it warms the returning gas.

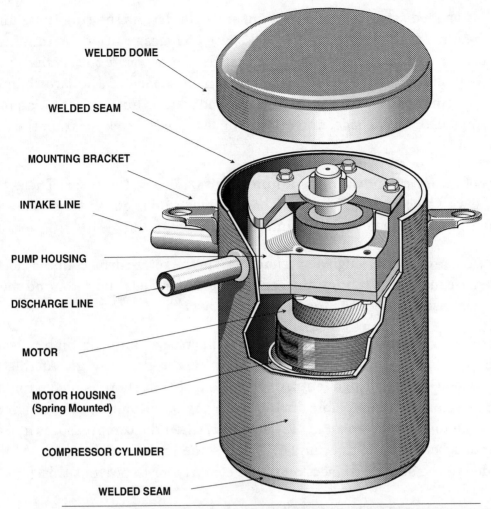

WELDED DOME

WELDED SEAM

MOUNTING BRACKET

INTAKE LINE

PUMP HOUSING

DISCHARGE LINE

MOTOR

MOTOR HOUSING
(Spring Mounted)

COMPRESSOR CYLINDER

WELDED SEAM

Figure 10. Basic components of a hermetically-sealed compressor.

Sound deadening devices are necessary on the smaller hermetic units for both the intake and exhaust openings. A device known as a muffler eliminates the gasping sound of the intake stroke and the puff noises of the exhaust gases.

6.1.0 ROTARY COMPRESSORS

Hermetically sealed rotary compressors are widely used for fractional-tonnage refrigeration applications. These rotary compressors may be divided roughly into two types. In the first type one or more stationary blades are used for sealing the suction from the discharge gases. The second type uses "sealing blades" which rotate with the shaft.

6.2.0 ROTARY COMPRESSOR WITH STATIONARY BLADE

The moving parts in a rotary compressor include a steel ring, a cam (eccentric) and a sliding barrier. The stationary parts consist of the motor which drives the shaft and the steel cylinder. The ring is precision machined so it fits over and may be turned on the cam. The outside rim of the ring fits inside the cylinder. As the shaft cam rotates, it moves the ring so that one point on its circumference is always in contact with the cylinder wall.

There is a crescent-shaped space between the ring and cylinder. As the ring turns and rolls on its rim against the inner wall of the cylinder, the space changes position. Within the cylinder head there is a suction port from the evaporator and a discharge port to the condenser. Between these two ports is a sliding barrier (sealing blade) which separates the two chambers thus formed. At the same time, the blade permits compressed refrigerant vapor at high pressure to be forced to the condenser on one side, and allows the low pressure vapor from the evaporator to enter the other side.

The second type of rotary compressor has a number of blades in the rotor. These blades are forced against the wall by centrifugal action. The cylinder head is off center so there is, again, a crescent-shaped space between the rotor and the cylinder.

The low pressure gas is drawn through a suction port. As the rotor turns counter-clockwise the gas is compressed because of the continually reduced space between the rotor and the cylinder. As the pressure increases, so does the temperature.

When the compressed gas reaches the discharge port, it passes into the high pressure dome because there is no space ahead of the rotor into which the gas may go. At this point the clearance is about .0001". This design feature, plus the fact that there is also a fine film of oil lubricating the parts, makes it impossible during operating conditions for the refrigerant vapor to leak from the high to the low pressure side. However, when the compressor is idle and there is no lubricating film between the rotor and cylinder, some vapor may leak out. A check valve should be provided in the suction connection to the compressor to prevent this.

6.3.0   IMPORTANCE OF LUBRICATION IN ROTARY COMPRESSORS

Proper operation of rotary compressors depends on maintaining a continuous film of oil on the cylinder, roller and blade surfaces. This oil feeds into the cylinder through the main bearings. A good lubricant for compressors has these properties:

- It must be free of moisture, wax and foam.
- It must have the correct viscosity for the specific refrigerant used.
- It must be free of impurities which cause carbon to form around exhaust valve.

Rotary compressors are quiet in operation, have limited vibration and may be used where a fairly high volume of refrigerant must be moved per ton of refrigeration.

6.4.0   CENTRIFUGAL COMPRESSORS

A centrifugal refrigeration system depends upon centrifugal force to compress the refrigerant vapor. The rotor (impeller) of a centrifugal compressor draws in vapor near the shaft and discharges at a high velocity at the outside edge of the impeller. The high velocity (inertia) is converted into pressure.

When the pressure drop is high, the compressor is built in stages. The discharge at one stage enters the suction inlet of the next until by the time the last stage is reached as much energy as possible has been used.

Centrifugal compressors are especially adapted for systems ranging as high as 5,000 tons and as low as 50 tons. They are also adaptable to temperature ranges between -130 degrees F to 50 degrees F.

Centrifugal compressors are reasonably efficient even when operating with loads as low as 20% of normal. Because of their high operating speeds, centrifugal compressors may be connected directly to a steam turbine drive. Smaller sizes are usually driven by electric motors, some of which are equipped with standard gear driven speed increasers.

6.5.0 METHODS OF CONTROLLING COMPRESSOR CAPACITY

The four common methods of controlling compressor capacity are, (1) by using a variable speed motor, (2) by bypassing the hot gas, (3) by bypassing the cylinders, (4) by cylinder unloading. The capacity of a compressor must be controlled because refrigerant loads are seldom constant. Operating under partial loads and low back pressures creates a condition where the coil may freeze or damage may result.

6.5.1 Variable Speed Motors

The capacity of a compressor is proportional to the speed of the driving motor. When the suction pressure of the refrigeration system is high, the motor speed and compressor capacity must be increased. For this reason, electric motors with two or more speeds are used. The motor speed may be selected according to the wiring. Variable-speed motors find limited application because the motors and their controls are expensive to use on large installations.

6.5.2 Hot Gas Bypass Method

The temperature or pressure of the refrigerant in a simple hot gas bypass compressor capacity control may be controlled by a solenoid stop valve in the bypass line. When a capacity reduction is needed, the solenoid opens and permits some of the hot gas discharge by the compressor to be returned the suction line.

For full compressor capacity the solenoid stop valve closes so that the gas from both banks of cylinders passes through the discharge line. The solenoid is opened for half-capacity operation and closed for full capacity operation.

When operating at reduced capacity for long periods of time, the cylinder heads become very hot, there are many lubricating problems and a great deal of noise. This method of controlling capacity is more practical where the reduction is of short duration and does not occur frequently.

6.5.3 Cylinder Bypass Method

The parts of a cylinder bypass system may be positioned either internally or externally. The cylinder bypass method is activated by either temperature or pressure controls.

The solenoid valve opens whenever the controller requires a capacity reduction. By this action, the discharge gases from one bank of cylinders goes directly to the suction line. The check valve does not permit any gas at high pressure to reach the isolated bank. Then, since the lines are large no high-pressure is created in the bypassed cylinders. As a result the bypassed cylinders have a suction pressure above and below the valve plate so the cylinders do no work. The horsepower required in this method decreases in proportion to capacity reduction.

6.5.4 Cylinder Unloading Method

A fourth method of satisfactorily controlling compressor capacity is known as "cylinder unloading". By this method the suction valves of some cylinders are held open, preventing compression. In an open position the piston draws gas from the suction manifold on the down stroke. On the up (return) stroke the piston returns the gas to the suction line without compressing it.

6.6.0 LUBRICATION OF COMPRESSORS

A number of conditions must be considered in the lubrication of compressors. For instance, the oil must remain fluid at low temperatures. This is necessary in systems where the refrigerant and oil are capable of being mixed and where some of the oil in circulation with the refrigerant works its way into the evaporator. Unless the oil remains fluid, the low temperatures cause it to "congeal". This may cause a low oil level in the compressor and decrease evaporator efficiency. Another requirement of the oil used in the compressor is that it must be free from moisture. Any large accumulation of moisture may also freeze in the expansion valve.

Methyl chloride and many of the "R" refrigerants are readily mixable with lubricating oil. For this reason a high viscosity oil must be used to compensate for its thinning out. However, this is not true for all refrigerants. Sulfur dioxide, for example, does not dilute oil. It is important, therefore, to follow the manufacturer's recommendation on the proper oil to use for the specific conditions required.

6.7.0 VISCOSITY OF OIL

The term "viscosity" refers to the resistance of oil to flow and is designated by a numeral. This viscosity numeral is determined by the number of seconds it takes for a specific quantity of oil at a given temperature to flow through the opening of a measuring device called a "viscosimeter" (viscosity meter). If a given quantity of oil at 100 degrees F takes 60 seconds to flow through the opening, it is said to have a viscosity of 60 at 100 degrees F. A heavier oil has a higher number; a lighter oil a lower number.

6.8.0   OIL SEPARATORS

An "Oil Separator" is a device used to separate oil from refrigerant gas, returning the oil to the compressor and allowing the refrigerant to continue on its circuit through the refrigerating system. It depends for its operation on a reduction of gas velocity in the super-heated state and is, therefore, located in the discharge line between the compressor and the condenser.

As the oil-laden refrigerant gas enters the oil separator its velocity is reduced. Since the oil particles have attained a greater inertia and are less inclined to change their direction of flow, the oil adheres to impingement screens allowing the gas to continue on its circuit through the refrigerating system.

The oil reservoir is that area in the base of the oil separator where oil is accumulated prior to its return to the compressor. When enough oil has accumulated to raise the float, the valve opens and since the pressure in the oil separator is greater than in the compressor crankcase, a positive oil return is accomplished.

Some refrigerating systems use low-side oil separators. These are built into the crankcase at the point where the suction line enters the compressor. Although one type of oil separator and a simple installation are illustrated in Fig. 11, there are many different types and sizes depending on the requirements of the refrigerating system.

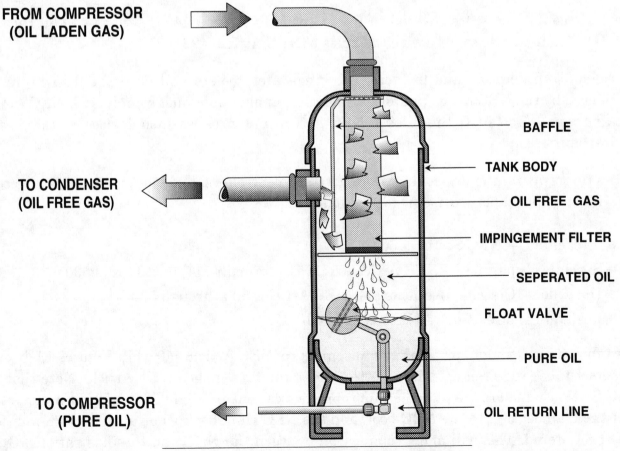

FROM COMPRESSOR (OIL LADEN GAS)

TO CONDENSER (OIL FREE GAS)

TO COMPRESSOR (PURE OIL)

BAFFLE

TANK BODY

OIL FREE GAS

IMPINGEMENT FILTER

SEPERATED OIL

FLOAT VALVE

PURE OIL

OIL RETURN LINE

Figure 11. Oil separator and practical application.

6.9.0   NEC REQUIREMENTS FOR COMPRESSORS

NEC Article 440 — *Air Conditioning and Refrigerating Equipment* — contains provisions for motor-driven equipment and for branch circuits and controllers for the equipment. It also takes into account the special considerations involved with sealed (hermetic-type) motor compressors, in which the motor operates under the cooling effect of the refrigeration.

It must be noted, however, that the rules of NEC Article 440 are in addition to, or are amendments to, the rules given in the basic NEC Article dealing with electric motors; that is, NEC Article 430. The basic rules of NEC Article 430 also apply to air conditioning and refrigerating equipment unless exceptions are indicated in NEC Article 440. This NEC Article (440) further clarifies the application of NEC rules to this type of equipment.

Where refrigeration compressors are driven by conventional motors (not the hermetic-type), the motors and controls are subject to NEC Article 430 — not NEC Article 440.

Other NEC Articles that will be covered in this module (besides NEC Articles 430 and 440) include:

- NEC Article 422 — Appliances
- NEC Article 424 — Space Heating Equipment

Room air conditioners are covered in Part G of NEC Article 440 (Sections 440-60 through 440-64), but must also comply with the rules of NEC Article 422.

Household refrigerators and freezers, drinking-water coolers and beverage dispensers are considered by the NEC to be appliances, and their application must comply with NEC Article 422 and must also satisfy the rules of NEC Article 440, because such devices contain sealed motor-compressors.

Hermetic refrigerant motor-compressors, circuits, controllers, and equipment must also comply with the applicable provisions for the following:

- Capacitors — NEC Section 460-9
- Special occupancies — NEC Articles 511, 513 through 517, Part D, and 530
- Hazardous (Classified) locations — NEC Articles 500 through 503
- Resistors and reactors — NEC Article 470

The table in Fig. 12 summarizes the requirements of NEC Article 440 while Figures 13 through 16 depict these requirements for a better understanding of the entire Article. Note that the NEC Section numbers are placed at appropriate locations in each illustration. It is suggested that the trainee refer to the NEC book and the NEC Handbook when such Section numbers are encountered. This will allow comparison between the NEC and the illustrations. Other NEC requirements are presented later in this module.

Application	NEC Regulation	NEC Section
Marking on hermetic compressors	Hermetic compressors must be provided with a nameplate containing manufacturer's name; trademark, or symbol; identifying designation; phase, voltage; frequency; rated-load current; locked-rotor current, and the words "thermally protect" if appropriate.	440-4
Marking on controllers	Controllers serving hermetically-sealed compressors must be marked with maker's name, trademark, or symbol; identifying designation; voltage; phase; full-load and locked-rotor current (or hp rating).	440-5
Ampacity and rating	Conductors for hermetically-sealed compressors must be sized according to NEC Tables 310-16 through 310-19 or calculated in accordance with NEC Section 310-15 as applicable.	440-6
Highest rated motor	The largest motor is considered to be the motor with the highest rated-load current.	440-7
Single machine	The entire HVAC system is considered to be one machine, regardless of the number of motors involved in the system.	440-8
Rating and interrupting capacity	Disconnecting means for hermetically-sealed compressors must be selected on the basis of the nameplate rated-load current or branch-circuit selection current, whichever is greater.	440-12
Cord-connected equipment	For cord connected equipment, an attachment plug and receptacle is permitted to serve as the disconnecting means.	440-13
Location	Disconnecting means must be located within sight of equipment. The disconnecting means may be mounted on or within the HVAC equipment.	440-14
Short-circuit and ground-fault protection	Amendments to NEC Article 240 are provided here for circuits supplying hermetically-sealed compressors against overcurrent due to short circuits and grounds.	440-21

Figure 12. Summary of NEC requirements for hermetically-sealed compressors.

Application	NEC Regulation	NEC Section
Rating of short-circuit and ground-fault protective device	Rating must not exceed 175% of the compressor rated-load current; if necessary for starting, device may be increased to a maximum of 225%.	440-22(a)
Compressor branch-circuit conductors	Branch-circuit conductors supplying a single compressor must have an ampacity not less than 125% of either the motor-compressor rated-load current or the branch-circuit selection current, whichever is greater.	440-32
	Conductors supplying more than one compressor must have conductors sized for the total load plus 25% of the largest motor's full-load amps.	440-33
Combination load	Conductors must be sufficiently sized for the other loads plus the required ampacity for the compressor as required in NEC Section 440-33.	440-34
Multimotor-load equipment	Conductors must be sized to carry the circuit ampacity marked on the equipment as specified in NEC Section 440-4(b).	440-35
Controller rating	Must have both a continuous-duty full-load current rating, and a locked-rotor current rating, not less than the nameplate rated-load current.	440-41
Application and Selection of controllers	Each motor-compressor must be protected against overload and failure to start by one of the means specified in NEC Section 440-52(a) (1) through (4).	440-52
Overload relays	Overload relays and other devices for motor overload protection, that are not capable of opening short circuits, must be protected by a suitable fuse or inverse time circuit breaker.	440-53
Equipment on 15- or 20-amp branch circuit; time-delay required	Short-circuit and ground-fault protective devices protecting 15- or 20-ampere branch circuits must have sufficient time delay to permit the motor compressor and other motors to start and accelerate their loads.	440-54

Figure 12. Summary of NEC requirements for hermetically-sealed compressors. *(Cont.)*

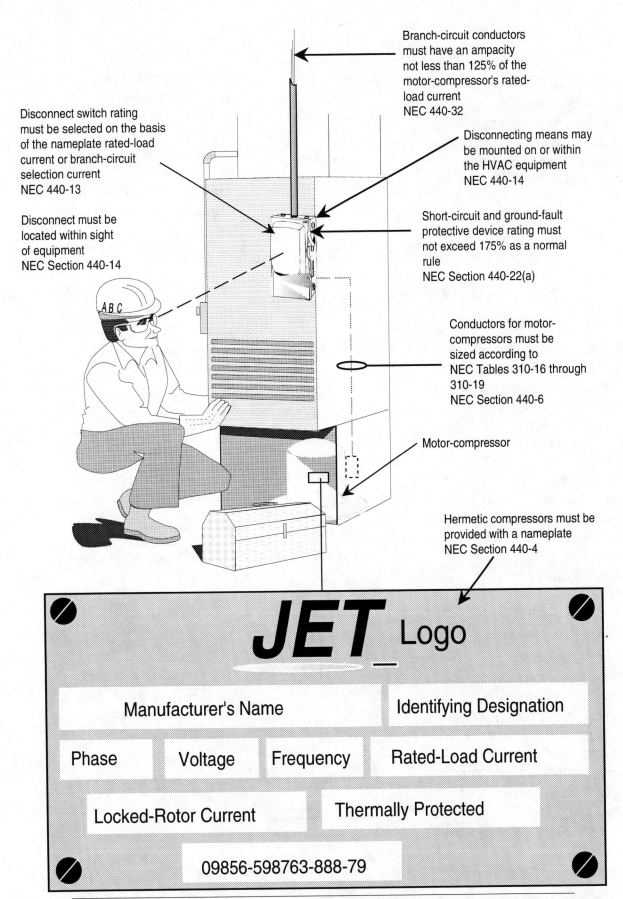

Branch-circuit conductors must have an ampacity not less than 125% of the motor-compressor's rated-load current
NEC 440-32

Disconnect switch rating must be selected on the basis of the nameplate rated-load current or branch-circuit selection current
NEC 440-13

Disconnecting means may be mounted on or within the HVAC equipment
NEC 440-14

Disconnect must be located within sight of equipment
NEC Section 440-14

Short-circuit and ground-fault protective device rating must not exceed 175% as a normal rule
NEC Section 440-22(a)

Conductors for motor-compressors must be sized according to NEC Tables 310-16 through 310-19
NEC Section 440-6

Motor-compressor

Hermetic compressors must be provided with a nameplate
NEC Section 440-4

JET Logo

Manufacturer's Name	Identifying Designation
Phase	Voltage	Frequency	Rated-Load Current
Locked-Rotor Current	Thermally Protected

09856-598763-888-79

Figure 13. NEC Article 440 deals with regulations governing motor-compressors.

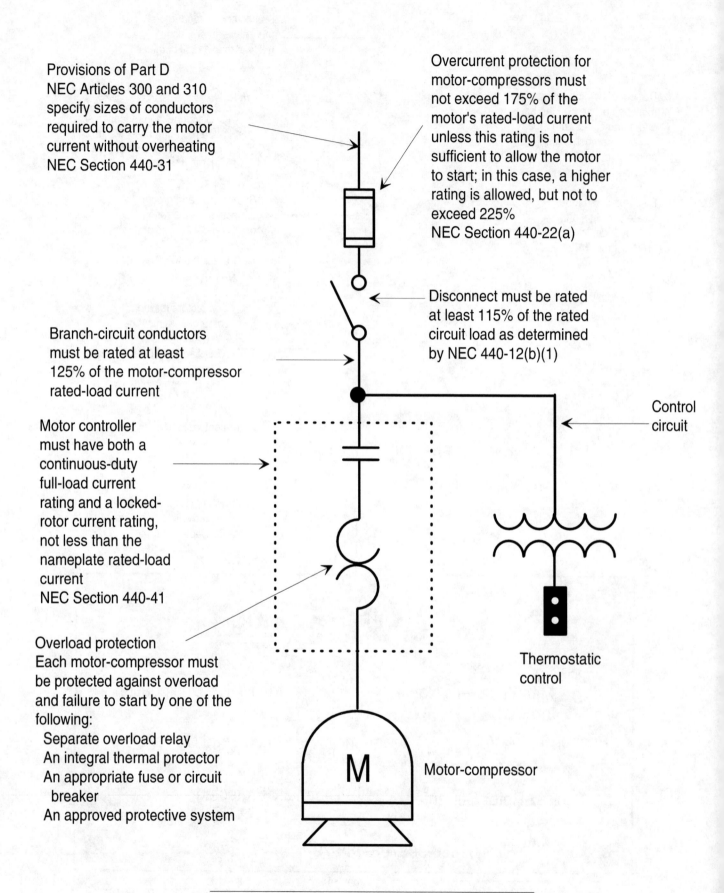

Provisions of Part D
NEC Articles 300 and 310
specify sizes of conductors
required to carry the motor
current without overheating
NEC Section 440-31

Overcurrent protection for
motor-compressors must
not exceed 175% of the
motor's rated-load current
unless this rating is not
sufficient to allow the motor
to start; in this case, a higher
rating is allowed, but not to
exceed 225%
NEC Section 440-22(a)

Disconnect must be rated
at least 115% of the rated
circuit load as determined
by NEC 440-12(b)(1)

Branch-circuit conductors
must be rated at least
125% of the motor-compressor
rated-load current

Control
circuit

Motor controller
must have both a
continuous-duty
full-load current
rating and a locked-
rotor current rating,
not less than the
nameplate rated-load
current
NEC Section 440-41

Overload protection
Each motor-compressor must
be protected against overload
and failure to start by one of the
following:
  Separate overload relay
  An integral thermal protector
  An appropriate fuse or circuit
    breaker
  An approved protective system

Thermostatic
control

M   Motor-compressor

Figure 14. Compressor branch and controls circuits.

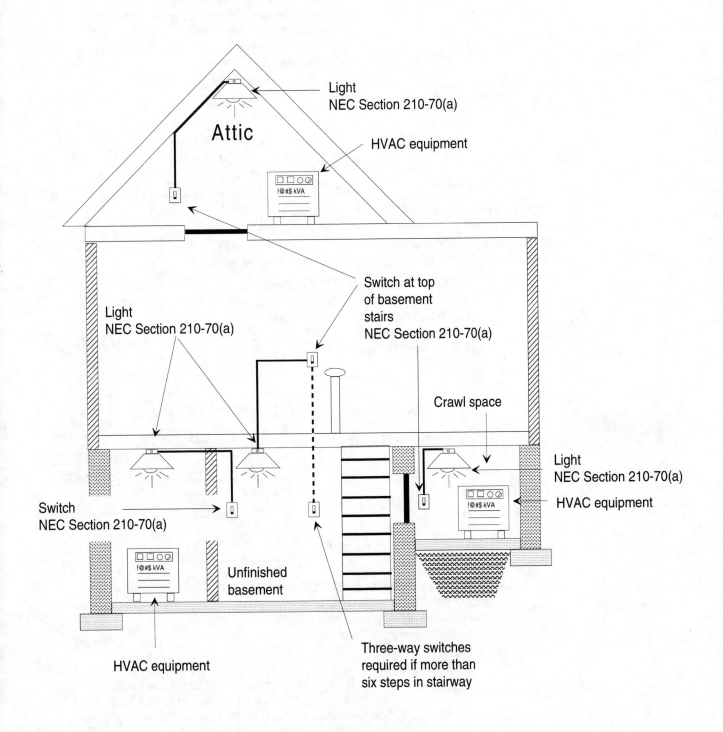

Light
NEC Section 210-70(a)

Attic

HVAC equipment

!@#$ kVA

Light
NEC Section 210-70(a)

Switch at top
of basement
stairs
NEC Section 210-70(a)

Crawl space

Light
NEC Section 210-70(a)

HVAC equipment

!@#$ kVA

Switch
NEC Section 210-70(a)

!@#$ kVA

Unfinished
basement

HVAC equipment

Three-way switches
required if more than
six steps in stairway

Figure 15. NEC requirements governing lighting and switches for HVAC equipment.

ELECTRICITY IN HVAC SYSTEMS — MODULE 20312

41

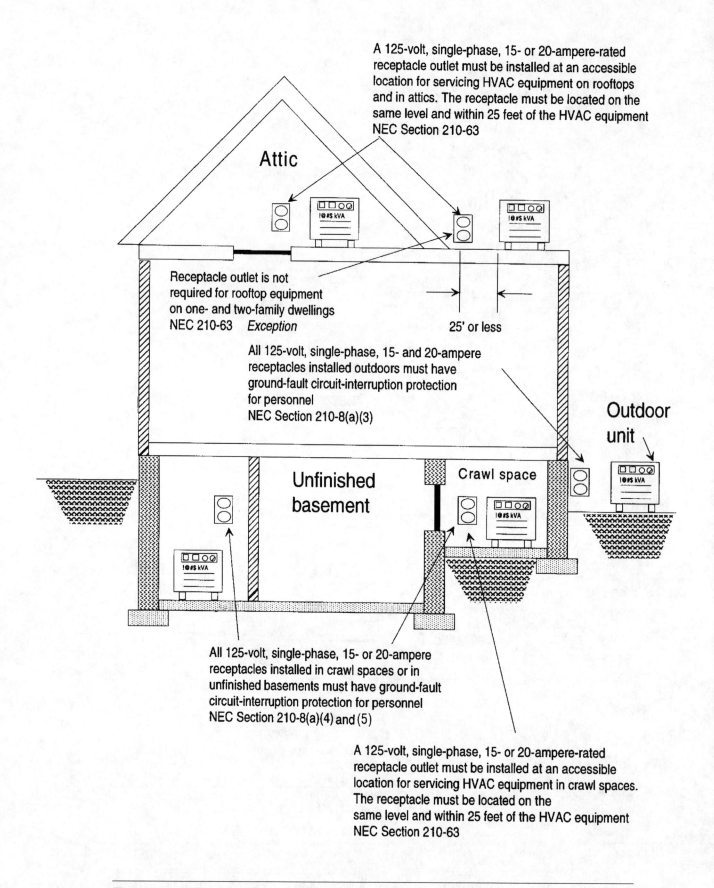

A 125-volt, single-phase, 15- or 20-ampere-rated receptacle outlet must be installed at an accessible location for servicing HVAC equipment on rooftops and in attics. The receptacle must be located on the same level and within 25 feet of the HVAC equipment NEC Section 210-63

Attic

Receptacle outlet is not required for rooftop equipment on one- and two-family dwellings NEC 210-63 *Exception*

25' or less

All 125-volt, single-phase, 15- and 20-ampere receptacles installed outdoors must have ground-fault circuit-interruption protection for personnel NEC Section 210-8(a)(3)

Outdoor unit

Unfinished basement

Crawl space

All 125-volt, single-phase, 15- or 20-ampere receptacles installed in crawl spaces or in unfinished basements must have ground-fault circuit-interruption protection for personnel NEC Section 210-8(a)(4) and (5)

A 125-volt, single-phase, 15- or 20-ampere-rated receptacle outlet must be installed at an accessible location for servicing HVAC equipment in crawl spaces. The receptacle must be located on the same level and within 25 feet of the HVAC equipment NEC Section 210-63

Figure 16. NEC requirements for locating 125-volt receptacles at HVAC equipment.

Some type of air filter is used on all types of forced-air heating and cooling systems—including air-to-air heat pumps. Individual room air conditioners also utilize air filters. The filter is placed in the return-air duct system to filter all air returning from the conditioned space. Air filters not only provide a cleaner environment, but also prolongs the life of the fan-coil unit in HVAC systems.

Air filters, as well as insect and bird screen, reduce the free intake area of the vent or ductwork and necessitates a larger overall area for compensation. Figure 17 shows the effective free area of three types of screen. Choose the appropriate one and multiply the free area by the percent of efficiency to obtain the size needed.

7.1.0   ELECTRONIC AIR FILTERS

In systems requiring high air-cleaning efficiency, electronic air filters are used. These devices operate on the principle of passing the airstream through an ionization field where a 12,000-volt potential imposes a positive charge on all airborne particles. The ionized particles are then passed between aluminum plates, alternately grounded and connected to a 6,000-volt source, and are precipitated onto the grounded plates. See Fig. 18.

The original design of the electronic air cleaner uses a water-soluble adhesive coating on the plates, which holds the dirt deposits until the plates require cleaning. The filter is then de-energized, and the dirt and adhesive films are washed off the plates. Fresh adhesive is applied before the power is turned on again. Newer versions of electronic air cleaners are designed so that the plates serve as agglomerators; the agglomerates of smaller particles are allowed to slough off the plates and to be trapped by viscous impingement or duty-type filters downstream of the electronic unit.

Designs of most electronic air cleaners are based on 500-feet per minute face velocity, with pressure losses at 0.20 in water gauge (w.g.) for the washable type and up to 1.0 in wag for the agglomerator type using dry-type after-filters. Power consumption is low, despite the 12,000-volt ionizer potential, because current flow is measured in milliamperes (mA).

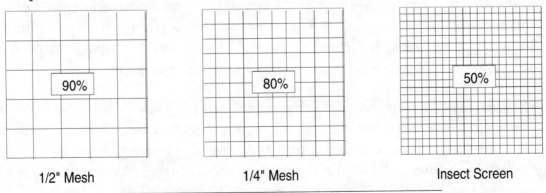

Figure 17. Examples of different-sized screen mesh.

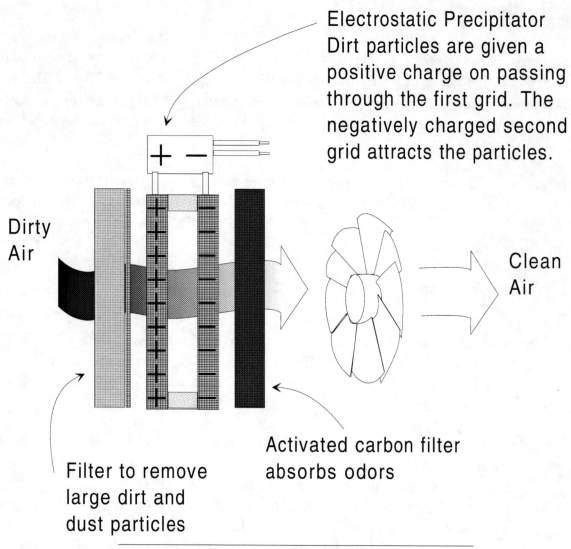

Electrostatic Precipitator
Dirt particles are given a positive charge on passing through the first grid. The negatively charged second grid attracts the particles.

Dirty Air

Clean Air

Activated carbon filter absorbs odors

Filter to remove large dirt and dust particles

Figure 18. Operating principles of electronic air cleaners.

8.0.0 ELECTRIC HEAT

Less than 30 years ago, electric heating units were used only for supplemental heat in small, seldom-used areas of the home, such as a laundry room or workshop, or in vacation homes on chilly autumn nights. Today, however, electric heat is used extensively in both new and renovated buildings—in residential, commercial, and industrial applications.

In addition to the fact that electricity is the cleanest fuel available, electric heat is usually the least expensive to install and maintain—although the fuel cost may be higher. Individual room heaters are very inexpensive compared to furnaces and ductwork required in oil and gas forced-air systems, no chimney is required, no utility room is necessary since there is no furnace or boiler, and the installation time and labor are less. Combine all these features and it is easy to see why electric heat currently ranks high on the list.

Several types of electric heating units are available (see Fig. 19) and a brief description of each is in order.

Electric baseboard heaters: Electric baseboard heaters are mounted on the floor along the baseboard, preferably on outside walls under windows for the most efficient operation. They are absolutely noiseless in operation and are the type most often used for heating residential occupancies and for use as supplemental heat in many commercial areas.

Electric baseboard heaters may be mounted on practically any surface (wood, plaster, drywall, and so on), but if polystyrene foam insulation is used near the unit, a ¾-inch (minimum) ventilated spacer strip must be used between the heater and the wall. In such cases, the heater should also be elevated above the floor or rug to allow ventilation to flow from the floor upward over the total heater space.

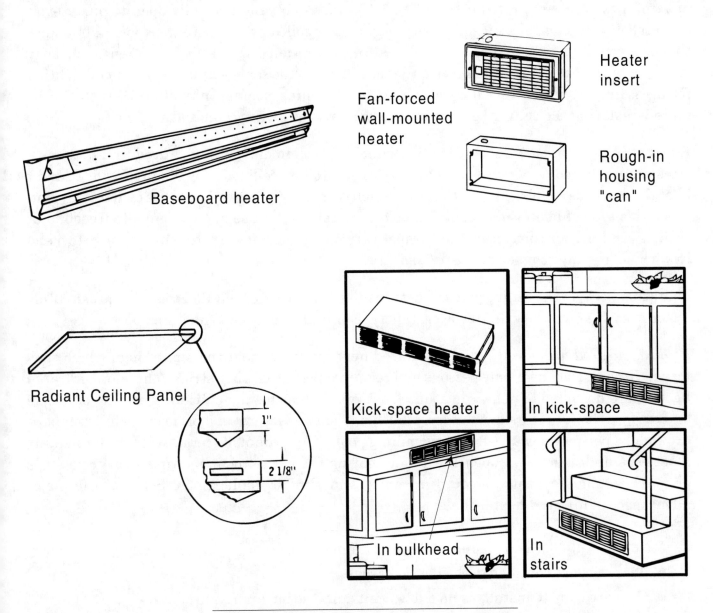

Figure 19. Several types of electric heating units.

One complaint received over the years about this type of heater has been wall discoloration directly above the heating units. When this problem occurred, the reason was almost always traced to one or more of the following:

- High wattage per square foot of heating element.
- Heavy smoking by occupants.
- Poor housekeeping.

Radiant ceiling heaters: Radiant ceiling heaters are often used in bathrooms and similar areas so that the entire room does not have to be overheated to meet the need for extra warmth after a bath or shower. They are also used in larger areas, such as a garage or basement, or for spot-warming a person standing at a workbench.

Most of these units are rated from 800 to 1,500 watts (W) and normally operate on 120-volt circuits. As with most electric units, they may be controlled by a remote thermostat, but since they are usually used for supplemental heat, a conventional wall switch is often used. They are quickly and easily mounted on an outlet box in much the same way as conventional lighting fixtures. In fact, where very low wattage is used, ceiling heaters may often be installed by merely replacing the ceiling lighting fixture with a light/heater combination.

Radiant heating panels: Radiant heating panels are commonly manufactured in 2-ft. by 4-ft. sizes and are rated at 500 watts. They may be located on ceilings or walls to provide radiant heat that spreads evenly through the room. Each room may be controlled by its own thermostat. Since this type of heater may be mounted on the ceiling, its use allows complete freedom for room decor, furniture placement, and drapery arrangement. Most are finished in beige to blend in with nearly any room or furniture color.

Units mounted on the ceiling give the best results when located parallel to and approximately 2 feet from the outside wall. However, this type of unit may also be mounted on walls.

Electric infrared heaters: Rays from infrared heaters do not heat the air through which they travel. Rather, they heat only persons and certain objects that they strike. Therefore, infrared heaters are designed to deliver heat into controlled areas for the efficient warming of people and surfaces both indoors and outdoors (such as to heat persons on a patio on a chilly night or around the perimeter of an outdoor swimming pool). This type of heater is excellent for heating a person standing at a workbench without heating the entire room, melting snow from steps or porches, sunlike heat over outdoor areas, and similar applications. Some of the major advantages of infrared heat include:

- No warm-up period is required. Heat is immediate.
- Heat rays are confined to the desired areas.
- They are easy to install, as no ducts, vents, and so on, are required.

When installing this type of heating unit, never mount the heater closer than 24 inches from vertical walls unless the specific heating unit is designed for closer installation. Read the manufacturer's instructions carefully.

NOTE   Infrared quartz lamps provide some light in addition to heat.

Forced-air wall heaters: Forced-air wall heaters are designed to bring quick heat into an area where the sound of a quiet fan will not be disturbing. Some are very noisy. Most of these units are equipped with a built-in thermostat with a sensor mounted in the intake air stream. Some types are available for mounting on high walls or even ceilings, but the additional force required to move the air to a usable area produces even more noise.

Floor insert convection heaters: Floor insert convection heaters require no wall space, as they fit into the floor. They are best suited for placement beneath conventional or sliding glass doors to form an effective draft barrier. All are equipped with safety devices, such as a thermal cutout to disconnect the heating element automatically in the event that normal operating temperatures are exceeded.

Floor insert convector heaters may be installed in both old and new homes by cutting through the floor, inserting the metal housing and wiring, according to the manufacturer's instructions. A heavy-gauge floor grille then fits over the entire unit.

Electric kick-space heaters: Modern kitchens contain so many appliances and so much cabinet space for the convenience of the owner that there often is no room to install electric heaters except on the ceiling. Therefore, a kick-space heater was added to the lines of electric heating manufacturers to overcome this problem.

For the most comfort, kick-space heaters should not be installed in such a manner that warm air blows directly on the occupant's feet. Ideally, the air discharge should be directed along the outside wall adjacent to normal working areas, not directly under the sink.

Radiant heating cable: Radiant heating cable provides an enormous heating surface over the ceiling or concrete floor so that the system need not be raised to to a high temperature. Rather, gentle warmth radiates downward (in the case of ceiling-mount cable) or upward (in the case of floor-mounted cable), heating the entire room or area evenly.

There is virtually no maintenance with a radiant heating system, as there are no moving parts and the entire heating system is invisible — except for the thermostat.

Combination heating and cooling units: One way to have individual control of each room or area in the home, as far as heating and cooling are concerned, is to install through-wall heating and cooling units. Such a system gives the occupants complete control of their environment with a room-by-room choice of either heating or cooling at any time of year at any temperature they desire. Operating costs are lower than for many other systems due to the high efficiency of room-by-room control. Another advantage is that if a unit should fail, the defective chassis can be replaced immediately or taken to a shop for repair without shutting down the remaining units in the buildings.

When selecting any electric heating units, obtain plenty of literature from suppliers and manufacturers before settling on any one type. In most cases you are going to get what you pay for, but most contractors and their personnel shop around at different suppliers before ordering the equipment. Delivery of any of these units may take some time, so once the brand, size, and supplier have been selected, the order should be placed well before the unit is actually needed.

Electric furnaces: Electric furnaces are becoming more popular, although they are somewhat surpassed by the all-electric heat pump. Most are very compact, versatile units designed for either wall, ceiling, or closet mounting. The vertical model can be flush mounted in a wall or shelf mounted in a closet; the horizontal design (Fig. 20) can be fitted into a ceiling (flush or recessed).

Central heating systems of the electrically energized type distribute heat from a centrally located source by means of circulating air or water. Compact electric boilers can be mounted on the wall of a basement, utility room, or closet with the necessary control and circuit protection, and will furnish hot water to convectors or to embedded pipes. Immersion heaters may be stepped in one at a time to provide heat capacity to match heat loss. The majority of electric furnaces are commonly available in sizes up to 24 kW for residential use. The larger boilers with proper controls can take advantage of lower off-peak electricity rates, where they prevail, by heating water during off-peak periods, storing it in insulated tanks, and circulating it to convectors or radiators to provide heat as needed.

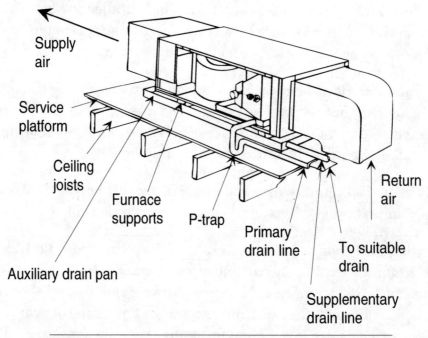

Figure 20. Horizontal application of an electric furnace.

Electric hot-water systems: A zone hydronic (hot-water) system permits selection of different temperatures in each zone of the home. Baseboard heaters located along the outer walls of rooms provide a blanket of warmth from floor to ceiling; the heating unit also supplies domestic hot water simultaneously, through separate circuits. A special attachment coupled to the hot-water unit can be used to melt snow and ice on walkways and driveways in winter, and a similar attachment can be used to heat, say, a swimming pool during the spring and fall seasons.

A typical hot-water system operating diagram is shown in Fig. 21, and is explained as follows: When a zone thermostat calls for heat, the appropriate zone valve motor begins to run, opening the valve slowly; when the valve is fully opened, the valve motor stops. At that time, the operating relay in the hydrostat is energized, closing contacts to the burner and the circulator circuits. The high-limit control contacts (a safety device) are normally closed so the burner will now fire and operate. If the boiler water temperature exceeds the high-limit setting, the high-limit contacts will open and the burner will stop, but the circulator will continue to run as long as the thermostat continues to call for heat. If the call for heat continues, the resultant drop in boiler water temperature—below the high-limit setting— will bring the burner back on. Thus, the burner will cycle until the thermostat is satisfied; then both the burner and circulator will shut off.

Hot-water boilers for the home are normally manufactured for use with oil, gas, or electricity. While a zoned hot-water system is comparatively costly to install, the cost is still competitive with the better hot-air systems. The chief disadvantage of hot-water systems is that they don't use ducts. If a central air conditioning system is wanted, a separate duct system must be installed along with the hot-water system.

Chillers, or refrigerated water, are sometimes used with commercial and industrial systems when hot-water is used for heating. The cycle is reversed in warmer months to provide cold water through the system to help cool the conditioned areas.

Heat pumps: The term heat pump, as applied to a year-round air conditioning system, commonly denotes a system in which refrigeration equipment is used in such a manner that heat is taken from a heat source and transferred to the conditioned space when heating is desired; heat is removed from the space and discharged to a heat sink when cooling and dehumidification are desired. Therefore, the heat pump is essentially a heat-transfer refrigeration device that puts the heat rejected by the refrigeration process to good use. A heat pump can do the following:

● Provide either heating or cooling.
● Change from one to the other automatically as needed.
● Supply both simultaneously if so desired.

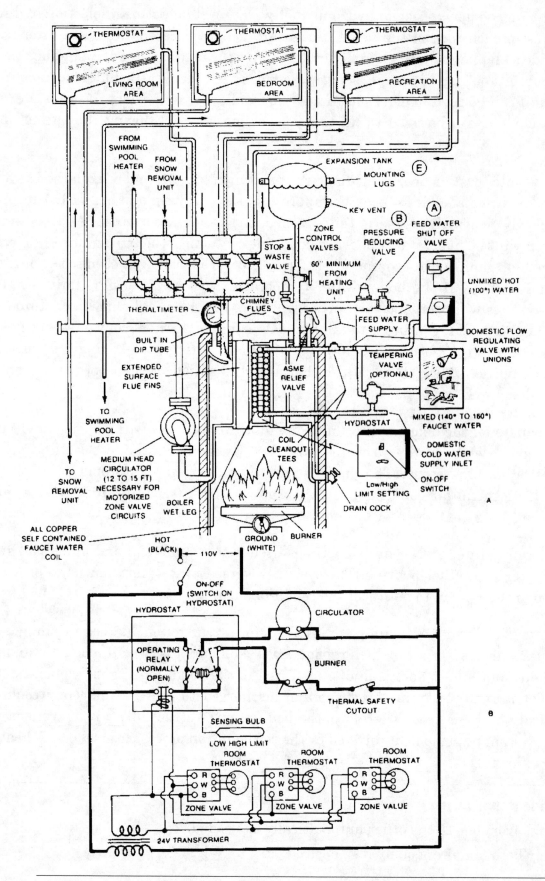

Figure 21. Typical hot-water system operating diagram with control-wiring diagram.

A heat pump has the unique ability to furnish more energy than it consumes. This uniqueness is due to the fact that electric energy is required only to move the heat absorbed by the refrigerant. Thus, a heat pump attains a heating efficiency of two or more to one; that is, it will put out an equivalent of 2 or 3 watts of heat for every watt consumed. For this reason, its use is highly desirable for the conservation of energy.

This module previously covered the operating principles of air-to-air heat pumps — the most popular type. However, in many cases, the water-to-air heat pump is even better. Water-to-air heat pumps, as opposed to air-to-air heat pumps, have the following advantages:

- The pumps can be located anywhere in the building since no outside air is connected to them.
- The outside air temperature does not affect the performance of the heat pump, as it does an air-to-air type of heat pump.
- Since the water source of the water-to-air heat pump will seldom vary more than a few degrees in temperature, a more consistent performance can be expected.

The schematic drawing in Fig. 22 shows how two heat pumps were connected (in parallel) for a large residence with an indoor swimming pool; the swimming-pool water was used as a means of heat exchange, and since this water was preheated to approximately 78°F, the efficiency of the heat pumps utilizing this same water approached the maximum.

During the warm months the heat pumps were reversed (cooling cycle) to cool the area. Again, the pool water was used as the heat exchange, this time acting as a "cooling tower." The air from these heat pumps was distributed by means of underfloor transit ducts with supply-air diffusers and return-air grilles raised above the floor level and mounted in a wainscot around the perimeter of the pool area. This was done to prevent water from entering the openings in the ductwork.

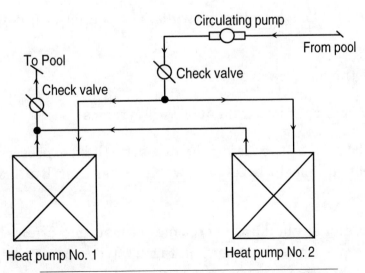

Figure 22. Diagram of water-to-air heat pumps.

8.1.0 HEATING CALCULATIONS

Most heating calculations for central systems will be performed by mechanical engineers and installed by mechanical contractors. Highly sophisticated calculations are required for systems of any consequence involving heat-loss and heat-gain calculations; cost analysis, psychrometrics, and other complicated computations that are outside the realm of this module. Electricians will normally install only the branch-circuit or feeder wiring, and sometimes install and connect the control wiring.

However, there may be times when electricians are required to size an electric heater for, say, a commercial or industrial toilet or perhaps an entire residential system. Therefore, the electrician should have a general knowledge of how simple heating calculations are performed.

The following are the basic goals of an electric heat installation:

- Adequate, dependable, and trouble-free service.
- Year-round comfort.
- Reasonable annual operating cost.
- Reasonable installation cost.
- Systems that are easy to service and maintain.

Heat-loss calculations must be made to ensure that heating equipment of proper capacity will be selected and installed. Heat loss is expressed in either Btu's per hour (abbreviated Btuh) or in watts (volt-amperes). Both are measures of the rate at which heat is transferred and are easily converted from one to the other:

$$Watts \ (volt\text{-}amperes) = \frac{Btuh}{3.4}$$

$$Btuh = watts \times 3.4$$

Basically, the calculation of heat loss through walls, roof, ceilings, windows, and floors requires three simple steps:

Step 1. Determine the net area in square feet.

Step 2. Find the proper heat-loss from appropriate tables.

Step 3. Multiply the area by the factor; the product will be expressed in Btuh. Since most electric heat equipment is rated in watts rather than Btuh, divide this product by 3.4 to convert to watts.

Calculations of heat loss for any building or area may be made more quickly and more efficiently by using a prepared form, such as that shown in Fig. 23. With spaces provided for all necessary data and calculations, the procedure becomes routine and simple.

Heating Load

1. Design Conditions	Dry Bulb (F)	Specific Humidity gr./lb.
Outside	0°F	
Inside	70°F	
Difference	70°F	

2. Transmission Gain (From appropriate Tables)

	Sq. Ft. ×	Factor ×	Dry Bulb Temp. Differ. =	Heat Load (Btuh)
Windows				
Walls				
Roof	5.4	0.09	70	340.2
Floor	5.4	0.20	70	756.0
Other				

3. Ventilation or Infiltration (from appropriate tables)

	CFM ×	Dry Bulb Temp. Diff. ×	Factor =	
Sensible Load	7.6	70	1.08 =	574.6
Humidification	CFM	Specific Humidity Diff. ×		
Load			0.67 =	

4. Duct Heat Loss (from appropriate tables)

Heat Loss_____ × Factor for insulation Thickness_____ × Duct Length (ft.) /100

=_____

5. Total Heating Load   =__1,670.8_____ Btuh

Figure 23. Typical heat-loss form.

Load estimate is based on design conditions inside the building and outside in the atmosphere surrounding the building. Outside design conditions are the maximum extremes of temperature occurring in a specific locality. The inside design condition is the degree of temperature and humidity that will give optimum comfort.

8.2.0 ELECTRIC BASEBOARD HEATERS

All requirements of the NEC apply for the installation of electric baseboard heaters, especially NEC Article 424 — *Fixed Electric Space Heating Equipment*. In general, electric baseboard heaters must not be used where they will be exposed to severe physical damage unless they are adequately protected from such possible damage. Heaters and related equipment installed in damp or wet locations must be approved for such locations and must be constructed and installed so that water cannot enter or accumulate in or on wired sections, electrical components, or duct work.

Baseboard heaters must be installed to provide the required spacing between the equipment and adjacent combustible material and each unit must be adequately grounded in accordance with NEC Section 424-14 and NEC Article 250.

Figure 24 summarizes the NEC regulations governing the installation of electric baseboard heaters, while Fig. 25 shows a residential floor plan layout for electric heat.

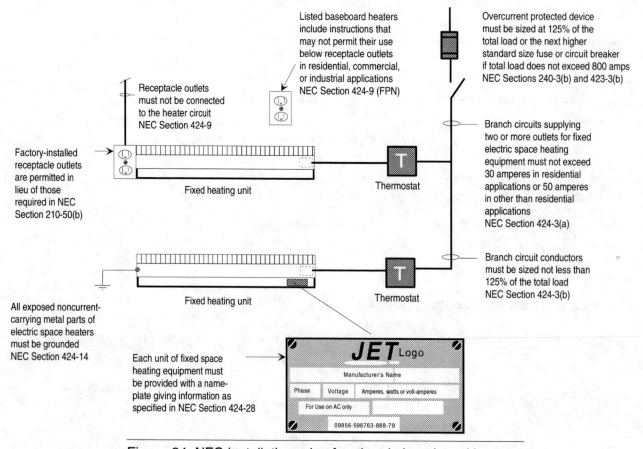

Figure 24. NEC installation rules for electric baseboard heaters.

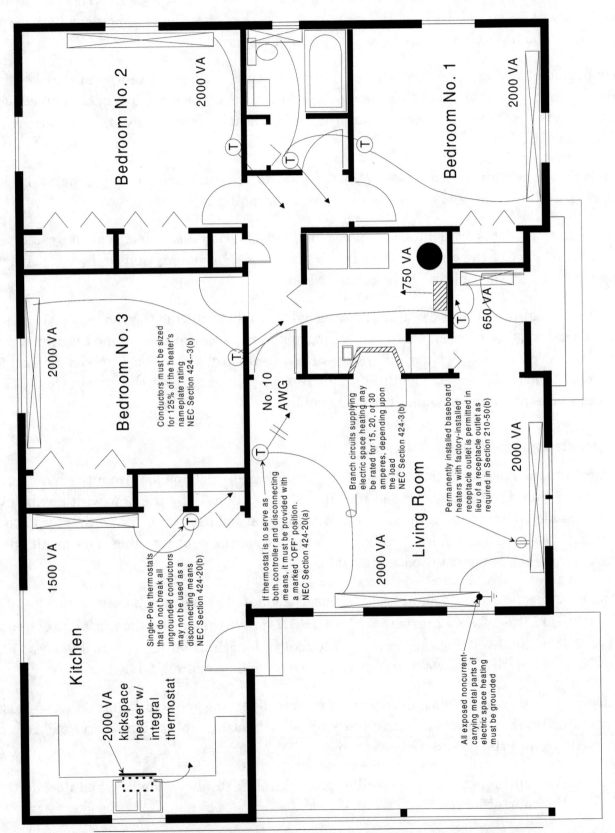

Figure 25. Floor-plan layout for a residential electric-heating application.

Within the figure, the following labels appear:

750 VA

2000 VA — Bedroom No. 2

2000 VA — Bedroom No. 1

750 VA

650 VA

2000 VA — Bedroom No. 3

Conductors must be sized for 125% of the heater's nameplate rating NEC Section 424-3(b)

No. 10 AWG

Branch circuits supplying electric space heating may be rated for 15, 20, of 30 amperes, depending upon the load NEC Section 424-3(b)

Permanently installed baseboard heaters with factory-installed receptacle outlet is permitted in lieu of a receptacle outlet as required in Section 210-50(b)

2000 VA — Living Room

2000 VA

Single-Pole thermostats that do not break all ungrounded conductors may not be used as a disconnecting means NEC Section 424-20(b)

If thermostat is to serve as both controller and disconnecting means, it must be provided with a marked "OFF" position. NEC Section 424-20(a)

All exposed noncurrent-carrying metal parts of electric space heating must be grounded

1500 VA

Kitchen

2000 VA kickspace heater w/ integral thermostat

8.3.0 ELECTRIC SPACE-HEATING CABLES

Radiant ceiling heat is acknowledged to be one of the greatest advances in structural heating since the Franklin stove, or so say the manufacturers, and thousands of homeowners all over the country who have chosen this type of heat for their homes.

The enormous heating surface precludes the necessity of raising air temperatures to a high degree. Rather, gentle warmth flows downward (or upward in the case of cable embedded in concrete floors) from the surfaces, heating the entire room or area evenly, and usually leaving no cold spots or drafts.

There is no maintenance with a radiant heating system as there are no moving parts, nothing to get clogged up, nothing to clean, oil, or grease, and nothing to wear out.

The installation of this system is within reach of even the smallest electrical contractor. The most difficult part of the entire project is the layout of the system; that is, how far apart to string the cable on the ceiling or in a concrete slab.

An ideal application of electric radiant heating cable would be during the renovation of an area within an existing residence, where the ceiling plaster is beginning to crack and this ceiling will be recovered with drywall or other type of plaster board. Or, perhaps the basement floor needs repair and three inches of additional concrete will be poured over the existing floor. These are ideal locations to install radiant heating cable.

8.3.1 Installation in Plaster Ceilings

To determine the spacing of the cable on a given ceiling, deduct one foot from the room length and one foot from the room width and multiply this new length by the new width, which will give the usable ceiling area in square feet. Multiply the square feet of the ceiling by 12 to get the ceiling area in inches before dividing by the length of the heating cable. The result will be the number of inches apart to space the cable.

For example, assume that a room 14×12 ft. has a calculated heat loss of 2,000 watts. Therefore, (14 ft. - 1 ft.) (12 ft. - 1 ft.) $\times$ 12 in. = $13 \times 11 \times 12$ = 1,716. We then look at manufacturers' tables and see that a 2,000-watt heating cable is 728 feet in length. Dividing the usable area (1,716 sq. in.) by this length (728), we find that the cable should be spaced 2.3 inches apart.

The drawing in Fig. 26 shows a floor plan of a typical heating cable installation as suggested by one manufacturer. However, nearly every brand of heating cable will be installed in exactly the same way, and the procedure is as follows:

Step 1.   Nail outlet box on inside wall approximately 5 ft. above the finished floor for your thermostat location.

Step 2.   Drill two holes in wall plate above this junction box location.

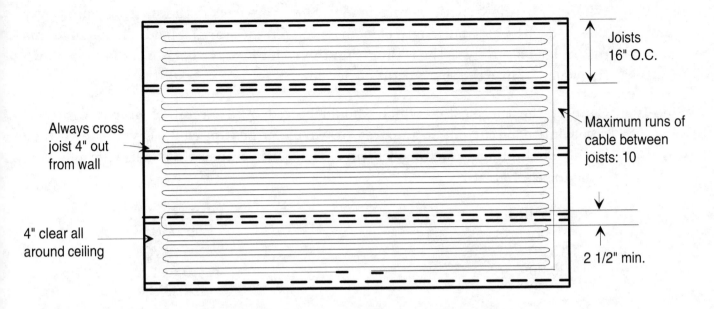

Joists 16" O.C.

Maximum runs of cable between joists: 10

Always cross joist 4" out from wall

4" clear all around ceiling

2 1/2" min.

Figure 26. Recommendations for heat cable layout.

Step 3.   Drill two holes through ceiling lath above thermostat junction box location.

Step 4.   Put the spool of heat cable on a nail, screwdriver, or any type of shaft you have at hand for unwinding the cable from the spool.

Step 5.   Cover the accessible end of the 8-ft. nonheating lead wire with nonmetallic loom. Loom should be long enough so at least 2 inches will go on ceiling surface and reach to the thermostat outlet box, but leave 6 inches of lead wire inside of this junction box for viewing the identification tags. Never, for any reason, remove those tags. Also do not cut or shorten the nonheating leads.

Step 6.   Run the accessible end (loom and all) of your cable through one of the holes in the ceiling, down the wall, through the plate, into the thermostat outlet or junction box.

Step 7.   Pull slack out of your nonheating leads and staple securely to the ceiling. Any excess nonheating lead should be covered with plaster the same as the heat cable. Do not staple or bend the cable.

Step 8.   Your next step will be to mark the ceiling. First mark a line all the way around the room 6 inches away from each wall; this accounts for the one foot you deducted from the width and length. A chalk line is best for this. Now take a yardstick, often given to regular customers at the local hardware store, and notch it for your calculated spacing. Run the cable along the ceiling 6 inches out from the wall to an outside wall, where you attach it in parallel spacing. In this example the spacing is $2\frac{1}{2}$ inches. Always keep the cable at least 2 inches from metal corner lath, or other metal reinforcing.

Step 9.   When you are down to the return lead wire, you are back to the starting wall. Cover this lead with the same length of loom as you did the starting lead. Then staple this return lead securely to the ceiling, run through the other hole in the ceiling, down the wall and into the thermostat outlet box.

Step 10.   Connect the thermostat, which should be fed by a circuit of proper wire size according to the current, in amperes, drawn by the heating cable. Divide the wattage by the rated voltage. Your answer will be the load in amperes. Then use the following table to size your wire.

AMPS	TYPE TW WIRE
15 or lower	10 AWG
15-20	8 AWG
20-30	6 AWG
30-50	4 AWG

Always make certain that the heating cable is connected to the proper voltage. A 120-volt cable connected to a 240-volt circuit will melt the cable, while a 240-volt heating cable connected to 120 volts will produce only 25 percent of the rated wattage of the cable. See Fig. 27 for a summary of NEC requirements governing the installation of electric space-heating cable.

8.3.2   Installing Concrete Cable

The heat loss is calculated in the same manner as for any other area. For best results, the heating cable should never be spaced less than $2\frac{1}{2}$ inches apart when installing in concrete floor except around outside walls, which may be spaced on $1\frac{1}{2}$-inch minimum centers for the first 2 feet out from the wall. The concrete thickness above the cable should be from $\frac{1}{2}$ to 1 inch.

The spacing of the cable is found exactly as previously shown for the spacing of the cable on the ceiling, and then the installation procedure is as follows:

Step 1.   Secure junction box on an inside wall approximately 5 feet from the finished floor to house the thermostat.

Step 2.   Install a piece of rigid conduit from the switch or thermostat junction box to house the nonheating leads, between the concrete slab and the switch or thermostat outlet box.

Step 3.   Approximately 6 inches of the lower end of the conduit should be embedded in the concrete and a smooth porcelain bushing should be on this end of the conduit to protect the nonheating leads where they leave the conduit.

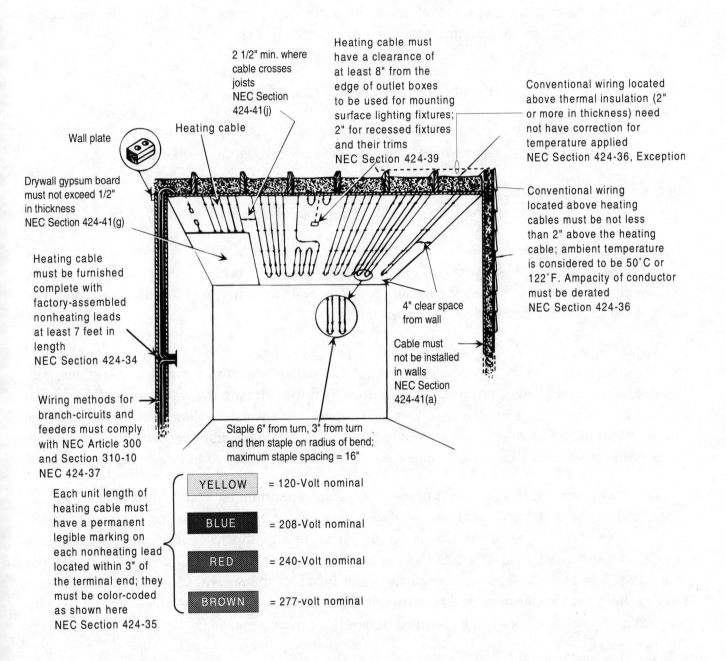

2 1/2" min. where cable crosses joists NEC Section 424-41(j)

Heating cable

Heating cable must have a clearance of at least 8" from the edge of outlet boxes to be used for mounting surface lighting fixtures; 2" for recessed fixtures and their trims NEC Section 424-39

Conventional wiring located above thermal insulation (2" or more in thickness) need not have correction for temperature applied NEC Section 424-36, Exception

Wall plate

Drywall gypsum board must not exceed 1/2" in thickness NEC Section 424-41(g)

Heating cable must be furnished complete with factory-assembled nonheating leads at least 7 feet in length NEC Section 424-34

Wiring methods for branch-circuits and feeders must comply with NEC Article 300 and Section 310-10 NEC 424-37

Each unit length of heating cable must have a permanent legible marking on each nonheating lead located within 3" of the terminal end; they must be color-coded as shown here NEC Section 424-35

Conventional wiring located above heating cables must be not less than 2" above the heating cable; ambient temperature is considered to be 50°C or 122°F. Ampacity of conductor must be derated NEC Section 424-36

4" clear space from wall

Cable must not be installed in walls NEC Section 424-41(a)

Staple 6" from turn, 3" from turn and then staple on radius of bend; maximum staple spacing = 16"

YELLOW = 120-Volt nominal

BLUE = 208-Volt nominal

RED = 240-Volt nominal

BROWN = 277-volt nominal

Figure 27. NEC regulations governing the installation of electric space-heating cable.

Step 4.   Place the spool of heating cable on a nail, screwdriver, or other shaft you have at hand for unwinding the cable from the spool.

Step 5.   Run the accessible end of the 8-ft. nonheating lead through the conduit to the thermostat junction box, leaving 6 inches extending out of the box. Never remove the identification tags or shorten the nonheating leads. They should be embedded in the concrete the same as the heating portion of the cable.

Step 6.   Run the cable along the floor 6 inches out from the wall to the outside or exposed wall, fastening the cable to the floor with staples or masking tape.

Step 7.   The cable usually is spaced $1\frac{1}{2}$ inches apart for the first 2 feet around the exposed wall and never less than $2\frac{1}{2}$ inches apart for the remaining area.

Step 8.   Run the return nonheating lead wire through the conduit to the thermostat junction box in the same manner as the starting nonheating lead was run in Step 5.

In general, the concrete slab should be prepared by applying a vapor barrier of 4 to 6 inches of gravel. Then pour 4 inches of vermiculite or other insulating concrete over the gravel after the outside edges of the slab have been insulated, in accordance with good building practice. The heating cable is then installed as described previously.

At this time, workers should inspect and test the cable before the final layer of concrete is installed, because once the concrete is poured, it's an expensive matter to repair. First, visually inspect the cable for any possible damage to the insulation during the application. Then, with a suitable ohmmeter, check for continuity and capacity of the cable. Concealed breaks may be found by leaving an ohmmeter connected to the cable, and then brushing the cable lightly with the bristles of a broom. Any erratic movement of the meter dial will indicate a fault.

During the pouring of the final coat of concrete, it is recommended that the ohmmeter be left connected to the heating cable leads to detect any possible damage to the cable during the pouring of the finish layer of ordinary concrete (do not use insulating concrete). If an ohmmeter is not available at the time, a 100-watt lamp may be connected in series with the cable to immediately detect any damage during the installation of the concrete. The lamp will glow as long as the circuit is complete and no damage occurs. However, if a break does occur, the lamp will go out and the break can be repaired before the concrete hardens.

Repairs to a broken cable are made by stripping the ends of the broken cable and rejoining the ends with a No. 14 AWG pressure-type connector provided and approved for this purpose. The splice must then be insulated with thermoplastic tape to a thickness equal to the insulation of the cable. Use any thermoplastic tape listed by the Underwriters' Laboratories as suitable for temperature of 176°F.

Once the finish layer of concrete sets, asphalt tile, linoleum tile, or linoleum can be laid on the concrete in the normal manner.

Besides the heating of interior spaces, electric-heating cable also has many other uses. It can be embedded in concrete or asphalt surfaces for the removal of ice and snow, provide freeze protection for water pipes exposed to cold weather, provide roof and gutter deicing, and heat soil in a hotbed or window box—keeping the temperature at a constant 70°F. The cost of heat cable is relatively inexpensive and the installation goes fast.

8.4.0 FORCED-AIR SYSTEMS

Electric forced-air heating systems are usually of three types:

- Heat pumps
- Electric furnaces
- Air conditioner with duct heaters

In the majority of installations utilizing central, forced-air systems, the system is usually designed by mechanical engineers and installed by mechanical contractors. Branch circuits, feeders, motor starters, and some control wiring is frequently performed by electrical workers. Consequently, every electrician should have a basic working knowledge of forced-air systems, along with applicable NEC installation requirements.

8.4.1 Duct Heaters

Duct heater is a term applied to any heater mounted in the air stream of a forced-air system where the air-moving (fan-coil) unit is not provided as an integral part of the equipment. Duct heaters are used in electric furnaces, combination electric heating/cooling systems, and most of the time in heat pumps to offer auxiliary heat when the pumps themselves cannot supply the demand.

In general, heaters installed in an air duct must be identified as suitable for the installation, and some means must be provided to assure uniform and adequate airflow over the face of the heater in accordance with the manufacturer's instructions. This latter requirement is normally accomplished by airflow controls and other components involving turning vanes, pressure plates, or other devices on the inlet side of the duct heater to assure an even distribution of air over the face of the heater.

Duct heaters installed closer than 4 feet to a heat pump or air conditioner must have both the duct heater and heat pump or air conditioner identified as suitable for such installation and must be so marked.

Duct heaters intended for use with elevated inlet air temperature must be identified as suitable for use at the elevated temperatures. Furthermore, duct heaters used with air conditioners or other air-cooling equipment that may result in condensation of moisture must be identified as suitable for use with air conditioners.

The NEC requires that all duct heaters are installed according to the manufacturer's instructions. Furthermore, duct heaters must be located with respect to building construction and other equipment so as to permit access to the heater. Sufficient clearance must be maintained to permit replacement of controls and heating elements and for adjusting and cleaning of controls and other parts requiring such maintenance. See Fig. 28.

Control requirements — including disconnecting means — are specified in NEC Sections 424-63, 424-64, and 424-65. Complete coverage of these requirements, along with HVAC controls in general, is presented in section 10.0.0 — *Control Wiring* — in this module. But in general, a fan-circuit interlock is one of the requirements. Such a control ensures that the fan circuit is energized when any heater circuit is energized. It would be a waste of energy, and perhaps also be a hazard, if the duct heaters became energized and no air flowed over or through them. However, the NEC permits a slight time- and temperature-delay before the fan may be energized. This prevents the system from blowing cold air into the conditioned space. In other words, such a control gives the duct heaters time to "warm-up" before air is induced in the system.

Furthermore, each duct heater must be provided with an automatic-reset limit control to deenergize the duct-heater circuit or circuits in case overheating or other faults occur. In addition, an integral independent supplementary control or controller shall be provided in each duct heater that will disconnect a sufficient number of conductors to interrupt current flow. This device must be manually resettable or replaceable.

The disconnect for duct heaters — the same as for other types of HVAC equipment — must be accessible and within sight of the controller. All control equipment must also be accessible.

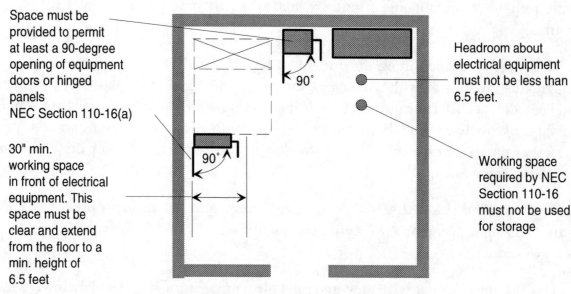

Space must be provided to permit at least a 90-degree opening of equipment doors or hinged panels NEC Section 110-16(a)

30" min. working space in front of electrical equipment. This space must be clear and extend from the floor to a min. height of 6.5 feet

Headroom about electrical equipment must not be less than 6.5 feet.

Working space required by NEC Section 110-16 must not be used for storage

Figure 28. Summary of NEC Section 110-16.

There are millions of room air conditioners in use throughout the United States and Canada. Consequently, the NEC deemed it necessary to provide Article 440 G, beginning with NEC Section 440-60, to ensure that such equipment will be installed so as not to provide a hazard to life or property. These NEC requirements apply to electrically energized room air conditioners that control temperature and humidity. In general, this section of the NEC considers room air conditioners (with or without provisions for heating) to be an alternating-current appliance of the air-cooled window, console, or in-wall (through-wall) type that is installed in a conditioned room or space and that incorporates a hermetic refrigerant motor-compressor(s). Furthermore, this NEC provision covers only equipment rated at 250 volts or less, single-phase, and such equipment is permitted to be cord- and attachment plug-connected.

Three-phase room air conditioners, or those rated at over 250 volts, are not covered under NEC Article 440 G. This type of equipment must be directly connected to a wiring method as described in NEC Chapter 3.

The majority of room air conditioners covered under NEC 440 G are connected by cord- and attachment-plug to receptacle outlets of general-purpose branch circuits. The rating of any such unit must not exceed 80% of the branch-circuit rating if connected to a 15-, 20-, or 30-ampere general-purpose branch circuit. The rating of cord-and-plug connected room air conditioners must not exceed 50% of the branch-circuit rating if lighting units and other appliances are also supplied.

Figures 29 and 30 depict the NEC application rules for room air conditioners. Note that the attachment plug and receptacle are allowed to serve as the disconnecting means. In some cases, the attachment plug and receptacle may also serve as the controller, or the controller may be

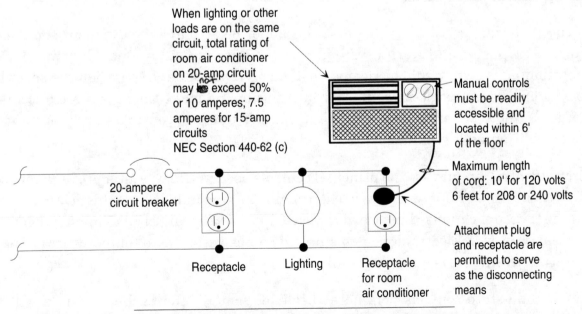

When lighting or other loads are on the same circuit, total rating of room air conditioner on 20-amp circuit may not exceed 50% or 10 amperes; 7.5 amperes for 15-amp circuits
NEC Section 440-62 (c)

Manual controls must be readily accessible and located within 6' of the floor

Maximum length of cord: 10' for 120 volts 6 feet for 208 or 240 volts

Attachment plug and receptacle are permitted to serve as the disconnecting means

20-ampere circuit breaker

Receptacle   Lighting   Receptacle for room air conditioner

Figure 29. Branch circuits for room air conditioners.

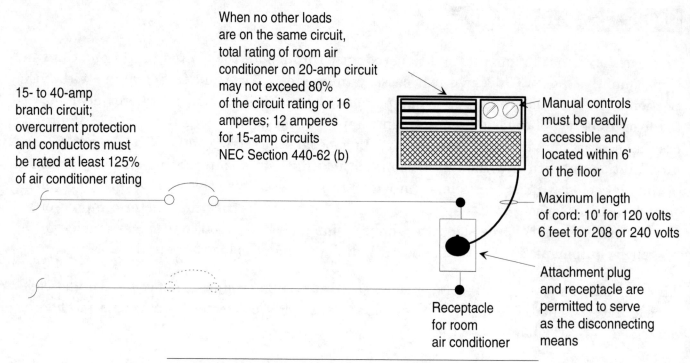

15- to 40-amp branch circuit; overcurrent protection and conductors must be rated at least 125% of air conditioner rating

When no other loads are on the same circuit, total rating of room air conditioner on 20-amp circuit may not exceed 80% of the circuit rating or 16 amperes; 12 amperes for 15-amp circuits NEC Section 440-62 (b)

Manual controls must be readily accessible and located within 6' of the floor

Maximum length of cord: 10' for 120 volts 6 feet for 208 or 240 volts

Attachment plug and receptacle are permitted to serve as the disconnecting means

Receptacle for room air conditioner

Figure 30. Fixed air conditioner connected to branch circuit.

a switch that is an integral part of the unit. The required overload protective device may be supplied as an integral part of the appliance and need not be included in the branch-circuit calculations.

Equipment grounding, as required by NEC Section 440-61, may be handled by the grounded receptacle.

10.0.0 CONTROL WIRING

All of the material described in this module cannot be accomplished without some means of control, if only to stop and start a system. For example, without a controlling mechanism, an air conditioning system would be turned on and run year-round. Even when the space became cool or hot enough, the system would continue to run and make the area more uncomfortable than if it had not been installed in the first place. Therefore, control is an important area of comfort conditioning for buildings.

Electronic control circuits are the principal controls used in HVAC systems of any consequence at the present time. Not too long ago, their main use was in highly sophisticated commercial and industry applications, but are now used in virtually all HVAC applications — from residential, upward. Electronic controls provide quick response to temperature changes, and temperature averaging is easily accomplished.

Although solid-state controls dominate the field, some of the earlier controls will also be covered, because many are still in use and will remain in use for some time to come.

10.1.0 INDOOR THERMOSTAT

HVAC controls can be electric, electronic or pneumatic. Almost limitless combinations of each type are possible. Therefore, it is not practical to attempt to describe all of the possible combinations in this module. Rather, fundamental terms, functions, operations, and simple maintenance and troubleshooting techniques are presented.

The indoor thermostat (Fig. 31) is the most easily recognized control device. Such controls can be found in practically every motel room in the United States. It is also a familiar ornament in every home utilizing central heating or cooling systems. Here's an overview of a thermostat used to control an electric heat pump.

A two-stage indoor changeover thermostat is the most popular type used to control heat pumps during the heating season. During the cooling season, it functions as a conventional air conditioning thermostat. The first stage of this thermostat controls the heat pump. The second stage (usually preset 1°F to 2°F below the first stage) allows the supplementary heat to be energized.

By placing the selector/switch on AUTO, the automatic changeover thermostat switches the system from heating to cooling or cooling to heating automatically. A manual thermostat also is available to offer the option to select manually the mode of operation for heating or cooling.

Emergency heat switch (EM.HT.): The indoor thermostat should have an emergency heat selector (EM.HT.) with a light to indicate full use of the supplementary heat when the heat pump is not in operation. If the heat pump should become inoperative, this switch enables bypassing the normal operation of the heat pump and heats the area by the supplementary

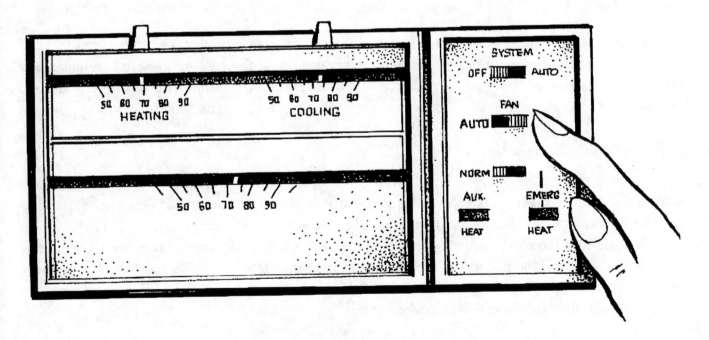

Figure 31. Typical indoor thermostat.

heat until the problem is corrected. An indicator light, usually red, mounted on the thermostat, will come on when the selector is in the emergency heat position. As soon as the unit has been repaired, return the switch to the normal operating position.

Supplementary heat light: Some thermostats have a light which indicates the supplementary heat is on to assist the heat pump in normal operation. When the outdoor temperature falls below the balance point of the heat pump and energizes the supplementary heat, the light will cycle on and off intermittently as the supplementary heat is activated.

The temperature range at which the light comes on will vary depending on the balance point.

10.1.1   Temperature Setting

Heating: The recommended setting for the heating cycle is 68°F. Once the thermostat is set, the best policy is to leave it alone. Raising the thermostat as little as 2°F may cause the supplementary heat to energize, thereby increasing the energy usage.

Night setback: Although night setback is recommended during the winter for most types of heating systems — to save energy and reduce costs — it is not generally recommended for a heat pump. When using a heat pump to raise the room temperature in the morning, the supplementary heat may come on using more energy than was saved during the night. However, thermostat temperature settings for weekend trips or vacations during the heating season should be reduced to save energy. The use of a standard automatic nighttime setback control with a heat pump system is not recommended.

Cooling: A setting of 78°F or higher is recommended for cooling. For each degree the temperature is set below 78°F, the cooling energy usage will increase approximately five percent.

Raising the temperature when the room or building is unoccupied is recommended to save energy. If the building will not be occupied for several days, the cooling system can be turned off altogether. However, frequent changes of the thermostat setting reduces the economical operation of the heat pump and tends to shorten the life of the compressor.

10.1.2   Fan Operation

Operation of the fan is the same for the automatic or manual changeover thermostat. The fan switch set in the AUTO position gives fan operation only when the unit is actually heating or cooling. In the ON position, the indoor fan will run continuously which may cost a little more. The ON position is recommended to obtain a more even temperature throughout the house. However, operating the fan in the ON position may allow the thermostat to be set at a lower temperature, or reduce the operating time of the compressor, which may offset the increased cost. It should also increase occupants' comfort.

10.1.3 Location

The indoor thermostat should be located on an interior wall in the central portion of the home, approximately five feet above the floor. It must be installed level. The location should be free from drafts, vibrations and any interior heat sources such as a lamp or a television set. Care should be taken to seal behind the wall where the thermostat wire or mounting box penetrates the wall.

10.2.0 BASIC PRINCIPLES OF CONTROL

The circuit in Fig. 32 shows a simple "control" for an electric resistance-type heater, consisting of a 120-volt, 2-wire circuit feeding the heater and a conventional single-pole toggle switch to interrupt the power supply. A 1-pole, 15-ampere circuit breaker is used to protect the circuit against short-circuits and ground-faults. When the toggle switch is in the ON position, the heater becomes energized; when in the OFF position, the heater is de-energized. Obviously, this is a manual control and leaves much to be desired. For example, once the space to be heated reaches the desired temperature, the toggle switch cannot sense the difference and continues to stay in the ON position until it is manually shut off. If the switch is left on, the space will continue to receive heat, even after the desired temperature has been reached.

A better solution is to replace the single-pole toggle switch with a thermostat control. A thermostat utilizing bimetal elements is shown in Fig. 33. The bimetal elements consist of thin strips of two different metals securely attached to each other. Since different metals expand and contract at different rates, the thin strips of metal actually curve toward or away from a given point when there is a change in temperature. In this way, the thermostat "makes" or "breaks" its contacts when the temperature changes — providing automatic control of the heater in question.

A floor plan of a typical year-round HVAC system is shown in Fig. 34, including the probable controls that will exist within the system. To get an overview of this system, let's take a look at the various controls and their relationship to the entire system.

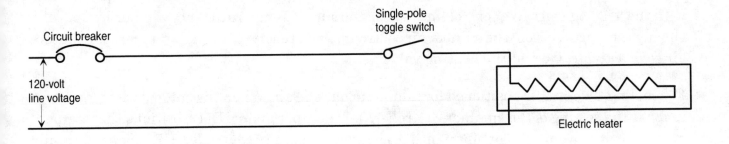

Figure 32. Simple heater control circuit.

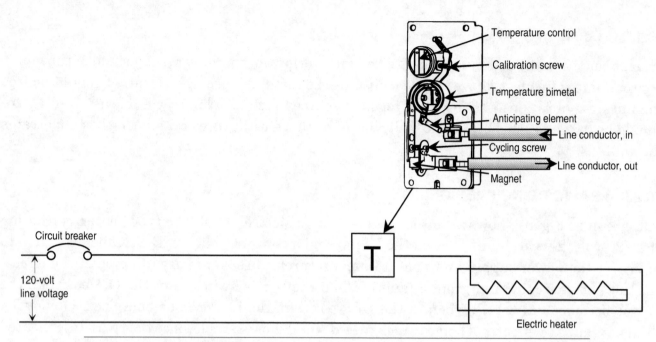

Figure 33. Bimetal elements in an electric thermostat and resulting control circuit.

The system is started by closing the fan starter switch. At the same time, an electric circuit provides the energy for operation of the controls. Notice that the outdoor and return air dampers are connected by a mechanical linkage so that the opening of one will close the other. Similarly, the face and bypass dampers for the cooling coil are interconnected.

When de-energized, the damper motors, D-1 and D-2, hold the outdoor air and exhaust dampers in the closed positions.

During the heating cycle, the space thermostat, T-1 operates the hot-water heating coil valve, V-1, to maintain room-air temperature. The insertion thermostat, T-2, positions the damper motor, D-1. The minimum quantity of outdoor air is determined by the switch, S. As the temperature of the air from the outside rises, T-2 moves D-1 so that more outside air and less return air is included in the mixture to be heated until 100% outside air is admitted at the top setting for T-2. Simultaneously, damper motor, D-2, must open the exhaust damper. The thermostat, T-4, is a low-limit controller to protect coils from freezing. It prevents the admission of too much cold outside air. A low-limit controller, T-6, in the supply duct is interconnected with the heating-coil valve control line to keep the supply temperature above a minimum value. During the warm season, the outdoor air controller, T-5, reacts to the high-temperature outdoor air to deactivate T-6 and the heating valve.

For cooling, switch S, is positioned for minimum outside air. When the outdoor air temperature reaches the setting of the summer controller, T-3, damper motor, D-1, positions the damper to admit the minimum outdoor air quantity. The space humidity controller, H-1, opens the cooling valve, V-2, whenever dehumidification is required. The cooling coil may be a direct-expansion or chilled-water coil. The controller, H-1, also sets the position of damper motor, D-3, so that the face and bypass dampers for the cooling coil allow the proper portion of air to pass over the

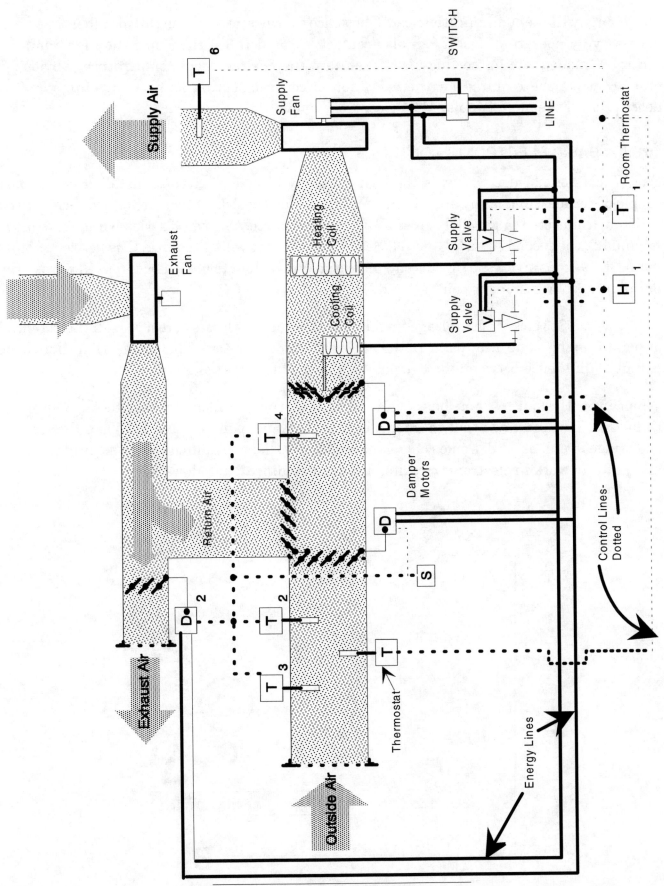

Figure 34. Typical HVAC control system.

coil to satisfy the dehumidification load. The space thermostat, T-1, can call for reheating when necessary by operating V-1. It can also control V-2 and D-3 if their operation for humidity control is inadequate for cooling air to maintain the desired space temperature. Obviously, other controls are required for the operation of central heating and cooling equipment. A description of the more popular HVAC controls follow.

10.3.0   BASIC ELECTRONIC CONTROLS

The Wheatstone-bridge circuit is the basis of many electronic circuits installed in the past couple of decades, and many are still in use. The bridge consists of two sets of two series-wired resistors, connected in parallel across a DC voltage source. One set of series-wired resistors is R1 and R2: the second set is R3 and R4. See Figure 35. The voltage source, E, is between points A and B. A sensitive electric current indicator (galvanometer), G, is connected across the parallel sets of resistors at points C and D.

When switch S is closed, the voltage from E flows through both sets of resistors. If the potential at point C is the same as at point D, the galvanometer reads zero. This means that there is no potential difference between the two points, and the bridge is balanced.

When any of the four resistors has a different voltage, the galvanometer registers a value other than zero. This indicates that there is a current between points C and D. When this occurs, the bridge is unbalanced. Some control manufacturers have modified this basic bridge circuit and put it to work in electronic circuits. A typical modification follows:

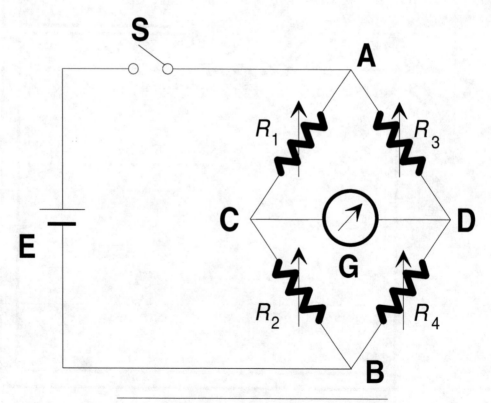

Figure 35. Basic Wheatstone-bridge circuit.

ELECTRICAL TRAINEE TASK MODULE 20312

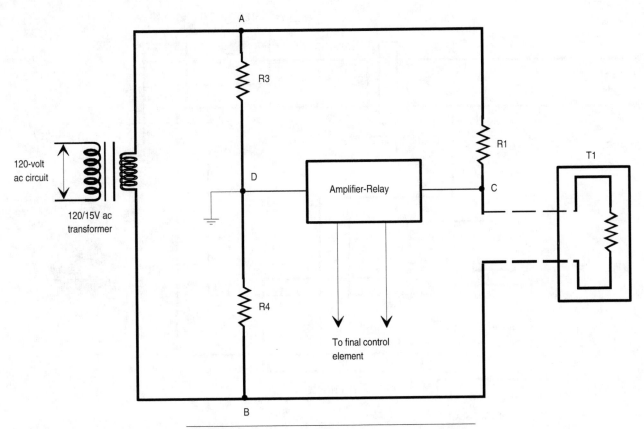

Figure 36. Modified Wheatstone-bridge circuit.

Figure 36 shows that the DC voltage has been replaced by an AC supply. Also, the switch is eliminated and the galvanometer is replaced by an amplifier-switching relay unit. Resistor R2 is replaced by a temperature-sensing element, T1.

In this modification, a resistance of 1,000 ohms is assigned to each of the three fixed resistors (R1, R2, R4) and to the thermostat element, T1. When conditions in the area to be conditioned are satisfied, the resistance in T1 does not change, and the voltage across the amplifier relay is zero. Because there is no voltage across this amplifier relay, the final control element (motor, valve, and the like) cannot be energized. The bridge circuit is balanced.

When the air temperature in the space changes, the thermostat element senses the change. This causes a corresponding change in the resistance at T1, and the bridge becomes unbalanced. Voltage now flows through the amplifier relay to the final control element. In a heating application, a drop in temperature in the space causes a decrease in resistance at T1. This results in voltage relationships in the bridge circuit that cause the amplifier relay to increase the voltage to the final control element, and heat is added to the space. As heat is added, resistance at T1 increases, and the resulting voltage change in the bridge circuit inactivates the heating action (shuts off the burner).

This modified bridge circuit can also be applied to cooling applications. Depending on whether the bridge voltage fed to the amplifier relay is in phase or out of phase with the supply voltage, the final control element is opened or closed.

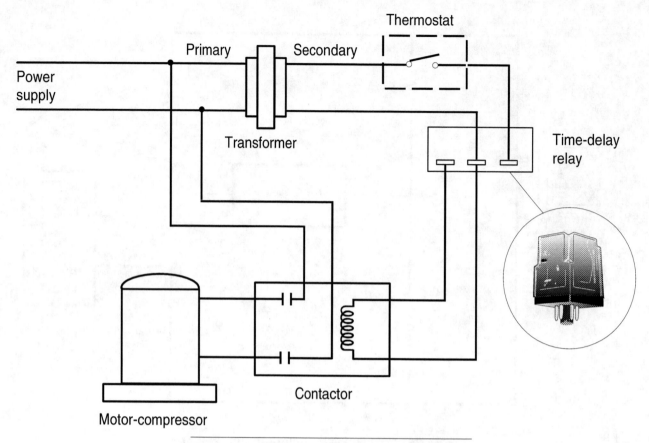

Figure 37. Solid state time-delay relay circuit.

10.4.0   TIME-DELAY CIRCUITS

The thermostat or other control for HVAC systems must respond to a gradual or average change in the space to be conditioned. An average change is produced by adding a tiny heater (timing device) near the thermostat temperature-sensing element in electrical controls. Electronic circuits produce an output signal delayed in time by a prescribed and controllable amount in relation to an input waveform in the transmission line or its lumped circuit approximation. More common forms of delay circuits are initiated by a controlling signal and produce an output, not necessarily related to the input in size or shape, at a later time. Usually the input pulses are recurrent at a specific rate, thus the time-displaced output signals are recurrent at the same rate. Such delay circuits are usually designed as linear delay circuits in the sense that a linear variation of some controlling element produces a delay that is a linear variation in time delay. See Fig. 37.

10.5.0   BASIC CONTROL COMPONENTS

Thermostat: In general, low-voltage room thermostats should be used for the best temperature control. The low-voltage thermostats respond much faster to temperature changes than the greater-mass line-voltage devices.

From a cost standpoint, the less-expensive installation of low-voltage wiring more than offsets the extra cost of the transformer. Also greater safety may be had with low-voltage thermostats.

The room thermostat is provided with a heat anticipator connected in series with the rest of the control circuit (Fig. 33 on page 68). These anticipators are made of a resistance-type material that produces heat in accordance with the current drawn through it. Heat anticipators are adjustable and are normally set to correspond with the current rating of the main gas valve, motor starter, or relay. The purpose of these devices is to make the room temperature more stable.

In operation, when the thermostat is calling for heat, the anticipator is also producing heat to the thermostat. This heating action causes the thermostat to become satisfied before the room actually reaches the set point of the thermostat. Thus, the thermostat stops the flame or de-energizes the motor starter, relay, etc., and the room temperature will not overshoot, or go too high.

Transformer: The transformer is a device used to reduce line voltage to a usable control voltage, usually 24 V. Transformers must be sized to provide sufficient power to operate the control circuit. Most are oversized sufficiently to provide enough power to operate an air conditioning control circuit also. This rating is usually 40 VA for small systems; greater for larger systems. Refer to Electrical Task Module 20307 for an in-depth coverage of control transformers.

Fan Control: The fan control is a temperature-actuated control that, when heated, will close a set of contacts to start the indoor fan motor. The sensing element of the fan control is positioned inside the heat exchanger where the temperature is the highest.

This control is actuated by a bimetal element that opens or closes the contacts on temperature change. The fan control is usually set to bring the fan on at about 100 degrees F in the heating mode.

In operation, the burner provides heat to the heat exchanger for a few seconds to warm the furnace before the fan is started. This operation is to prevent blowing cold air into the room on furnace start-up. When the thermostat is satisfied, the main burner stops providing heat, but the fan continues to operate until the temperature in the furnace has been reduced, thus removing any excess heat in the furnace. See Figs. 38 and 39.

Limit Control: The limit control is also a heat-actuated switch with a bimetal sensing element positioned inside the heat exchanger. This is a safety control that is wired into the primary side of the transformer. If the temperature inside the furnace reaches approximately 200 degrees F, the power to the transformer will be shut off, which also stops all power in the temperature-control circuit.

Main Gas Valve: The main gas valve is the device that acts on demand from the thermostat to either admit gas to the main burners or to stop the gas supply. This valve has many functions. It has a gas pressure regulator, a pilot safety, a main gas cock, a pilot gas cock, and the main gas solenoid all in a single unit — the combination gas control.

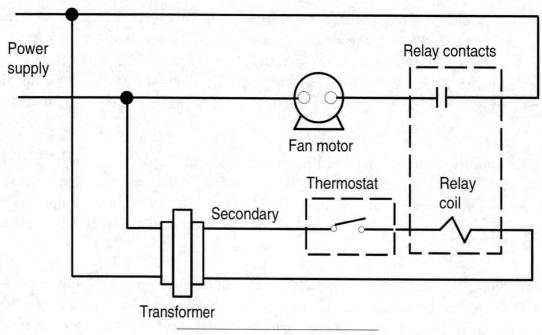

Figure 38. Wiring for a fan relay.

As the thermostat calls for heat, energizing the solenoid coil, the valve lever opens the cycling valve. The inlet gas now flows through the control orifice past the cycling valve. At this point, gas flow is in two directions, as follows:

1. Part of the flow is to the back of the diaphragm by means of internal passageways. The resulting increase in pressure pushes the main valve to the open position, compressing the diaphragm springs lightly.

2. Part of the flow is through the seat of the master regulator into the valve outlet by means of internal passageways. This causes the master regulator to begin its function.

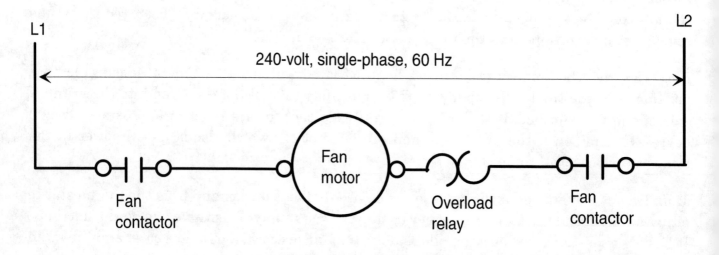

Figure 39. Fan control circuit.

The gas valve remains in this position and the master regulator continues to regulate until the relay coil is de-energized, at which time the cycling valve seals off.

As the cycling valve closes, the regulator spring causes the seat of the master regulator to close off. The function of the bypass orifice is to permit gas in the passageways to escape into the outlet of the valve, thereby causing the main gas valve to close.

Pilot Safety: There are two types of pilot safety controls, 90 percent safe, and 100 percent safe. These two names refer to the amount of gas cut off when the pilot light is unsafe. The 100-percent safe device is incorporated in the combination main gas valve. The 90-percent safe device incorporates a set of contacts that open the control circuit during an unsafe condition.

These units are used in conjunction with a thermocouple to keep the control contacts closed or the valve open during normal operation. If, at any time, the control "drops out" (the contacts open) the reset button must be manually reset before operation of the furnace can be resumed.

Thermocouple: The thermocouple is a device that uses the difference in metals to provide electron flow. The hot junction of the thermocouple is put in the pilot flame where the dissimilar metals are heated. When heat is applied to the welded junction, a small voltage is produced. This small voltage is measured in millivolts (mV) and is the power used to operate the pilot safety control. The output of a thermocouple is approximately 30 mV. This simple device can cause many problems if the connections are not kept clean and tight.

The Fire-Stat: The fire-stat is a safety device mounted in the fan compartment to stop the fan if the return air temperature reaches about 60 degrees F. It is a bimetal-actuated, normally closed switch that must be manually reset before the fan can operate.

The reason for stopping the fan when the high return-air temperatures exist is to prevent agitation of any open flame in the house, thus helping to prevent the spreading of any fire that may be present.

Furnace Wiring: There are three different circuits and three different voltages in a modern furnace 24-volt control system. The following diagrams will illustrate each of these circuits:

- The fan or circulator circuit (Fig. 39)
- The temperature control circuit (Fig. 40)
- The pilot safety circuit, 30 mV (Fig. 41)

When all three of these circuits are connected together (Fig. 42), we have a modern furnace 24-V control system.

The control of electric furnaces is much the same as that just described; however, there are no pilot safety devices, and the main gas valve is replaced with relays that actuate to complete the electrical circuit to the heating elements.

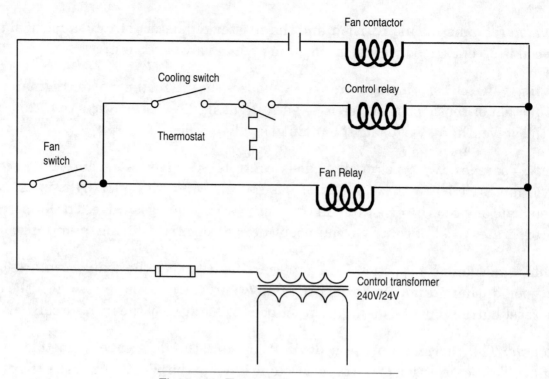

Figure 40. Temperature control circuit.

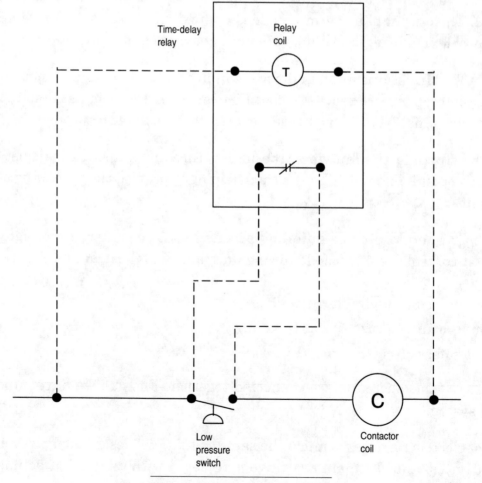

Figure 41. Pilot safety circuit.

The variety of functions performed by a heating system is limited only by the use of controls. The more a service technician knows about controls, the easier his or her job will be in servicing such equipment.

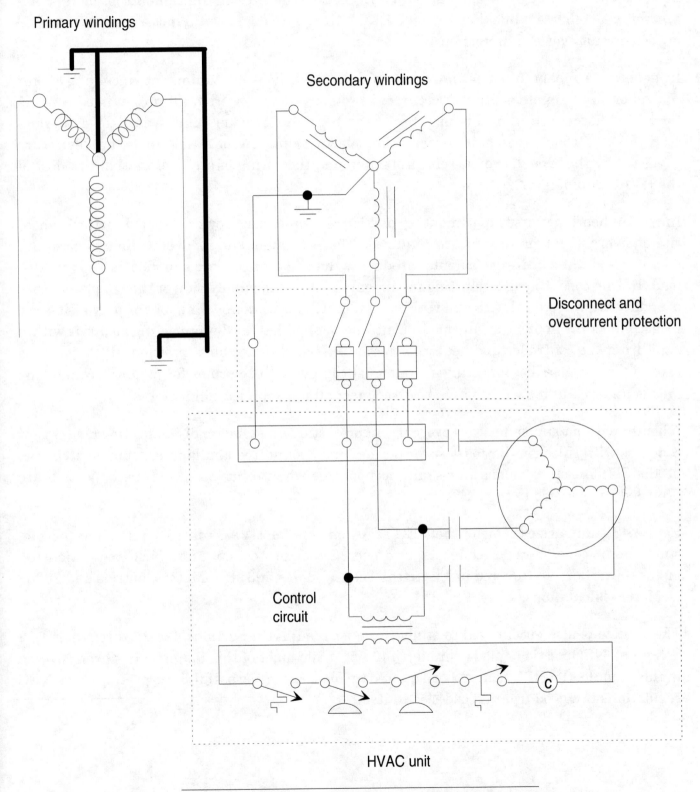

Figure 42. Major components for an HVAC control circuit.

10.6.0 NEC REQUIREMENTS FOR HVAC CONTROLS

Several NEC Sections cover the requirements for HVAC controls. For example, Part C of NEC Article 424 deals with the control and protection of fixed electric space heating equipment (NEC Sections 424-19 through 24-22); Part E of NEC Article 440 covers requirements for motor-compressor controllers, while Part F of NEC Article 440 deals with motor-compressor and branch-circuit overload protection.

In general, a means must be provided to disconnect heating equipment, including motor controllers and supplementary overcurrent protective devices, from all ungrounded conductors. The disconnecting means may be a switch, circuit breaker, unit switch, or a thermostatically controlled switching device. The selection and use of a disconnecting device are governed by the type of overcurrent protection and the rating of any motors that are part of the HVAC equipment.

In certain heating units, supplementary overcurrent protective devices other than the branch-circuit overcurrent protection are required. These supplementary overcurrent devices are usually used when heating elements rated more than 48 amperes are supplied as a subdivided load. In this case, the disconnecting means must be on the supply side of the supplementary overcurrent protective device and within sight of it. This disconnecting means may also serve to disconnect the heater and any motor controllers, provided the disconnecting means is within sight from the controller and heater, or it can be locked in the open position. If the motor is rated over ⅛ HP, a disconnecting means must comply with the rules for motor disconnecting means unless a unit switch is used to disconnect all ungrounded conductors.

A heater without supplementary overcurrent protection must have a disconnecting means that complies with rules similar to those for permanently connected appliances. A unit switch may be the disconnecting means in certain occupancies when other means of disconnection are provided as specified in the NEC.

Figure 43 summarizes some of the NEC requirements for HVAC controls, including thermostats, motor controllers, disconnects, and overcurrent protection. More NEC regulations on motor controllers are covered in Electrical Task Modules 20311 — Motor Controls and 20309 — Motor Calculations.

The trainee is also encouraged to study (not just read) NEC Sections 424-19 through 424-22 as well as NEC Section 440-41 through 440-55. While most of this material has been covered in this module, the NEC has not been quoted (in most cases) verbatim. Interpreting these NEC regulations is a good training exercise in itself.

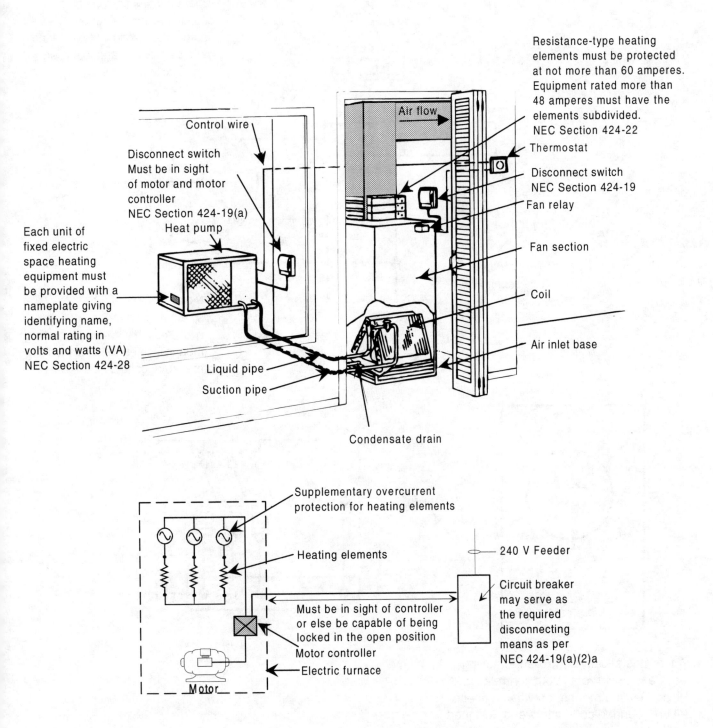

Resistance-type heating elements must be protected at not more than 60 amperes. Equipment rated more than 48 amperes must have the elements subdivided. NEC Section 424-22

Air flow

Control wire

Disconnect switch Must be in sight of motor and motor controller NEC Section 424-19(a)

Heat pump

Each unit of fixed electric space heating equipment must be provided with a nameplate giving identifying name, normal rating in volts and watts (VA) NEC Section 424-28

Liquid pipe

Suction pipe

Condensate drain

Thermostat

Disconnect switch NEC Section 424-19

Fan relay

Fan section

Coil

Air inlet base

Supplementary overcurrent protection for heating elements

Heating elements

240 V Feeder

Must be in sight of controller or else be capable of being locked in the open position

Motor controller

Electric furnace

Motor

Circuit breaker may serve as the required disconnecting means as per NEC 424-19(a)(2)a

Figure 43. Summary of NEC requirements for HVAC controls.

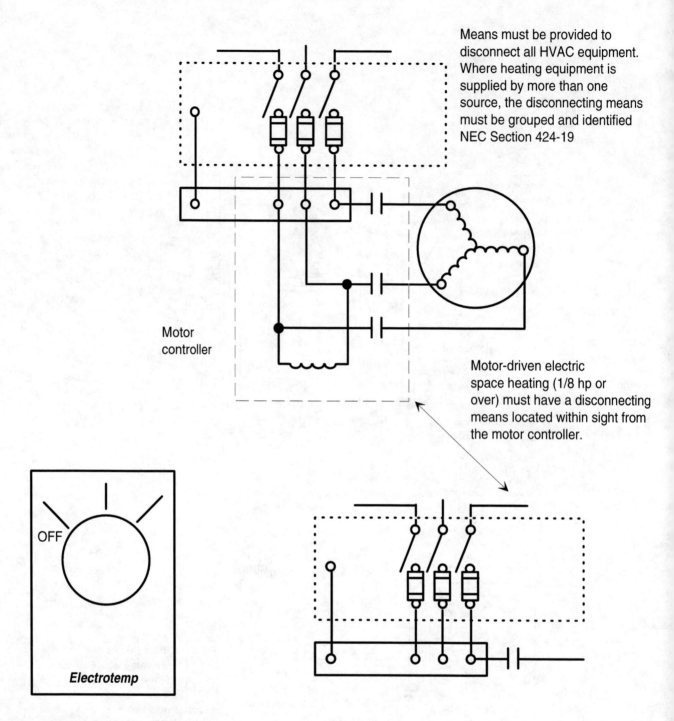

Means must be provided to disconnect all HVAC equipment. Where heating equipment is supplied by more than one source, the disconnecting means must be grouped and identified NEC Section 424-19

Motor controller

Motor-driven electric space heating (1/8 hp or over) must have a disconnecting means located within sight from the motor controller.

OFF

Electrotemp

A thermostatically controlled switching device may serve as both controller and disconnecting means, provided it opens all ungrounded conductors and designed so that the circuit cannot be energized when manually placed in the OFF position. The device must have an OFF marking.

Figure 43. Summary of NEC requirements for HVAC controls. *(Cont.)*

Troubleshooting heating and cooling systems cover a wide range of electrical and mechanical problems, from finding a short circuit in the power supply line, through adjusting a pulley on a motor shaft, to tracing loose connections in complex control circuits. However, in nearly all cases, the technician can determine the cause of the trouble by using a systematic approach, checking one part of the system at a time in the right order.

Every heating and cooling system's problems can be solved, and it is the purpose of this section to show the trainee exactly how to go about solving the more conventional ones in a safe and logical manner.

The data in Fig. 44 are arranged so that the problem is listed first. The possible causes of this problem are listed in the order in which they should be checked. Finally, solutions to the various problems are given, including step-by-step procedures where it is felt that they are necessary.

To better illustrate the use of these solutions to heating and cooling equipment problems, assume that an air-conditioner fan or blower motor is operating, but the compressor motor is not. Glance down the left-hand column in the troubleshooting charts until you locate the problem titled "Compressor Motor and/or Condenser Motor will not start, but Blower Motor Operates." Begin with the first item under "Probable cause" which tells you to check the thermostat system switch to make sure it is set to "Cool." Finding that the switch is set in the proper position, you continue on to next item; that is, "Check Temperature Setting." You may find that the temperature setting is above the room temperature so the system is not calling for cooling. Set the thermostat below room temperature, and the cooling unit will function.

This example is, of course, very simple, but most of the heating and cooling problems can be just as simple if a systematic approach to troubleshooting is used.

This section gives troubleshooting and repair procedures for common HVAC function problems. For more information, also see Module 20311 — Motor Controls. Motor controls are used extensively in HVAC applications, and the information contained in Module 20311 can be very helpful to the service technician.

Manufacturers of HVAC equipment also provide troubleshooting and maintenance manuals for their equipment. These manuals can be one of the most helpful "tools" imaginable for troubleshooting specific HVAC equipment. When unpacking equipment, controls, and other components for the system, always save any manuals or instructions that accompany the items. File them in a safe place so that you and other maintenance personnel can readily find them. Many electricians like to secure these manuals on the inside of a cabinet door within the equipment. This way, they will always be available when needed.

Malfunction	Probable Cause	Corrective Action
Compressor motor and condenser motor will not start, but fan/coil unit (blower motor) operates normally	Check the thermostat system switch to ascertain that it is set to "COOL."	Make necessary adjustments to settings.
	Check the thermostat to make sure that it is set below room temperature.	Make necessary adjustments.
	Check the thermostat to see if it is level. Most thermostats must be mounted level; any deviation will ruin their calibration.	Remove cover plate, place a spirit level on top of the thermostat base, loosen the mounting screws, and adjust the base until it is level; then tighten the mounting screws.
	Check all low-voltage connections for tightness.	Tighten.
	Make a low-voltage check with a voltmeter on the condensate float switch; the condensate may not be draining.	The float switch is normally found in the fan/coil unit. Repair or replace.
	Low air flow could be causing the trouble, so check the air filters.	Clean or replace.
	Make a low-voltage check of the antifrost control.	Replace if defective.
	Check all duct connections to the fan-coil unit.	Repair if necessary.
Compressor, condenser, and fan/coil unit motors will not start	Check the thermostat system switch setting to ascertain that it is set to "COOL."	Adjust as necessary.
	Check thermostat setting to make sure it is below room temperature.	Adjust as necessary.
	Check thermostat to make sure it is level.	Correct as required.
	Check all low-voltage connections for tightness.	Tighten.

Figure 44. HVAC troubleshooting chart.

Malfunction	Probable Cause	Corrective Action
Compressor, condenser, and fan/coil unit motors will not start	Check for a blown fuse or tripped circuit breaker.	Determine the cause of the open circuit and then replace the fuses or reset the circuit breaker.
	Make a voltage check of low-voltage transformer.	Replace if defective.
	Check the electrical service against minimum requirements; that is, for correct voltage, amperage, etc.	Update as necessary.
Condensing unit cycles too frequently, contactor opens and closes on each cycle and blower motor operates	Check condensate drain.	Repair or replace.
	Check all low-voltage wiring connections for tightness.	Tighten.
	Defective blower motor.	Test amperage reading while motor is running. Do not confuse the full-load (starting) amperes shown on the motor nameplate with the actual running amperes. The latter should be about 25% less. If the amperage varies considerably from that on the nameplate, check the motor for bad bearings, defective winding insulation, etc.
	Low voltage.	Test circuit for proper voltage.
Inadequate cooling with condensing unit and blower running continuously	Check all low-voltage connections (control wiring) against the wiring diagram furnished with the system.	Correct if necessary. Then check for leaks in the refrigerant lines.
	Check all joints in the supply and return ductwork.	Make all joints tight.

Figure 44. HVAC troubleshooting chart. *(cont.)*

Malfunction	Probable Cause	Corrective Action
Inadequate cooling with condensing unit and blower running continuously	The equipment could be undersized. Check heat gain calculations against the output of the unit.	Correct structural deficiencies with insulation, awnings, etc., or install properly sized equipment.
Condensing unit cycles but blower motor does not run	Check all low voltage connections against the wiring diagram furnished with the system.	Correct if necessary.
	Check all-low voltage connections for tightness.	Tighten.
	Make a voltage check on the blower relay.	Replace if necessary.
	Make electrical and mechanical checks on blower motor. Check for correct voltage at motor terminals. Mechanical problems could be bad bearings or a loose blower wheel. Bearing trouble can be detected by turning the blower wheel by hand (with current off), and checking for excessive wear, roughness, or seizure.	Repair or replace defective components.
Continuous short cycling of blower coil unit and insufficient cooling	Make electrical and mechanical checks.	Repair or replace motor if necessary.
Sweating at blower coil output or at electric duct heater outlet	Check to see if the insulation is installed properly.	Insulate properly.
	Inspect the joints at the duct heater or blower coil receiving collar.	Seal properly.
Thermostat calls for heat but blower motor will not operate	Check all low-voltage connections against the wiring diagram furnished with the system.	Correct if necessary.

Figure 44. HVAC troubleshooting chart. *(cont.)*

Malfunction	Probable Cause	Corrective Action
Thermostat calls for heat but blower motor will not operate	Check all low-voltage connections for tightness.	Tighten.
	Check all low-voltage against the unit's nameplate.	Correct if necessary.
	Check all line-voltage connections for tightness.	Tighten.
	Check for blown fuses or a tripped circuit breaker in the line.	Determine the reason for the open circuit and replace the fuses or reset circuit breaker.
	Check the low-voltage transformer.	Replace if defective.
	Make a low-voltage check on the magnetic relay.	Repair or replace if necessary.
	Make electrical and mechanical checks on the blower motor.	Repair or replace the motor if defective.
Thermostat calls for heat blower motor operates but delivers cold air	Make a visual and electrical check on the heating elements.	If not operating, continue on to next step.
	Make an electrical check on the heater limit switch—first disconnecting all power to the unit—using an ohmmeter to check continuity between the two terminals of the switch.	If the limit switch is open, repair or replace.
	Make an electrical check on the time-delay relay. Most are rated at 24 volts and have one set of normally-open auxiliary contacts for pilot duty.	If the relay heater coil is open or grounded.
	Make an electrical check on the magnetic relay.	Repair or replace if defective.
	Check the electric service-entrance and related circuits against the minimum recommendations.	Upgrade if necessary.

Figure 44. HVAC troubleshooting chart. *(cont.)*

Malfunction	Probable Cause	Corrective Action
Thermostat calls for heat and blower motor operates continuously, system delivers warm air but the thermostat is not satisfied	Check all joints in the ductwork for air leaks.	Make all defective joints tight.
	Check all duct joints and blower outlets for tightness.	Seal where necessary.
	Make a visual and electrical check of the electric heating element.	Repair or replace if necessary.
	Make an electrical check on the heater limit switch as described previously.	Repair or replace.
	Make an electrical check on the heater limit switch.	Repair or replace.
	Check the heating element against the blower unit for the possibility of a mismatch.	Replace if incompatible.
	Check your heat loss calculations. The equipment could be undersized.	If so, correct structural deficiencies by installing more insulation, storm windows and doors, etc., or install properly sized equipment.
Blower unit operates properly and delivers air but thermostat is not satisfied	Check all joints in the ductwork for air leaks.	Repair if necessary.
	Check the air filter.	Clean or replace if necessary. Also check the number of air outlets for adequacy and make sure they are balanced.
	Check for undersized equipment.	Correct structural deficiencies or install properly sized equipment.
Electric heater cycles on limit switches but blower motor does not operate	Make an electrical check on the magnetic relay.	Repair or replace if defective.

Figure 44. HVAC troubleshooting chart. *(cont.)*

Malfunction	Probable Cause	Corrective Action
Electric heater cycles on limit switches but blower motor does not operate	Make electrical and mechanical checks on the blower motor.	Repair or replace if defective.
	Check the line connection against the wiring diagram furnished with the system.	Make any necessary changes.
Excessive air noise at terminator	Duct or outlet undersized; air velocity too great.	Increase size of duct and/or outlet.
	Make external static pressure check.	Correct restrictions in system if necessary.
	Check for properly balanced system.	Make corrections if necessary.
Excessive noise at return air grille	Check the return duct to make sure it has a 90 degree bend.	Correct if necessary.
	Make a visual check of the blower unit to ascertain that all shipping blocks and angles have been removed.	Remove if necessary.
	Check blower motor assembly suspension and fasteners.	Tighten if necessary.
Excessive vibration at blower unit	Visually check for vibration isolators (which isolate the blower coil from the structure).	If missing, install as recommended by the manufacturer.
	Visually check to ascertain that shipping blocks and angles have been removed from the blower unit.	Remove if necessary.
	Check blower motor assembly suspension and fasteners.	Tighten if necessary.

Figure 44. HVAC troubleshooting chart. *(cont.)*

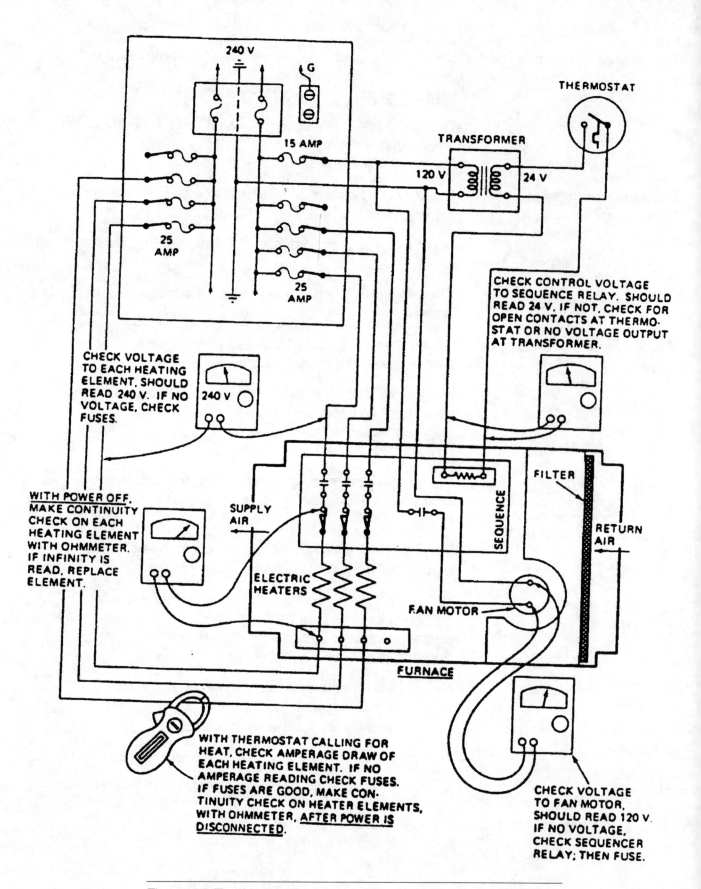

Figure 45. Troubleshooting diagram for an electric heating system.

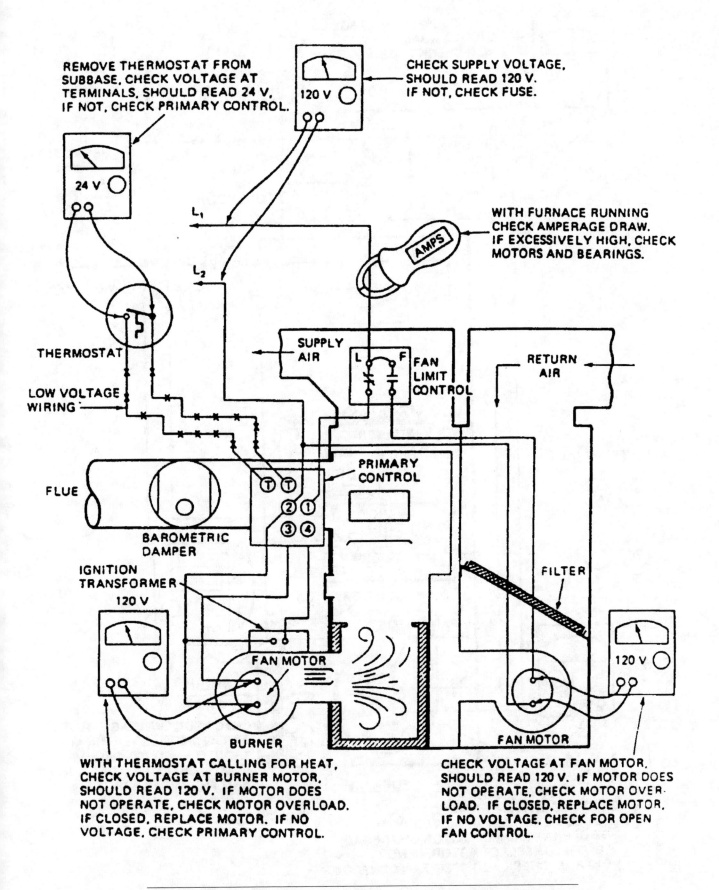

REMOVE THERMOSTAT FROM SUBBASE, CHECK VOLTAGE AT TERMINALS, SHOULD READ 24 V. IF NOT, CHECK PRIMARY CONTROL.

CHECK SUPPLY VOLTAGE, SHOULD READ 120 V. IF NOT, CHECK FUSE.

120 V

24 V

L_1

L_2

WITH FURNACE RUNNING CHECK AMPERAGE DRAW. IF EXCESSIVELY HIGH, CHECK MOTORS AND BEARINGS.

AMPS

THERMOSTAT

LOW VOLTAGE WIRING

SUPPLY AIR

L F

FAN LIMIT CONTROL

RETURN AIR

FLUE

BAROMETRIC DAMPER

T T

2 1

3 4

PRIMARY CONTROL

FILTER

IGNITION TRANSFORMER

120 V

FAN MOTOR

BURNER

FAN MOTOR

120 V

WITH THERMOSTAT CALLING FOR HEAT, CHECK VOLTAGE AT BURNER MOTOR, SHOULD READ 120 V. IF MOTOR DOES NOT OPERATE, CHECK MOTOR OVERLOAD. IF CLOSED, REPLACE MOTOR. IF NO VOLTAGE, CHECK PRIMARY CONTROL.

CHECK VOLTAGE AT FAN MOTOR, SHOULD READ 120 V. IF MOTOR DOES NOT OPERATE, CHECK MOTOR OVER-LOAD. IF CLOSED, REPLACE MOTOR. IF NO VOLTAGE, CHECK FOR OPEN FAN CONTROL.

Figure 46. Oil burner troubleshooting; thermal primary safety control.

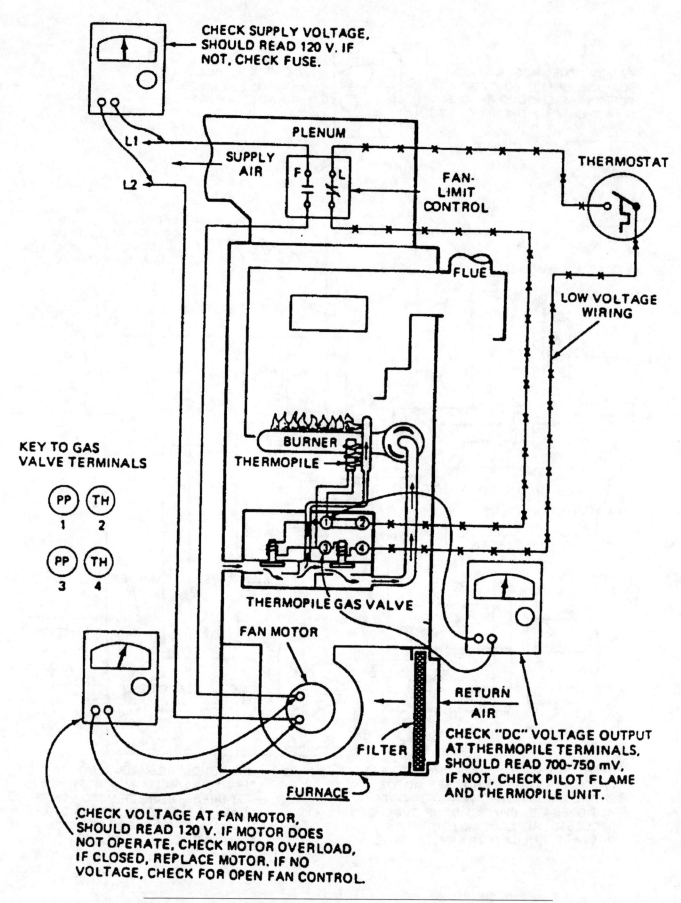

CHECK SUPPLY VOLTAGE.
SHOULD READ 120 V. IF
NOT, CHECK FUSE.

PLENUM

THERMOSTAT

SUPPLY AIR

FAN-LIMIT CONTROL

FLUE

LOW VOLTAGE WIRING

KEY TO GAS VALVE TERMINALS

PP 1   TH 2

PP 3   TH 4

BURNER

THERMOPILE

THERMOPILE GAS VALVE

FAN MOTOR

RETURN AIR

FILTER

CHECK "DC" VOLTAGE OUTPUT
AT THERMOPILE TERMINALS.
SHOULD READ 700-750 mV.
IF NOT, CHECK PILOT FLAME
AND THERMOPILE UNIT.

FURNACE

CHECK VOLTAGE AT FAN MOTOR,
SHOULD READ 120 V. IF MOTOR DOES
NOT OPERATE, CHECK MOTOR OVERLOAD.
IF CLOSED, REPLACE MOTOR. IF NO
VOLTAGE, CHECK FOR OPEN FAN CONTROL.

Figure 47. Troubleshooting gas-fired furnace; thermopile system.

The old saying, "an ounce of prevention..." certainly holds true for heating and cooling systems. A correctly installed system that is maintained according to the manufacturer's recommendations will give years of trouble-free service at minimum cost.

Maintenance procedures and frequency will depend on the type of system, but the information given here covers most residential and commercial HVAC systems in general. For further information, consult the Owner's Handbook accompanying the equipment. If none is available, obtain a copy from the equipment supplier or write directly to the manufacturer.

Besides saving on repairs, good maintenance of heating and cooling equipment will ensure that the equipment operates at maximum efficiency, which will save fuel and reduce other operating expenses.

Cleaning: Heating and cooling equipment should be cleaned at regular intervals in order to maintain operating efficiency, lengthen the life of the equipment, and minimize energy consumption and operating costs.

Begin by removing lint and dust with a cloth and brush from all finned convector-type heaters, and then vacuum them. This includes electric and hot-water baseboard heaters, wall, and unit heaters. Air conditioning evaporator and condenser coils should be vacuumed, scrubbed with a liquid solvent or detergent, and flushed.

Air filters in forced-air systems collect dust, as is their purpose. Periodic inspections will tell you how often they should be cleaned or replaced. Permanent metal mesh and electrostatic air filters should be washed and treated; throw-away filters should be replaced with the same size and type.

Vacuum or wipe off lint and dust from all supply and return grilles, diffusers, and registers, using a detergent solution if necessary. While you're doing this, also remove any dirt on the dampers and check the levers for proper operation.

Motors should also be vacuumed to remove dust and lint, but you should first turn off the power supply. Once free of dust and lint, wipe their exterior surfaces clean with a rag and reconnect the power.

Propeller fans and blower wheels are especially susceptible to dust deposits and should be cleaned often. Again make certain that the power is turned off—to prevent losing a finger—and remove the dirt deposits with a liquid solvent.

Every periodic inspection of any HVAC system should include a check of the condensate drain. Wash the drain pan with a mild detergent and flush out the drain line. All loose particles of dirt should be brushed from the evaporator coil and a fin comb should be used to open all clogged air passages in the coil. If the coil is extremely dirty, a small pressurized sprayer may be used with a strong dishwasher detergent to flush the coil. Always rinse with clean water after using detergent.

Lubrication: Adequate, regular lubrication ensures efficient operation, long equipment life, and minimum maintenance cost, but never over-lubricate! Observe the manufacturer's instructions when lubricating all bearings, rotary seals, and movable linkages of:

- Motors—Direct or belt-drive type.
- Shafts—Fans, blower wheel, and damper.
- Pumps—Water circulating.
- Motor Controllers—Sequential and damper operators.

Periodic Inspection: This check list should be followed during periodic inspections of most heating and cooling systems in order to minimize maintenance expense and to save as much fuel as possible.

1. Drive Belts

Examine for:

Proper tension and alignment
Sidewall wear
Deterioration, cracks
Greasy surfaces
Safety guards in position and secure

2. "V" Pulleys

Examine for:

Alignment
Wear of "V" wall
Tightness of pulley and set screws

3. Fan Blade and Blower Wheel Assemblies

Examine for:

Metal fatigue cracks
Tightness of hubs and set screws
Balance
Safety guards in position and secure

4. Electrical Components

Examine, burnish, or replace electrical contacts of:

Magnetic contactors and relays
Control switches, thermostats, timers, etc.

Examine wire and terminals for:

Corrosion and looseness at switches, relays, thermostats, controllers, fuse clips, capacitors, etc.

Examine motor capacitors for:

Case swelling
Electrolyte leakage

When checking the electrical system examine all components for evidence of overheating and insulation deterioration.

13.0.0   SAFETY

Experience in the electrical industry has shown that all electrical workers and technicians expose themselves to danger at times; new workers because of inexperience, older ones because of over-confidence and habits of work which they form. A few may even assume an attitude that it is a sign of weakness, or an indication of inexperience to observe safety precautions. However, to eliminate or reduce the incidence of accidents of any nature, it is extremely important that safety precautions be observed. Everyone involved in the electrical industry must continually be alert to this objective and its importance. There is no halfway mark.

The Core Curricula Modules have excellent coverage of safety precautions to use while on the job. However, since this module covers rotary equipment such as fans and motors, a brief review of some pertinent safety precautions are in order.

13.1.0   PERSONAL CLOTHING

The clothing of electricians will vary tremendously from job to job—depending on the type of work encountered. For example, technicians performing final control connections and testing in a nearly-finished office building with the environmental system in operation will probably wear ordinary slacks and sport shirt; technicians installing the motor controls on the deck of a high-rise building during cold weather will probably wear insulated underwear, sweaters, and coveralls. However, in all cases, clothing worn by electricians should not be frayed, torn or otherwise unsuitable for the work and the job conditions. See Fig. 48.

All shoes worn on the job should be heavy soled and in good repair. Many manufacturing plants require that all workers wear "hard-toe" shoes. This is a good type of shoe to wear on any project as they can save a lot of mashed toes—as in the case when a bundle of rigid conduit or a piece of heavy equipment is accidentally dropped. It is also a good idea to wear shoes containing steel plates in the soles. Workers stepping on nails driven into wooden boards is a common accident on most construction projects. The steel-plated shoes will prevent the majority of these accidents. Furthermore, the bottom of the sole should be rubber or other insulated material to prevent electric shocks when working around energized circuits. For example, electrical technicians are frequently called upon to replace damaged electrical components on 120- and 240-volt systems. If the electrician is standing on, say, a grounded concrete floor and accidentally brushes against an energized conductor or other component, the electric current will discharge through the person's body to ground, and will cause possible body injury or at least an uncomfortable shock. With rubber insulated shoes, the person will not be grounded (provided no other part of his body is in contact with a grounded object) and the electric current will not pass through his body.

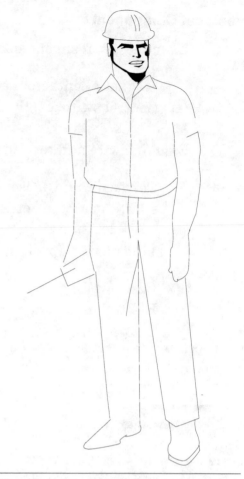

Figure 48. Proper clothing is essential.

When working in damp locations or in water, the electrical technician must be especially cautious of their footwear. Most contractors provide rubber boots for their workers when such conditions exist. The boots should always be worn and if a pair is not furnished by the employer, the individual workers should obtain a pair for their own personal safety.

Electricians working on energized equipment and circuits should avoid wearing unnecessary metal articles such as key chains, rings, metal hard hats and the like; all of these objects can come in contact with live electrical equipment or circuits and cause injury or death to the worker.

Gloves will save the technician's hands from many minor injuries such as blisters, cuts from metal burrs on equipment, splinters, and so on. However, gloves should never be worn around rotating machinery such as drill motors, fan-coil units, and the like, as they can easily become entangled in a drill bit or fan blade and injure the worker's hand severely.

All workers exposed to the hazards of falling objects or working in tight quarters should wear an approved type of fiber or plastic hard hat. A metal hard hat should never be worn by electrical workers. See Fig. 49 for a summary of clothing that should (and should not) be worn by electrical workers.

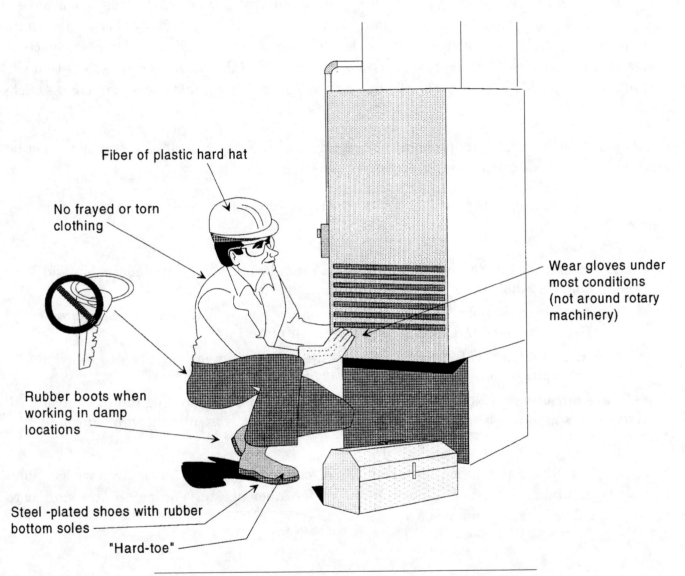

Figure 49. Summary of proper clothing for electrical workers.

Summary

The general rules for the installation of fixed heating equipment are presented in NEC Article 424. Installation of the units must be made in an approved manner.

In addition to the requirements of installation, all fixed heating equipment, including any associated motor controllers and supplementary overcurrent protective devices, must have a disconnecting means. This disconnecting means may be a switch, circuit breaker, unit switch, or a thermostatically-controlled switching device. The selection and use of a disconnecting device are governed by the type of overcurrent protection and the rating of any motors that are part of the unit. The following is a brief summary of NEC requirements covered in NEC Article 424:

Fixed electric space heating equipment with conductors having over 60° C insulation must be marked and marking must be visible after installation.

Fixed electric heating equipment exposed to severe physical damage must be adequately protected.

- Heaters installed in damp or wet conditions must be approved for the location and be installed so that water cannot enter or accumulate.
- Heating equipment must be installed with required spacing from combustible material or be acceptable for direct contact with such material.
- Exposed metal parts of fixed electric space heating equipment must be grounded as specified in the Code.
- Means must be provided to disconnect heating equipment, including motor controllers and supplementary overcurrent protective devices, from all ungrounded conductors.

Note: See NEC for specific rules about the disconnecting means for heaters with supplementary overcurrent protection. Rules for other heating equipment are similar to those for any appliance.

References

For more advanced study of topics covered in this Task Module, the following works are suggested:

National Electrical Code Handbook, Latest Edition, NFPA, Quincy, MA
American Electricians Handbook, Latest Edition, Croft, McGraw-Hill, New York, NY

SELF-CHECK REVIEW/PRACTICE QUESTIONS

1. Under normal atmospheric pressure, what temperature (in degrees Fahrenheit) does water boil?

   a. 110°F
   b. 120°F
   c. 212°F
   d. 240°F

2. At what temperature does common ammonia boil?

   a. -21°F
   b. -11°F
   c. -16°F
   d. -29°F

3. What is the common name given the liquid control valve in a refrigeration system?

   a. Expansion valve
   b. LCV
   c. Four-way valve
   d. Oil-recovery valve

4. What is the function of the condenser in a refrigeration system?

   a. Change the liquid to a gas
   b. Reject the heat in the refrigerant to the surrounding air or to a water supply
   c. Change the gas to a liquid
   d. Change from cooling to heating or vice-versa

5. When a liquid boils, what state does it take?

   a. Solid
   b. Liquid
   c. Gas
   d. Atom

6. When installing a disconnecting means for an HVAC system, containing a motor controller, which of the following is a true statement?

   a. The disconnect must be within 60 feet of the motor controller
   b. The disconnect must be within 75 feet of the motor and controller
   c. The disconnect must not be rated more than 80% of the motor's FLA
   d. The disconnect must be within sight of the motor and controller

7. When using a 240-volt, line-voltage thermostat as an electric space-heater controller, which of the following also qualifies as the equipment disconnect.

   a. One of the ungrounded conductors must be disconnected when the thermostat is in the OFF position
   b. Two ungrounded conductors must be disconnected when the thermostat is in the OFF position
   c. All ungrounded conductors must be disconnected when the thermostat is in the OFF position
   d. Only the grounded conductor needs to be disconnected

8. Why should gloves not be worn around rotary machinery?

   a. They can become entangled in a drill bit or fan and cause injury to the hands
   b. They prevent the worker from "getting the feel" of his or her work
   c. They prevent accurate work
   d. Leather will contaminate the refrigerant

9. Which of the following is not a good material for electrician's hard hats?

   a. Fiberglass
   b. Plastic
   c. Insulating fiber
   d. Sheet metal

10. When an air conditioning system does not start, what should be one of the first things checked?

   a. The refrigerant
   b. The conductor insulation
   c. The overcurrent-protective devices
   d. The fan limit switch

PERFORMANCE/LABORATORY EXERCISE

1. Remove the cover from an electric bimetal thermostat and identify the following:

    a. Temperature adjustment controls.

    b. Calibration screw.

    c. Temperature bimetal.

    d. Anticipating element.

    e. Cycling screw.

    f. Line contacts.

2. Connect a single-pole, line-voltage thermostat to a 500-watt, 120-volt electric heater; attach the line side to a cord-and-plug assembly, plug unit into a 120-volt receptacle, and check thermostat for proper functioning.

3. Connect the lines to the various control components in Fig. 50.

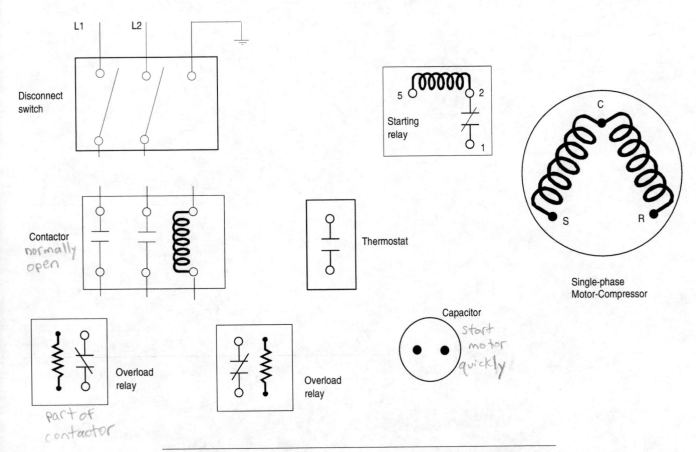

Figure 50. Wiring diagram for performance/laboratory exercise.

Answers to Self-Check Questions

1. c

2. d

3. a

4. b

5. c

6. d

7. c

8. a

9. d

10. c

Hazardous Locations

Module 20313

Electrical Trainee Task Module 20313

HAZARDOUS LOCATIONS

Objectives

Upon completion of this module, the trainee will be able to:

1. Identify the various classifications of hazardous locations.
2. Select and install branch circuits and feeders in specific hazardous locations.
3. Select motors for specific hazardous locations.
4. Select wiring methods for Class I hazardous locations.
5. Select wiring methods for Class II hazardous locations.
6. Select wiring methods for Class III hazardous locations.
7. Select and install lighting fixtures for specific hazardous locations.
8. Install wiring in specific hazardous locations.
9. Follow NEC requirements for specific hazardous locations.

Prerequisites

Successful completion of the following Task Modules is required before beginning study of this Task Module: Core Curricula, Electrical Levels 1 and 2, Electrical Level 3, Modules 20301 through 20312.

Required Student Materials

1. Trainee Task Module
2. Copy of the latest edition of the National Electrical Code

COURSE MAP

This course map shows all of the *Wheels of Learning* task modules in the third level of the Electrical curricula. The suggested training order begins at the bottom and proceeds up. Skill levels increase as a trainee advances on the course map. The training order may be adjusted by the local Training Program Sponsor.

Course Map: Electrical, Level 3

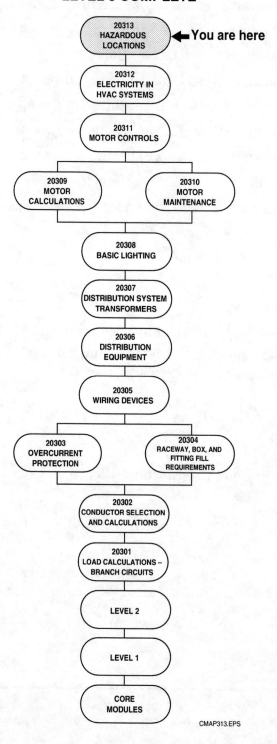

LEVEL 3 COMPLETE

20313
HAZARDOUS
LOCATIONS ← You are here

20312
ELECTRICITY IN
HVAC SYSTEMS

20311
MOTOR CONTROLS

20309
MOTOR
CALCULATIONS

20310
MOTOR
MAINTENANCE

20308
BASIC LIGHTING

20307
DISTRIBUTION SYSTEM
TRANSFORMERS

20306
DISTRIBUTION
EQUIPMENT

20305
WIRING DEVICES

20303
OVERCURRENT
PROTECTION

20304
RACEWAY, BOX, AND
FITTING FILL
REQUIREMENTS

20302
CONDUCTOR SELECTION
AND CALCULATIONS

20301
LOAD CALCULATIONS –
BRANCH CIRCUITS

LEVEL 2

LEVEL 1

CORE
MODULES

CMAP313.EPS

TABLE OF CONTENTS

Section	Topic	Page

Trade Terms Introduced in This Module

approved: Acceptable to the authority having jurisdiction.

conduit: A tubular raceway such as electrical metallic tubing (EMT); rigid metal conduit, rigid nonmetallic conduit, etc.

conduit body: A separate portion of a conduit or tubing system that provides access through a removable cover(s) to the interior of the system at a junction of two or more sections of the system or at a terminal point of the system.

Boxes such as FS and FD or larger cast or sheet metal boxes are not classified as conduit bodies.

Condulet: A trade name for conduit body.

dustproof: So constructed or protected that dust will not interfere with its successful operation.

dusttight: So constructed that dust will not enter the enclosing case under specified test conditions.

equipment: A general term including material, fittings, devices, appliances, fixtures, apparatus, and the like used as a part of, or in connection with, an electrical installation.

explosionproof: Designed and constructed to withstand an internal explosion without creating an external explosion or fire.

explosionproof apparatus: Apparatus enclosed in a case that is capable of withstanding an explosion of a specified gas or vapor that may occur within it, and also capable of preventing the ignition of a specified gas or vapor surrounding the enclosure by sparks, flashes, or explosion of the gas or vapor within, and which operates at such an external temperature that a surrounding flammable atmosphere will not be ignited thereby.

hazardous (classified) location: A location in which ignitable vapors, dust, or fibers may cause fire or explosion.

identified: (As applied to equipment.) Recognizable as suitable for the specific purpose, function, use, environment, application, etc., where described in a particular Code requirement.

sealing compound: The material poured into an electrical fitting to seal and minimize the passage of vapors.

seal-off: Fittings required in conduit systems to prevent the passage of gases, vapors, or flames from one portion of the electrical installation to another through the conduit. Sometimes referred to as *seals*.

NEC Articles 500 through 504 (and 505, the international version) of the NEC cover the requirements of electrical equipment and wiring for all voltages in locations where fire or explosion hazards may exist due to flammable gases or vapor, flammable liquids, combustible dust, or ignitable fibers or flyings. Locations are classified depending on the properties of the flammable vapors, liquids, gases or combustible dusts or fibers that may be present, as well as the likelihood that a flammable or combustible concentration or quantity is present.

Any area in which the atmosphere or a material in the area is such that the arcing of operating electrical contacts, components, and equipment may cause an explosion or fire is considered as a hazardous location. In all such cases, explosionproof equipment, raceways, and fittings are used to provide an explosionproof wiring system.

Hazardous locations have been classified in the NEC into certain class locations. Various atmospheric groups have been established on the basis of the explosive character of the atmosphere for the testing and approval of equipment for use in the various groups. However, it must be understood that considerable skill and judgment must be applied when deciding to what degree an area contains hazardous concentrations of vapors, combustible dusts or easily ignitable fibers and flyings. Furthermore, many factors—such as temperature, barometric pressure, quantity of release, humidity, ventilation, distance from the vapor source, and the like—must be considered. When information on all factors concerned is properly evaluated, a consistent classification for the selection and location of electrical equipment can be developed.

The appendices of this module list all of the flammable gases and combustible dusts which have been classified by the National Fire Protection Association (NFPA), along with their ignition temperatures. This information will prove invaluable when various atmospheres are encountered in your career as an electrician.

NEC Article 505 allows classification and application to international standards but it will be applied only under engineering supervision. In addition, few American made products are listed for this application. As a result, this module will not cover NEC Article 505 in any detail.

1.1.0 CLASS I LOCATIONS

Class I atmospheric hazards are divided into two Divisions (1 and 2) and also into four groups (A, B, C, and D). The Divisions are summarized in the paragraphs to follow while the groups are summarized in the table in Fig. 4.

Those locations in which flammable gases or vapors may be present in the air in quantities sufficient to produce explosive or ignitable mixtures are classified as Class I locations. If these gases or vapors are present under normal operations, under frequent repair or maintenance operations, or where breakdown or faulty operation of process equipment might also cause simultaneous failure of electrical equipment, the area is designated as Class I, Division 1.

Examples of such locations are interiors of paint spray booths where volatile, flammable solvents are used, inadequately ventilated pump rooms where flammable gas is pumped, anesthetizing locations of hospitals (to a height of 5 feet above floor level), and drying rooms for the evaporation of flammable solvents. See Fig. 1.

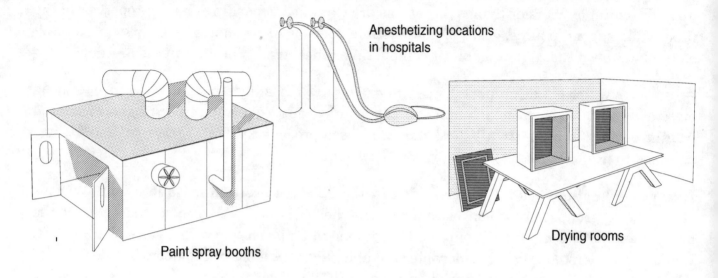

Anesthetizing locations
in hospitals

Drying rooms

Paint spray booths

Figure 1. Typical NEC Class I locations.

Class I, Division 2 covers locations where flammable gases, vapors or volatile flammable gases, vapors or volatile liquids are handled either in a closed system, or confined within suitable enclosures, or where hazardous concentrations are normally prevented by positive mechanical ventilation. Area adjacent to Division 1 locations, into which gases might occasionally flow, would also belong in Division 2.

1.2.0 CLASS II LOCATIONS

Class II locations are those that are hazardous because of the presence of combustible dust. Class II, Division 1 locations are areas where combustible dust, under normal operating conditions, may be present in the air in quantities sufficient to produce explosive or ignitable mixtures; examples are working areas of grain-handling and storage plants and rooms containing grinders or pulverizers. Class II, Division 2 locations are areas where dangerous concentrations of suspended dust are not likely, but where dust accumulations might form. See Fig. 2.

Besides the two Divisions (I and II), Class II atmospheric hazards cover three groups of combustible dusts. The groupings are based on the resistivity of the dust. Group E is always Division 1. Group F, depending on the resistivity, and Group G may be either Division 1 or 2. Since the NEC is considered the definitive classification tool and contains explanatory data about hazardous atmospheres, refer to NEC Section 500-6 for exact definitions of Class II, Divisions 1 and 2. Again, these groups are summarized in the table in Fig. 4.

ELECTRICAL TRAINEE TASK MODULE 20313

1.3.0   CLASS III LOCATIONS

These locations are those areas that are hazardous because of the presence of easily ignitable fibers or flyings, but such fibers and flyings are not likely to be in suspension in the air in these locations in quantities sufficient to produce ignitable mixtures. Such locations usually include some parts of rayon, cotton, and textile mills, clothing manufacturing plants, and woodworking plants. See Fig. 3.

In Class I and Class II locations the hazardous materials are further divided into groups; that is, Groups A, B, C, D in Class I and Groups E, F, and G in Class II. These groups are summarized in Fig. 4. For a more complete listing of flammable liquids, gases and solids, see *Classification of Gases, Vapors and dusts for Electrical Equipment in Hazardous (Classified) Locations*, NFPA Publication No. 497M.

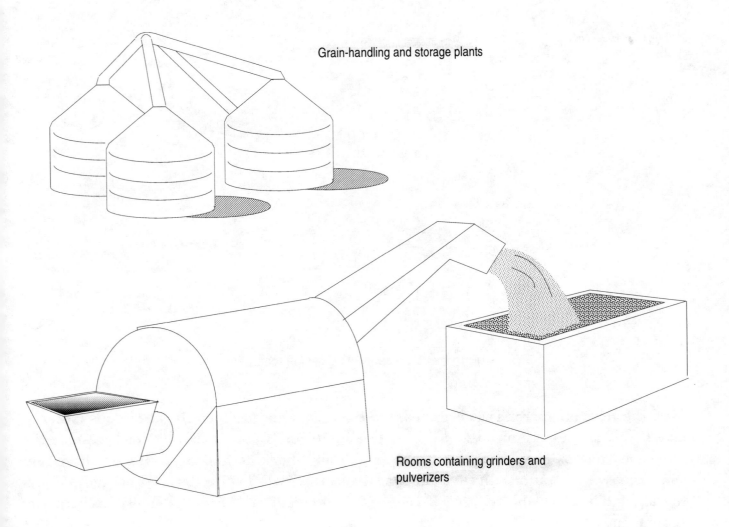

Grain-handling and storage plants

Rooms containing grinders and pulverizers

Figure 2. Typical NEC Class II locations.

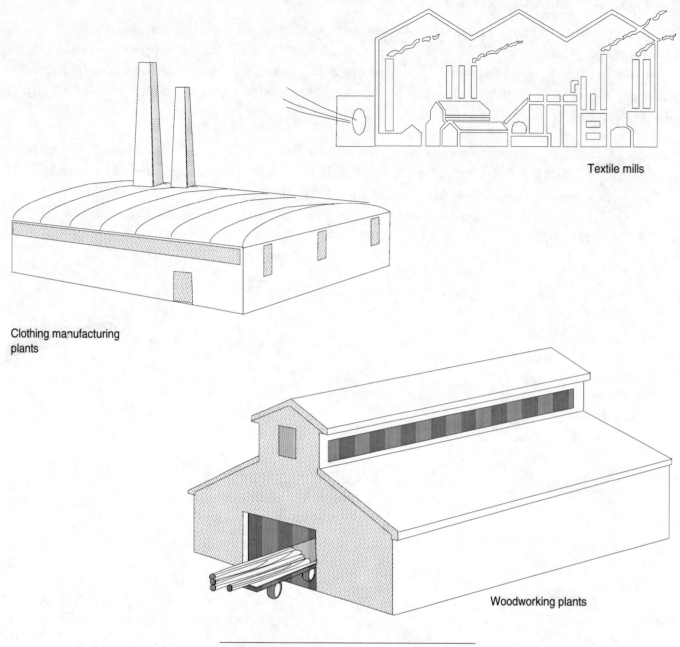

Textile mills

Clothing manufacturing plants

Woodworking plants

Figure 3. Typical NEC Class III locations.

Once the class of an area is determined, the conditions under which the hazardous material may be present determines the division. In Class I and Class II, Division 1 locations, the hazardous gas or dust may be present in the air under normal operating conditions in dangerous concentrations. In Division 2 locations, the hazardous material is not normally in the air, but it might be released if there is an accident or if there is faulty operation of equipment.

The information in Figures 5 through 10 gives a summary of the various classes of hazardous locations, as defined by the NEC.

Class	Division	Group	Typical Atmosphere/Ignition Temps.	Devices Covered	Temperature Measured	Limiting Value
I	1	A	Acetylene (305C, 581F)	All electrical devices and wiring	Maximum external temperature in 40C ambient	See Sect. 500-3 of NEC
Gases, vapors	Normally hazardous	B	1,3-Butadiene (420C, 788F)			
			Ethylene Oxide (429C, 804F)			
			Hydrogen (520C, 968F)			
			Manufactured Gas (containing more than 30% hydrogen by volume)			
			Propylene Oxide (449C, 840F)			
		C	Acetaldehyde (175C, 347F)			
			Diethyl Ether (160C, 320F)			
			Ethylene (450C, 842F)			
			Unsymmetrical Dimethyl Hydrazine (UDMH) (249C, 480F)			
		D	Acetone (465C, 869F)			
			Acrylonitrile (481C, 898F)			
			Ammonia (498C, 928F)			
			Benzene (498C, 928F)			
			Butane (288C, 550F)			
			1-Butanol (343C, 650F)			
			2-Butanol (405C, 761F)			
			n-Butyl Acetate (421C, 790F)			
			Cyclopropane (503C, 938F)			
			Ethane (472C, 882F)			
			Ethanol (363C, 685F)			
			Ethyl Acetate (427C, 800F)			
			Ethylene Dichloride (413C, 775F)			
			Gasoline (280-471C, 536-880F)			
			Heptane (204C, 399F)			
			Hexane (225C, 437F)			
			Isoamyl Alcohol (350C, 662F)			
			Isoprene (220C, 428F)			
			Methane (630C, 999F)			
			Methanol (385C, 725F)			
			Methyl Ethyl Ketone (404C, 759F)			
			Methyl Isobutyl Ketone (449C, 840F)			
			2-Methyl-1-Propanol (416C, 780F)			
			2-Methyl-2-Propanol (478C, 892F)			
			Naphtha (petroleum) (288C, 550F)			
			Octane (206C, 403F)			
			Pentane (243C, 470F)			
			1-Pentanol (300C, 572F)			

Figure 4. Summary of hazardous atmospheres.

Class	Division	Group	Typical Atmosphere/Ignition Temps.	Devices Covered	Temperature Measured	Limiting Value
			Propane (450C, 842F)			
			1-Propanol (413C, 775F)			
			2-Propanol (399C, 750F)			
			Propylene (455C, 851F)			
			Styrene (490C, 914F)			
			Toluene (480C, 896F)			
			Vinyl Acetate (402C, 756F)			
			Vinyl Chloride (472C, 882F)			
			Xylenes (464-529C, 867-984F)			
I	2	A	Same as Division 1	Lamps, resistors, coils, etc., other than arcing devices. (see Div. 1)	Max. internal or external temp. not to exceed the ignition temperature in degrees Celsius (°C) of the gas or vapor involved	See Sect. 500-3
Gases, vapors	Not normally hazardous	B	Same as Division 1			
		C	Same as Division 1			
		D	Same as Division 1			
II Combustible dusts	1 Normally hazardous	E	Atmospheres containing combustible metal dusts regardless of resistivity, or other combustible dusts of similarly hazardous characteristics having resistivity of less than 10^2 ohm-centimeter	Devices not subject to overloads (switches, meters).	Max. external temp. in 40C ambient with a dust blanket	Shall be less than ignition temperature of dust but not more than: No overload: E—200C (392F) F—200C (392F) G—165C (329F) Possible overload in operation: Normal E—200C (392F) F—150C (302F) G—120C (248F) Abnormal E—200C (392F) F—200C (392F) G—165C (329F)
		F	Atmospheres containing carbonaceous dusts having resistivity between 10^2 and 10^8 ohm-centimeter			
		G	Atmospheres containing combustible dusts having resistivity of 10^8 ohm-centimeter or greater			
	2	F	Atmospheres containing carbonaceous dusts having resistivity of 10^5 ohm-centimeter or greater	Lighting fixtures	Max. external temp under conditions of use	Same as Division 1
	Not normally hazardous	G	Same as Division 1			

Figure 4. Summary of hazardous atmospheres. (*Cont.*)

Components	Characteristics	NEC Section
Boxes, fitting	Explosionproof	501-4(a)
Seal offs	Approved for purpose	501-5(c)
Wiring methods	Rigid metal conduit, steel intermediate metal conduit, Type MI cable and, under certain conditions, rigid PVC and MC cable	501-4(a)
Receptacles	Explosionproof	501-12
Lighting fixtures	Explosionproof	501-9
Panelboards	Explosionproof	501-6(a)
Circuit breakers	Class I enclosure	501-6(a)
Fuses	Class I	501-6(a)
Switches	Class I enclosure	501-6(a)
Motors	Class I, totally enclosed or submerged	501-8(a)
Motor controls	Class I, Division 1	501-10(a)
Liquid-filled transformers	Installed in approved vault	501-2(a)
Dry-type transformers	Class I, Division 1 enclosures	501-7(a)
Utilization equipment	Class I, Division 1	501-10(a)
Flexible connections	Class I explosionproof	501-4
Portable lamps	Explosionproof	501-9(a)
Generators	Class I, totally enclosed or submerged	501-8(a)
Alarm systems	Class I, Division 1	501-14

Figure 5. Application rules for Class I, Division 1 locations.

2.0.0 PREVENTION OF EXTERNAL IGNITION/EXPLOSION

The main purpose of using explosionproof fittings and wiring methods in hazardous areas is to prevent ignition of flammable liquids or gases and to prevent an explosion.

2.1.0 SOURCES OF IGNITION

In certain atmospheric conditions when flammable gases or combustible dusts are mixed in the proper proportion with air, any source of energy is all that is needed to touch off an explosion.

Components	Characteristics	NEC Section
Boxes, fitting	Do not have to be explosionproof unless current interrupting contacts are exposed	501-4(b)
Seal offs	Approved for purpose	501-5(c)
Wiring methods	Rigid metal conduit, steel intermediate metal conduit, or Types MI, MC, MV, TC, ITC, PLTC cables, or enclosed gasketed busways or wireways	501-4(b)
Receptacles	Explosionproof	501-12
Lighting fixtures	Protected from physical damage	501-9(b)
Panelboards	General purpose with exceptions	501-6(b)
Circuit breakers	Class I enclosure	501-6(b)
Fuses	Class I	501-6(b)
Switches	Class I enclosure	501-6(b)
Motors	General purpose unless motor has sliding contacts, switching contacts or integral resistance devices; if so, use Class I	501-8(b)
Motor controls	Class I, Division 1	501-10(b)
Liquid-filled transformers	General purpose	501-2(b)
Dry-type transformers	Class I, General purpose except switching mechanism Division 1 enclosures	501-7(b)
Utilization equipment	Class I, Division 1	501-10(b)
Flexible connections	Class I explosionproof	501-11
Portable lamps	Explosionproof	501-9(b)
Generators	Class I, totally enclosed or submerged	501-8(b)
Alarm systems	Class I, Division 1	501-14(b)

Figure 6. Application rules for Class I, Division 2 locations.

Components	Characteristics	NEC Section
Boxes, fittings	Class II boxes required when using taps, joints or other connections; otherwise use dust tight boxes with no openings	502-4(a)(1)
Wiring methods	Rigid metal conduit, steel intermediate metal conduit, or Types MI and, under certain conditions, MC cables.	502-4(a)
Receptacles	Class II	502-13(a)
Lighting fixtures	Class II	502-11(a)
Panelboards	Dust-ignition proof	502-6(a)
Circuit breakers	Dust-ignition proof enclosure	502-6(a)
Fuses	Dust-ignition proof enclosure	502-6(a)
Switches	Dust-ignition proof enclosure	502-6(a)
Motors	Class II, Division 1 or totally enclosed	502-8(a)
Motor controls	Dust-ignition proof	502-6(a)
Liquid-filled transformers	Install in vault	502-2(a)
Dry-type transformers	Class II, vault	502-2(a)
Utilization equipment	Class II, Division 1	502-10(a)
Flexible connections	Extra-hard usage cord, liquidtight, and others	502-4(a)
Portable lamps	Class II	502-11(a)
Generators	Class II, Division 1 or totally enclosed	501-8(a)

Figure 7. Application rules for Class II, Division 1 hazardous locations.

Components	Characteristics	NEC Section
Boxes, fittings	Use tight covers to minimize entrance of dust	502-4(b)(1)
Wiring methods	Rigid metal conduit, steel intermediate metal conduit, or Types MI, MC, TC, ITC, PLTC, cables, or enclosed dust-tight busways or wireways	502-4(b)
Receptacles	Exposed live parts are not allowed	502-13(b)
Lighting fixtures	Class II	502-11(b)
Panelboards	Dust-tight enclosures	502-6(b)
Circuit breakers	Dust-tight enclosures	502-6(b)
Fuses	Dust-tight enclosure	502-6(b)
Switches	Dust-tight enclosure	502-6(b)
Motors	Class II, Division 1 or totally enclosed	502-8(b)
Motor controls	Dust-tight enclosures	502-6(b)
Liquid-filled transformers	Install in vault	502-2(b)
Dry-type transformers	Class II vault	502-2(b)
Utilization equipment	Class II	502-10(b)
Flexible connections	Extra-hard usage cord, liquidtight and others	502-4(b)
Portable lamps	Class II	502-11(b)
Generators	Class II, Division 1 or totally enclosed	501-8(b)

Figure 8. Application rules for Class II, Division 2 hazardous locations.

Components	Characteristics	NEC Section
Boxes, fittings	Use tight covers to minimize entrance of dust	503-3(a) & (b)
Wiring methods	Rigid metal conduit, steel intermediate metal conduit, EMT, or Types MI and MC, cable, or enclosed dust-tight busways or wireways	503-4(a) & (b)
Receptacles	Tight with no openings	503-11
Lighting fixtures	Tight enclosure with no openings	503-9
Panelboards	Dust-tight enclosures	503-4
Circuit breakers	Dust-tight enclosures	503-4
Fuses	Tight metal enclosure with no openings	503-4
Switches	Dust-tight enclosures	503-4
Motors	Totally enclosed	503-6
Motor controls	Dust-tight enclosures	503-4
Liquid-filled transformers	Install in vault	503-2
Dry-type transformers	Class II vault	503-2
Utilization equipment	Class II	503-8
Flexible connections	Extra-hard usage cord, and other flexible conduit/fittings	503-3
Portable lamps	Unswitched, guarded with tight enclosure for lamp	503-9
Generators	Totally enclosed	503-6

Figure 9. NEC Application rules for Class III, Divisions 1 and 2 hazardous locations.

One prime source of energy is electricity. Equipment such as switches, circuit breakers, motor starters, pushbutton stations, or plugs and receptacles, can produce arcs or sparks in normal operation when contacts are opened and closed. This could easily cause ignition.

Other hazards are devices that produce heat, such as lighting fixtures and motors. Here surface temperatures may exceed the safe limits of many flammable atmospheres.

Finally, many parts of the electrical system can become potential sources of ignition in the event of insulation failure. This group would include wiring (particularly splices in the wiring), transformers, impedance coils, solenoids, and other low-temperature devices without make-or-break contacts.

Non-electrical hazards such as sparking metal can also easily cause ignition. A hammer, file or other tool that is dropped on masonry or on a ferrous surface can cause a hazard unless the tool is made of non-sparking material. For this reason, portable electrical equipment is usually made from aluminum or other material that will not produce sparks if the equipment is dropped.

Electrical safety, therefore, is of crucial importance. The electrical installation must prevent accidental ignition of flammable liquids, vapors and dusts released to the atmosphere. In addition, since much of this equipment is used outdoors or in corrosive atmospheres, the material and finish must be such that maintenance costs and shutdowns are minimized.

2.2.0   COMBUSTION PRINCIPLES

Three basic conditions must be satisfied for a fire or explosion to occur:

- A flammable liquid, vapor or combustible dust must be present in sufficient quantity.
- The flammable liquid, vapor or combustible dust must be mixed with air or oxygen in the proportions required to produce an explosive mixture.
- A source of energy must be applied to the explosive mixture.

In applying these principles, the quantity of the flammable liquid or vapor that may be liberated and its physical characteristics must be recognized.

Vapors from flammable liquids also have a natural tendency to disperse into the atmosphere, and rapidly become diluted to concentrations below the lower explosion limit particularly when there is natural or mechanical ventilation.

> *CAUTION:*   The possibility that the gas concentration may be above the upper explosion limit does not afford any degree of safety, as the concentration must first pass through the explosive range to reach the upper explosion limit.

3.0.0   EXPLOSIONPROOF EQUIPMENT

Each area that contains gases or dusts that are considered hazardous must be carefully evaluated to make certain the correct electrical equipment is selected. Many hazardous atmospheres are Class I, Group D, or Class II, Group G. However, certain areas may involve other groups, particularly Class I, Groups B and C. Conformity with the NEC requires the use of fittings and enclosures approved for the specific hazardous gas or dust involved.

The wide assortment of explosionproof equipment now available makes it possible to provide adequate electrical installations under any of the various hazardous conditions. However, the electrician must be thoroughly familiar with all NEC requirements and know what fittings are available, how to install them properly, and where and when to use the various fittings.

For example, some workers are under the false belief that a fitting rated for Class I, Division 1 can be used under any hazardous conditions. However, remember the groups! A fitting rated for, say, Class I, Division 1, Group C cannot be used in areas classified as Groups A or B. On the other hand, fittings rated for use in Group A may be used for any group beneath A; fittings rated for use in Class I, Division 1, Group B can be used in areas rated as Group B areas or below, but not vice-versa.

WARNING!   Never interchange fittings or covers between one hazardous area and another. Such items must be rated for the appropriate Class, Division, and Group.

Explosionproof fittings are rated for both classification and groups. All parts of these fittings, including covers, are rated accordingly. Therefore, if a Class I, Division 1, Group A fitting is required, a Group B (or below) fitting cover must not be used. The cover itself must be rated for Group A locations. Consequently, when working on electrical systems in hazardous locations, always make certain that fittings and their related components match the condition at hand.

3.1.0 INTRINSICALLY SAFE EQUIPMENT

Intrinsically safe equipment is equipment and wiring that are incapable of releasing sufficient electrical energy under normal or abnormal conditions to cause ignition of a specific hazardous atmospheric mixture in its most easily ignited concentration.

The use of intrinsically safe equipment is primarily limited to process control instrumentation, since these electrical systems lend themselves to the low energy requirements.

Installation rules for intrinsically safe equipment are covered in NEC Article 504. In general, intrinsically safe equipment and its associated wiring must be installed so they are positively separated from the non-intrinsically safe circuits because induced voltages could defeat the concept of intrinsically safe circuits. Underwriters' Laboratories Inc. and Factory Mutual list several devices in this category.

3.2.0 EXPLOSIONPROOF CONDUIT AND FITTINGS

A floor plan for a hazardous area is shown in Fig. 10. In hazardous locations where threaded metal conduit is required, the conduit must be threaded with a standard conduit cutting die (Fig. 11) that provides ¾-inch taper per foot. The conduit should be made up wrench tight to prevent sparking in the event fault current flows through the raceway system (NEC Section

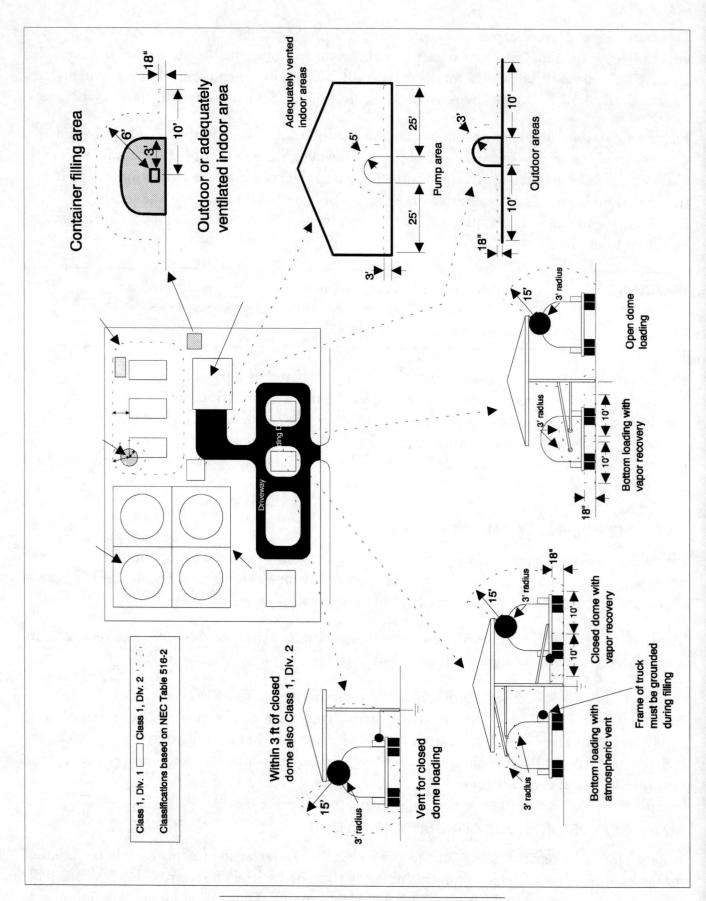

Figure 10. Floor plan of a hazardous location.

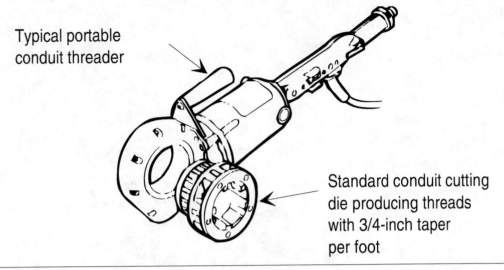

Typical portable conduit threader

Standard conduit cutting die producing threads with 3/4-inch taper per foot

Figure 11. Conduit must be threaded with a standard cutting die that provides ¾-inch taper per foot.

500-2). All boxes, fittings, and joints shall be threaded for connection to the conduit system and shall be an approved, explosionproof type (Fig. 12). Threaded joints must be made up with at least five threads fully engaged. Where it becomes necessary to employ flexible connectors at motor or fixture terminals (Fig. 13), flexible fittings approved for the particular class location shall be used.

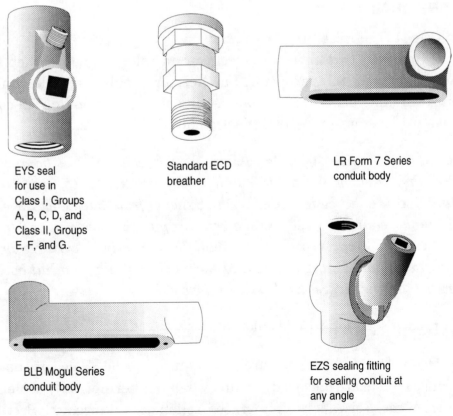

EYS seal for use in Class I, Groups A, B, C, D, and Class II, Groups E, F, and G.

Standard ECD breather

LR Form 7 Series conduit body

BLB Mogul Series conduit body

EZS sealing fitting for sealing conduit at any angle

Figure 12. Typical fittings approved for hazardous areas.

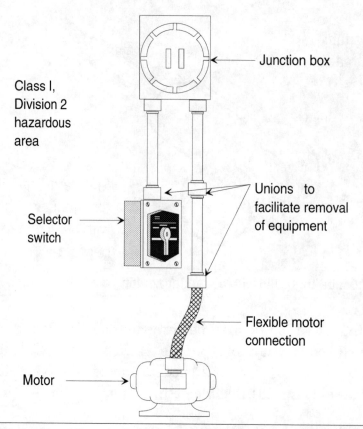

Class I,
Division 2
hazardous
area

Junction box

Unions to
facilitate removal
of equipment

Selector
switch

Flexible motor
connection

Motor

Figure 13. Explosionproof flexible connectors are frequently used for motor terminations.

3.3.0 SEALS AND DRAINS

Seal-off fittings (Fig. 14) are required in conduit systems to prevent the passage of gases, vapors, or flames from one portion of the electrical installation to another at atmospheric pressure and normal ambient temperatures. Furthermore, seal-offs (seals) limit explosions to the sealed-off enclosure and prevents precompression of "pressure piling" in conduit systems. For Class I, Division 1 locations, the NEC [Section 501-5(1)] states:

In each conduit run entering an enclosure for switches, circuit breakers, fuses, relays, resistors, or other apparatus which may produce arcs, sparks, or high temperatures, seals shall be installed with 18 inches from such enclosures. Explosionproof unions, couplings, reducers, elbows, capped elbows and conduit bodies similar to "L," "T," and "cross" types shall be the only enclosures or fittings permitted between the sealing fitting and the enclosure. The conduit bodies shall not be larger than the largest trade size of the conduits.

There is, however, one exception to this rule:

Conduits 1½ inches and smaller are not required to be sealed if the current-interrupting contacts are enclosed within a chamber hermetically sealed against the entrance of gases or vapors, immersed in oil in accordance with Section 501-6 of the NEC, or enclosed within a factory-sealed explosionproof chamber within an enclosure approved for the location and marked "factory sealed" or equivalent.

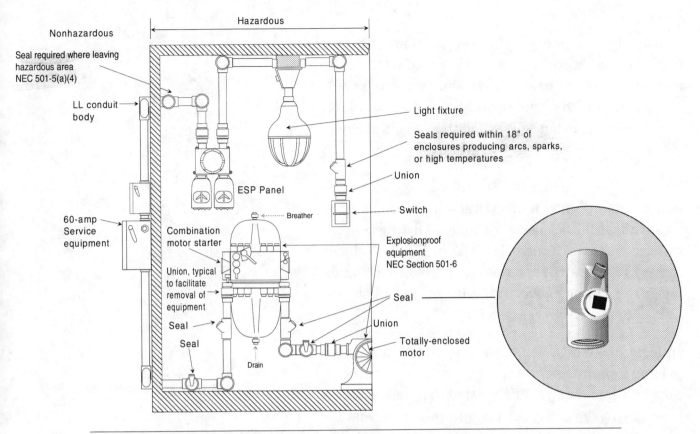

Nonhazardous

Hazardous

Seal required where leaving hazardous area NEC 501-5(a)(4)

LL conduit body

60-amp Service equipment

Combination motor starter

Union, typical to facilitate removal of equipment

Seal

Seal

Drain

Breather

ESP Panel

Light fixture

Seals required within 18" of enclosures producing arcs, sparks, or high temperatures

Union

Switch

Explosionproof equipment NEC Section 501-6

Seal

Union

Totally-enclosed motor

Figure 14. Seals must be installed at various locations in Class I, Division 1 locations.

Seals are also required in Class II locations under the following condition (NEC Section 502-5):

- Where a raceway provides communication between an enclosure that is required to be dust-ignitionproof and one that is not.

A permanent and effective seal is one method of preventing the entrance of dust into the dust-ignitionproof enclosure through the raceway. A horizontal raceway, not less than 10 feet long, is another approved method, as is a vertical raceway not less than 5 feet long and extending downward from the dust-ignitionproof enclosure.

Where a raceway provides communication between an enclosure that is required to be dust-ignitionproof and an enclosure in an unclassified location, seals are not required.

Where sealing fittings are used, all must be accessible.

While not an NEC requirement, many electrical designers and workers consider it good practice to sectionalize long conduit runs by inserting seals not more than 50 to 100 feet apart, depending on the conduit size, to minimize the effects of "pressure piling."

In general, seals are installed at the same time as the conduit system. However, the conductors are installed after the raceway system is complete and *prior* to packing and sealing the seal-offs.

3.3.1 DRAINS

In humid atmospheres or in wet locations, where it is likely that water can gain entrance to the interiors of enclosures or raceways, the raceways should be inclined so that water will not collect in enclosures or on seals but will be led to low points where it may pass out through integral drains.

Frequently, the arrangement of raceway runs makes this method impractical—if not impossible. In such instances, special drain/seal fittings should be used, such as Crouse-Hinds Type EZDs as shown in Fig. 15. These fittings prevent harmful accumulations of water above the seal and meets the requirements of NEC Section 501-5(f).

In locations which usually are considered dry, surprising amounts of water frequently collect in conduit systems. No conduit system is airtight; therefore, it may "breathe." Alternate increases and decreases in temperature and/or barometric pressure due to weather changes or due to the nature of the process carried on in the location where the conduit is installed will cause "breathing."

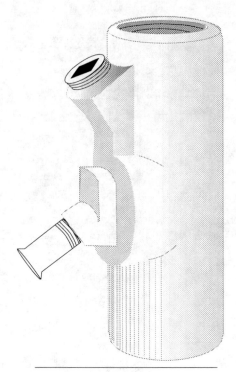

Figure 15. Typical drain seal.

Outside air is drawn into the conduit system when it "breathes in." If this air carries sufficient moisture, it will be condensed within the system when the temperature decreases and chills this air. The internal conditions being unfavorable to evaporation, the resultant water accumulation will remain and be added to by repetitions of the breathing cycle.

In view of this likelihood, it is good practice to insure against such water accumulations and probable subsequent insulation failures by installing drain/seal fittings with drain covers or fittings with inspection covers even though conditions prevailing at the time of planning or installing do not indicate their need.

3.4.0 SELECTION OF SEALS AND DRAINS

The primary considerations for selecting the proper sealing fittings are as follows:

- Select the proper sealing fitting for the hazardous vapor involved; that is, Class 1, Groups A, B, C, or D.
- Select a sealing fitting for the proper use in respect to mounting position. This is particularly critical when the conduit runs between hazardous and nonhazardous areas. Improper positioning of a seal may permit hazardous gases or vapors to enter the system beyond the seal, and permit them to escape into another portion of the

hazardous area, or to enter a nonhazardous area. Some seals are designed to be mounted in any position; others are restricted to horizontal or vertical mounting.

- Install the seals on the proper side of the partition or wall as recommended by the manufacturer.

- Installation of seals should be made *only* by trained personnel in strict compliance with the instruction sheets furnished with the seals and sealing compound.

- It should be noted that NEC Section 501-5(c)(4) prohibits splices or taps in sealing fittings.

- Sealing fittings are listed by UL for use in Class I hazardous locations with Chico A compound only. This compound, when properly mixed and poured, hardens into a dense, strong mass which is insoluble in water, is not attacked by chemicals, and is not softened by heat. It will withstand with ample safety factor, pressure of the exploding trapped gases or vapor.

- Conductors sealed in the compound may be approved thermoplastic or rubber insulated type. Both may or may not be lead covered.

3.5.0 TYPES OF SEALS AND FITTINGS

Certain seals, such as Crouse-Hinds EYS seals, are designed for use in vertical or nearly vertical conduit in sizes for ½- through 1-inch. Other styles are available in sizes ½-inch through 6 inches for use in vertical or horizontal conduit. In horizontal runs, these are limited to face-up openings. This, and other types of seals are shown in Fig. 16.

Seals ranging in sizes from 1¼-inch through 6-inches have extra large work openings, and separate filling holes, so that fiber dams are easy to make. However, the overall diameter of these fittings are scarcely greater than that of unions of corresponding sizes, permitting close conduit spacing.

Crouse-Hinds EZS seals are for use with conduit running at any angle, from vertical through horizontal.

EYD drain seals provide continuous draining and thereby prevent water accumulation. EYD seals are for vertical conduit runs and range in size from ½-inch to 4-inches inclusive. They are provided with one opening for draining and filling, a rubber tube to form drain passage and a drain fitting.

EZD drain seals provide continuous draining and thereby prevent water accumulation. The covers should be positioned so that the drain will be at the bottom. A set screw is provided for locking the cover in this position.

EZD fittings are suitable for sealing vertical conduit runs between hazardous and nonhazardous areas, but must be installed in the hazardous area when it is above the nonhazardous area. They must be installed in the nonhazardous area when it is above the hazardous area.

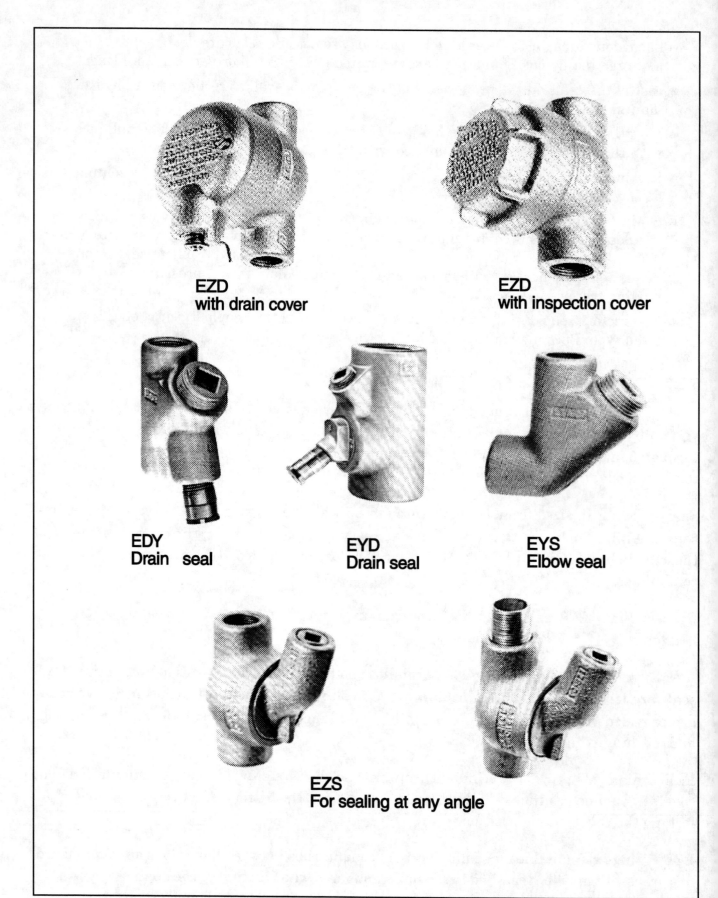

EZD
with drain cover

EZD
with inspection cover

EDY
Drain seal

EYD
Drain seal

EYS
Elbow seal

EZS
For sealing at any angle

Figure 16. Various types of seals and related components.

EZD drain seals are designed so that the covers can be removed readily, permitting inspection during installation or at any time thereafter. After the fittings have been installed in the conduit run and conductors are in place, the cover and barrier are removed. After the dam has been made in the lower hub opening with packing fiber, the barrier must be replaced so that the sealing compound can be poured into the sealing chamber.

EZD inspection seals are identical to EZD drain seals to provide all inspection, maintenance and installation advantages except that the cover is not provided with an automatic drain. Water accumulations can be drained periodically by removing the cover (when no hazards exist). The cover must be replaced immediately.

3.6.0   SEALING COMPOUNDS AND DAMS

Poured seals should be made only by trained personnel in strict compliance with the specific instruction sheets provided with each sealing fitting. Improperly poured seals are worthless.

Sealing compound shall be approved for the purpose; it shall not be affected by the surrounding atmosphere or liquids; and it shall not have a melting point of less than 200 degrees F. (93 degrees C.). The sealing compound and dams must also be approved for the type and manufacturer of fitting. For example, Crouse-Hinds CHICO® sealing compound is the only sealing compound approved for use with Crouse-Hinds ECM sealing fittings.

To pack the seal-off, remove the threaded plug or plugs from the fitting and insert the fiber supplied with the packing kit. Tamp the fiber between the wires and the hub before pouring the sealing compound into the fitting. Then pour in the sealing cement and reset the threaded plug tightly. The fiber packing prevents the sealing compound (in the liquid state) from entering the conduit lines.

Most sealing-compound kits contain a powder in a polyethylene bag within an outer container. To mix, remove the bag of powder, fill the outside container, and pour in the powder and mix.

> *CAUTION!*   Always make certain that the sealing compound is compatible for use with the packing material, brand and type of fitting, and also with the type of conductors used in the system.

In practical applications, there may be dozens of seals required for a particular installation. Consequently, after the conductors are pulled, each seal in the system is first packed. To prevent the possibility of overlooking a seal, one color of paint is normally sprayed on the seal hub at this time. This indicates that the seal has been packed. When the sealing compound is poured, a different color paint is once again sprayed on the seal hub to indicate a finished job. This method permits the job supervisor to visually inspect the conduit run, and if a seal is not painted the appropriate color, he or she knows that proper installation on this seal was not done; therefore, action can be taken to correct the situation immediately.

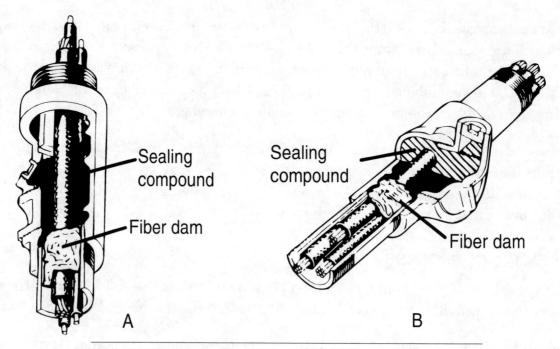

Figure 17. Seals made with fiber dams and sealing compound.

The seal-off fittings in Fig. 17 are typical of those used. The type in Fig. 17A is for vertical mounting and is provided with a threaded, plugged opening into which the sealing cement is poured. The seal-off in Fig. 17B has an additional plugged opening in the lower hub to facilitate packing fiber around the conductors to form a dam for the sealing cement.

The following procedures are to be observed when preparing sealing compound:

- Use a clean mix vessel for every batch. Particles of previous batches or dirt will spoil the seal.

- Recommended proportions are by volume—usually two parts powder to one part clean water. Slight deviations in these proportions will not affect the result.

- Do not mix more than can be poured in 15 minutes after water is added. Use cold water. Warm water increases setting speed. Stir immediately and thoroughly.

- If batch starts to set do not attempt to thin it by adding water or by stirring. Such a procedure will spoil seal. Discard partially set material and make fresh batch. After pouring, close opening immediately.

- Do not pour compound in sub-freezing temperatures, or when these temperatures will occur during curing.

- See that compound level is in accordance with the instruction sheet for that specific fitting.

Most other explosionproof fittings are provided with threaded hubs for securing the conduit as described previously. Typical fittings include switch and junction boxes, conduit bodies, union and connectors, flexible couplings, explosionproof lighting fixtures, receptacles, and panel-board and motor starter enclosures. A practical representation of these and other fittings is shown in Figs. 18 through 22.

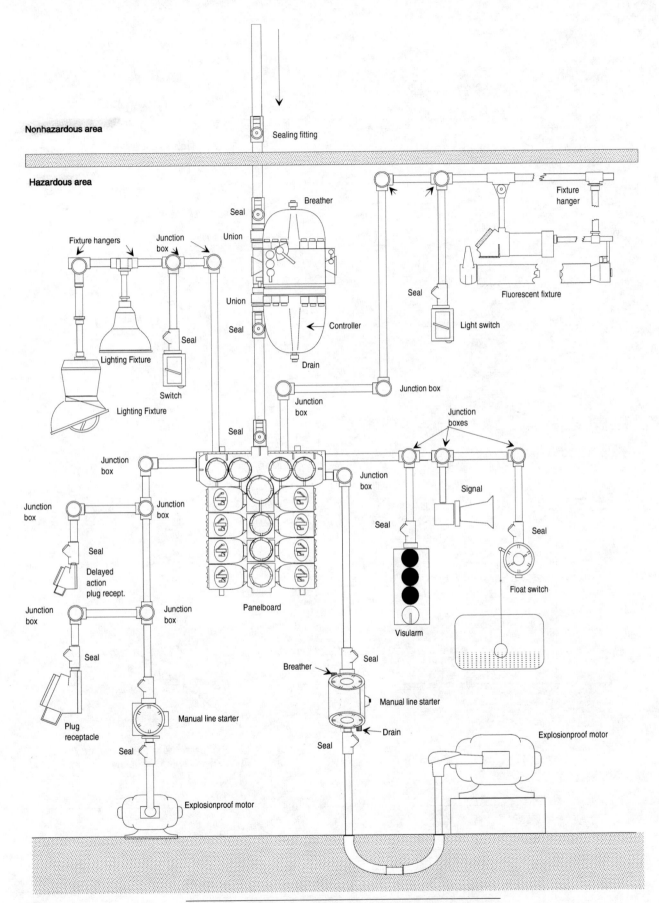

Nonhazardous area

Sealing fitting

Hazardous area

Fixture hangers

Junction box

Seal

Union

Breather

Fixture hanger

Union

Seal

Controller

Seal

Light switch

Fluorescent fixture

Lighting Fixture

Seal

Switch

Drain

Junction box

Lighting Fixture

Seal

Junction box

Junction boxes

Junction box

Junction box

Signal

Junction box

Junction box

Seal

Seal

Seal

Junction box

Delayed action plug recept.

Float switch

Junction box

Junction box

Visularm

Seal

Panelboard

Plug receptacle

Seal

Breather

Seal

Manual line starter

Explosionproof motor

Manual line starter

Seal

Drain

Seal

Explosionproof motor

Explosionproof motor

Figure 18. Class I, Division 1 electrical installation.

HAZARDOUS LOCATIONS — MODULE 20313

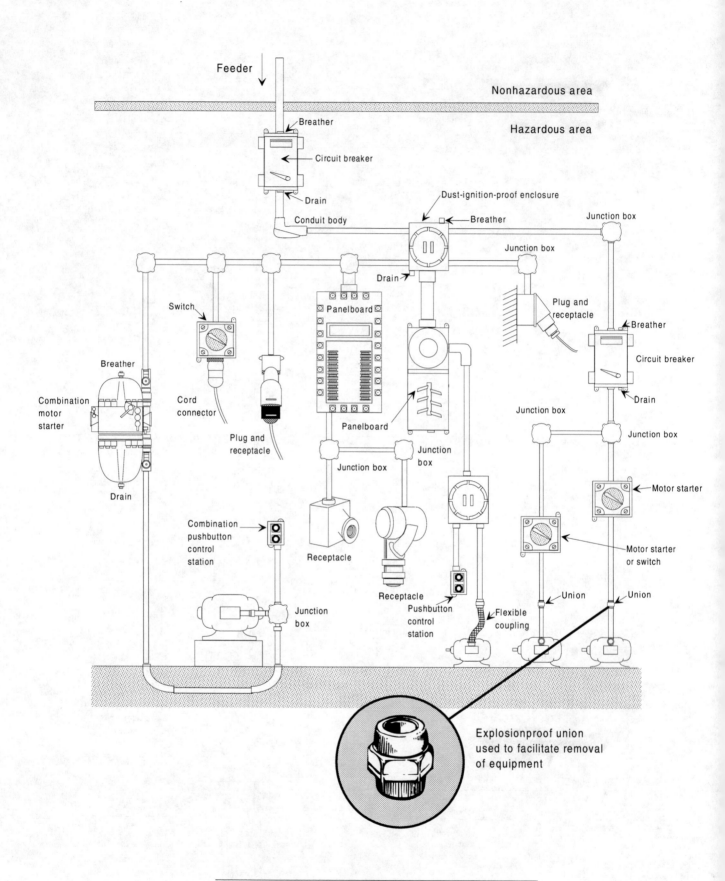

Figure 19. Class II, Division 1 electrical installation.

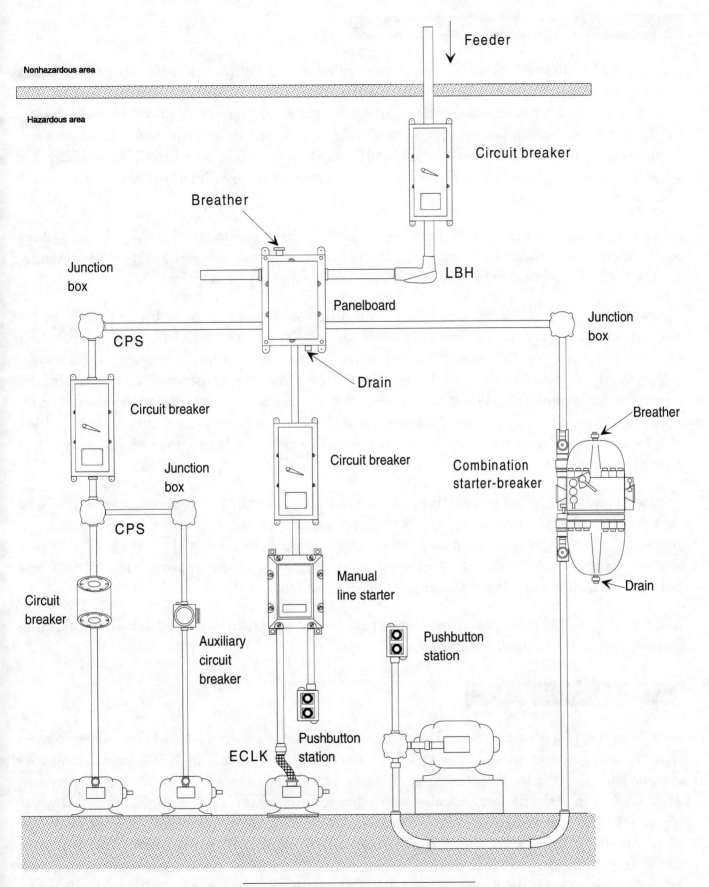

Nonhazardous area

Hazardous area

Feeder

Circuit breaker

Breather

Junction box

LBH

CPS

Panelboard

Junction box

Drain

Circuit breaker

Junction box

Breather

CPS

Combination starter-breaker

Circuit breaker

Manual line starter

Drain

Circuit breaker

Auxiliary circuit breaker

Pushbutton station

ECLK

Pushbutton station

Pushbutton station

Figure 20. Class II power installation.

4.0.0   GARAGES AND SIMILAR LOCATIONS

Garages and similar locations where volatile or flammable liquids are handled or used as fuel in self-propelled vehicles (including automobiles, buses, trucks, and tractors) are not usually considered critically hazardous locations. However, the entire area up to a level 18 inches above the floor is considered a Class I, Division 2 location, and certain precautionary measures are required by the NEC. Likewise, any pit or depression below floor level shall be considered a Class I, Division 2 location, and the pit or depression may be judged as Class I, Division I location if it is unvented.

Normal raceway (conduit) and wiring may be used for the wiring method above this hazardous level, except where conditions indicate that the area concerned is more hazardous than usual. In this case, the applicable type of explosionproof wiring may be required.

Approved seal-off fittings should be used on all conduit passing from hazardous areas to nonhazardous areas. The requirements set forth in NEC Sections 501-5 and 501-5(b)(2) shall apply to horizontal as well as vertical boundaries of the defined hazardous areas. Raceways embedded in a masonry floor or buried beneath a floor are considered to be within the hazardous area above the floor if any connections or extensions lead into or through such an area. However, conduit systems terminating to an open raceway, in an outdoor unclassified area, shall not be required to be sealed between the point at which the conduit leaves the classified location and enters the open raceway.

Figure 21 shows a typical automotive service station with applicable NEC requirements. Note that space in the immediate vicinity of the gasoline-dispensing island is denoted as Class I, Division I, to a height of 4 feet above grade. The surrounding area, within a radius of 20 feet of the island, falls under Class I, Division 2, to a height of 18 inches above grade. Bulk storage plants for gasoline are subject to comparable restrictions.

A summary of NEC rules governing the installation of electrical wiring at and about gasoline dispensing pumps is shown in Fig. 22.

5.0.0   AIRPORT HANGARS

Buildings used for storing or servicing aircraft in which gasoline, jet fuels, or other volatile flammable liquids or gases are used fall under Article 513 of the NEC. In general, any depression below the level of the hangar floor is considered to be a Class I, Division I location. The entire area of the hangar including any adjacent and communicating area not suitably cut off from the hangar is considered to be a Class I, Division 2 location up to a level of 18 inches above the floor. The area within 5 feet horizontally from aircraft power plants, fuel tanks, or structures containing fuel is considered to be a Class I, Division 2 hazardous location; this area extends upward from the floor to a level 5 feet above the upper surface of wings and of engine enclosures. See Fig. 23.

ELECTRICAL TRAINEE TASK MODULE 20313

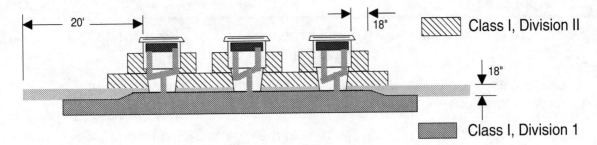

Gasoline dispensing units

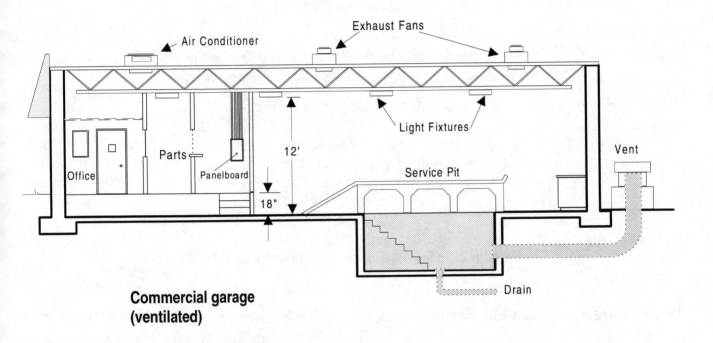

**Commercial garage
(ventilated)**

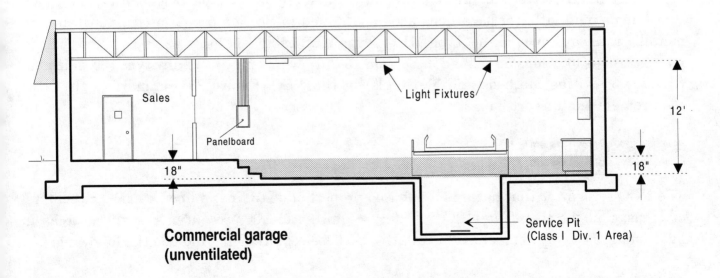

**Commercial garage
(unventilated)**

Figure 21. Commercial garage classifications.

Application	NEC Regulation	NEC Section
Equipment in hazardous locations	All wiring and components must conform to the rules for Class I locations	514-3
Equipment above hazardous locations	All wiring must conform to the rules for such equipment in commercial garages	514-4
Gasoline dispenser	A disconnecting means must be provided for each circuit leading to or through a dispensing pump to disconnect all conductors including the grounded neutral. An approved seal (seal-off) is required in each conduit entering or leaving a dispenser	514-6(a)
Grounding	Metal portions of all noncurrent-carrying parts of dispensers must be effectively grounded	
Underground wiring	Underground wiring must be installed within 2 feet of ground level—in rigid metal or IMC. If underground wiring is buried 2 feet or more, rigid nonmetallic conduit may be used along with the types mentioned above; Type MI cable may also be used in some cases	514-8

Figure 22. NEC application rules of service stations.

Adjacent areas in which hazardous vapors are not likely to be released, such as stock rooms and electrical control rooms, should not be classed as hazardous when they are adequately ventilated and effectively cut off from the hangar itself by walls or partitions. All fixed wiring in a hangar not within a hazardous area as defined in Section 513-2 must be installed in metallic raceways or shall be Type MI or Type ALS cable; the only exception is wiring in nonhazardous locations as defined in Section 13-2(d), which may be of any type recognized in Chapter 3 (Wiring Methods and Materials) in the NEC. Figure 23 summarizes the NEC requirements for airport hangars.

6.0.0 THEATERS

The NEC recognizes that hazards to life and property due to fire and panic exist in theaters, cinemas, and the like. The NEC therefore requires certain precautions in these areas in addition to those for commercial installations. These requirements include the following:

● Proper wiring of motion picture projection rooms (Article 540).

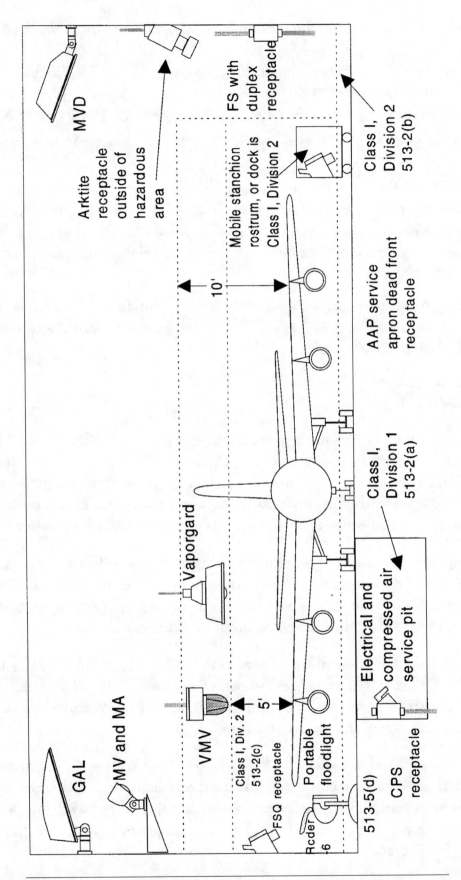

Figure 23. Sections of airport hangar showing hazardous locations.

- Heat-resistant, insulated conductors for certain lighting equipment (Section 520-43(b)).

- Adequate guarding and protection of the stage switchboard and proper control and overcurrent protection of circuits (Section 520-22).

- Proper type and wiring of lighting dimmers [Sections 520-52(e)] and 520-25.

- Use of proper types of receptacles and flexible cables for stage lighting equipment (Section 520-45).

- Proper stage flue damper control (Section 520-49).

- Proper dressing-room wiring and control (Sections 520-71, 72, and 73).

- Fireproof projection rooms with automatic projector port closures, ventilating equipment, emergency lighting, guarded work lights, and proper location of related equipment (Article 540).

Outdoor or drive-in motion picture theaters do not present the inherent hazards of enclosed auditoriums. However, the projection rooms must be properly ventilated and wired for the protection of the operating personnel.

7.0.0   HOSPITALS

Hospitals and other health-care facilities fall under Article 517 of the NEC. Part B of Article 517 covers the general wiring of health-care facilities. Part C covers essential electrical systems for hospitals. Part D gives the performance criteria and wiring methods to minimize shock hazards to patients in electrically susceptible patient areas. Part E covers the requirements for electrical wiring and equipment used in inhalation anesthetizing locations.

With the widespread use of x-ray equipment of varying types in health-care facilities, electricians are often required to wire and connect equipment such as discussed in Article 660 of the NEC. Conventional wiring methods are used, but provisions should be made for 50- and 60-ampere receptacles for medical x-ray equipment (Section 660-3b).

Anesthetizing locations of hospitals are deemed to be Class I, Division 1, to a height of 5 feet above floor level. Gas storage rooms are designated as Class I, Division 1, throughout. Most of the wiring in these areas, however can be limited to lighting fixtures only—locating all switches and other devices outside of the hazardous area.

The NEC recommends that wherever possible electrical equipment for hazardous locations be located in less hazardous areas. It also suggests that by adequate, positive-pressure ventilation from a clean source of outside air the hazards may be reduced or hazardous locations limited or eliminated. In many cases the installation of dust-collecting systems can greatly reduce the hazards in a Class II area.

8.0.0   PETRO/CHEMICAL HAZARDOUS LOCATIONS

Most manufacturing facilities involving flammable liquids, vapors, or fibers must have their wiring installations conform strictly to the NEC as well as governmental, state, and local ordnances. Therefore, the majority of electrical installations for these facilities are carefully designed by experts in the field—either the plant in-house engineering staff or else an independent consulting engineering firm.

Industrial installations dealing with petroleum or some types of chemicals are particularly susceptible to many restrictions involving many governmental agencies. Electrical installations for petro/chemical plants will therefore have many pages of electrical drawings and specifications which first go through the gambit for approval from all the agencies involved. Once approved, these drawings and specifications must be following exactly because any change whatsoever must once again go through the various agencies for approval.

9.0.0   MANUFACTURERS' DATA

Manufacturers of explosionproof equipment and fittings expend a lot of time, energy, and expense in developing guidelines and brochures to ensure that their products are used correctly and in accordance with the latest NEC requirements. The many helpful charts, tables, and application guidelines are invaluable to anyone working on projects involving hazardous locations. Therefore, it is recommended that the trainee obtain as much of this data as possible. Once obtained, study this data thoroughly. Doing so will enhance your qualifications for working in hazardous locations of any type. Manufacturers' data is usually available to qualified personnel (electrical workers) at little or no cost and can be obtained from local distributors of electrical supplies, or directly from the manufacturer. You may write to the following companies requesting their catalogs and brochures dealing with explosionproof equipment.

Appleton Electric Co.
1701 Wellington Ave.
Chicago, IL 60657

Automatic Switch Co.
50-60 Hanover Rd.
Florham Park, NJ 07932

Crouse-Hinds ECM
PO Box 4999
Syracuse, NY 19221

Electric Panelboard Co., Inc.
Box 2483010
Pixley Industrial Pkwy.
Rochester, NY 14624

Hope Electrical Products Co., Inc.
39 Long Ave.
Hillside, NJ 07205

Killark Electric Mfg. Co.
Box 5325
3940 M.L. King Dr.
Saint Louis, MO 63115

Spring City Electrical Mfg. Co.
Box A
Hall & Main Sts.
Spring City, PA 19475

RAB Electric Mfg. Co.
170 Ludlow Ave.
Northvale, NJ 07647

Square D Co.
330 Weakley Rd.
Smyrna, TN 37167

Summary

Any area in which the atmosphere or a material in the area is such that the arcing of operating electrical contacts, components, and equipment may cause an explosion or fire is considered as a hazardous location. In all such cases, explosionproof equipment, raceways, and fittings are used to provide an explosionproof wiring system.

The wide assortment of explosionproof equipment now available makes it possible to provide adequate electrical installations under any of the various hazardous conditions. However, the electrician must be thoroughly familiar with all NEC requirements and know what fittings are available, how to install them properly, and where and when to use the various fittings.

Many factors—such as temperature, barometric pressure, quantity of release, humidity, ventilation, distance from the vapor source, and the like—must be considered. When information on all factors concerned is properly evaluated, a consistent classification for the selection and location of electrical equipment can be developed.

References

For a more advanced study of topics covered in this Task Module, the following works are suggested:

American Electricians Handbook, Latest Edition, Croft, McGraw-Hill, New York, NY
National Electrical Code Handbook, Latest Edition, NFPA, Quincy, MA
Code Digest, Latest Edition, Crouse-Hinds, Syracuse, NY
Manufacturers' Catalogs, See Section 8.0.0 in this module

SELF-CHECK REVIEW/PRACTICE QUESTIONS

1. How many classifications of hazardous atmospheres are listed in the NEC?

    a. 1
    b. 2
    c. 3
    d. 4

2. How many Divisions are there for each classification?

    a. 1
    b. 2
    c. 3
    d. 4

3. How many Groups are listed under Class I, Division 1?

    a. 1
    b. 2
    c. 3
    d. 4

4. Which of the following are the Groups listed under Class II, Division 1 locations?

    a. A, B, C, and D
    b. E, F, G
    c. H, I, J
    d. L, M, N

5. Which of the following best describes the purpose of packing fiber in a sealing fitting?

    a. To identify the type of seal and the sealing compound to use
    b. To prevent any liquids, gases or vapors from passing through the fitting
    c. To prevent flammable vapors from mixing with the sealing compound
    d. To provide a dam to contain the sealing compound until it hardens

6. When rigid metal conduit is required in hazardous locations, at what taper must the threads be cut?

    a. ½-inch taper per foot
    b. ¾-inch taper per foot
    c. 1-inch taper per foot
    d. 1¼-inch taper per foot

7. When installing switches or other arc-producing apparatus in Class 1, Division 1 locations, within what distance of the switch (or other apparatus) must sealing fittings be installed?

    a. 6 inches
    b. 12 inches
    c. 18 inches
    d. 24 inches

8. What is the main purpose of a union in conduit runs?

    a. To facilitate installation and removal of equipment
    b. To ensure grounding continuity
    c. To seal the system from flammable gases or vapors
    d. To form tighter joints in the system

9. Which of the following best describes the time to pour in the sealing compound in a sealing fitting?

    a. Within five minutes after it is installed in a raceway system
    b. After the conduit system and seals are installed and the conductors and packing fiber has been installed
    c. Prior to installing the packing fiber
    d. After the threaded plug is set tight

10. When installing circuit breakers in Class II, Division 1 hazardous locations, what type of enclosure must be used?

    a. Dust-ignition proof
    b. Dust-tight
    c. Tight metal enclosure with no openings
    d. A standard enclosure with several openings

PERFORMANCE/LABORATORY EXERCISE

1. Using two ¾″ rigid metal conduit nipples, a sealing fitting (with ¾″ threaded hubs), 3 pieces of No. 12 AWG THHN conductors, and a packing fiber/sealing kit, perform the following operations:

   a. Secure one conduit nipple in each end of the seal.

   b. Make sure the required amount of threads are engaged.

   c. Pull the three THHN conductors through the nipples and seal, so that about 6 inches is protruding from each nipple.

   d. Pack the fiber as per instructions furnished with the sealing kit.

   e. Mix the sealing compound as per instruction furnished with the kit.

   f. Position unit in required position and pour in sealing compound.

   g. When sealing compound starts to set, have your instructor examine your work.

2. Identify, as to Groups (A, B, C, etc.), a minimum of three different covers for explosionproof fittings.

3. Remove the inspection cover on an explosionproof fitting as furnished by your instructor and check for moisture.

Answers to Self-Check Questions

1. c

2. b

3. d

4. b

5. d

6. b

7. c

8. a

9. b

10. a

Appendix I   Substances Used in Business and Industry

Class I Group	Substance	°F	°C	°F	°C	Lower	Upper	Vapor Density (Air Equals 1.0)
C	Acetaldehyde	347	175	-38	-39	4.0	60	1.5
D	Acetic Acid	967	464	103	39	4.0	19.9 @200°F	2.1
D	Acetic Anhydride	600	316	120	49	2.7	10.3	3.5
D	Acetone	869	465	-4	-20	2.5	13	2.0
D	Acetone Cyanohydrin	1270	688	165	74	2.2	12.0	2.9
D	Acetonitrite	975	524	42	6	3.0	16.0	1.4
A	Acetylene	581	305	gas	gas	2.5	100	0.9
B(C)	Acarolein (inhibited)	455	235	-15	-26	2.8	31.0	1.9
D	Acrylic Acid	820	438	122	50	2.4	8.0	2.5
D	Acrylonitrite	898	481	32	0	3.0	17	1.8
D	Adiponitrite	—	—	200	93	—	—	—
C	Allyl Alcohol	713	378	70	21	2.5	18.0	2.0
D	Allyl Chloride	905	485	-25	-32	2.9	11.1	2.6
B(C)	Allyl Glycidyl Ether	—	—	—	—	—	—	—
D	Ammonia	928	498	gas	gas	15	28	0.6
D	n-Amyl Acetate	680	360	60	16	1.1	7.5	4.5
D	sec-Amyl Acetate	—	—	89	32	—	—	4.5
D	Aniline	1139	615	158	70	1.3	11	3.2
D	Benzene	928	498	12	-11	1.3	7.9	2.8
D	Benzyl Chloride	1085	585	153	67	1.1	—	4.4
B(D)	1,,3-Butadiene	788	420	gas	gas	2.0	12.0	1.9
D	Butane	550	288	gas	gas	1.6	8.4	2.0
D	1-Butanol	650	343	98	37	1.4	11.2	2.6
D	2-Butanol	761	405	75	24	1.7 @ 212°F	9.8 @ 212°F	2.6
D	n-Butyl Acetate	790	421	72	22	1.7	7.6	4.0
D	iso-Butyl Acetate	790	421	—	—	—	—	—
D	sec-Butyl Acetate	—	—	88	31	1.7	9.8	4.0
D	t-Butyl Acetate	—	—	—	—	—	—	—
D	n-Butyl Acrylate (inhibited)	559	293	118	48	1.5	9.9	4.4
C	n-Butyl Formal	—	—	—	—	—	—	—
B(C)	n-Butyl Glycidyl Ether	—	—	—	—	—	—	—
C	Butyl Mercaptan	—	—	35	2	—	—	3.1
D	t-Butyl Toluene	—	—	—	—	—	—	—
D	Butylamine	594	312	10	-12	1.7	9.8	2.5
D	Butylene	725	385	gas	gas	1.6	10.0	1.9
C	n-Butyraldehyde	425	218	-8	-22	1.9	12.5	2.5
D	n-Butyric Acid	830	443	161	72	2.0	10.0	3.0
?	Carbon Disulfide	194	90	-22	-30	1.3	50.0	2.6

Figure 24. Gases and vapors used in business and industry.

Class I Group	Substance	°F	°C	°F	°C	Lower	Upper	Vapor Density (Air Equals 1.0)
C	Carbon Monoxide	1128	609	gas	gas	12.5	74.0	1.0
C	Chloroacetaldehyde	—	—	—	—	—	—	—
D	Chlorobenzene	1099	593	82	28	1.3	9.6	3.9
C	1-Chloro-1-Nitropropane	—	—	144	62	—	—	4.3
D	Chloroprene	—	—	-4	-20	4.0	20.0	3.0
D	Cresol	1038-1110	559-599	178-187	81-86	1.1-1.4	—	—
C	Crotonaldehyde	450	232	55	13	2.1	15.5	2.4
D	Cumene	795	424	96	36	0.9	6.5	4.1
D	Cyclohexane	473	245	-4	-20	1.3	8.0	2.9
D	Cyclohexanol	572	300	154	68	—	—	3.5
D	Cyclohexanone	473	245	111	44	1.1 @ 212°F	9.4	3.4
D	Cyclohexene	471	244		-7	—	—	2.8
D	Cyclopropane	938	503	gas	gas	2.4	10.4	1.5
D	p-Cymene	817	436	117	47	0.7 @ 212°F	5.6	4.6
C	C n-Decaldehyde	—	—	—	—	—	—	—
D	n-Decanol	550	288	180	82	—	—	5.5
D	Decene	455	235	â	7	—	—	4.84
D	Diacetone Alcohol	1118	603	148	64	1.8	6.9	4.0
D,o-Dichlorobenzene		1198	647	151	66	2.2	9.2	5.1
D	1,,1-Dichloroethane	820	438	22	-6	5.6	—	—
D	1,,2-Dichloroethylene	860	460	36	2	5.6	12.8	3.4
C	1,,1-Dichloro-1-Nitroethane	—	—	168	76	—	—	5.0
D	1,,3-Dichloropropene	—	—	95	35	5.3	14.5	3.8
C	Dicyclopentadiene	937	503	90	32	—	—	—
D	Diethyl Benzene	743-842	395-450	133-135	56-57	—	—	4.6
C	Diethyl Ether	320	160	-49	-45	1.9	36.0	2.6
C	Diethylamine	594	312	-9	-23	1.8	10.1	2.5
C	Diethylaminoethanol	—	—	—	—	—	—	—
C	Diethylene Glycol Monobutyl Ether	442	228	172	78	0.85	24.6	5.6
C	Diethylene Glycol Monomethyl Ether	465	241	205	96	—	—	—
D	Di-isobutyl Ketone	745	396	120	49	0.8 @ 200°F	7.1 @ 200°F	4.9
D	Di-isobutylene	736	391	23	-5	0.8	4.8	3.9
C	Di-isopropylamine	600	316	30	-1	1.1	7.1	3.5

Figure 24. Gases and vapors used in business and industry. (*Cont.*)

Class I Group	Substance	°F	°C	°F	°C	Lower	Upper	Vapor Density (Air Equals 1.0)
C	N-N-Dimethyl Aniline	700	371	145	63	—	—	4.2
D	Dimethyl Formamide	833	455	136	58	2.2 @ 212°F	15.2	2,5
D	Dimethyl Sulfate	370	188	182	83	—	—	4.4
C	Dimethylamine	752	400	gas	gas	2.8	14.4	1.6
C	1,,4-Dioxane	356	180	54	12	2.0	22	3.0
D	Dipentene	458	237	113	45	0.7 @ 302°F	6.1 @ 302°F	4.7
C	Di-n-propylamine	570	299	63	17	—	—	3.5
C	Dipropylene Glycol Methyl Ether	—	—	185	85	—	—	5.11
D	Dodecene	491	255	—	—	—	—	—,
C	Epichlorohydrin	772	411	88	31	3.8	21.0	3.2
D	Ethane	882	472	gas	gas	3.0	12.5	1.0
D	Ethanol	685	363	55	1.3	3.3	19	1.6
D	Ethyl Acetate	800	427	24	-4	2.0	11.5	3.0
D	Ethyl Acrylate (inhibited)	702	372	50	10	1.4	14	3.5
D	Ethyl sec-Amyl Ketone	—	—	—	—	—	—	—
D	Ethyl Benzene	810	432	70	21	1.0	6.7	3.7
D	Ethyl Butanol	—	—	—	—	—	—	—
D	Ethyl Butyl Ketone	—	—	115	46	—	—	4.0
D	Ethyl Chloride	966	519	-58	-50	3.8	15.4	2.2
D	Ethyl Formate	851	455	-4	-20	2.8	16.0	2.6
D	2-Ethyl Hexanol	448	231	164	73	0.88	9.7	4.5
D	2-Ethyl Hexyl Acrylate	485	252	180	82	—	—	—
C	Ethyl Mercaptan	572	300		-18	2.8	18.0	2.1
C	n-Ethyl Morpholine	—	—	—	—	—	—	—
C	2-Ethyl-3-Propyl Acrolein	—	—	155	68	—	—	4.4
D	Ethyl Silicate	—	—	125	52	—	—	7.2
D	Ethylamine	725	385		-18	3.5	14.0	1.6
C	Ethylene	842	450	gas	gas	2.7	36.0	1.0
D	Ethylene Chlorohydrin	797	425	140	60	4.9	15.9	2.8
D	Ethylene Dichloride	775	413	56	13	6.2	16	3.4
C	Ethylene Glycol Monobutyl Ether	460	238	143	62	1.1 @ 200°F	12.7 @ 275°F	4.1
D	Ethylene Glycol Monobutyl Ether Acetate	645	340	160	71	0.88 @ 200°F	8.54 @ 275°F	—
C	Ethylene Glycol Monoethyl Ether	455	235	110	43	1.7 @ 200°F	15.6 @ 200°F	3.0
C	Ethylene Glycol Monoethyl Ether Acetate	715	379	124	52	1.7	—	4.72

Figure 24. Gases and vapors used in business and industry. (*Cont.*)

Class I Group	Substance	°F	°C	°F	°C	Lower	Upper	Vapor Density (Air Equals 1.0)
D	Ethylene Glycol Monomethyl Ether	545	285	102	39	1.8 @ STP	14 @ STP	2.6
B(C)	Ethylene Oxide	804	429	-20	-28	3.0	100	1.5
D	Ethylenediamine	725	385	93	34	4.2	14.4	2.1
C	Ethylenimine	608	320	12	-11	3.6	46.0	1.5
C	2-Ethylhexaldehyde	375	191	112	44	0.85 @ 200°F	7.2 @ 275°F	4.4
B	Formaldehyde (Gas)	795	429	gas	gas	7.0	73	1.0
D	Formic Acid (90%)	813	434	122	50	18	57	—
D	Fuel Oils	410-765	210-407	100-336	38-169	0.7	5	—
C	Furfual	600	316	140	60	2.1	19.3	3.3
C	Furfuryl Alcohol	915	490	167	75	1.8	16.3	3.4
D	Gasoline	536-880	280-471	-36 to -50	-38 to -46	1.2-1.5	7.1-7.6	3-4
D	Heptane	399	204	2.5	-4	1.05	6.7	3.5
D	Heptene	500	260			—	—	3.39
D	Hexane	437	225	-7	-22	1.1	7.5	3.0
D	Hexanol	—	—	145	63			3.5
D	2-Hexanone	795	424	77	25	—	8	3.5
D	Hexenes	473	245		-7	—	—	3.0
D	sec-Hexyl Acetate	—	—	—	—	—	—	—
C	Hydrazine	74-518	23-270	100	38	2.9	9.8	1.1
B	Hydrogen	968	520	gas	gas	4.0	75	0.1
C	Hydrogen Cyanide	1000	538	0	-18	5.6	40.0	0.9
C	Hydrogen Selenide	—	—	—	—	—	—	—
C	Hydrogen Sulfide	500	260	gas	gas	4.0	44.0	1.2
D	Isoamyl Acetate	680	360	77	25	1.0 @ 212°F	7.5	4.5
D	Isoamyl Alcohol	662	350	109	43	1.2	9.0 @ 212°F	3.0
D	Isobutyl Acrylate	800	427	86	30	—		4.42
C	Isobutyraldehyde	385	196	-1	-18	1.6	10.6	2.5
C	Isodecaldehyde	—	—	185	85	—	—	5.4
D	Iso-octyl Alcohol	—	—	180	82	—	—	
C	Iso-octyl Aldehyde	387	197	—	—	—	—	—
D	Isophorone	860	460	184	84	0.8	3.8	—
D	Isoprene	428	220	-65	-54	1.5	8.9	2.4
D	Isopropyl Acetate,860	460	35	2	1.8 @ 100°F	8	3.5	
D	Isopropyl Ether	830	443	-18	-28	1.4	7.9	3.5
C	Isopropyl Glycidyl Ether	—	—	—	—	—	—	—
D	Isopropylamine	756,402	-35	-37	—	—	2.0	
D	Kerosene	410	210	110-162	43-72	0.7	5	—
D	Liquefied Petroleum Gas	761-842	405-450	—	—	—	—	—

Figure 24. Gases and vapors used in business and industry. (*Cont.*)

ELECTRICAL TRAINEE TASK MODULE 20313

Class I Group	Substance	°F	°C	°F	°C	Lower	Upper	Vapor Density (Air Equals 1.0)
B	Manufactured Gas (containing more than 30% H							
D	Methyl Isobutyl Ketone	840	440	64	18	1.2 @ 200°F. 8.0 @ 200°F 3.5		
D	Methyl Isocyanate	994	534	19	-7	5.3	26	1.97
C	Methyl Mercaptan	—	—	—	—	3.9	21.8	1.7
D	Methyl Methacrylate	792	422	50	10	1.7	8.2	3.6
D	2-Methyl-1-Propanol	780	416	82	28	1.7 @ 123°F	10.6 @ 202°F	2.6
D	2-Methyl-2-Propanol	892	478	52	11	2.4	8.0	2.6
D	alpha-Methyl Styrene	1066	574	129	54	1.9	6.1	—
C	Methylacetylene	—	—	gas	gas	1.7	—	1.4
C	Methylacetylene-Propadiene (stabilized)	—	—	—	—	—	—	—
D	Methylamine	806	430	gas	gas	4.9	20.7	1.0
D	Methylcyclohexane	482	250	25	-4	1.2	6.7	3.4
D	Methylcyclohexanol	565	296	149	65	—	—	3.9
D	0-Methylcyclohexanone	—	—	118	48	—	—	3.9
D	Monoethanolamine	770	410	185	85	—	—	2.1
D	Monoisopropanolamine	705	374	171	77	—	—	2.6
C	Monomethyl Aniline	900	482	185	85	—	—	3.7
C	Monomethyl Hydrazine	382	194	17	-8	2.5	92	1.6
C	Morpholine	590	310	98	37	1.4	11.2	3.0
D	Naphtha (Coal Tar)	531	277	107	42	—	—	—
D	Naphtha (Petroleum)	550	288		-18	1.1	5.9	2.5
D	Nitrobenzene	900	482	190	88	1.8 @ 200°F —	4.3	
C	Nitroethane	778	414	82	28	3.4	—	2.6
C	Nitromethane	785	418	95	35	7.3	—	2.1
C	1-Nitropropane	789	421	96	36	2.2	—	3.1
C	2-Nitropropane	802	428	75	24	2.6	11.0	3.1
D	Nonane	401	205	88	31	0.8	2.9	4.4
D	Nonene	—	—	78	26	—	—	4.35
D	Nonyl Alcohol	—	—	165	74	0.8 @ 212°F	6.1 @ 212°F	5.0
D	Octane	403	206	56	13	1.0	6.5	3.9
D	Octene	446	230	70	21	—	—	3-9
D	n-Octyl Alcohol	—	—	178	81	—	—	4.5

Figure 24. Gases and vapors used in business and industry. (*Cont.*)

Class I Group	Substance	°F	°C	°F	°C	Lower	Upper	
D	Pentane	470	243	-40	-40	1.5	7.8	2.5
D	1-Pentanol	572	300	91	33	1.2	10.0 @ 212°F	3.0
D	2-Pentanone	846	452	45	7	1.5	8.2	3.0
D	1-Pentene	527	275	0	-18	1.5	8.7	2.4
D	Phenylhydrazine	—	—	190	88	—	—	—
D	Propane	842	450	gas	gas	2.1	9.5	1.6
D	1-Propanol	775	413	74	23	2.2	13.7	2.1
D	2-Propanol	750	399	53	12	2.0	12.7 @ 200°F	2.1
D	Propiolactone	—	—	165	74	2.9	—	2.5
C	Propionaldehyde	405	207	-22	-30	2.6	17	2.0
D	Propionic Acid	870	466	126	52	2.9	12.1	2.5
D	Propionic Anhydride	545	285	145	63	1.3	9.5	4.5
D	n-Propyl Acetate	842	450	55	13	1.7 @ 100°F	8	3.5
C	n-Propyl Ether	419	215	70	21	1.3	7.0	3.53
B	Propyl Nitrate	347	175	68	20	2	100	—
D	Propylene	851	455	gas	gas	2.0	11.1	1.5
D	Propylene Dichloride	1035	557	60	16	3.4	14.5	3.9
B(C)	Propylene Oxide	840	449	-35	-37	2.3	36	2.0
D	Pyridine	900	482	68	20	1.8	12.4	2.7
D	Styrene	914	490	88	3.1	1.1	7.0	3.6
C	Tetrahydrofuran	610	321	6	-14	2.0	11.8	2.5
D	Tetrahydronaphthalene	725	385	160	71	0.8 @ 212°F	5.0 @ 302°F 4.6	
C	Tetramethyl Lead	—	—	100	38	—	—	6.5
D	Toulene	896	480	40	4	1.2	7.1	3.1
D	Tridecene	—	—	—	—	—	—	—
C	Triethylamine	480	249	16	-9	1.2	7.1	3.1
D	Triethylbenzene	—	—	181	83	—	—	5.6
D	Tripropylamine	—	—	105	41	—	—	—
D	Turpentine	488	253	95	35	0.8	—	—
D	Undecene	—	—	—	—	—	—	—
C	Unsymmetrical Dimethyl Hydrazine (UDMH)	480	249	5	-15	2	95	2.0
C	Valeraldehyde	432	222	54	12	—	—	3.0
D	Vinyl Acetate	756	402	18	-8	2.6	13.4	3.0
D	Vinyl Chloride	882	472	gas	gas	3.6	33.0	2.2
D	Vinyl Toluene	921	494	120	49	—	11.0	4.1
D	Vinylidene Chloride	1058	570	-19	-28	6.5	15.5	3.4
D	Xylenes	867-984	464-529	81-90	27-32	1.0-1.1	7.0	3.7

Figure 24. Gases and vapors used in business and industry. (*Cont.*)

ELECTRICAL TRAINEE TASK MODULE 20313